Seismic Design of Building Structures

A Professional's Introduction to Earthquake Forces and Design Details

Eleventh Edition

Michael R. Lindeburg, PE
with Kurt M. McMullin, PhD, PE

Professional Publications, Inc. • Belmont, California

SEISMIC DESIGN OF BUILDING STRUCTURES: A PROFESSIONAL'S INTRODUCTION TO EARTHQUAKE FORCES AND DESIGN DETAILS

Eleventh Edition

Current printing of this edition: 3

Printing History

date	edition number	printing number	update
Oct 2014	11	1	New edition. Code updates. Copyright update.
Jul 2015	11	2	Minor corrections. Minor cover updates.
Feb 2017	11	3	Minor corrections. Minor cover updates.

Printed in the United States of America.

PPI
1250 Fifth Avenue
Belmont, CA 94002
(650) 593-9119
ppi2pass.com

ISBN: 978-1-59126-470-5

Library of Congress Control Number: 2014950240

F E D C B A

Table of Contents

5 Response of Structures

6 Seismic Building Codes

7 Diaphragm Theory

8 General Structural Design

9 Details of Seismic-Resistant Concrete Structures

10 Details of Seismic-Resistant Steel Structures

11 Details of Seismic-Resistant Masonry Structures

12 Details of Seismic-Resistant Wood Structures

13 Tilt-Up Construction

14 Special Design Features

15 Practice Problems

Appendices

Preface

Seismic Design of Building Structures is a concise introduction to basic seismic concepts and principles. It is primarily intended to be used as a review manual for the California Civil Seismic Principles Exam. However, this book is also appropriate to use if you are studying for the NCEES Structural Engineering (SE) Examination and the Architect Registration Examination (ARE).

This book has been updated to conform to the 2012 *International Building Code* (IBC), the 2010 *Minimum Design Loads for Buildings and Other Structures* (ASCE/SEI7), the 2011 *Building Code Requirements for Structural Concrete* (ACI 318), and the 2011 *Building Code Requirements for Masonry Structures* (ACI 530). Though the California Civil Seismic Principles Exam is ostensibly based on the 2013 *California Building Code* (CBC), Volume II of the CBC contains the same material and section numbers as IBC-2012 Volumes I and II. I have included many important tables and figures from these codes and standards that support the topics presented. Since these tables and figures may have been abridged and/or reformatted, this book is not a replacement for the source documents.

The 2012 IBC defers almost entirely to ASCE/SEI7. This eliminates some of the conflicts that were present in the past, such as the handling of (what used to be called) special seismic load combinations. In addition, the IBC's acceleration maps are now consistent with the 2008 USGS maps; and Next Generation Attenuation (NGA) relationships are used exclusively for the western United States.

Other important changes related to the use of risk-targeted ground motion have resulted in some terminology changes. For example, the term "risk-targeted maximum considered earthquake (MCE_R) ground motion response acceleration" has replaced "maximum considered earthquake (MCE_R) ground motion." And "risk category" has replaced the term "occupancy category" in the 2010 ASCE/SEI7.

Examinees often comment that the depth of structural analysis experience and code-related knowledge expected of them in the California seismic exam far exceeds anything they will ever be faced with in practice. Accordingly, the exam has shifted toward a mixture of technical, code-detailed, calculation-intensive questions, and conceptual questions requiring general familiarity with provisions, concepts, and procedures. This book is intended to prepare you for both types of questions. This book transcends the issue of selecting between LRFD and ASD. The subject of seismic design, as described in this book and as used in the exam, basically stops after forces have been calculated and distributed to stories and members. Designing members to support those forces is not part of the California Civil Seismic Principles Exam or the ARE (although it is certainly part of the NCEES SE exam).

This book uses both customary U.S. and SI units. All of the examples and problems give you a choice of units: You can choose to work solely in SI, solely in customary U.S., or in both (for twice the practice). This book places the customary U.S. values first to accommodate the unit system typically used in the exams and in conventional seismic design.

Over the past 25 years, examinees have shaped this book considerably. I will be grateful if you, too, take the time to tell me how you think this book can be improved. You may contact me via the web at **ppi2pass.com/errata**. I will humbly accept any suggestions you think might help future examinees.

Michael R. Lindeburg, PE

Acknowledgments

A lot of people contributed to producing this eleventh edition of *Seismic Design of Building Structures*. As with the ninth and tenth editions, most of the technical changes were completed by Kurt M. McMullin, PhD, PE, professor of civil engineering at San Jose State University and a member of the American Society of Civil Engineers and the American Society for Engineering Education. He thoroughly reviewed the book, updated all of the code section references, rewrote sections, and revised numerous problems and examples. He also served as PPI's "go-to" person during the prepress preparation of the book.

At PPI, the task of making the vision of a new edition a reality fell to a Product Development and Implementation Department team that consisted of Magnolia Molcan, editorial project manager; Ellen Nordman, lead editor; David Chu, Nicole Evans, Hilary Flood, Tyler Hayes, Julia Lopez, Scott Marley, Heather Turbeville, and Ian A. Walker, copy editors; Ralph Arcena, EIT, engineering intern; Tom Bergstrom, technical illustrator and production associate; Kate Hayes, production associate; Cathy Schrott, production services manager; and Sarah Hubbard, director of product development and implementation.

Previous editions drew upon the expertise of many individuals over several decades. Their contributions and my appreciation are documented in the acknowledgements of thousands of copies of previous editions still in circulation.

I acknowledge and thank the International Code Council (ICC) for its permission to include so many *International Building Code* (IBC) figures and tables. These figures and tables enhance the understanding and use of the code-related provisions of this book.

I acknowledge and thank the American Society of Civil Engineers (ASCE) for its permission to include content from *Minimum Design Loads for Buildings and Other Structures* (ASCE/SEI7). That permission means a lot of illustrative content has been included, resulting in less cross-referencing while you are reading.

I also acknowledge the Simpson Strong-Tie Company for kindly giving its permission to reproduce various pages of its "Connectors for Wood Construction," 2013 catalog. These pages illustrate typical seismic appliances and support some of the practice problems in this book. Pages from the Simpson Strong-Tie Company product catalog are included mainly to support solving example and practice problems. The inclusion of these pages is for illustrative purposes, not for the purpose of endorsement or exclusive recommendation. Check the Simpson Strong-Tie Company's website, strongtie.com, for the latest catalog.

Other publishers which granted permission to reproduce important supporting tables and figures from their books are credited where their material appears.

A few problems in *Seismic Design* have their basis in California engineering licensing exam problems, from the days when such exams were released to the public after being administered. The problems are used with permission of the California Board for Professional Engineers, Land Surveyors, and Geologists.

Many changes in this new edition were needed simply because of the new seismic provisions in IBC-2012 and ASCE/SEI7-2011. However, in addition to bringing the code-related sections up to date, this edition incorporates improvements and errata suggested by readers. These readers took the time to tell me how to make the book better. Hopefully, I'll hear from you, as well.

As I did in the previous edition, I include the following paragraph because I will always remember and love my family.

"Finally, I acknowledge the unwavering support of my family which, as usual, has had to put up with my habitual writing. My wife, Elizabeth, and my two daughters, Jenny and Katie, lost the family time that went into this book. They may be beginning to understand how I think and what is important to me. If not, they never seem to complain."

Michael R. Lindeburg, PE

How to Use This Book

If you are the type of person who never reads instructions, here are my "Quickstart" suggestions on how to get the most from *Seismic Design of Building Structures* during your exam preparation.

1. Get copies of the IBC and ASCE/SEI7 and their errata.
2. Start reading this book from the first chapter. Don't skip around, because the book builds on concepts.
3. Read slowly; a page or two a day is plenty. Look up every code section in the referenced sources.
4. Work through all of the example and practice problems.
5. Put lots of tabs on the building code tables.
6. Use the indexes extensively.
7. Don't forget to take it with you to your exam.

Now, beyond those suggestions, how you will use this book depends on why you obtained it. *Seismic Design* was specifically written for engineering exam review. As the exams vacillated each year between areas of emphasis, the scope and depth of this book also increased. And so, this book now covers a lot of bases, and it can be used for other exams (e.g., SE and ARE) and general familiarity. However, even though its scope and depth have increased enormously over the years, I suspect this book will remain typecast in its leading role—that of an engineering exam review book. Therefore, I am writing this section assuming that you are using *Seismic Design* for that purpose.

Although this book develops subjects gradually, gently, and linearly, it crams innumerable concepts onto every page. If you don't have a seismic background, you'll pretty much have to start at page one and work your way through the book, page by page. I don't assume that you know anything. I've skipped almost all of the higher-order mathematics. And I've tried to write and edit the material to provide instruction that is intuitive. However, you'll still need to go slowly.

I know that some engineering examinees will buy this book just to take it into the open-book exam, rather than reading and working through it. During the exam, they plan to use the index a lot, hoping to hit the jackpot. If that describes you, I predict you won't pass your exam using that plan. The subject of seismic design is different from most everything else civil engineers do. Even the language is different. What's the difference between a "drag strut" and a "collector"? (Answer: There is no difference.) What's the difference between a "space frame" and a "ductile moment-resisting space frame"? (Answer: There can be plenty of difference.) You can read the words, but without having studied the material in this book, they won't mean anything to you. You'll just be wasting your time if you think you're a hotshot engineer who can get by on good looks and this book's good index.

The California Civil Seismic Principles Exam is based on the *California Building Code* (CBC). The seismic design principles in the CBC are essentially those of the IBC and ASCE/SEI7. Due to the cost of the CBC, its two-book format, and its references to the IBC and ASCE/SEI7, few examinees purchase it. The IBC and ASCE/SEI7 are suitable substitutes.

For the open-book engineering exams, you are much more likely to need to pull a number out of the IBC or ASCE/SEI7 than to read a code section. Not surprisingly, the only place where a complete compilation of all of the constants and other numerical values can be found is in the building codes themselves. The numerous tables and figures in this book give a false impression of completeness. True, every detail needed to solve an example or practice problem is probably contained somewhere in this book. However, the actual exam will not be so kind. It's always a shame to lose points simply because you couldn't perform a simple table lookup in "the code." So, as the suggestions at the top of this section say, if you are taking an open-book engineering exam, buy the IBC and ASCE/SEI7.

The California seismic exam places a lot of emphasis on design and detailing of connections. What is a good connection? What used to be considered a good connection? What are the modes of failure of connections? And, what can be done to strengthen a connection? Learning to recognize adequacy in a variety of concrete, steel, masonry, and timber (including structural panel) connections is an important part of the exam. The example and practice problems in this book have been selected to make, or emphasize, certain points. But these problems do not try to mimic exam complexity. PPI has produced two other seismic problem-oriented books, *Seismic Design Solved Problems* and *Seismic Principles Practice Exams for the California Civil Seismic Exam*, by Majid Baradar, to provide hundreds of additional practice problems in exam format. You can get extensive practice by working problems in these two well-organized books.

ABOUT THE CALIFORNIA CIVIL SEISMIC PRINCIPLES EXAM

The California Civil Seismic Principles Exam is a computer-based test (CBT) administered at a Prometric testing center. There are Prometric testing centers throughout the United States. The exam is offered twice a year, around the dates of the NCEES professional engineering exams, during a relatively short (e.g., three-week) period defined by the California Board for Professional Engineers, Land Surveyors, and Geologists.

The exam is $2^1/_2$ hours in length and is taken in one single sitting. The exam consists of 55 questions, all of which are multiple-choice, each with four options. The average time allowed per question is about three minutes, but you can spend as much or as little time as you want on any problem. "Too little time," is a common comment from examinees. Some of the questions are conceptual, some are theoretical, some are practical, some are straight lookup, and some require simple calculations.

Navigation through the exam using the CBT interface is fairly standard. There is a list of all questions indicating which have been answered, marked for review, or skipped. You can skip a question or mark it for later review. You can navigate to a specific question. You don't have to move sequentially through the exam, although you can if you wish to. You can return to any question and change your answer.

A timer is shown on the screen indicating how much time remains. There is no significant lag time between the problems. In some cases, you may have to toggle back and forth between a question and an on-screen illustration if the illustration takes up too much screen space.

Creature comforts are typical of CBT testing centers. Procedures used to ensure security may be objectionable and/or silly to some examinees. Desktop space is small, sufficient for only one book at a time. You can take a restroom break, but the clock doesn't stop running. You simply raise your hand and wait for a proctor to come and release you. You can have a bottle of water. The chair will be adjustable and comfortable. Noise-reducing earplugs are allowed.

The total number of points on the exam and the minimum passing score are not public knowledge. The percentage of people who pass varies considerably: between 30% and 40%, typically. Many examinees receive specialized instruction in seismic design (i.e., they take a specialized course), and the reported passing percentage factors in those "knowledgeable" examinees. There is only a moderate correlation between the people who pass the NCEES eight-hour exam, and the people who pass the California seismic exam.

The *Special Civil Seismic Principles Test Plan*, which defines the fundamental principles, tasks, and elements of knowledge that you need to know, has been made public by the California board. If you want to know what is needed in order to pass, you can print out the test plan from the website of the California Board for Professional Engineers, Land Surveyors, and Geologists. A link to the test plan is provided at **ppi2pass.com/CAspecial**. What you will find out is that you need to know everything in this book and then some. You cannot learn much from the test plan other than that the scope of the exam is huge. The areas of emphasis may change from exam to exam. You just have to take whatever they throw at you when you show up for your exam.

Your results are mailed to you; grading is not instantaneous. You will never learn your actual score. The only result reported is "Pass" or "Fail." If you fail, you will receive a Diagnostic Report that indicates your performance level (i.e., Proficient, Marginal, or Deficient) in each of the five content areas of the test plan.

ABOUT THE NCEES STRUCTURAL ENGINEERING (SE) EXAM

The NCEES Structural Engineering (SE) Examination is offered in two parts. The first part—vertical forces (gravity/other) and incidental lateral—takes place on a Friday. The second part—lateral forces (wind/earthquakes)—takes place on a Saturday. Each part comprises a breadth section and a depth section. The breadth sections in the morning are each 4 hours and contain 40 multiple-choice problems that cover a range of structural engineering topics specific to vertical and lateral forces. The depth sections in the afternoon are also each 4 hours, but instead of multiple-choice problems, they contain essay (design) problems. You may choose either the bridges or the buildings depth section, but you must work the same depth section across both parts of the exam. That is, if you choose to work buildings for the lateral forces part, you must also work buildings for the vertical forces part. Both breadth and depth sections use customary U.S. units.

According to NCEES, the vertical forces (gravity/other) and incidental lateral depth section in buildings covers loads, lateral earth pressures, analysis methods, general structural considerations (e.g., element design), structural systems integration (e.g., connections), and foundations and retaining structures. The depth section in bridges covers gravity loads, superstructures, substructures, and lateral loads other than wind and seismic. It may also require pedestrian bridge and/or vehicular bridge knowledge.

The lateral forces (wind/earthquake) depth section in buildings covers lateral forces, lateral force distribution, analysis methods, general structural considerations (e.g., element design), structural systems integration (e.g., connections), and foundations and retaining structures. The depth section in bridges covers gravity loads, superstructures, substructures, and lateral forces. It may also require pedestrian bridge and/or vehicular bridge knowledge. For more information regarding the NCEES SE exam, go to the NCEES website, ncees.org.

ABOUT THE ARCHITECT REGISTRATION EXAM

The Architect Registration Examination (ARE) is composed of seven divisions that test various areas of architectural knowledge and problem-solving ability: programming, planning, and practice; site planning and design; schematic design; structural systems; building systems; building design and construction systems; and construction documents and services. Most divisions include both multiple-choice and graphic/design-type problems. Seismic force problems are part of the structural systems division, but experienced test-takers will tell you that there is quite a bit of overlap among the divisions so it's best to be prepared for seismic problems in any division. Like with the California seismic exam, you have to take what they throw at you, and run with it.

You may schedule any division of the ARE at any time and may take the divisions in any order. Divisions can be taken one at a time, to spread out preparation time and exam costs, or can be taken together in any combination. However, all seven divisions of the ARE must be passed within a single five-year period. If you have not completed the ARE within five years, the divisions you passed more than five years ago are no longer credited, and the content must be retaken. For more information about the ARE, see **ppi2pass.com/areinfo**.

Codes Used in This Book

book abbreviation	title
ACI 318	American Concrete Institute: *Building Code Requirements for Structural Concrete*, 2011
ACI 530	American Concrete Institute: *Building Code Requirements for Masonry Structures*, 2011
AISC	American Institute of Steel Construction: *Steel Construction Manual*, 14th ed., 2011
AISC 341	American Institute of Steel Construction: *Seismic Provisions for Structural Steel Buildings*, 2010
AISC 360	American Institute of Steel Construction: *Specification for Structural Steel Buildings*, 2010
ASCE/SEI7	American Society of Civil Engineers: *Minimum Design Loads for Buildings and Other Structures*, 2010
IBC	International Code Council: *International Building Code*, 2012
NDS®	American Wood Council: *National Design Specification® for Wood Construction ASD/LRFD*, 2012, including *National Design Specification Supplement*, 2012
SDPWD	American Wood Council: *Special Design Provisions for Wind and Seismic*, 2008

Nomenclature

Unless defined otherwise in the text, the following symbols are used in this book. Consistent units are presented. However, in some cases (such as modulus of elasticity and drift), it is customary to report values in smaller units (e.g., lbf/in^2 and in).

symbol	term	units U.S.	units SI
a	acceleration	ft/sec^2	m/s^2
a	link beam distance	ft	m
a_p	in-structure component amplification factor	–	–
A	amplitude	ft	m
A	area	ft^2	m^2
A_B	base area of a structure	ft^2	m^2
A_B	ground floor area	ft^2	m^2
A_c	combined effective area of shear wall	ft^2	m^2
A_e	minimum cross-sectional area	ft^2	m^2
A_i	area of shear wall i	ft^2	m^2
A_x	floor area of diaphragm immediately below story x	ft^2	m^2
A_x	torsional amplification factor	–	–
ARS	acceleration response spectrum	–	–
b	link beam distance	ft	m
b	parallel wall length	ft	m
b	width	ft	m
B	damping coefficient	lbf-sec/ft	N·s/m
B	seismic parameter	–	–
C	chord force	lbf	N
C	numerical coefficient	–	–
C	seismic parameter	mi^{-2}	km^{-2}
C_d	deflection amplification factor	–	–
C_e	snow exposure coefficient	–	–
C_q	pressure coefficient	–	–
C_s	base shear coefficient	–	–
$C_{s,\text{min,SDCEF}}$	minimum seismic coefficient for seismic design category E and F	–	–
C_t	numerical coefficient	–	–
C_t	building period coefficient based on structural system	–	–
C_t	snow thermal coefficient	–	–
C_w	alternative building period coefficient for masonry or concrete shear wall structures	–	–
d	depth	ft	m
d_i	thickness of layer i	ft	m
d_s	total thickness of cohesionless soil layers	ft	m
D	column depth	ft	m
D	dead load effect	lbf or ft-lbf	N or N·m
D	diameter	ft	m
D	fault slip	ft	m
D_e	length of a shear wall	ft	m

symbol	term	units U.S.	SI
D_i	length of a shear wall i	ft	m
D_p	relative displacement that a component must be designed to accept	in	mm
e	eccentricity	ft	m
E	seismic load effect	lbf	N
E	energy released	ft-lbf	J
E	modulus of elasticity	lbf/ft^2	Pa
E_h	horizontal earthquake load	lbf	N
E_m	seismic load effect including the overstrength factor	lbf	N
E_v	vertical earthquake load	lbf	N
EPA	effective ground acceleration	ft/sec^2	m/s^2
f	frequency	Hz	Hz
f	stress	lbf/ft^2	Pa
f'_m	masonry compressive strength	lbf/ft^2	Pa
F	force	lbf	N
F	story shear	lbf	N
F_p	design seismic forces on a part of the structure	lbf	N
F_{px}	design seismic forces on a diaphragm	lbf	N
g	acceleration of gravity, 32.2 (9.81)	ft/sec^2	m/s^2
g_c	gravitational constant, 32.2	ft-lbm/ lbf-sec^2	n.a.
G	shear modulus	lbf/ft^2	Pa
h	height	ft	m
h	story height	ft	m
h_n	height above base to level n	ft	m
h_r	structure roof elevation with respect to grade	ft	m
h_x	element or component attachment elevation with respect to grade	ft	m
H	horizontal force	lbf	N
H	story height	ft	m
I	moment of inertia	ft^4	m^4
I_e	seismic importance factor	–	–
I_p	seismic importance factor for an element or component	–	–
I_s	snow importance factor	–	–
I_w	wind importance factor	–	–
J	polar moment of inertia	ft^4	m^4
k	exponent for vertical distribution calculation	–	–
k	stiffness (spring constant)	lbf/ft	N/m
K	adjustment coefficient	–	–
l_d	development length	ft	m
L	length	ft	m
L	live load	lbf or ft-lbf	N or N·m
m	mass	lbm	kg
M	moment	ft-lbf	N·m
M	Richter magnitude	–	–
MCE	maximum considered earthquake	g's	g's
MCE_R	risk-targeted maximum considered earthquake	g's	g's
n	cycle number	–	–
n	exponent	–	–
n	uppermost story of a structure	–	–
N	number of earthquakes	–	–
N	standard penetration resistance of soil layer	blows/ft	blows/m
N_{ch}	standard penetration resistance of cohesionless soil layer	blows/ft	blows/m

symbol	term	units U.S.	units SI
OTM	overturning moment	ft-lbf	N·m
p_f	flat roof snow load	lbf/ft^2	Pa
p_g	ground snow load	lbf/ft^2	Pa
p_s	sloped roof snow load	lbf/ft^2	Pa
P	design wind pressure	lbf/ft^2	Pa
P	magnitude of forcing function	lbf	N
P	sum of dead and live loads	lbf	N
PGA	peak ground acceleration	ft/sec^2	m/s^2
PI	plasticity index	–	–
q	shear flow	lbf/ft	N/m
q_z	wind pressure	lbf/ft^2	Pa
Q_E	effect of horizontal seismic force	lbf or ft-lbf	N or N·m
r	radius or moment arm	ft	m
r_{max}	maximum element story shear ratio	–	–
R	electrical resistance	ohms	ohms
R	response modification factor	–	–
R	rigidity	various	various
R_{abs}	absolute rigidity	lbf/ft	N/m
R_{obs}	observed rigidity	1/ft	1/m
R_p	component response modification factor	–	–
R_{rel}	relative rigidity	–	–
R_{tab}	tabulated rigidity	1/ft	1/m
s	distance	ft	m
s_u	average undrained shear strength	lbf/ft^2	Pa
S	snow load	lbf/ft^2	Pa
S_1	mapped acceleration parameter	g's	g's
S_a	spectral acceleration	g's or ft/sec^2	g's or m/s^2
S_d	spectral displacement	ft	m
S_{D1}	design spectral response acceleration parameter at 1 sec period	g's	g's
S_{DS}	design spectral response acceleration parameter at short period	g's	g's
S_{M1}	MCE spectral response acceleration for a 1 sec period	g's	g's
S_{MS}	MCE spectral response acceleration for short periods	g's	g's
S_S	spectral response acceleration for short periods	g's	g's
S_{v}	spectral velocity	ft/sec	m/s
SC_f	seismic coefficient of a floor of a structure	–	–
SR	slip rate	ft/yr	m/yr
SUG	seismic use group	–	–
t	thickness	ft	m
t	time	sec	s
T	fundamental period of vibration	sec	s
T_a	approximate fundamental period of vibration of a structure	sec	s
T_s	site period	sec	s
u	displacement determined from code level seismic design force	in	mm
U	strain energy (per unit volume)	$ft\text{-}lbf/ft^3$	J/m^3
U	ultimate capacity	lbf or ft-lbf	N or N·m
v	shear stress	lbf/ft^2	Pa
v	velocity	ft/sec	m/s
v_s	shear wave velocity	ft/sec	m/s
V	base shear	lbf	N
V	basic wind speed	mi/hr	m/s
w	load per unit length	lbf/ft	N/m

symbol	term	units	
		U.S.	SI
w_i	weight of level i	lbf	N
w_{mc}	moisture content	–	–
W	load due to wind pressure	lbf or ft-lbf	N or N·m
W	nail and spike withdrawal design values	lbf/in	N/m
W	weight	lbf	N
W_p	weight of an element or component	lbf	N
x	exponent for fundamental period of vibration calculation	–	–
x	position or excursion	ft	m
y	height over which wind acts	ft	m
Y	number of years	–	–
z	the story under consideration	–	–
Z	ductility reduction factor	–	–
Z	lateral design force	lbf	N
α	lateral angular drift of story or building	rad	rad
β	magnification factor	–	–
Γ	participation factor	–	–
δ	decay decrement	–	–
δ	displacement	ft	m
δ_i	horizontal displacement at level i	in	mm
$\delta_{\max}$	maximum elastic deflection	in	mm
δ_M	maximum inelastic response displacement	in	mm
δ_{MT}	total inelastic separation	ft	m
δ_x	deflection of story x at the center of mass	in	mm
δ_{xe}	deflection of story x at the center of mass calculated using linear elastic analysis	in	mm
Δ	design story drift	in	mm
Δ	story drift	in	mm
Δ_M	maximum inelastic response displacement	in	mm
Δ_S	story displacement	in	mm
Δ_{SR}	story drift ratio	–	–
Δ_x	design story drift at story x	in	mm
ϵ	strain	–	–
ζ	damping ratio	–	–
θ	link rotation of eccentric braced frame	rad	rad
θ	stability coefficient	–	–
λ	Lamé's constant	–	–
μ	ductility factor	–	–
ν	Poisson's ratio	–	–
ρ	redundancy factor	–	–
ρ	reinforcement ratio	–	–
ρ	redundancy coefficient for structure	–	–
ρ	density	lbm/ft^3	kg/m^3
ρ_x	redundancy coefficient for story x	–	–
σ	stress	lbf/ft^2	Pa
τ	short period of time	sec	s
ϕ	mode shape factor	–	–
ω	angular frequency	rad/sec	rad/s
Ω_O	seismic system overstrength factor	–	–

subscripts

0	initial or calibration
α	allowable
ave	average
b	bending
d	damped or displacement
e	elastic
f	foundation
E	elastic strain
H	hysteresis
i	inertial or level number
j	mode number
n	cycle number
p	part or portion (component)
P	P-wave
R	resilience
st	static
sx	between levels $x-1$ and x
S	S-wave
t	top or total
T	toughness
v	velocity
x	level number
x	with respect to x-axis
y	with respect to y-axis

1 Basic Seismology

1. THE NATURE OF EARTHQUAKES

An earthquake is an oscillatory, sometimes violent movement of the ground's surface that follows a release of energy in the earth's crust. This energy can be generated by a sudden dislocation of segments of the crust, a volcanic eruption, or a manufactured explosion. Most of the destructive earthquakes, however, are caused by dislocations of the crust.

When subjected to geologic forces from plate tectonics, the crust initially strains (i.e., bends and shears) elastically. For pure axial loading, Hooke's law gives the stress that accompanies this strain.

$$\sigma = E\epsilon \quad \text{[axial loading]} \qquad 1.1$$

As rock is stressed, it stores *strain energy*, *U*. The elastic strain energy per unit volume for pure axial loading is[1]

$$U = \frac{\sigma\epsilon}{2} \quad \text{[axial loading]} \qquad 1.2$$

When the stress exceeds the ultimate strength of the rocks, the rocks break and quickly move (i.e., they "snap") into new positions. In the process of breaking, the strain energy is released and *seismic waves* are generated. This is the basic description of the *elastic rebound theory* of earthquake generation.

Seismic waves travel from the source of the earthquake (known as the *hypocenter* or *focus*) to more distant locations along the surface of and through the earth. The wave velocities depend on the nature of the waves and the material through which the waves travel. (See Sec. 1.14.) Some of the vibrations are of high enough frequency to be audible, while others are of low frequency with periods of many seconds and, thus, are inaudible.

2. EARTHQUAKE TERMINOLOGY

The *epicenter* of an earthquake is the point on the earth's surface directly above the *focus* (also known as the *hypocenter*). The location of an earthquake is commonly described by the geographic position of its epicenter and its focal depth. The *focal depth* of an earthquake is the depth from the earth's surface to the focus. These terms are illustrated in Fig. 1.1.

Figure 1.1 *Earthquake Terminology*

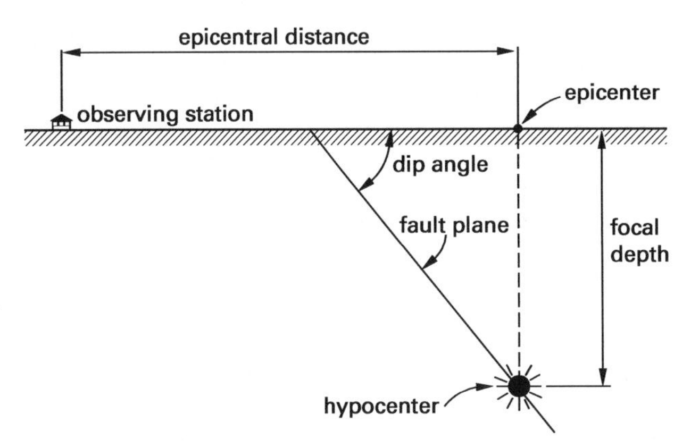

Earthquakes with focal depths of less than approximately 40 mi (60 km) are classified as *shallow earthquakes*. Very shallow earthquakes are caused by the fracturing of brittle rock in the crust or by internal strain energy that overcomes the friction locking opposite sides of a fault. California earthquakes are typically shallow.[2] *Intermediate earthquakes*, whose causes are not fully understood, have focal depths ranging from 40 mi to 190 mi (60 km to 300 km). *Deep earthquakes* may have focal depths of up to 450 mi (700 km).

The slip propagates from the hypocenter along the fault plane with a velocity up to that of the outward-radiating seismic shear wave front—about 1.8 mi/sec

[1]When the rock is stressed in shear, an analogous term for shear strain energy can be written. Both energy forms can be present simultaneously.

[2]Since there is no deep subduction zone (see Fig. 1.3), earthquakes in California typically occur at depths of less than 10 mi (15 km).

(3 km/s)—until the entire affected segment is in motion. (See Sec. 1.14 for a description of shear waves.)

3. GLOBAL SEISMICITY

Most earthquakes occur in areas bordering the Pacific Ocean.[3] This circum-Pacific belt, nicknamed the *ring of fire*, includes the Pacific coasts of North America and South America, the Aleutian Islands, Japan, Southeast Asia, and Australia. The reason for such a concentration is explained by plate tectonics theory. (See Sec. 1.5.)

The United States has experienced less destruction than other countries located in this earthquake zone. This is partly due to the country's relatively young age (and subsequently, the young age of its buildings) and its attention to earthquake-resistant construction methods. However, millions of Americans still live in potential earthquake areas such as in parts of the western United States, and particularly in California.[4] Because nuclear reactors, dams, schools, hospitals, and high-rise buildings are planned and built in locations of high seismicity, there remains an urgent need for greater attention to the mitigation of earthquake-induced damage.

4. CONTINENTAL DRIFT

The validity of *continental drift* (that the continents are moving relative to one another) was reasonably established during the 1930s, but was not universally accepted until the 1950s, when the emerging science of *paleomagnetism* provided new supporting evidence. Many rocks, such as volcanic rock solidified from molten lava, contain tiny grains of magnetic minerals such as *magnetite*. When these minerals are formed, they retain the magnetic orientation of the earth's magnetic field at the time of their formation. The magnetic orientations of rocks confirm the same ancient locations of the continents suggested by paleoclimatology and other geologic criteria. In fact, fossilized records of past climates (the subject of the field of *paleoclimatology*) indicate that the continents have been moving slowly about the globe for hundreds of millions of years. For example, the same 300-million-year-old fossilized deposits are found in India and in the Arctic.

An enormous amount of geophysical data was gathered during the 1950s and 1960s, particularly from oceanographic research vessels such as the *Glomar Challenger*. A system of interconnecting submarine ridges, called *mid-ocean ridges*, was discovered circling the earth. These ridges are located approximately midway between continents that are moving apart (e.g., between Africa and South America). New oceanic crust is formed at the ridges and is added to the plates moving apart. This is known as *seafloor spreading*.

Great submarine trenches were also located, particularly along the convex oceanic sides of the volcanic arcs that make up the Pacific ring of fire. Inclined zones of earthquakes dip down from these trenches as deep as 450 mi (700 km) into the mantle beneath and behind the volcanic arcs. Oceanic crust is formed at spreading ridges behind the moving plates.

The ocean crust, known to consist of alternating belts of highly and weakly magnetic oceanic crust material, represents magnetic records as new crust forms in the gaps behind separating plates. The symmetrical belts record the ambient magnetism on opposite sides of the newly formed ridges.

As crust is created, it is also destroyed elsewhere at the same rate as oceanic plates dip down at the trenches and slide deep into the mantle along the seismic zones. However, there is a global balance between crust formation and destruction. The formation of plate material in the Atlantic Ocean is compensated by absorption of plate material, primarily in the Pacific Ocean.

5. PLATE TECTONICS

Most earthquakes are a manifestation of the fragmentation of the earth's outer shell (known as the *lithosphere*) into various large and small plates. (The academic field that studies plate motion is known as *plate tectonics*.) There are seven very large plates, each consisting of both oceanic and continental portions. There are also a dozen or more small plates, not all of which are shown in Fig. 1.2.

Figure 1.2 *Lithosphere Plates*

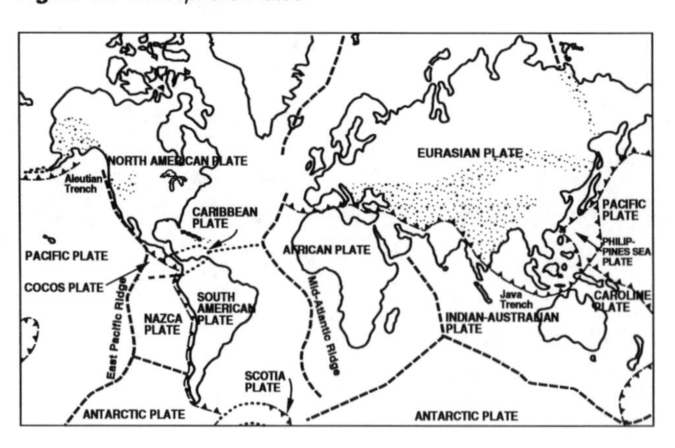

Each plate is approximately 50 mi to 60 mi (80 km to 100 km) thick, with some parts thicker or thinner than others. Thinner parts deform by elastic bending and brittle breakage. Thicker parts yield plastically. Beneath the plate is a viscous layer on which the entire plate slides. The plates themselves tend to be internally rigid, interacting only at the edges.

[3]The other major concentration of earthquakes is in a much smaller east-west belt that runs between Asia and the Mediterranean.

[4]It is interesting that the largest earthquakes on the North American continent in the history of the United States probably occurred in the east (the 1811 and 1812 New Madrid, Missouri, and the 1886 Charleston, South Carolina, earthquakes, the latter of which had an estimated Richter magnitude of 7.2–8.1). However, earthquakes in these regions are much less frequent than earthquakes in the western United States.

These plates move relative to one another with steady velocities that approach 0.4 ft/yr (0.13 m/yr).[5] Although plate velocities are slow by human standards, they are extremely rapid geologically. For example, a motion of 0.15 ft/yr (0.05 m/yr) adds up to 30 mi (50 km) in only 1 million years. Some plate motions have been continuous for 100 million years.

Depending on location, the plates can be moving apart, colliding slowly to build mountain ranges, or slipping laterally past or sliding over and under one another.

6. SUBMARINE RIDGES

When plates pull apart, hot material from the deeper mantle wells up to fill the gap. Some of the mantle material appears as lava in volcanoes. Most solidifies beneath the ocean surface, forming a *submarine ridge.* The ridge is high relative to the ocean bottom because the mantle material is hot and low in density.

As the plates move apart, the ridge material gradually cools and contracts, and its surface sinks. Ridges generally form step-like alterations in height perpendicular to the direction of plate motion. Strike-slip faults form parallel to the direction of plate motion. (See Sec. 1.10.)

7. SUBMARINE TRENCHES

Where plates converge, one dips down and slides beneath the other in a process known as *subduction.* Generally, an oceanic plate slides, or subducts, beneath a continental plate (as is happening along the west coast of South America) or beneath another oceanic plate (as is happening along the east side of the Philippine Sea plate). A trench is formed where the subducting plate dips down. The sediment from the ocean floor is scraped off against the front edge of the top plate. This is illustrated in Fig. 1.3.[6]

Far back under the top plate, inclined zones of earthquakes reach down into the mantle. The average depth of these zones is approximately 80 mi (125 km), but the zones can approach 450 mi (700 km) in depth. The hypocenters of earthquakes in these zones indicate the trajectory of the subducted plate.

A belt of volcanoes typically occurs above this earthquake zone, roughly paralleling the plate edges. (The Pacific Ocean coastal region of South America is typical of such an area.) Rock melting, which ultimately produces the volcanoes, starts when water combined in the crystalline structures of various minerals, or otherwise trapped, is removed by the increase in pressure on the subducted plate. The water loss lowers the net energy required to melt the remaining rock.

[5]There are grounds for suggesting that the African plate may be fixed relative to the deep mantle. If so, it is the only major plate that is fixed.

[6]Details and dimensions are those for western Java and the Java trench system, but other systems are similar.

Figure 1.3 *Zone of Subduction*

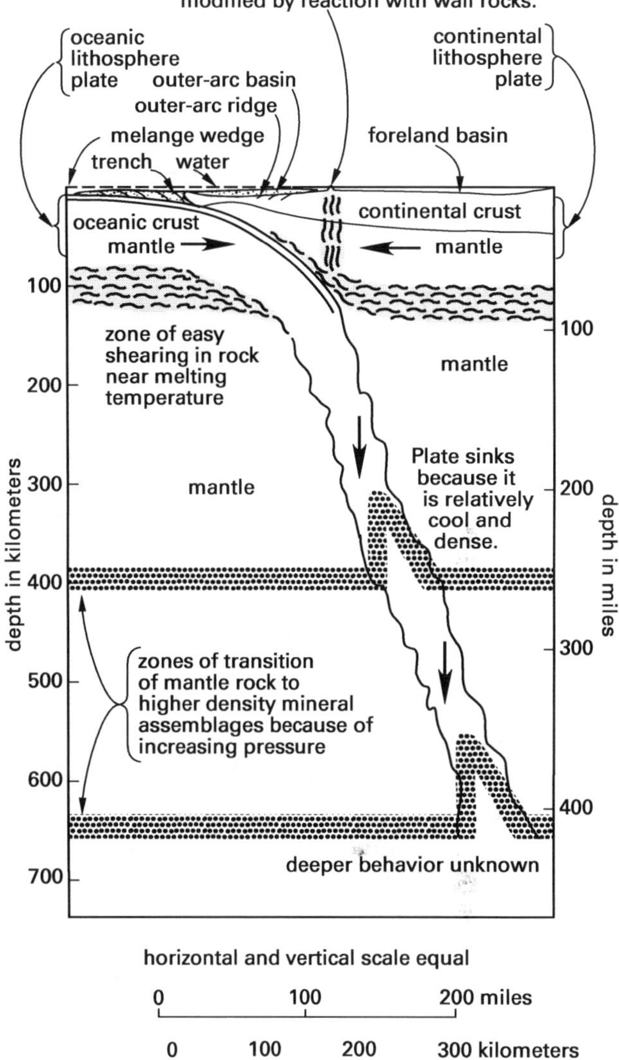

8. EARTHQUAKE ENERGY RELEASE[7]

Shallow earthquakes represent sudden slippages and are accompanied by a release of elastic strain energy stored in the rock over a long period. It is not completely clear whether deep mantle subduction-zone earthquakes are accompanied by similar elastic releases (i.e., the elastic rebound theory) or are merely abrupt contractions of part of the subducting plate into rock of higher density (i.e., an aspect of the *dilatational theory*). However, recent research seems to indicate the former explanation is more appropriate than the latter.

Only a fraction of the energy released in an earthquake actually appears in seismic waves. Most of the released

[7]See also Sec. 2.5.

strain energy is reabsorbed locally by the moving, deforming, and heating of the rock. The fraction absorbed increases irregularly with the increasing size of earthquakes. Minor earthquakes generally do not represent a sufficient release of energy to dissipate the strain energy and prevent great earthquakes, although a slow creep along a fault can provide a partial release. A great earthquake, however, does not necessarily release all of the strain energy, either.

Great earthquakes occur primarily along convergent (subducting) plate boundaries.[8] Submerged ridges (where plates are spreading apart) are so hot at relatively shallow depths that the solid rock above them cannot store enough elastic strain energy to produce great earthquakes. The infrequent large earthquakes that do occur in these ridge systems are mostly on the longer strike-slip faults. (See Sec. 1.10.)

9. SEISMIC SEA WAVES

When the seafloor suddenly rises up during a great earthquake, water also rises with it and, then, rushes away to find a level surface. If the floor drops, water rushes in. An enormous mass of water is suddenly set in motion, and a complex sloshing back and forth between continents continues for many hours. The result is a train of surface-water waves, each of which is known as a *seismic sea wave* (also known as a *tidal wave*), or, in Japanese, a *tsunami*. The most pronounced sudden changes in seafloor depth, and hence the greatest sea waves, result from shallow subduction-zone earthquakes.

As with any surface wave or surge wave, the velocity of a tsunami depends primarily on the ocean depth.[9] The greater the depth, the greater the energy content of the tsunami. In deep ocean, waves travel at about 500 mi/hr (800 km/h). The waves at sea may be an hour apart and perhaps only 1 ft (0.3 m) in height. Combined with a wave period of 5 min to 60 min, they are virtually undetectable. As a wave approaches land, however, the wave velocity decreases due to increased friction with the increasingly shallow seafloor. As the wave velocity decreases, the wave height increases.

Only normal (dip-slip) and thrust (reverse) faults produce tsunamis. Where seafloor topography and orientation are optimal for tsunami formation (where there is a gently sloping seafloor and where the slope is parallel to wave direction), the wave can form a wall of water more than 50 ft (15 m) in height. Such a wave can cause enormous destruction when it rushes onto shore. Nearby coastal points, where the bottom configuration is much different (i.e., more abrupt in depth change), may see the same wave pass as only a rapid surge and withdrawal of water.

[8]The great 1906 San Francisco earthquake, however, was not a subduction-zone earthquake.

[9]This is greatly simplified. There has been much research on the effect of depth and other aspects of tsunami generation and propagation. A good survey of the subject is contained in Murty (1977).

10. FAULTS

A fault is a fracture in the earth's crust along which two blocks slip relative to each other. One crustal block may move horizontally in one direction while the opposite block moves horizontally in the opposite direction. Alternatively, one block may move upward while the other moves downward.

One of the ways movement along faults can occur is by sudden displacement, or *slip*, of the crust or rock along a fault. During the 1906 San Francisco earthquake, the ground was displaced as much as 21 ft (6.5 m) in northern California along the San Andreas Fault. By comparison, the 1989 Loma Prieta earthquake had a maximum displacement of approximately 6 ft (2 m).

Most of the faults in California are vertical or near-vertical breaks. Movement along these breaks is predominantly horizontal in the northerly or northwesterly direction.[10] With *right-lateral movement* (such as the movement of earthquakes in the San Andreas system), a block on the opposite side of the fault (relative to an observer) moves to the right. Conversely, the block moves to the left in a *left-lateral fault*. Lateral movement is produced by *strike-slip* (*wrench*) *faults*.

A fault in which the movement is vertical is called a *dip-slip fault*. In a *normal fault*, the hanging wall moves down relative to the foot wall. In a *reverse fault*, also known as a *thrust fault*, the hanging wall moves up relative to the foot wall.

Along many faults movement is both horizontal and vertical. Such faults are named by combining the names of each kind of movement. For example, Fig. 1.4 shows a left-lateral normal fault. The term *oblique fault* is also used.

A few *reverse faults* have been active in California. The planes of such faults are inclined to the earth's surface. The rocks above the fault plane have been thrust upward and over the rocks below the fault plane. The Arvin-Tehachapi earthquake of 1952 was caused by the White Wolf reverse fault. The San Fernando earthquake of 1971 was caused by a sudden rupture along a reverse fault at the foot of the San Gabriel Mountains.

11. CALIFORNIA FAULTS

The most earthquake-prone areas in the United States are those that are adjacent to the San Andreas Fault system of California, as well as the fault system that separates the Sierra Nevada from the Great Basin. Many of the individual faults of these major systems are known to have been active during the past 200 years. Others are believed to have been active since the end of the last great ice advance about 10,000 years ago.

[10]Notable exceptions are the Garlock and Big Pine left-lateral faults, which trend westerly.

Figure 1.4 Types of Faults

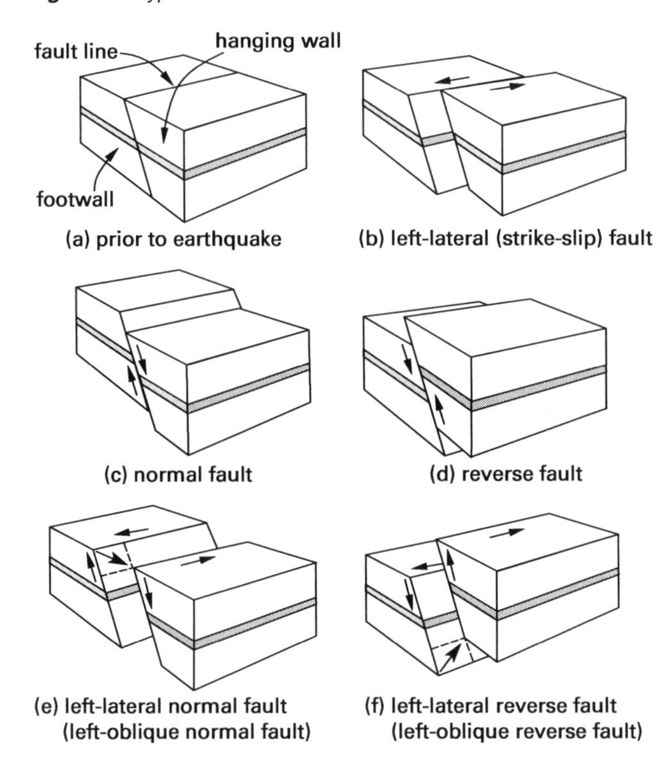

During the past 200 years, many of the faults shown in Fig. 1.5 have experienced either sudden slip or slow creep. Activity of other faults, however, can only be inferred from geologic and topographic relations that indicate the faults have been active during the past several thousand years. Such activity suggests that these faults could slip or creep again. (See Sec. 1.13.)

Earthquakes in California are relatively shallow and are clearly related to movement along active faults. Many California earthquakes have produced surface rupture.

12. SAN ANDREAS FAULT

The San Andreas Fault is the major fault in a network of faults that cuts through rocks of the California coastal region. This right-lateral fault is a huge fracture more than 600 mi (950 km) long. It extends almost vertically into the earth to a depth of at least 20 mi (30 km). In detail, it is a complex zone of crushed and broken rock from only a few feet wide to a mile wide. Many smaller faults branch from and join the San Andreas Fault.

A linear trough in the surface of the earth reveals the presence of the San Andreas Fault over much of its length. From the air, the linear arrangements of lakes, bays, and valleys is apparent. On the ground, the fault zone can be recognized by long, straight escarpments, narrow ridges, and small, undrained ponds formed by the settling of small areas of rock. However, people on the ground usually do not realize when they are on or near the fault.

Geologists who have studied the fault between Los Angeles and San Francisco have suggested that the total accumulated displacement along the fault may be as much as 350 mi (550 km). Similarly, geological study of a segment of the fault between the Tejon Pass and the Salton Sea has revealed geologically similar terrains on opposite sides of the fault separated by 150 mi (250 km). This indicates that the separation is a result of movement along the San Andreas and branching San Gabriel faults.

Since 1934, earthquake activity along the San Andreas Fault system has been concentrated in three areas: (1) an off-shore area at the northernmost tip of the fault known as the *Mendocino fracture zone*, (2) the area along the fault between San Francisco and Parkfield, and (3) the southernmost fault section roughly bounded by Los Angeles and the border with Mexico. Creep, slip, and moderate earthquakes have occurred on a regular basis in these areas.

The two zones between these three active areas have had almost no earthquakes or known slip since the great earthquakes of 1857 in the southern segment and 1906 in the northern segment. This implies that these two zones of the San Andreas Fault system are temporarily locked and that strain energy is building. The lack of seismic activity in the locked sections could mean that the sections are subject to less frequent but larger fault movements and, correspondingly, more severe earthquakes.

13. CREEP

In addition to fault rupture, a second type of fault movement known as *creep* can occur. Creep is characterized by continuous or intermittent movement without noticeable earthquakes. Fault creep occurring on portions of the Hayward, Calaveras, and San Andreas faults has produced cumulative offsets ranging from mere millimeters to almost 1 ft (0.3 m) in curbs, streets, and railroad tracks.

The offsets observed seem consistent with the creep rate measured. Precise surveying shows a slow drift approaching 2 in/yr (5 cm/yr) in some places along the San Andreas Fault. Over 350 mi (550 km) of offset has occurred during the past 100 million years.

14. SEISMIC WAVES

Seismic waves are of three types: compression, shear, and surface waves. (See Fig. 1.6.) Compression and shear waves travel from the hypocenter through the earth's interior to distant points on the surface. Only compression waves, however, can pass through the earth's molten core. Because *compression waves* (also known as *longitudinal waves*) travel at great speeds (19,000 ft/sec, or 5800 m/s, in granite) and ordinarily

Figure 1.5 *Active California Faults*

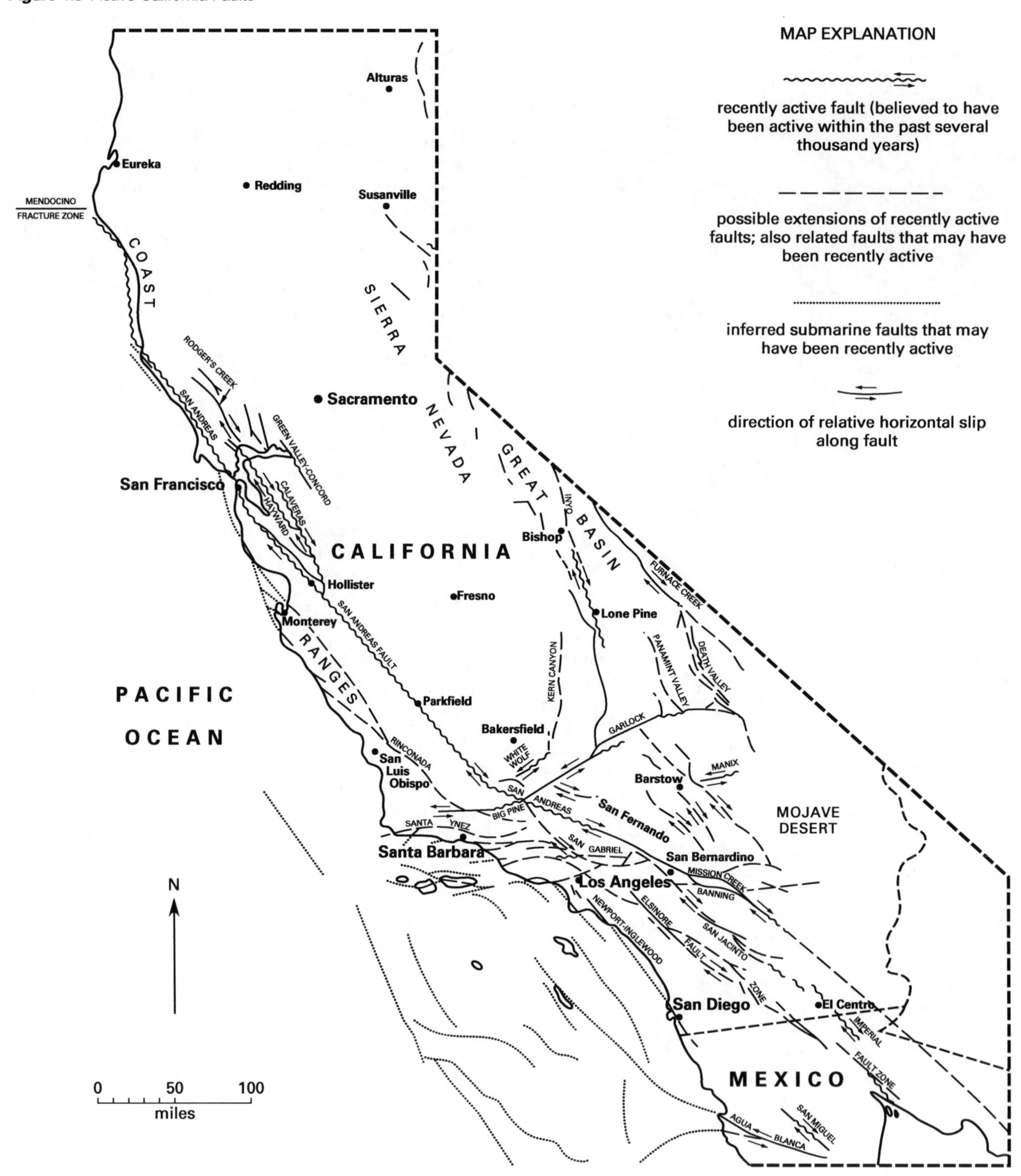

Figure 1.6 *Types of Seismic Waves*

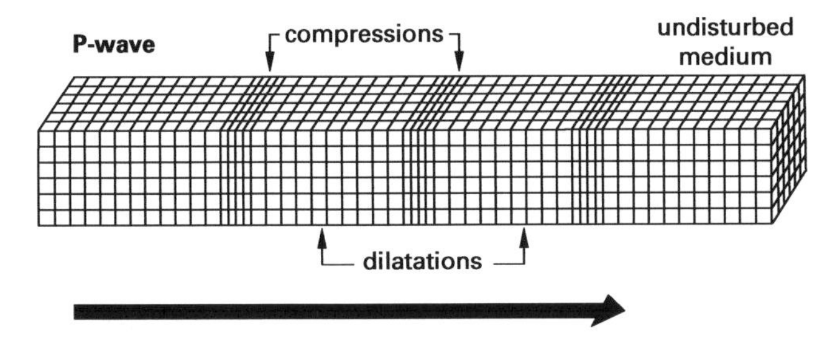

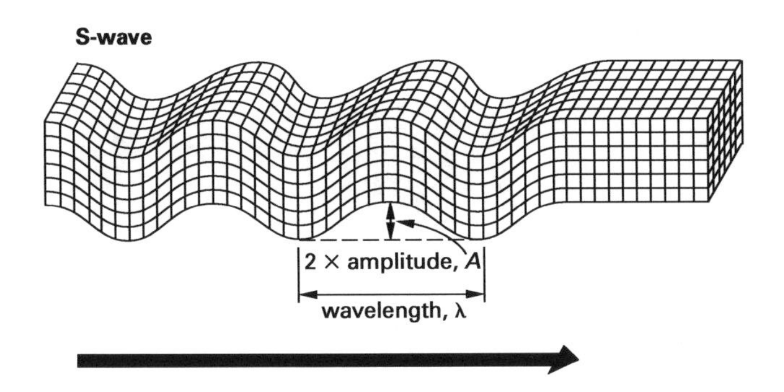

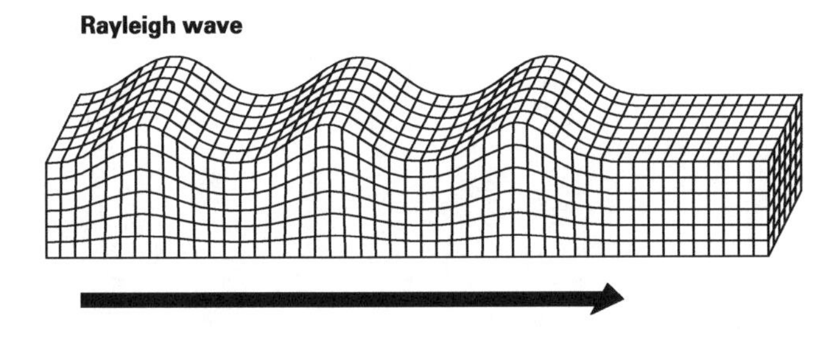

reach the surface first, they are known as *P-waves* (for "primary waves").[11] P-wave velocity is given by Eq. 1.3.

$$v_P = \sqrt{\frac{(\lambda + 2G)g_c}{\rho}} \quad \text{[U.S.]} \quad 1.3(a)$$

$$v_P = \sqrt{\frac{\lambda + 2G}{\rho}} \quad \text{[SI]} \quad 1.3(b)$$

In Eq. 1.3, λ is Lamé's constant,

$$\lambda = \frac{G(E - 2G)}{3G - E} = \frac{\nu E}{(1+\nu)(1-2\nu)}$$

Shear waves (also known as *transverse waves*) do not travel as rapidly (10,000 ft/sec, or 3000 m/s, in granite) through the earth's crust and mantle as do compression waves. Because they ordinarily reach the surface later, they are known as *S-waves* (for "secondary waves"). Instead of affecting material directly behind or ahead of their lines of travel, shear waves displace material at right angles to their path. Equation 1.4 gives the velocity of S-waves. While S-waves travel more slowly than P-waves, they transmit more energy and cause the majority of damage to structures.

$$v_S = \sqrt{\frac{Gg_c}{\rho}} \quad \text{[U.S.]} \quad 1.4(a)$$

$$v_S = \sqrt{\frac{G}{\rho}} \quad \text{[SI]} \quad 1.4(b)$$

Surface waves, also known as *R-waves* (for "Rayleigh waves") or *L-waves* (for "Love waves"), may or may not form. They arrive after the primary and secondary waves. In granite, R-waves move at approximately 9000 ft/sec (2700 m/s).

15. LOCATING THE EPICENTER

The first indication of an earthquake will often be a sharp "thud" signaling the arrival of the compression wave front. This will be followed by the shear waves and, then, the ground roll caused by the surface waves. The times separating the arrivals of the compression and shear waves at various seismometer stations can be used to locate the epicenter's position and depth.

The distance, s, from a seismometer to the epicenter can be determined from the wave velocities and the observed time between the arrival of the compression (P-) and shear (S-) waves.[12]

$$t_S - t_P = \left(\frac{1}{v_S} - \frac{1}{v_P}\right)s \quad 1.5$$

The epicenter and hypocenter correspond to the locations of initial fault slip but do not necessarily coincide with the center of energy release. For small and medium earthquakes (i.e., Richter magnitude $M < 6$), the points of initial fault slip and energy release are relatively close. For larger earthquakes, however, hundreds of kilometers can separate the two.

[11]The term *dilatation* is used to describe negative compression (i.e., the "expansion" of rock from its normal density). (See Fig. 1.6.)

[12]It is not always possible to accurately determine the difference in arrival times from the seismometer record.

2 Earthquake Characteristics

1. INTENSITY SCALE

The *intensity* of an earthquake (not to be confused with *magnitude*, see Sec. 2.3) is based on the damage and other observed effects on people, buildings, and other features. Intensity varies from place to place within the disturbed region. An earthquake in a densely populated area that results in many deaths and considerable damage may have the same magnitude as a shock in a remote area that does nothing more than frighten the wildlife. Large magnitude earthquakes that occur beneath the oceans may not even be felt by humans.

An intensity scale consists of a series of responses, such as people awakening, furniture moving, and chimneys being damaged. Although numerous intensity scales have been developed, the scale encountered most often in the United States is the *Modified Mercalli Intensity scale*, developed in 1931 by the American seismologists Harry Wood and Frank Neumann.[1]

The Modified Mercalli scale, given in Table 2.1, consists of 12 increasing levels of intensity (expressed as roman numerals following the initials MM) that range from imperceptible shaking to catastrophic destruction. The lower numbers of the intensity scale generally are based on the manner in which the earthquake is felt by people. The higher numbers are based on observed structural damage. The numerals do not have a mathematical basis and, therefore, are more meaningful to nontechnical people than to those in technical fields.

Table 2.1 *Modified Mercalli Intensity Scale*

intensity	observed effects of earthquake
I	Not felt except by very few under especially favorable conditions.
II	Felt only by a few persons at rest, especially by those on upper floors of buildings. Delicately suspended objects may swing.
III	Felt quite noticeably by persons indoors, especially in upper floors of buildings. Many people do not recognize it as an earthquake. Standing vehicles may rock slightly. Vibrations similar to the passing of a truck. Duration estimated.
IV	During the day, felt indoors by many, outdoors by a few. At night, some awakened. Dishes, windows, doors disturbed; walls make cracking sound. Sensation like heavy truck striking building. Standing vehicles rock noticeably.
V	Felt by nearly everyone; many awakened. Some dishes, windows broken. Unstable objects overturned. Pendulum clocks may stop.
VI	Felt by all, many frightened. Some heavy furniture moved. A few instances of fallen plaster. Damage slight.
VII	Damage negligible in buildings of good design and construction; slight to moderate in well-built ordinary structures; considerable damage in poorly built structures. Some chimneys broken.
VIII	Damage slight in specially designed structures; considerable damage in ordinary substantial buildings, with partial collapse. Damage great in poorly built structures. Fallen chimneys, factory stacks, columns, monuments, walls. Heavy furniture overturned.
IX	Damage considerable in specially designed structures; well-designed frame structures thrown out of plumb. Damage great in substantial buildings, with partial collapse. Buildings shifted off foundations.
X	Some well-built wooden structures destroyed; most masonry and frame structures with foundations destroyed. Rails bent.
XI	Few, if any, masonry structures remain standing. Bridges destroyed. Rails bent greatly.
XII	Damage total. Lines of sight and level are distorted. Objects thrown into air.

[1]The original Mercalli scale was developed in 1902 by the Italian seismologist and volcanologist of the same name. The *Rossi-Forel scale* (its ten values are used in Fig. 2.1 to describe the 1906 San Francisco earthquake) was developed in the 1880s.

2. ISOSEISMAL MAPS

It is possible to compile a map of earthquake intensity over a region. Data for such an *isoseismal map* can be obtained by observation or, in some cases, by questions answered by residents after an earthquake. (See Fig. 2.1.)

Figure 2.1 *Isoseismal Map of 1906 San Francisco Earthquake (based on the Rossi-Forel Intensity scale)*

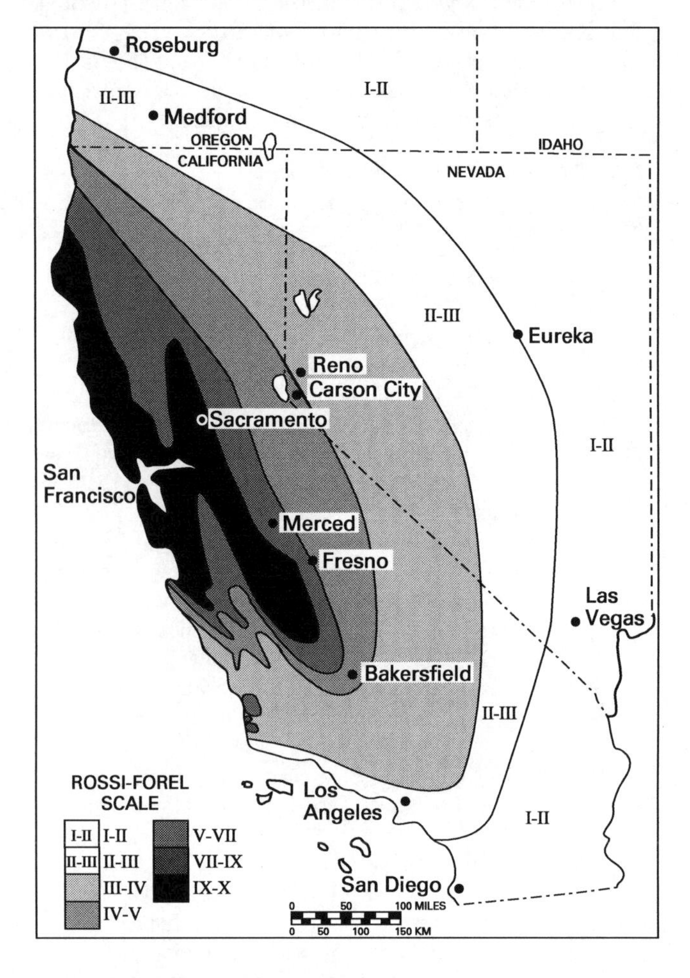

Source: *Earthquake Engineering*, Robert L. Wiegel, ed., copyright © 1970 by Prentice-Hall, Inc. Redrawn from map 23 in *Report of the State Earthquake Investigation Committee*, Atla, Carnegie Institution of Washington, publication 87 (1908).

3. RICHTER MAGNITUDE SCALE

In 1935, Charles F. Richter of the California Institute of Technology developed the Richter magnitude scale to measure earthquake strength. The magnitude, M, of an earthquake is determined from the logarithm to base ten of the amplitude recorded by a seismometer. (See Sec. 2.4 and Sec. 2.5.) Adjustments are included in the magnitude to compensate for the variation in the distance between the various seismometers and the epicenter. Because the Richter magnitude is a logarithmic scale, each whole number increase in magnitude represents a tenfold increase in measured amplitude.

Richter magnitude is expressed in whole numbers and decimal fractions. For example, a magnitude of 5.3 might correspond to a *moderate earthquake*. A *strong earthquake* might be rated at 7.3. *Great earthquakes* have magnitudes above 7.5. Earthquakes with magnitudes of 2.0 or less are known as *microearthquakes*. While recorded on seismometers, microearthquakes are rarely felt by people.

Earthquakes of 4.5 and below have little potential to cause structural damage. Several thousand seismic events with magnitudes of approximately 4.5 or greater occur each year and are strong enough to be recorded by seismometers all over the world. Great earthquakes, such as the 1906 San Francisco earthquake and the 1964 Alaskan earthquake, occur, on the average, once each year someplace in the world.

The magnitude of an earthquake depends on the length and breadth of the *fault slip*, as well as on the amount of slip. The largest examples of fault slip recorded in California accompanied the earthquakes of 1857, 1872, and 1906—all of which had estimated magnitudes over 8.0 on the Richter scale.

Although the Richter scale has no lower or upper limit (i.e., it is "open ended"), the largest known shocks have had magnitudes in the 8.7 to 8.9 range.[2] The actual factor limiting energy release—and hence Richter magnitude—is the strength of the rocks in the earth's crust.[3]

Because of the physical limitations of the faults and crust in the area, earthquakes larger than 8.5 in Southern California are considered to be highly improbable.

4. RICHTER MAGNITUDE CALCULATION

The Richter magnitude, M, is now often designated M_L to signify local magnitude, and is calculated from the maximum amplitude, A, of the seismometer trace, as illustrated in Fig. 2.2. A_0 is the seismometer reading produced by an earthquake of standard size (i.e., a *calibration earthquake*). Generally, A_0 is 3.94×10^{-5} in (0.001 mm).

$$M = \log_{10}\frac{A}{A_0} \qquad 2.1$$

Equation 2.1 assumes that a distance of 62 mi (100 km) separates the seismometer and the epicenter. For other distances, the nomograph of Fig. 2.3 and the following

[2]Depending on the sensitivity of the seismograph, earthquakes with negative Richter magnitudes can occur.

[3]It is said that Richter magnitudes much higher than 9 would correspond to an energy release sufficient to destroy the earth itself. While this may theoretically be true for much larger Richter magnitudes (due to the logarithmic nature of the measurement), the limited strength of the rock itself serves to ensure that such a doomsday earthquake will never occur.

Figure 2.2 *Typical Seismometer Amplitude Trace*

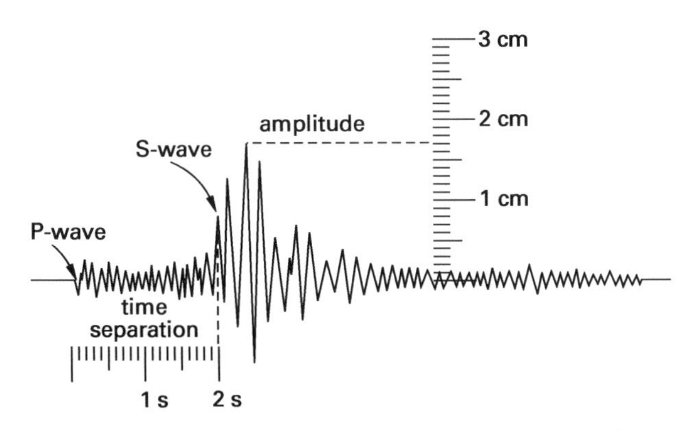

Figure 2.3 *Richter Magnitude Correction Nomograph*

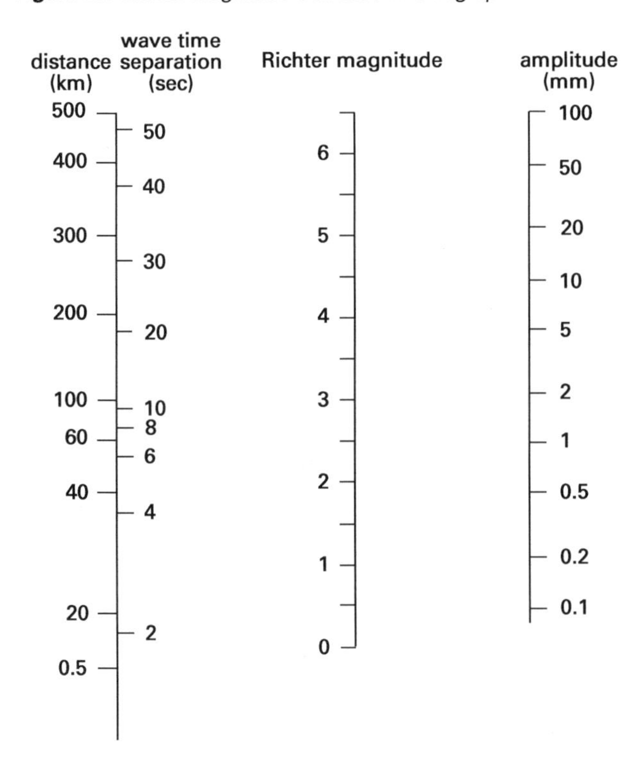

procedure can be used to calculate the magnitude. Due to the lack of reliable information on the nature of the earth between the observation point and the earthquake epicenter, an error of 5 mi to 20 mi (10 km to 40 km) in locating the epicenter is not unrealistic.

step 1: Determine the time between the arrival of the P- and S-waves.

step 2: Determine the maximum amplitude of oscillation.

step 3: Connect the arrival time difference on the left scale and the amplitude on the right scale with a straight line.

step 4: Read the Richter magnitude on the center scale.

step 5: Read the distance separating the seismometer and the epicenter from the left scale.

Whereas one seismometer can determine the approximate distance to the epicenter, it takes three seismometers to determine and verify the location of the epicenter.

5. ENERGY RELEASE AND MAGNITUDE CORRELATION

Since the Richter magnitude is a logarithmic quantity, an increase of 1.0 in the magnitude represents a tenfold increase in recorded shaking amplitude. However, the energy release of an earthquake scales with the $^3/_2$ power of the shaking amplitude. For every increase in Richter magnitude of 1.0, the energy release increases by $(10^{1.0})^{3/2} = 31.6$. A magnitude increase of 2.0 is equivalent to an energy increase of approximately $(10^{2.0})^{3/2} = 1000$.

Once the Richter magnitude, M, is known, an approximate relationship can be used to calculate the energy, E, radiated. Most of the relationships are of the form of Eq. 2.2.

$$\log_{10} E = \log_{10} E_0 + aM \qquad 2.2$$

In 1956, Gutenberg and Richter determined the approximate correlation to be as given in Eq. 2.3. E is the energy in ergs. (See App. A for conversions to other units.) Although there have been other relationships developed, Eq. 2.3 has been verified against data from underground explosions and is the primary correlation cited.

$$\log_{10} E = 11.8 + 1.5M \qquad 2.3$$

The radiated energy is less than the total energy released by the earthquake. The difference goes into heat generation and other nonelastic effects, which are not included in Eq. 2.3. Little is known about the amount of total energy release.

The fact that a fault zone has experienced an earthquake offers no assurance that enough stress has been relieved to prevent another earthquake. As indicated by the logarithmic relationship between seismic energy and Richter magnitude, a small earthquake (of magnitude 5, for example) would radiate approximately only $^1/_{32}$ of the energy of an earthquake just one magnitude larger (of magnitude 6, for example). Therefore, it would take 32 small earthquakes to release the same energy as an earthquake one magnitude larger.

6. LENGTH OF ACTIVE FAULT

Equation 2.4 correlates the Richter magnitude, M, with the approximate total fault length, L in kilometers, involved in an earthquake. Such correlations are very site dependent, and even then, there is considerable scatter in such data. Equation 2.4 should be considered only representative of the general (approximate) form of the correlation.

$$\log_{10}L = 1.02M - 5.77 \qquad 2.4$$

7. LENGTH OF FAULT SLIP

Equation 2.5 (as derived by King and Knopoff in 1968) correlates the Richter magnitude, M, and the fault length, L in meters, with the approximate length of vertical or horizontal fault slip, D (for *displacement*) in meters.[4] As with Eq. 2.4, this correlation should be considered representative of the general relationship.

$$\log_{10}(LD^2 \times 10^6) = 1.90M - 2.65 \qquad 2.5$$

8. PEAK GROUND ACCELERATION

The *peak* (maximum) *ground acceleration*, PGA, is easily measured by a seismometer (see Sec. 2.14) or accelerometer (see Sec. 2.15) and is one of the most important characteristics of an earthquake.[5] The PGA can be given in various units, including ft/sec^2, in/sec^2, or m/s^2. However, it is most common to specify the PGA in "g's" (i.e., as a fraction or percent of gravitational acceleration).

$$\text{PGA} = \frac{a_{\text{ft/sec}^2}}{32.2} \times 100\% \qquad \text{[U.S.]} \qquad 2.6(a)$$

$$\text{PGA} = \frac{a_{\text{in/sec}^2}}{386} \times 100\% \qquad \text{[U.S.]} \qquad 2.6(b)$$

$$\text{PGA} = \frac{a_{\text{m/s}^2}}{9.81} \times 100\% \qquad \text{[SI]} \qquad 2.6(c)$$

Significant ground accelerations in California include 1.25 g (1971 San Fernando earthquake, Pacoima dam site), 0.50 g (1966 Parkfield earthquake), 0.65 g (1989 Loma Prieta earthquake), and 1.85 g (1992 Cape Mendocino earthquake).

Equation 2.7 (as determined by Gutenberg and Richter in 1956) is one of many approximate relationships between the Richter magnitude, M, and the PGA at the epicenter. Of course, the ground acceleration (in rock) will decrease as the distance from the epicenter increases, and for this reason, relationships called *attenuation equations* have been developed. (See Sec. 2.13.)

Blume's 1965 equation (see Eq. 2.7) for California earthquakes depends on the epicentral distance (R' in kilometers), the focal depth (h in kilometers), and a specific site factor, b.

$$\text{PGA}_{\text{gravities}} = \frac{y_o}{1 + \left(\frac{R'}{h}\right)^2}$$

$$\log y_o = -(b+3) + 0.81M - 0.027M^2 \qquad 2.7$$

Attenuation equations are very site dependent. Since Eq. 2.7 was developed, newer studies have resulted in better correlations in different formats and for many different locations, but they are all based on limited data. Such studies regularly result in revisions of the seismic provisions of building codes.

Table 2.2 is a commonly cited correlation between magnitude, PGA, and duration of *strong-phase shaking* (see Sec. 2.14) in the vicinity of the epicenter of California earthquakes.[6] The values of acceleration in the table are somewhat on the high side. Ground acceleration in observed earthquakes is usually lower.

Table 2.2 *Approximate Peak Ground Acceleration and Duration of Strong-Phase Shaking (California earthquakes)*

magnitude	maximum acceleration (g)	duration (sec)
5.0	0.09	2
5.5	0.15	6
6.0	0.22	12
6.5	0.29	18
7.0	0.37	24
7.5	0.45	30
8.0	0.50	34
8.5	0.50	37

9. CORRELATION OF INTENSITY, MAGNITUDE, AND ACCELERATION WITH DAMAGE

Although there are some empirical relationships, no exact correlations of intensity, magnitude, and acceleration with damage are possible since many factors contribute to seismic behavior and structural performance. For example, seismic damage depends on the care that was taken at the time of building design and construction. Buildings in villages in undeveloped countries fare much worse than high-rise buildings in developed countries in earthquakes of equal magnitudes. This damage causes a corresponding lack of correlation between intensity and magnitude.

However, within a geographical region with consistent design and construction methods, fairly good correlation

[4]King, Chi-Yu, and L. Knopoff, "Stress Drop in Earthquakes," *Bulletin of the Seismological Society of America*, 58 (1968): 249.

[5]It is possible for an earthquake to exceed the range of the accelerometer or seismometer, in which case the PGA will not be recorded.

[6]Table 2.2 gives the impression of high correlation even though the correlation is actually low. For example, the 1989 Loma Prieta earthquake magnitude was approximately 7.1, but the peak ground acceleration was 0.65 g.

exists between structural performance and ground acceleration, because the Mercalli intensity scale is based specifically on observed damage. Table 2.3 gives an approximate relationship between the Mercalli intensity scale and ground acceleration.

Table 2.3 *Approximate Relationship Between Mercalli Intensity and Peak Ground Acceleration*

MMI	PGA (g)
IV	0.03 and below
V	0.03–0.08
VI	0.08–0.15
VII	0.15–0.25
VIII	0.25–0.45
IX	0.45–0.60
X	0.60–0.80
XI	0.80–0.90
XII	0.90 and above

10. VERTICAL ACCELERATION

The shear (transverse) waves are at right angles to the compression (longitudinal) waves. (See Sec. 1.14.) Since there is nothing constraining the shear waves to a horizontal direction, it is not surprising that the S-wave (shear wave) can be broken down into horizontal and vertical components. When necessary, these are identified as SH-waves and SV-waves for horizontal and vertical shear waves, respectively.

Vertical ground acceleration is known to occur in almost all earthquakes. The peak vertical acceleration is usually approximately one-third of the peak horizontal acceleration, but often reaches a ratio of two-thirds. Combined with resonance site effects, vertical forces can become substantial. Furthermore, forces from all three coordinate directions combine into a resultant force that can easily exceed the yield (and, sometimes, the ultimate) strength of a member.

Previous design codes were generally based on horizontal acceleration alone. This practice was justified by assuming that structures with horizontal seismic resistance will automatically have adequate vertical seismic resistance. One of the reasons this assumption was accepted is that factors of safety should have been applied during the building design to ensure that a member is able to withstand a force equal to one gravity downward.

Experience has shown, however, that disregarding details to resist vertical forces can be a serious problem. Columns and walls in compression, cantilever beams, overhangs, and prestressed concrete structures that have not been designed according to specific seismic provisions are particularly susceptible to damage by vertical accelerations because they have little factor of safety against upward vertical acceleration. *Transfer girders*, horizontal members that support exterior perimeter columns in *tube buildings* (see Sec. 5.18), are definitely sensitive to vertical acceleration. For that reason, ASCE/SEI7 requires designing for the maximum effect of horizontal and vertical earthquake forces [ASCE/SEI7 Sec. 12.4.2].

11. PROBABILITY OF OCCURRENCE[7]

The probability that an earthquake of magnitude M or greater will occur in a specific region in any given year is given approximately by Eq. 2.8.[8] B is a seismic parameter that has been approximately determined as 2.1 for the entire state of California, and 0.48 for 100,000 mi^2 (260 000 km^2) of Southern California. While Eq. 2.8 does not place any upper bound on M, the probability of exceeding 8.5 is effectively zero.

$$p\{M\} = e^{-M/B} \qquad 2.8$$

The expected number of earthquakes having magnitude greater than M during Y years is given by Eq. 2.9. Equation 2.9 is known as a *recurrence formula*. For northern California, $C = 76.7\ \text{yr}^{-1}$ and $B = 0.847$, approximately. For the San Francisco area, $C = 19{,}700\ \text{yr}^{-1}$ and $B = 0.463$, approximately.

$$N = CYe^{-M/B} \qquad 2.9$$

12. FREQUENCY OF OCCURRENCE

For a specific area, an equation for the expected number, N, of earthquakes of a given magnitude, M, per year will be of the form

$$\log_{10}N = a - bM \qquad 2.10$$

For the south-central segment of the San Andreas fault, a and b have values of 3.3 and 0.88, respectively. Taking the entire world as a whole, the approximate relationship (up to approximately $M = 8.2$) is

$$\log_{10}N = 7.7 - 0.9M \qquad 2.11$$

Table 2.4 gives the expected number of earthquakes of any given magnitude per 100 years in California. (The table does not give the frequency over any particular location in the state.) Table 2.4 cannot be derived exactly from Eq. 2.11 because adjustments have been made to account for California's increased seismicity. (See Fig. 2.4.)

13. ATTENUATION OF GROUND MOTION

Ground motion at a site is related to the seismic energy received at that site, and when the propagation path is through rock, the amount of energy decreases the

[7]Wiegel (1970) contains a more complete presentation of this subject.
[8]The form of Eq. 2.8 is easily derived from a Poisson distribution, which is commonly used to calculate the probability of an infrequent event.

Table 2.4 Approximate Expected Frequency of Occurrence of Earthquakes (per 100 years)

magnitude	number
4.75–5.25	250
5.25–5.75	140
5.75–6.25	78
6.25–6.75	40
6.75–7.25	19
7.25–7.75	7.6
7.75–8.25	2.1
8.25–8.75	0.6

farther a site is from the epicenter.[9] This decrease is known as *attenuation*. Some of the factors affecting attenuation include path line, path length, focal depth, geological formations, properties of the crustal rock, and orientation of the fault.

Unfortunately, the geology and local conditions affect the actual values so much that little more than generalizations such as the following are possible about the rate of attenuation.[10]

- Intensity generally decreases with distance from the epicenter.
- There is little attenuation in the vicinity of the epicenter.
- Higher-frequency components of the seismic wave attenuate faster than slower components do.

Figure 2.4 Expected Number of Earthquakes per Year (world)

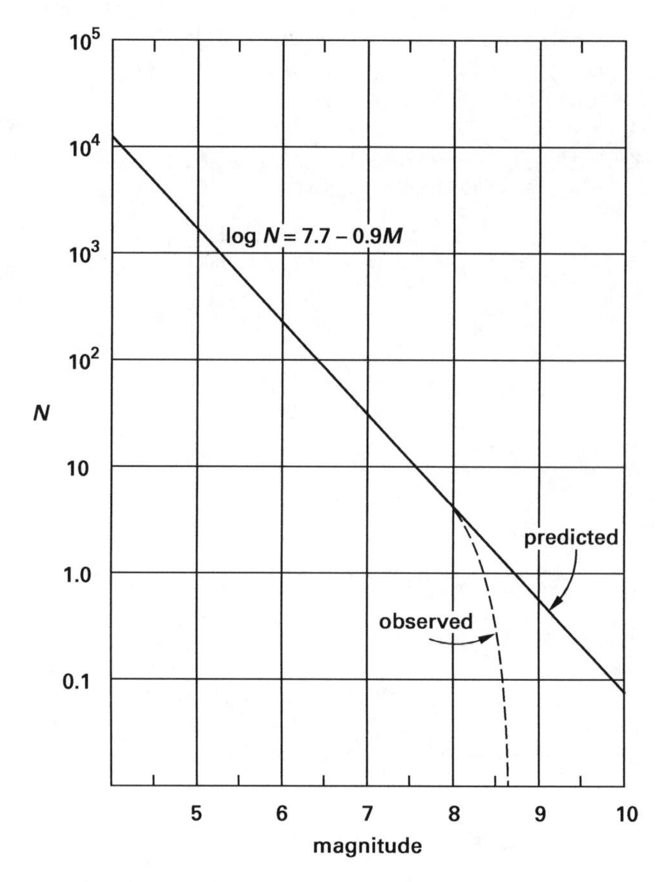

14. SEISMOMETER

Seismic waves (see Sec. 1.14) travel through the earth and are recorded on seismometers. A *seismometer* is the detecting and recording part of a larger apparatus known as a *seismograph*. Seismometers are pendulum-type devices that are mounted on the ground and measure the displacement of the ground with respect to a stationary reference point. Since a seismometer usually records motion in only one orthogonal direction, three seismometers are needed to record all components of ground motion. Figure 2.2 illustrates a typical seismometer trace, known as a *seismogram*. Appendix F is the actual seismogram of the 1940 El Centro earthquake.

Notice that while seismic activity usually continues for some time after the start of the earthquake, the major movement occurs in a concentrated period known as the *strong phase*. The longer the earthquake shakes, the more seismic energy is absorbed by buildings; thus, the duration of strong-phase shaking greatly affects the damage inflicted.[11]

Seismometers record the varying amplitude of ground oscillations beneath the instrument. Sensitive seismometers greatly magnify these ground motions and can detect strong earthquakes occurring anywhere in the world. The time, location, and magnitude of an earthquake can be determined from the data recorded by seismometer stations.

Since a seismometer is a spring-mass-dashpot device, it will magnify or distort earthquakes with frequencies in certain ranges. The ratio of actual damping to critical damping can be changed to minimize such distortion. Good seismometer design calls for a damping ratio of between 0.6 and 0.7 with a natural period of vibration

[9]While the overall wave energy attenuates when the transmission path is through rock, the damage at a site does not necessarily decrease with distance. There are other factors that can concentrate the energy that reaches the site, as was proved by the Mexico City (1985) and Loma Prieta (1989) earthquakes. This is analogous to the decrease in the intensity of sunlight with distance from the sun. While the light may be diffused when it reaches the earth's surface, it can be sufficiently concentrated with a lens to kindle a fire.

[10]Numerous attenuation relationships have been published, particularly for sites and faults in California. However, there is little similarity between the correlations. Nevertheless, for sites located on firm soil, the attenuation laws are quite useful for predicting expected ground motions from future earthquakes.

[11]For example, the 1985 Chile earthquake (magnitude 7.8) had almost 80 sec of strong ground motion. There were approximately 60 sec of strong ground motion in the 1985 Mexico earthquake (magnitude 8.1). Both of these earthquakes resulted in significant loss of life and destruction. By comparison, the strong-phase motion of the 1940 El Centro earthquake (magnitude 7.1) lasted only 10 sec, and the 1989 Loma Prieta earthquake (magnitude 7.1) had a mere 5 sec. There is great debate and no consensus on whether or not long-duration earthquakes can occur in California.

smaller than the smallest period to be measured.[12] (See Sec. 4.8 for damping ratio.)

15. ACCELEROMETER

An *accelerometer* (*accelerograph*) is a seismometer mounted in buildings for the purpose of recording large accelerations.[13] For this reason, they are also known as *strong motion seismometers*. The large swings accelerometers record typically exceed the scale limits of most seismometers. An accelerometer located in a building does not run continually. It is triggered by a P-wave (see Sec. 1.14) and runs for a fixed period of time.

16. OTHER SEISMIC INSTRUMENTS

A *tiltmeter* installed in the ground works on the same principle as a carpenter's level. The slightest movement of a bubble floating in a spherical dome is electronically detected to reveal tilting of the ground.

The strain (deformation) of rock under pressure can be measured by a *magnetometer*. Such strain changes the magnetic permeability of the rock, resulting in a local change in the magnetic field of the earth. *Strain gauges* measure how much the earth deforms.

Dilatometers measure the earth's dilations. A dilatometer is a closed, fluid-filled tube approximately 10 ft (3 m) long that is buried in the ground. Changes in the earth's "squeeze" are detected and measured by a pressure sensor or gauge at the top of the tube.

Scintillation counters are installed in wells to measure the amount of radioactive radon gas in the water. Minute amounts of radon are released into well water by rocks under stress.

Changes in the resistance of rock can be measured by a *resistivity gauge* and are indications of density and water content changes. Both density and water content change during periods of fluctuating stress.

A *creepmeter* measures minute gradual movement along a fault. In the past, such a meter relied on a wire stretched across a fault. Movement of the fault increased the tension in the wire. Current creepmeters use laser technology.

A *gravimeter* responds to variations in the local force of gravity. Such variations are the result of changes in underground rock density.

A *laser* can measure the round-trip travel time of a light beam between two points. When the relative positioning of the two points changes as a direct result of an earthquake, the travel time also changes.

17. EARTHQUAKE PREDICTION

While reliable long-term earthquake prediction remains elusive, short-term predictions based on the observation or measurement of various precursors (*premonitory signs*) seem possible. Most of the measuring devices mentioned in Sec. 2.14 through Sec. 2.16 can be adapted to instantaneous reporting. It may be possible to correlate sudden and unexpected changes in behavior (i.e., creep rate, tilt, accumulating strain, elevation, fluid pressure, seismic wave speed, electrical conductivity, and magnetic susceptibility) with the probability of an impending earthquake. Thus far, however, no reliable indicators have been found.

18. OTHER EARTHQUAKE CHARACTERISTICS

In addition to the peak ground acceleration, two other characteristics that contribute significantly to the effects of an earthquake are its duration of strong shaking (motion) and frequency content.[14] Roughly speaking, the longer the duration of strong shaking, the greater the energy that can be imparted to a structure. Since various parts of a structure can absorb only a limited amount of elastic strain energy, a longer earthquake has a greater chance of driving structural performance into inelastic behavior.

The shaking in the 1989 Loma Prieta earthquake lasted approximately 10 sec to 15 sec. It is generally believed that a longer duration (i.e., another 20 sec) would have resulted in significantly greater damage.

The effects of resonance on all types of machinery and structures are well known. Basically, if a regular disturbing force is applied at the same frequency as the natural frequency (see Sec. 4.6), the oscillation of the structure can be greatly magnified. In such cases, the effects of damping are minimal. While earthquakes are never as regular as a sinusoidal waveform, there is usually a predominant waveform that is roughly regular.[15]

[12]These design principles are incorporated into the Wood-Anderson seismometer, which has a damping ratio of approximately 0.8 (almost critical) and a natural period of 0.8 sec (i.e., a natural frequency of 1.25 Hz).

[13]The distinction between the terms *seismometer* and *accelerometer* is not always made. However, it is important to recognize that seismometers typically run continually and record displacement, while accelerometers are triggered by P-waves and record acceleration.

[14]This characteristic is defined as the duration of strong shaking, not the overall duration of the earthquake.

[15]Fourier analysis can be used to separate the dominant frequencies (i.e., the frequencies at which most seismic energy arrives at a site) of a specific earthquake. However, this is rarely done for most seismic design projects.

3 Effects of Earthquakes on Structures

1. SEISMIC DAMAGE

Structural damage due to an earthquake is not solely a function of the earthquake ground motion. The primary factors affecting the extent of damage are

- *earthquake characteristics*, such as (a) peak ground acceleration, (b) duration of strong shaking, (c) frequency content, and (d) length of fault rupture
- *site characteristics*, such as (a) distance between the epicenter and structure, (b) geology between the epicenter and structure, (c) soil conditions at the site, and (d) natural period of the site
- *structural characteristics*, such as (a) natural period and damping of the structure, (b) age and construction method of the structure, and (c) seismic provisions (i.e., detailing) included in the design

2. GEOGRAPHIC DISTRIBUTION OF SEISMIC RISK

In order to design a structure to withstand the effects of an earthquake, it is necessary to determine the expected earthquake magnitude. While extensive mathematical models could be developed for each location, seismic codes have evolved a simplified model based on the *geographic distribution of seismic risk*. The national maps referenced by the 2012 *International Building Code* (IBC) are shown in Fig. 3.1 and Fig. 3.2.[1]

Another interpretation of the significance of the different regions is to correlate them to the effects of an earthquake and the Modified Mercalli Intensity scale as in Table 3.1.

Table 3.1 *Effects of an Earthquake by Location*

location	effect
Minnesota	no damage
Ohio	minor damage corresponding to MM intensities V and VI; distant earthquakes may damage structures with fundamental periods greater than 1.0 sec
South Carolina	moderate damage corresponding to MM intensity VII
Oregon coast	major damage corresponding to MM intensity VIII
California coast	major damage corresponding to MM intensity VIII and higher

3. RISK MICROZONES

Certain limited areas, referred to as *microzones*, consistently experience higher ground accelerations than do surrounding areas.[2] This tendency is primarily attributed to the site (i.e., soil) conditions in the microzone, as the soil profile affects the peak ground acceleration, frequency content, and duration of strong motion. Inasmuch as seismic damage is at least partially related to ground acceleration, knowledge of such microzonification is essential. An initial attempt at the *microzonification* concept has been incorporated in the *International Building Code*, and it is implicit in the design process.

Mexico City has incorporated microzones into its rebuilding plan following the devastating earthquake in 1985. Considering the significant variations in damage in different areas of the San Francisco Bay Area during the Loma Prieta earthquake, implementing microzonification has begun. (For example, unreinforced buildings in Chinatown were not damaged, while similarly constructed buildings in the Marina district were. Portions of the Cypress structure on clay collapsed, but other portions built on firmer soil did not.)

[1]The maps shown in Fig. 3.1 and Fig. 3.2 are subject to change with different versions of the seismic code.

[2]For example, in the 1989 Loma Prieta earthquake, peak ground accelerations in San Francisco did not generally exceed 0.09 g. However, a 31-story instrumented office building in Emeryville experienced a horizontal acceleration of 0.26 g at the ground. Peak accelerations at the Bay Bridge are believed to have ranged between 0.22 g and 0.33 g. The Golden Gate Bridge experienced 0.24 g, while 0.33 g was recorded at the San Francisco Airport. Similar ground accelerations were recorded near the collapsed Interstate 880 Cypress structure.

Figure 3.1 *Geographic Distribution of Seismic Risk (0.2 sec spectral response acceleration) [IBC Fig. 1613.3.1(1)]*

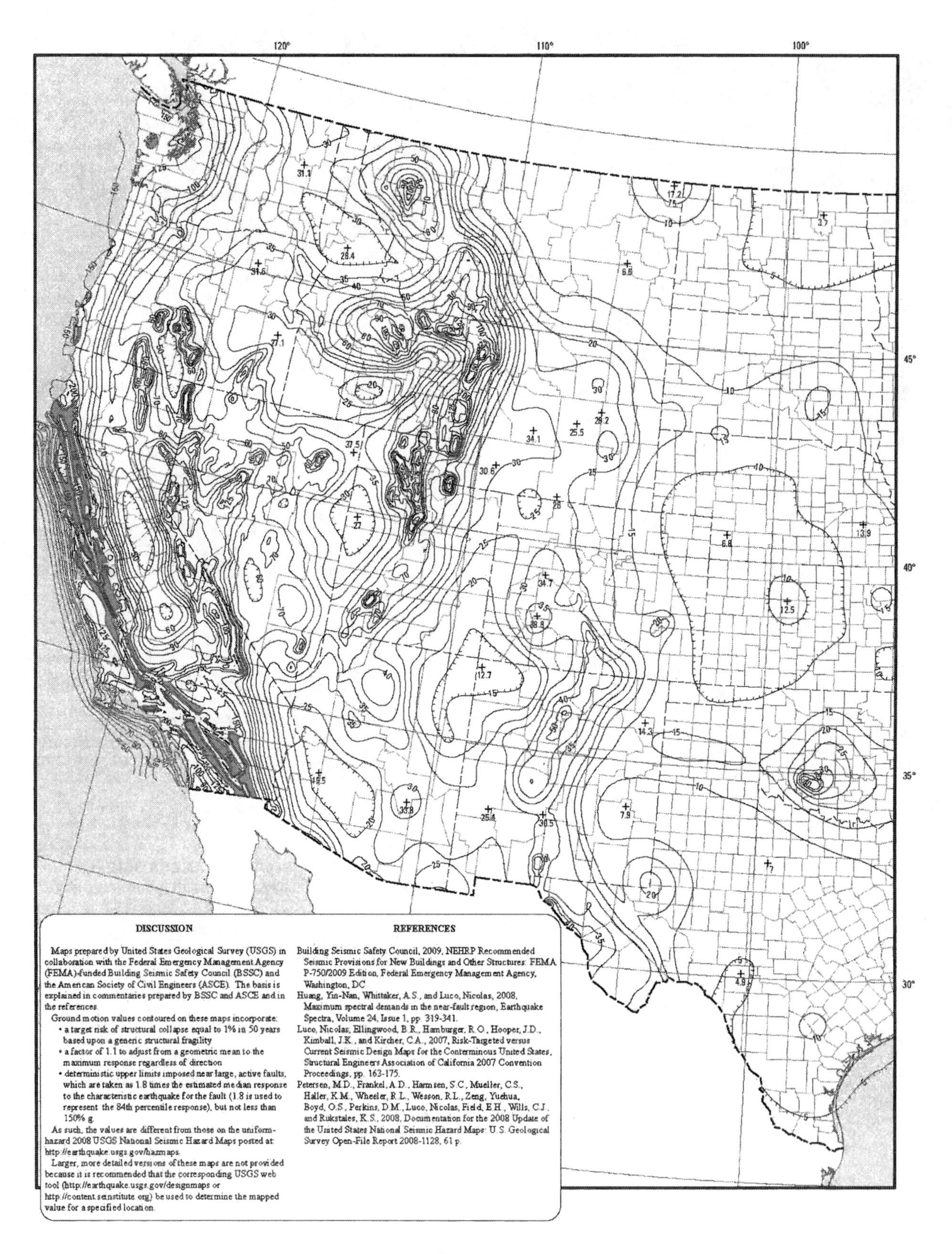

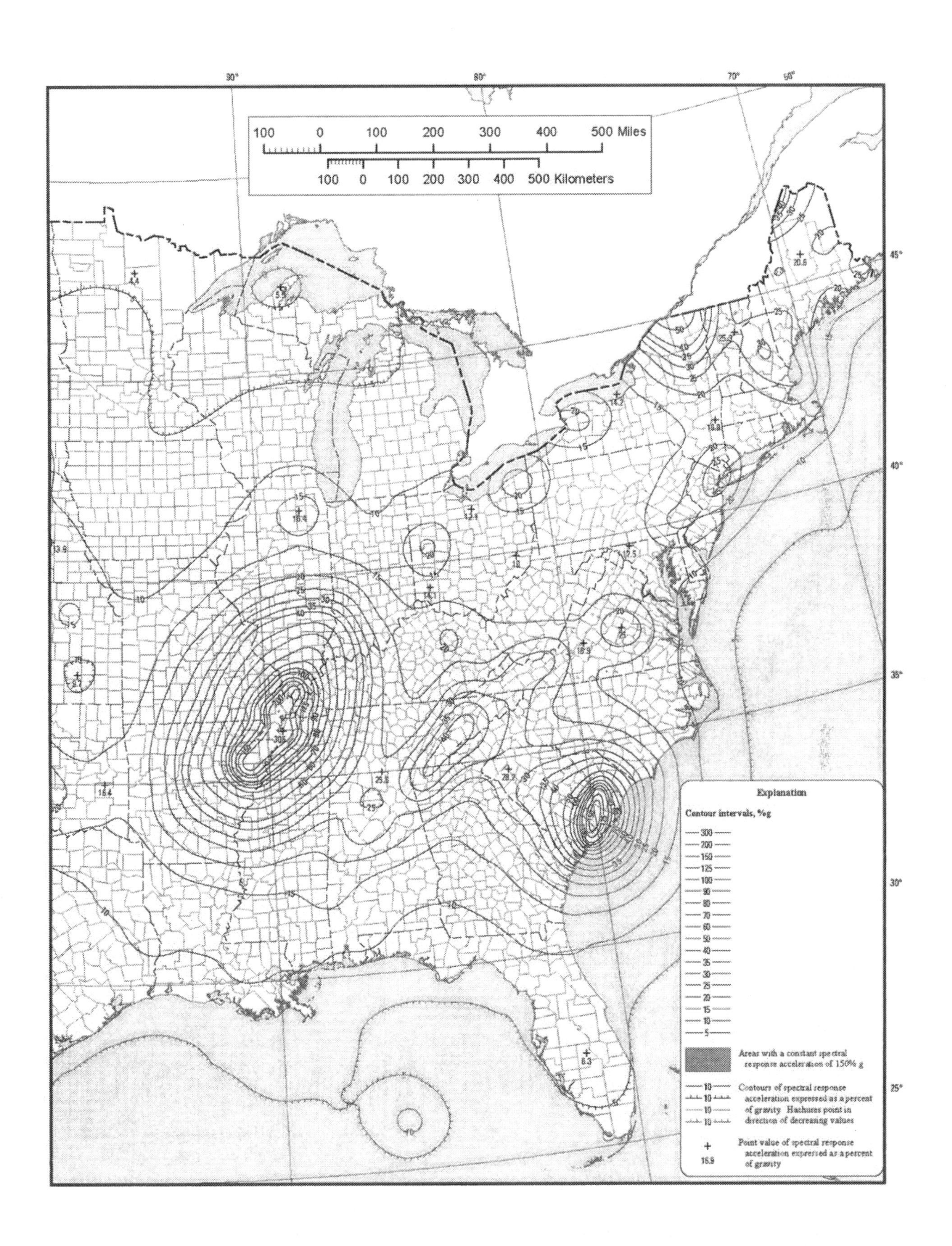
100 0 100 200 300 400 500 Miles
100 0 100 200 300 400 500 Kilometers
Explanation
Contour intervals, %g
Areas with a constant spectral response acceleration of 150% g
Contours of spectral response acceleration expressed as a percent of gravity. Hachures point in direction of decreasing values
Point value of spectral response acceleration expressed as a percent of gravity

Figure 3.2 *Geographic Distribution of Seismic Risk (1 sec spectral response acceleration) [IBC Fig. 1613.3.1(2)]*

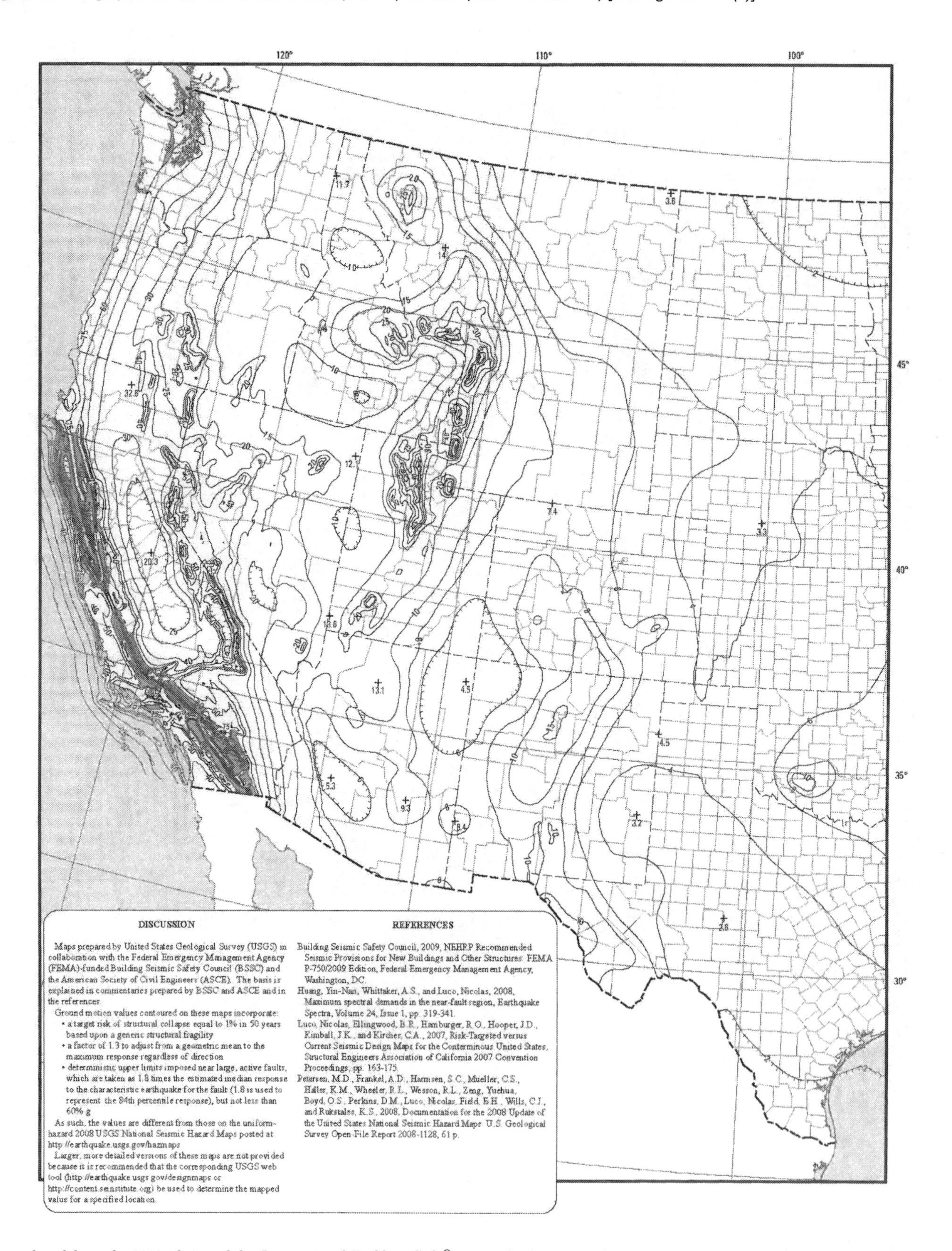

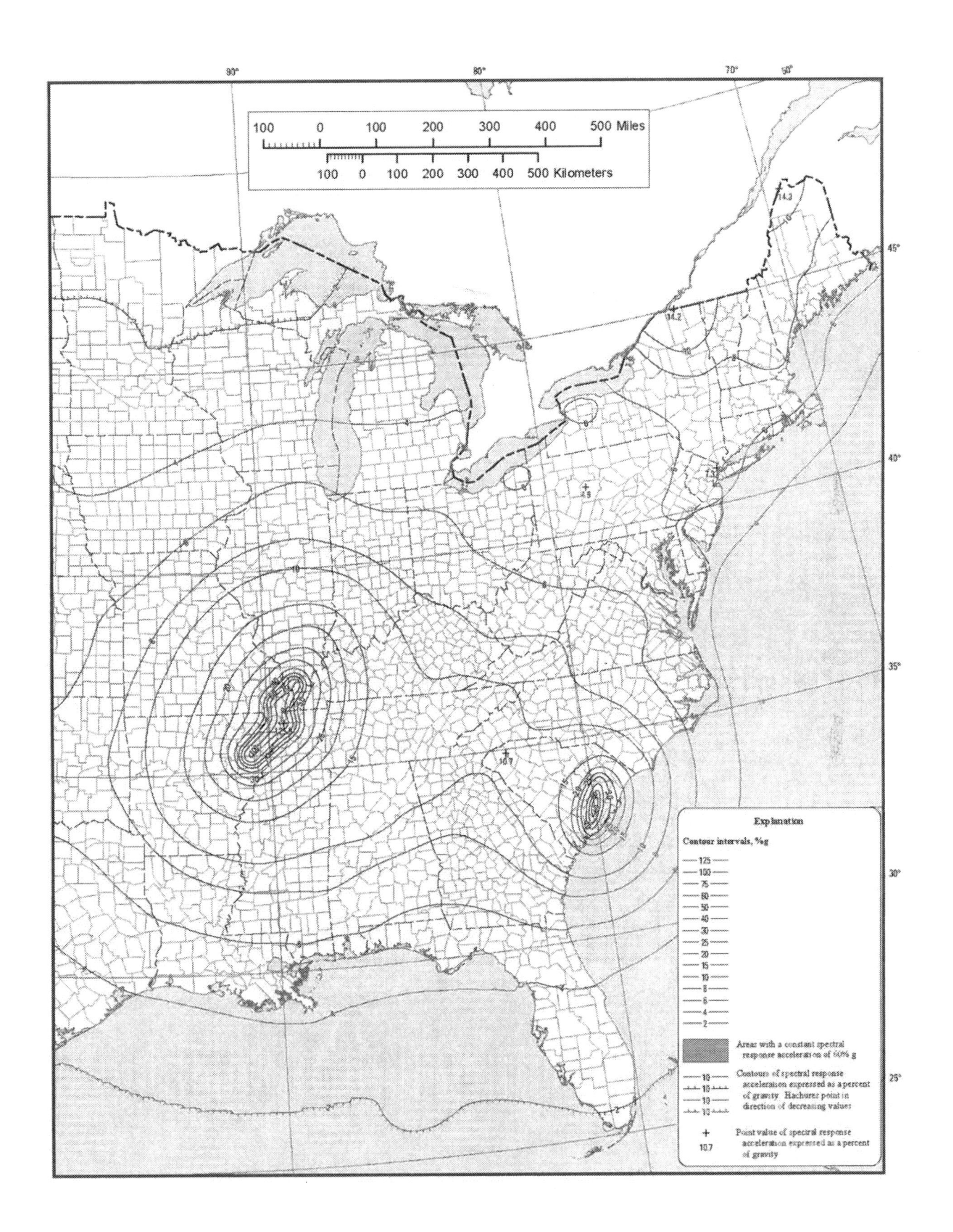
100 0 100 200 300 400 500 Miles
100 0 100 200 300 400 500 Kilometers
Explanation
Contour intervals, %g
Areas with a constant spectral response acceleration of 60% g
Contours of spectral response acceleration expressed as a percent of gravity. Hachures point in direction of decreasing values
Point value of spectral response acceleration expressed as a percent of gravity

4. PROBABLE AND MAXIMUM CONSIDERED EARTHQUAKE GROUND MOTIONS

A *maximum probable earthquake* ground motion at a site is the largest earthquake shaking that has a significant probability of occurring within the lifetime of a structure due to earthquakes from all sources. Since the potential losses of property and life are great, the probability does not have to be very large to be significant.

The *maximum considered earthquake* (MCE) ground motion at a site is the maximum possible earthquake ground motion based on the mapped spectral response acceleration, modified by site soil profile effects. The maximum considered earthquake is difficult to evaluate.

In the IBC, the *risk-targeted maximum considered earthquake* (MCE_R) *ground motion response acceleration*, S_1, is determined from maps based on location [IBC Sec. 202]. It requires the building to be oriented such that an earthquake produces the largest maximum response to horizontal ground motions, and that the acceleration be adjusted for targeted risk. MCE_R ground motion response acceleration was previously known as *MCE ground motion*.

5. EFFECTIVE PEAK GROUND ACCELERATION

The *effective peak ground acceleration*, EPA (or A_a as it is used in some documents), is nominally equal to the maximum ground acceleration associated with the design earthquake, but is more of a contrived design parameter set by code than a feature of an actual earthquake.[3] As a code provision, the EPA depends on the region and corresponds to the maximum considered earthquake ground motion in the IBC. (See Sec. 6.12.)

As initially explained in the 1990 *Blue Book* commentary, the EPA is derived from the log-tripartite graph given in ATC 3-06 scaled downward from the spectral acceleration by dividing it by a spectral amplification factor. (See Sec. 3.9 for spectral acceleration.) The value of the spectral amplification factor depends on the amount of damping present in the building and the probability of the earthquake's occurrence. For example, for 5% damping and a hazard level (probability of occurrence) of 10% in a 50-year period, the spectral amplification factor is 2.5.

The method of deriving the EPA is subject to continuing study and analysis.

[3]The terms *effective peak ground acceleration* (EPA) and *effective peak velocity* (EPV) were originally defined in the commentary of ATC 3-06.

6. SITE PERIOD

The *site* (*soil*) *period* is now recognized as a significant factor contributing to structural damage.[4] When a site has a natural frequency of vibration that corresponds to the predominant earthquake frequency, site movement can be greatly magnified. This is known as *resonance*. (See Sec. 4.14.) Thus, the buildings can experience ground motion much greater than would be predicted from only the seismic energy release.

Determining the actual site period is no easy matter. Since the site period can be computed precisely from widely available formulas and still be grossly inaccurate, such determinations are best left to experts familiar with the area.

Other soil characteristics, including density, bearing strength, moisture content, compressibility (i.e., tendency to settle), and sensitivity (i.e., tendency to liquefy), are additional factors not addressed by the seismic code. These factors, nevertheless, must be considered in structural design.

7. SOIL LIQUEFACTION

Liquefaction occurs in soils, particularly in soils of saturated cohesionless particles such as sand, and is a sudden drop in shear strength. This is experienced as a drop in bearing capacity. In effect, the soil turns into a liquid, allowing everything it previously supported to sink. It is not necessary for the soil to be located on a cliff or other escarpment for liquefaction to occur. Perfectly flat soil layers can become major mud puddles if the conditions are right.[5]

Continued cycles of reversed shear in saturated sand can cause *pore water pressure* to increase, which in turn decreases the *effective stress* and *shear strength*.[6] When the shear strength drops to zero, the sand liquefies.

Conditions most likely to contribute to or indicate a potential for liquefaction include (1) a lightly loaded sand layer within 49 ft to 66 ft (15 m to 20 m) of the surface, (2) uniform particles of medium size, (3) a saturated condition below a water table, and (4) a low penetration-test value.

8. BUILDING PERIOD

When a lightly damped building is displaced laterally by an earthquake, wind, or other force, it will oscillate back

[4]The resonance-induced damages of the 1985 Mexico City earthquake and the 1989 Loma Prieta earthquake are prime examples.
[5]Dramatic examples of liquefaction occurred in the 1964 earthquakes in Alaska and Niigata, Japan.
[6]These are standard terms used in soils and foundations handbooks.

and forth with a regular *period*, or *natural period*. (This building period should not be confused with the site period mentioned in Sec. 3.6 or with the period of the earthquake.)

The natural period of modern buildings can seldom, if ever, be calculated from simple vibrational theory. Five other methods, listed as follows, can be used when knowledge of a building period is needed.

1. Analytical models based on finite element analysis (FEA) and other modeling techniques can be used.
2. A scale model of the building can be constructed and the natural period extrapolated from measurements on the model. (This is seldom done, however.)
3. If the building has been constructed, actual measurements can be taken.
4. Empirical relations (such as are incorporated in ASCE/SEI7 Sec. 12.8.2.1) can be used. (See Sec. 6.26.)
5. Rayleigh's method can be used.

9. SPECTRAL CHARACTERISTICS

Despite some inherent regularity, earthquake seismograms are quite "noisy." It is difficult to determine how a building behaves at all times during an earthquake consisting of many random pulses. It is also unnecessary in many cases to know the entire time-history response of the building, since the maximum seismic force on (and, hence, damage in) a structure depends partially on the effective peak acceleration experienced, not on lower accelerations that might have occurred during the earthquake.

The maximum acceleration[7] that is experienced by a single-degree-of-freedom vibratory system (see Sec. 4.2) is known as the *spectral acceleration*, S_a.[8] Similarly, the maximum displacement and velocity are known as the *spectral displacement*, S_d, and *spectral velocity*, S_v, respectively.

[7]The maximum building acceleration should not be confused with the effective ground acceleration. The building acceleration is typically higher than the ground acceleration. The ratio of building to ground acceleration depends on the building period, a concept that is discussed elsewhere in this book. For infinitely stiff buildings (with zero natural periods), the ratio is 1. The spectral acceleration from typical California design earthquakes (i.e., those used as a basis in establishing the IBC provisions) for a 10% damped building located on rocks or other firm soil is approximately 2.0 to 2.5 times the peak ground acceleration. (See Sec. 3.5.)

[8]Another name is *spectral pseudo acceleration*—"pseudo" because the value does not correspond exactly to the maximum acceleration.

10. SEISMIC FORCE

The theoretical maximum seismic force (referred to as "base shear" in the IBC), V, on a structure of mass m (weight W) is given by Newton's second law ($F=ma$).[9]

$$V=\frac{mS_a}{g_c}=\frac{WS_a}{g} \quad \text{[U.S.]} \qquad 3.1(a)$$

$$V=mS_a=\frac{WS_a}{g} \quad \text{[SI]} \qquad 3.1(b)$$

Design base shear, as defined in the IBC, takes on the form of $V=CS_aW$.

11. RELATIONSHIP BETWEEN SPECTRAL VALUES

Equation 3.2 indicates that the spectral displacement, velocity, and acceleration can be derived from one another if the natural frequency (in rad/sec) of vibration, ω, is known.[10] Equation 3.2 is exact for the case of an undamped, single-degree-of-freedom system in simple harmonic motion but is approximate otherwise (i.e., with damping and for multiple-degree-of-freedom systems). (See Sec. 4.2.)

$$|S_d|=\left|\frac{S_v}{\omega}\right|=\left|\frac{S_a}{\omega^2}\right| \quad \text{[undamped SDOF]} \qquad 3.2$$

[9]The total building dead load (dead weight) is used in the calculation of the base shear. This practice should not be confused with a calculation of diaphragm force, which omits half of the ground floor dead load. Except for a warehouse, no live load is included. While these may seem like arbitrary provisions, the IBC is specific in including and excluding certain fractions of the building weight and live load. (See Sec. 6.29.)

[10]Equation 3.2 is easily derived for the case of sinusoidal oscillation. Starting with a sinusoidal position equation, $x(t)=A\sin\omega t$, the first derivative (i.e., the velocity equation) is $v(t)=\omega A\cos\omega t$, whose maximum amplitude is ω multiplied by the amplitude of the position function. The maximum acceleration amplitude is similarly determined.

4 Vibration Theory

1. TWO APPROACHES TO SEISMIC DESIGN

There are two greatly different approaches to seismic design, both of which are "correct" in their own ways. In a *dynamic analysis*, the overall building and story stiffnesses and rigidities are calculated. (See Sec. 4.5.) A specific design earthquake, including magnitude and loading history, is selected and applied to a mathematical model (consisting of lumped masses, damping, and spring stiffness) of the building. The solution may rest heavily on vibrational theory, finite element analysis, and other advanced structural techniques requiring computer analysis. The response of the system (including the displacement and acceleration functions) is calculated and used to determine the forces in each member as a function of time. This method is almost always used for critical structures such as dams and power plants.

There are a number of factors that can render the dynamic approach inappropriate. The building itself may be too simple or too standardized to warrant the rigorous approach of the design analysis. Conversely, the building may be too complex and have too many degrees of freedom to model mathematically. Also, in the initial design phases, the member sizes and locations may not be known, making it difficult to estimate stiffnesses and rigidities.[1] The dynamic approach is inappropriate, too, when the design earthquake is not known. Additionally, the analysis may be beyond the financial or computational abilities of the engineering firm performing the design. And, finally, unless the building is particularly irregular as defined in ASCE/SEI7 Sec. 12.3.2, there may be no code requirement to perform a dynamic analysis.

The alternative to a dynamic analysis is a *static analysis*. The *equivalent lateral (seismic) force* is calculated as simply some fraction of the dead weight. Chapter 16 of the IBC and Chap. 12 of the ASCE/SEI7 codifies this analysis so that there is no need to know the design earthquake.

In the chapters and sections that follow, these two methods are at times discussed separately and, at other times, aspects of each method are combined. Although considerably different in approach, the static method is based on engineering logic that can, in many cases, be traced back to vibration theory.

2. SIMPLE HARMONIC MOTION[2]

Ideal vibrational systems that consist of springs and masses and that are not acted upon by external disturbing forces (after an initial displacement) are known as *simple harmonic oscillators*. During steady-state motion, such oscillators move in a repetitive sinusoidal pattern known as *simple harmonic motion*. Simple harmonic motion is characterized by the absence of a continued disturbing force and a lack of frictional damping.

Examples of simple harmonic oscillators are a mass hanging on an ideal spring (as shown in Fig. 4.1(a)), a pendulum on a frictionless pivot, and a slab supported on two massless cantilever springs (as shown in Fig. 4.1(b)).

The number of variables needed to define the position of all parts of a system is known as its *degrees of freedom*. If the oscillator is constrained to move in one dimension only, or alternatively, if one linear or angular variable is sufficient to describe the position of the

[1]Although there are some real design programs, most "design" programs are actually analysis programs that require the user to input information about the locations and characteristics of the structural members.

[2]There is no need actually to develop the differential equations of oscillatory motion for a building. However, this section introduces some of the terms and concepts related to structural dynamics.

Figure 4.1 *Simple Harmonic Oscillator*

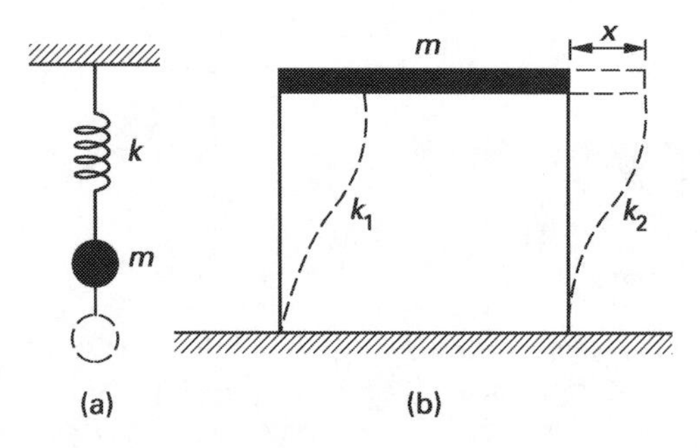

oscillator, the system is known as a *single-degree-of-freedom* (SDOF) *system*. The moving mass in an SDOF system is usually concentrated at one point and is known as a *lumped mass*.

Oscillation of the SDOF system shown in Fig. 4.1 is initiated by displacing and releasing the mass. The displacement, x, is measured from the equilibrium position. Once the system has been displaced and released, no further external force acts on it. Because there is no friction once it is set in motion, the mass remains in motion indefinitely, as shown in Fig. 4.2.

Figure 4.2 *Time Response of an Undamped Simple Harmonic Oscillator*

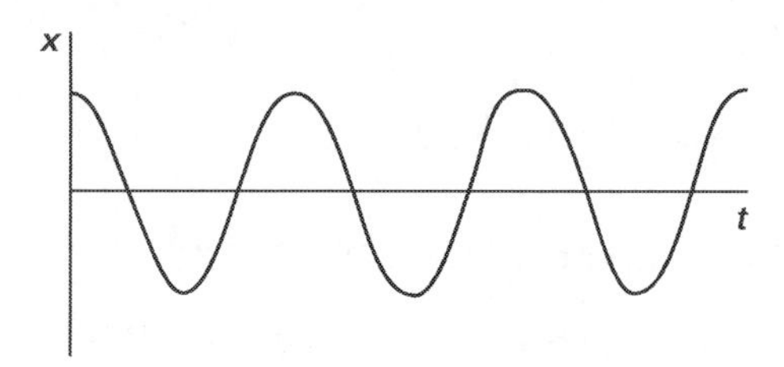

3. STIFFNESS AND FLEXIBILITY

When a force, F, acts on an ideal linear spring, *Hooke's law* predicts the magnitude of the spring deflection, x. In Eq. 4.1, k is the *stiffness* or *spring constant* in lbf/ft (N/m). The stiffness is the force that must be applied in order to deflect the spring a distance of one unit.

$$F = kx \quad \text{[Hooke's law]} \qquad 4.1$$

Referring to the mass-spring system that is shown in Fig. 4.1(a), the spring is undeflected until the mass is attached to it. After the mass is attached, the spring will deflect an amount known as the *static deflection*, x_{st}.

$$W = \frac{mg}{g_c} = kx_{st} \qquad \text{[U.S.]} \qquad 4.2(a)$$

$$W = mg = kx_{st} \qquad \text{[SI]} \qquad 4.2(b)$$

The stiffness, k, of a beam can be calculated as the ratio of applied force to deflection from the beam deflection tables that are typically in every mechanics of materials textbook. Table 4.1 summarizes some of these terms.

Compliance (*flexibility*) is the reciprocal of stiffness. It is the deflection obtained when a unit force is applied. Therefore, its units are ft/lbf (m/N).

Table 4.1 *Deflection and Stiffness for Various Systems (due to bending moment alone)*

system	maximum deflection, x	stiffness, k
F, h, x	$\frac{Fh}{AE}$	$\frac{AE}{h}$
F, h, x	$\frac{Fh^3}{3EI}$	$\frac{3EI}{h^3}$
F, h, x	$\frac{Fh^3}{12EI}$	$\frac{12EI}{h^3}$
w, h, x (w is load per unit length)	$\frac{wL^4}{8EI}$	$\frac{8EI}{L^3}$
F, I_1, I_2, h, x	$\frac{Fh^3}{12E(I_1+I_2)}$	$\frac{12E(I_1+I_2)}{h^3}$
F, x, L	$\frac{FL^3}{48EI}$	$\frac{48EI}{L^3}$
w, L, x (w is load per unit length)	$\frac{5wL^4}{384EI}$	$\frac{384EI}{5L^3}$
F, x, L	$\frac{FL^3}{192EI}$	$\frac{192EI}{L^3}$
w, x, L (w is load per unit length)	$\frac{wL^4}{384EI}$	$\frac{384EI}{L^3}$

4. RIGIDITY

An understanding of structural rigidity is essential to applying seismic design principles. Most people are comfortable using the terms "rigid," "stiff," and "inflexible" as synonyms. However, engineers actually need to be able to quantify rigidity, since some structures are more rigid than others. Rigidity has traditionally been defined as "the reciprocal of deflection," although rigidity is seldom used in that context. Unfortunately for someone learning seismic design principles for the first time, the term "rigidity" has at least four different incompatible usages; and the codes and literature, which are written for practitioners, not students, do not distinguish between the usages and are often inconsistent. Without a clear understanding of the usages, the rigidity variable, R, is easily misused.

Absolute rigidity, R_{abs}, is 100% synonymous with stiffness. Hooke's law states that the deflection of a structure is proportional to the applied force, and the constant of proportionality is the *stiffness*, k. For example, Eq. 4.1 gives the relationship between stiffness and the deflection, x, of a spring acted upon by a force, F.

For seismic design use, Hooke's law would be rewritten as $V = R_{abs}\delta$ or $V = R_{abs}\Delta$, where V is the applied shear force and either δ or Δ is used to represent the deflection. Notice that R_{abs} has units of force/length (e.g., kips/in). Either the context or the units can be used to identify when the word "rigidity" should be taken to mean "stiffness." The usefulness of this definition of rigidity is that the deflection can be determined for any disturbing force (i.e., for any earthquake). The absolute rigidity of a member can be used to find the deflection from any force.

Hooke's law shows that rigidity has something to do with deflection, but the "reciprocal of deflection" definition does not follow from the mathematics. In order to use this definition, the force would have to be disregarded. For analysis (not for design) work, this definition could be used to compare the behavior (response) of different designs exposed to the same disturbing force. Basically, the force can be disregarded if it is a constant, and even then, the reciprocal of deflection can only be used to rate, rank, or describe the observed behavior of a specific structure under those specific conditions. Under those conditions, the *observed rigidity* could be defined as

$$R_{obs} = \frac{1}{\delta} \qquad 4.3$$

Observed rigidity can be distinguished from any other definition of rigidity by its units, which are 1/length (e.g., 1/in). Unfortunately, observed rigidity is often reported without units, which requires the practitioner to assume units or decide the units based on logic. Observed rigidity isn't particularly useful, as it can't be used to determine the applied force or the stiffness of a structure or the structure's response to a different disturbing force. Observed stiffness can be determined and reported (with units), but by itself, the observed stiffness of a single wall or structure can't be used for anything.

The *tabulated rigidity*, R_{tab}, combines the concepts of stiffness and observed rigidity. Tabulated rigidity is the reciprocal of deflection as calculated from mechanics of materials principles for an imaginary structure with arbitrary properties. For manual solutions, tabulated rigidity, as the name implies, is looked up in a table. Values can be calculated instead, but the calculations must consider the contributions of both moment and shear to deflection, and the equations are somewhat more complex than the simple elastic beam equations. For the convenience of anyone solving a seismic design problem manually, tabulated rigidity is most easily read from a table.

Not unexpectedly, the deflection depends on the applied force, the structure's material, and its dimensions, so the deflection equation contains terms for modulus of elasticity, E, moment of inertia, I, and length. The actual values of these terms are unknown, so values of tabulated rigidity represent the reciprocals of deflection for an arbitrary structure based on a convenient but arbitrary force, dimensions, and material properties. For example, a thickness of 1 in, a force of 100,000 lbf, and a modulus of elasticity of 1,000,000 psi are used to calculate the tabulated rigidities in most tabulations. Other values could be used, but these are the most common. Since convenient but arbitrary values are used, the deflection (and its reciprocal, the tabulated rigidity) clearly cannot be applied to any specific structure.

Since tabulated rigidity has the same units as observed rigidity, 1/length, tabulated rigidity suffers from the same limitation as observed rigidity (i.e., it is largely descriptive). However, by itself, it is even less useful than the $1/\delta$ calculation for observed rigidity (which was based on an observed deflection of an actual structure); it is based on arbitrary values of an imaginary structure. Individual values of tabulated rigidity can't be applied to any specific structure. It would be impossible, for example, to obtain a value of R_{tab} for use in determining the deflection of a shear wall.

Tabulated rigidity's usefulness comes from an understanding of how seismic design is practiced. Seismic designers essentially know the force that the building must withstand. They know the expected ground acceleration, and as the design progresses, they have increasingly good estimates of the building's final mass. $F = ma$ and its code-related equivalent ($V = CW$) tell them what total force the building will be exposed to. This force is resisted by individual pieces of the structure. Seismic designers don't design buildings; they design the pieces of buildings—the walls, columns, roofs, floors, and connections. Seismic designers need to know how much of the applied force each piece resists, and they need a way to distribute the total earthquake force among the various pieces. They do this by determining the relative strength of each piece. The relative strength of a piece is the strength relative to all the force-resisting members. The term "relative" implies more than one force-resisting member.

The relative strength of a particular component (e.g., one of several shear walls) is referred to as the *relative rigidity*, R_{rel}. A single member may have its own relative rigidity, but the relative rigidity can't be determined without knowledge of all the force-resisting members. The relative rigidity of a particular component, i, among n components that share the earthquake force, V, is calculated from a ratio of the tabulated rigidities.

$$R_{\text{rel},i} = \frac{R_{\text{tab},i}}{\sum_{i=1}^{n} R_{\text{tab},i}} \quad 4.4$$

Even though the individual values of tabulated rigidity are based on arbitrary values of force, material properties, and dimensions, these arbitrary values cancel out since they are represented in both the numerator and denominator values. Only the relevant dimensions (base and height) used to look up the R_{tab} values implicitly remain. The force resisted by component i is

$$V_i = R_{\text{rel},i} V = \frac{R_{\text{tab},i}}{\sum_{i=1}^{n} R_{\text{tab},i}} V \quad 4.5$$

Relative rigidity is unitless. It is a fraction less than 1.0—the fraction of the total earthquake force resisted by component i. The sum of all relative rigidities within a system of force-resisting members is 1.0.

$$\sum_{i=1}^{n} R_{\text{rel},i} = 1 \quad 4.6$$

Even though relative rigidity shares the common name "rigidity," it is inconsistent with all three other definitions of rigidity. It can't directly be used to calculate the deflection of an individual wall.

Unfortunately, unlike what is presented in this book, common usage does not distinguish between absolute rigidity, observed rigidity, and tabulated rigidity. In truth, the terms "absolute rigidity," "observed rigidity," and "tabulated rigidity" were made up specifically for this section; they don't exist in practice. And, although the term "relative rigidity" is used in practice, the term is sometimes applied incorrectly to the tabulated values. The correct definition and usage of the term "rigidity" must usually be determined from the context, units, and value. (A relative rigidity could never have a value greater than 1.0.) This determination is essential. None of the four types of rigidity can be substituted for the others in usage.

5. TABULATED PIER RIGIDITY

Tables of tabulated rigidity are useful when allocating forces to resisting members. Both moment and shear contribute to the deflection experienced by a resisting member (e.g., a shear wall).[3] Consider the wall shown in Fig. 4.3(a). This wall is fixed at the top and bottom and bends in double curvature since the top and bottom must remain vertical. Such a wall is known as a *fixed pier*. The deflection due to both shear and flexure of a fixed pier is given by Eq. 4.7.

$$x_{\text{fixed}} = \frac{Fh^3}{12EI} + \frac{1.2Fh}{AG} \quad \text{[fixed pier]} \quad 4.7$$

$$A = td \quad 4.8$$

$$I = \frac{td^3}{12} \quad 4.9$$

A wall that is fixed at the bottom, but free to rotate at the top, bends in simple curvature and is known as a *cantilever pier*. The deflection of a cantilever wall due to both effects is

$$x_{\text{cantilever}} = \frac{Fh^3}{3EI} + \frac{1.2Fh}{AG} \quad \text{[cantilever pier]} \quad 4.10$$

For concrete, $E \approx 3 \times 10^6$ psi (2.1×10^7 kPa) and $G \approx 0.4E$. For masonry, $E \approx 1 \times 10^6$ psi (6.9×10^6 kPa) and $G \approx 0.4E$. For steel, $E \approx 3 \times 10^7$ psi (2.1×10^8 kPa) and $G \approx 1.2 \times 10^7$ psi (8.3×10^7 kPa). For wood, $E \approx 1 \times 10^6$ psi to 1.8×10^6 psi (6.9×10^6 kPa to 12×10^6 kPa). In buildings where all members consist of the same material and all walls have the same thickness (for example, a masonry-walled building or an all-concrete building), the deflection is traditionally calculated with arbitrary values of applied force, modulus of elasticity, and wall thickness. This is permitted when distributing the applied lateral loads to vertical members because the load "taken" by each member is proportional to the member's relative rigidity. Since the shear that is distributed to each vertical member (i.e., each pier) is in proportion to its relative rigidity and does not depend on the actual rigidity, the deflections can be calculated with arbitrary values of total shear, F, and wall thickness, t. Equation 4.11 and Eq. 4.12 use $F = 100{,}000$ lbf (445 000 N), $t = 1.0$ in (25 mm), $E = 1 \times 10^6$ psi (6.9×10^6 kPa), and arbitrary units. These values are used to calculate the tabulated rigidity values in App. D and App. E, which, in turn, are used to calculate the relative rigidity of each vertical member.

$$R_{\text{tab,fixed}} = \frac{1}{0.1\left(\frac{h}{d}\right)^3 + 0.3\left(\frac{h}{d}\right)} \quad 4.11$$

$$R_{\text{tab,cantilever}} = \frac{1}{0.4\left(\frac{h}{d}\right)^3 + 0.3\left(\frac{h}{d}\right)} \quad 4.12$$

[3]Common beam deflection equations, such as those presented in Table 4.1, usually disregard the effect of shear. However, shear contributes to deflection when the ratio of height to depth is low. In general, shear deflection should not be neglected, unless beam spans are long.

Figure 4.3 Fixed and Cantilever Piers

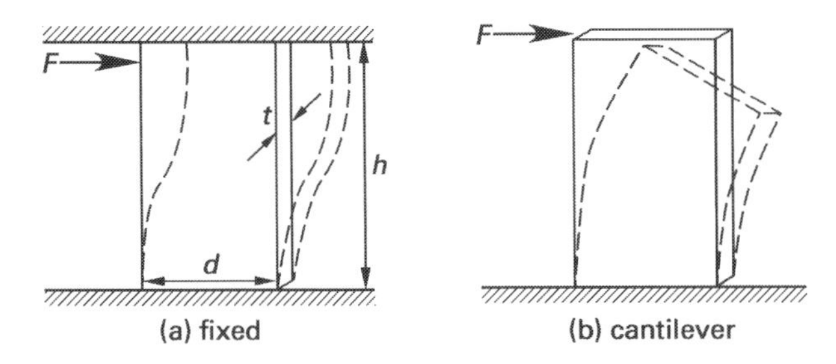

6. NATURAL PERIOD AND FREQUENCY

The time for a complete cycle of oscillation of an SDOF system is known as the *fundamental* or *natural period*, T, usually expressed in seconds. The reciprocal of the natural period is the *linear natural frequency*, f, usually called *natural frequency*, *fundamental frequency*, or just *frequency*, and expressed in Hz (i.e., cycles per second). It is important to distinguish between the natural frequency of a system (building, oscillator, etc.) and the frequency of an applied force. The natural frequency, f, in Eq. 4.13 has nothing to do with an external force.

$$f = \frac{1}{T} \qquad 4.13$$

The natural frequency can also be expressed in radians per second (rad/sec), in which case it is known as the *circular frequency*, *angular natural (fundamental) frequency*, or just *angular frequency*, ω.

$$\omega = 2\pi f = \frac{2\pi}{T} \qquad 4.14$$

It is easy to derive the natural frequency for the case of a simple harmonic oscillator.[4] For a mass on a spring,

$$\omega = \sqrt{\frac{kg_c}{m}} = \sqrt{\frac{kg}{W}} \qquad \text{[U.S.]} \quad 4.15(a)$$

$$\omega = \sqrt{\frac{k}{m}} \qquad \text{[SI]} \quad 4.15(b)$$

Substituting k from Hooke's law (see Eq. 4.1) and recognizing that the mass, m, can be calculated from the weight, W, an expression is derived for the natural frequency in terms of the static deflection, x_{st}, calculated in Sec. 4.3.

$$\omega = \frac{2\pi}{T} = \sqrt{\frac{Fg_c}{x_{st}m}} = \sqrt{\frac{Fg}{x_{st}W}} \qquad \text{[U.S.]} \quad 4.16(a)$$

$$\omega = \frac{2\pi}{T} = \sqrt{\frac{F}{x_{st}m}} \qquad \text{[SI]} \quad 4.16(b)$$

[4]This is done in virtually every physics, dynamics, and earthquake book, but not here.

Since Eq. 4.16 can be used to calculate the natural period, it is tempting to substitute the maximum allowable code drift (i.e., 2.5% of the total building height; see Sec. 6.40) for the static deflection in order to calculate the natural building period.[5] Such a substitution would require no structural analysis at all, but implies that the building will have maximum flexibility permitted by the code and will remain elastic when this drift is achieved. One problem with this approach is that it assumes the maximum allowable drift to be the same for all geographic regions, although the flexibility actually depends on the location since flexibility is affected by the building's seismic resistance. While the lateral forces on the building differ, the maximum drift and, thus, the period, do not. Obviously, the building period cannot be calculated in this way.

Example 4.1

A 5 lbm (2.27 kg) mass hangs from two ideal springs as shown. The block is constrained so that it does not rotate. What is the natural period of vibration?

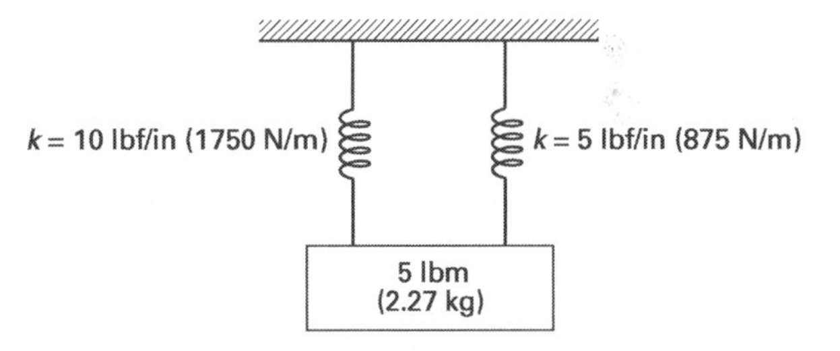

Customary U.S. Solution

Both springs must deflect in order for the mass to move. The total composite spring constant is

$$k_t = k_1 + k_2 = \left(5\ \frac{\text{lbf}}{\text{in}} + 10\ \frac{\text{lbf}}{\text{in}}\right)\left(12\ \frac{\text{in}}{\text{ft}}\right) = 180\ \text{lbf/ft}$$

From Eq. 4.14 and Eq. 4.15, the natural period is

$$T = 2\pi\sqrt{\frac{m}{g_c k}} = 2\pi\sqrt{\frac{5\ \text{lbm}}{\left(32.2\ \frac{\text{ft-lbm}}{\text{lbf-sec}^2}\right)\left(180\ \frac{\text{lbf}}{\text{ft}}\right)}}$$
$$= 0.185\ \text{sec}$$

SI Solution

Both springs must deflect in order for the mass to move. The total composite spring constant is

$$k_t = k_1 + k_2 = 1750\ \frac{\text{N}}{\text{m}} + 875\ \frac{\text{N}}{\text{m}}$$
$$= 2625\ \text{N/m}$$

[5]The IBC permits the drift limits to be exceeded when it is demonstrated that greater drift can be tolerated by both structural elements and nonstructural elements.

From Eq. 4.14 and Eq. 4.15, the natural period for vertical translation is

$$T = 2\pi\sqrt{\frac{m}{k}} = 2\pi\sqrt{\frac{2.27 \text{ kg}}{2625 \frac{\text{N}}{\text{m}}}}$$

$$= 0.185 \text{ s}$$

Example 4.2

A small water tank is supported on a slender column as shown. Neglecting the weight of the column, calculate the natural period of vibration.

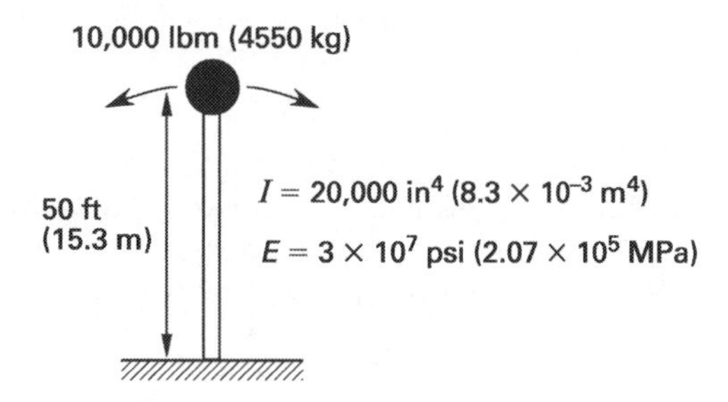

Customary U.S. Solution

Consider the water tower to be a cantilever beam. The stiffness is the force required to deflect the tank 1 ft laterally.

From Table 4.1,

$$k = \frac{3EI}{h^3} = \frac{(3)\left(3 \times 10^7 \frac{\text{lbf}}{\text{in}^2}\right)(20{,}000 \text{ in}^4)}{(50 \text{ ft})^3\left(12 \frac{\text{in}}{\text{ft}}\right)^2}$$

$$= 1 \times 10^5 \text{ lbf/ft}$$

From Eq. 4.14 and Eq. 4.15, the period is

$$T = 2\pi\sqrt{\frac{m}{g_c k}} = 2\pi\sqrt{\frac{10{,}000 \text{ lbm}}{\left(32.2 \frac{\text{ft-lbm}}{\text{lbf-sec}^2}\right)\left(1 \times 10^5 \frac{\text{lbf}}{\text{ft}}\right)}}$$

$$= 0.35 \text{ sec}$$

SI Solution

Consider the water tower to be a cantilever beam. The stiffness is the force required to deflect the tank 1 m laterally.

From Table 4.1,

$$k = \frac{3EI}{h^3}$$

$$= \frac{(3)(2.07 \times 10^5 \text{ MPa})\left(10^6 \frac{\text{Pa}}{\text{MPa}}\right)(8.3 \times 10^{-3} \text{ m}^4)}{(15.3 \text{ m})^3}$$

$$= 1.44 \times 10^6 \text{ N/m}$$

From Eq. 4.14 and Eq. 4.15, the natural period is

$$T = 2\pi\sqrt{\frac{m}{k}} = 2\pi\sqrt{\frac{4550 \text{ kg}}{1.44 \times 10^6 \frac{\text{N}}{\text{m}}}}$$

$$= 0.35 \text{ s}$$

7. DAMPING

Damping is the dissipation of energy from an oscillating system, primarily through friction. (See Fig. 4.4.) The kinetic energy is transformed into heat. All structures have their own unique ways of dissipating kinetic energy, and in certain designs, mechanical systems known as *dampers* (see Sec. 14.4) can be installed to increase the damping rate.[6]

Figure 4.4 *Oscillator with Damping*

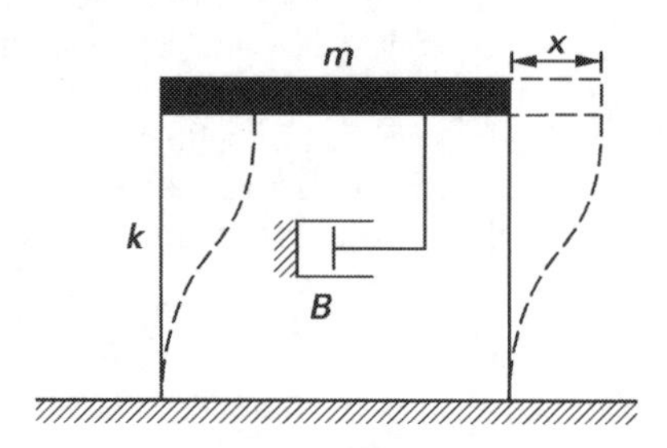

There are several sources of damping. *External viscous damping* is caused by the structure moving through surrounding air (or water, in some cases). It is generally small in comparison to other sources of damping. *Internal viscous damping*, commonly the only type of damping actually modeled, is related to the viscosity of the structural material. It is proportional to velocity. (See Eq. 4.17.) *Body-friction damping*, also known as *Coulomb friction*, results from friction between members in contact. It includes friction at connection points. Sections of opposed cracked masonry walls rubbing back and forth against one another are very effective body-friction dampers. Another source of damping, *radiation damping*, occurs as a structure vibrates and becomes a source of energy itself. Some of the energy is reradiated through the foundation back into the ground. Finally, *hysteresis damping* occurs when the structure yields during reversals of the load. (See Sec. 5.7.)

For internal viscous damping, the frictional damping force opposing motion is given by Eq. 4.17. The exponent n in Eq. 4.17 is usually taken as 1.0 for slow-moving systems and 2.0 for fast-moving systems. However, even these values are idealizations. The coefficient B in Eq. 4.17 is known as the *damping coefficient*.

$$F_{\text{damping}} = Bv^n \qquad 4.17$$

[6]A damper is similar in design to a shock absorber and is often depicted as a plunger moving through a pot of viscous fluid. In modeling, dampers are also known as *dashpots*, although this term is more common among mechanical engineers.

8. DAMPING RATIO

An oscillating system with a small amount of damping will continue to oscillate, although the amplitude of the oscillations will decay. Many cycles and a long time may elapse before the system eventually reaches the motionless equilibrium position. Figure 4.5 shows this type of system, known as an *underdamped system.*

Figure 4.5 *Underdamped Motion (Moderate Damping)*

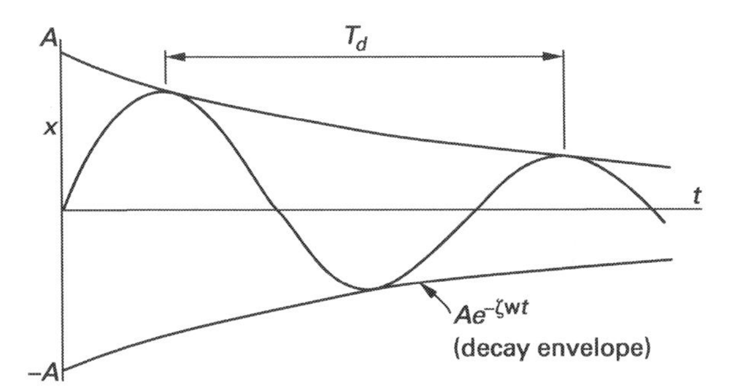

Conversely, a system may have a large amount of damping. When displaced, such an *overdamped system* seems to "hang in space," taking an extremely long time to return to the motionless equilibrium position, as shown in Fig. 4.6.[7]

Figure 4.6 *Overdamped Motion*

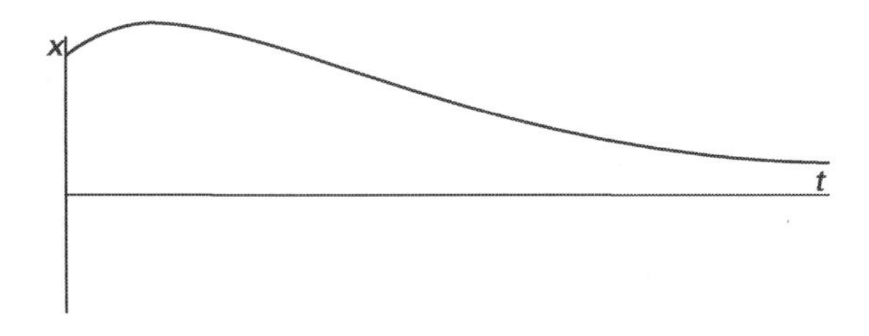

Both the underdamped and overdamped cases bring the system back to the equilibrium position only after a long time. Figure 4.7 shows the one particular amount of damping, known as *critical damping,* that brings the system to equilibrium in a minimum time without oscillation. In this case, the damping coefficient, B, is known as the *critical damping coefficient,* B_{critical}.

Figure 4.7 *Critically Damped Motion*

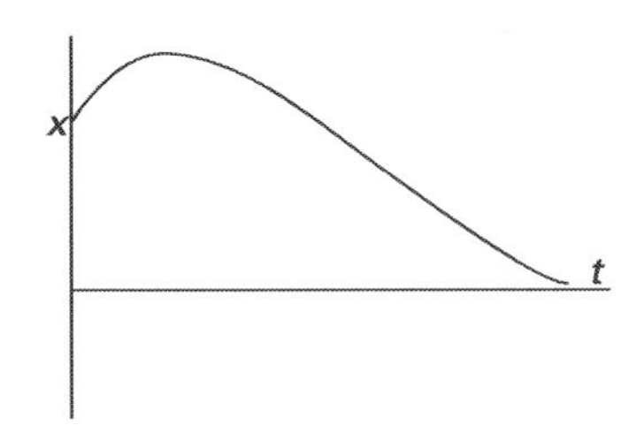

[7]An example of an overdamped system is a door with a slow-closing device that will not permit the door to slam shut. Instead, the door approaches the fully closed position slowly.

Most systems are not critically damped. The ratio of the actual damping coefficient to the critical damping coefficient is known as the *damping ratio,* ζ.

$$\zeta = \frac{B}{B_{\text{critical}}} \qquad 4.18$$

9. DECAY ENVELOPE

For small and moderate amounts of damping (i.e., the underdamped case), the oscillation will be bounded by a *decay envelope* as illustrated in Fig. 4.5. The equation of the decay envelope is given by Eq. 4.19.

$$x = Ae^{-\zeta\omega t} \qquad 4.19$$

The ratio of one cycle's amplitude to the subsequent cycle's amplitude is the *decay decrement.* The natural logarithm of the decay decrement is the *logarithmic decrement,* δ.

$$\delta = \ln\left(\frac{x_n}{x_{n+1}}\right) = \frac{2\pi\zeta}{\sqrt{1-\zeta^2}} \qquad 4.20$$

10. DAMPING RATIO OF BUILDINGS

The exact damping ratio, ζ, of an actual structure is difficult to determine. Furthermore, the damping ratio increases during large swings. Available data on actual structures suggest the values given in Table 4.2. There is little evidence to support damping ratios in real structures that exceed 15%.

Although the damping ratio is essentially constant for a given building, the damping ratio of a particular building type or construction material appears to depend on the natural period of the building. Buildings with natural periods of less than 1.0 sec may have damping ratios two to three times higher than buildings with similar construction but natural periods greater than 1.0 sec. While generalizations that do not consider all factors are possible, it appears that the building's damping ratio, period, and construction method are all related.

11. DAMPED PERIOD OF VIBRATION

The period of oscillation of a system will be slightly greater with damping than without it, since the damping slows down the movement. Equation 4.21 and Eq. 4.22 give the damped frequency and period. Most buildings have only small amounts of damping. Therefore, the damped and undamped periods are almost identical.

$$\omega_d = \omega\sqrt{1-\zeta^2} \qquad 4.21$$

$$T_d = \frac{2\pi}{\omega_d} \qquad 4.22$$

Table 4.2 Typical Damping Ratios

type of construction	ζ
steel frame welded connections with flexible walls	0.02
steel frame welded connections with normal floors and exterior cladding	0.05
steel frame bolted connections with normal floors and exterior cladding	0.10
concrete frame flexible internal walls	0.05
concrete frame flexible internal walls with exterior cladding	0.07
concrete frame concrete or masonry shear walls	0.10
concrete or masonry shear wall	0.10
wood frame and shear wall	0.15

12. FORCED SYSTEMS

A *forced system* is an oscillatory system that is supplied energy on a regular, irregular, or random basis. The force that supplies the energy is known as a *forcing function.* Forcing functions can be constant (i.e., a *step function*), applied and quickly removed (i.e., an *impulse function*), sinusoidal, or random.

An example of a regularly forced system is a flexible floor supporting an out-of-balance motor. When turning, the motor will generate a force at a frequency proportional to the motor's rotational speed. An example of a randomly forced system is a structure acted on by wind or seismic forces. In the latter case, there is little or no regularity to the applied forces.

It is not significant whether a lateral force (e.g., seismic force or wind) is applied to a building directly, or whether the base moves out from under the building (e.g., as in an earthquake). In the latter case, the equivalent lateral force is an inertial force, but it is just as effective at displacing the building relative to its base as any direct force is.

The system response (i.e., the behavior of a building) to a force depends on the nature of the forcing function. Unfortunately, earthquakes are never simple sinusoids and buildings have more than a single degree of freedom (see Sec. 4.17), so the determination of system response is time consuming and complex. Computers and numerical techniques, however, greatly simplify the analysis.[8]

[8]It may not always be a simple matter, however, to interpret the results of the analysis.

13. MAGNIFICATION FACTOR

It is not difficult to show that when a sinusoidal forcing function with the form $F(t) = P\sin\omega_f t$ is applied to a system with stiffness k, the steady-state response will be of the form of Eq. 4.23.

$$x(t) = \beta\left(\frac{P}{k}\right)\sin\omega_f t \qquad 4.23$$

In Eq. 4.23, P/k is the *static deflection*, x_{static} (see Sec. 4.3), that is experienced if a constant force P is applied to the system. β is a dynamic *magnification factor* that depends on all other characteristics of the system.[9]

14. RESONANCE

For a given system, the dynamic magnification factor, β, can be less than or greater than unity, depending on the ratio of the natural and forcing frequencies. Figure 4.8 illustrates how the magnification factor varies for different frequency ratios. At one point, corresponding to where the forcing function frequency equals the natural frequency of the system, the magnification factor is very large (theoretically infinite for undamped systems). Such a condition is known as *resonance.* The ratio ω_f/ω must be greater than $\sqrt{2}$ for β to drop below $|1.0|$.

Figure 4.8 Undamped Magnification Factor

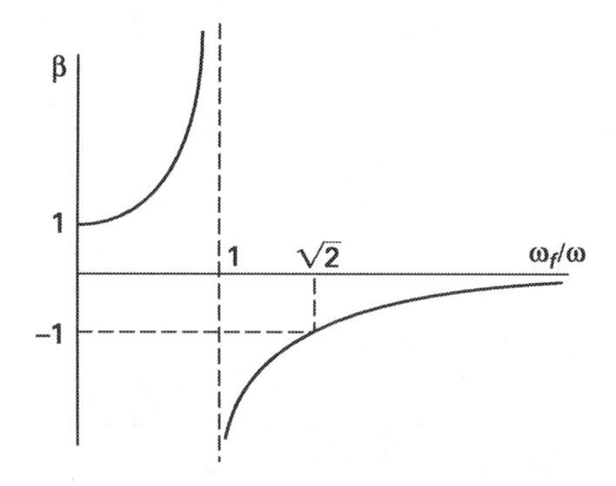

The 1985 Mexico City magnitude 8.1 earthquake occurred on September 19, with a 7.5 aftershock occurring the next day. Approximately 400 buildings were destroyed, and 700 were damaged. The death toll was over 5000. The earthquake consisted of (approximately) twenty 0.18 g pulses coming every 2 sec (the natural period of the ground). This coincided with the period for buildings in the 7 to 20 story range. The resulting resonance-related yielding was the primary cause of structural failure. Quality of construction was not a major factor in the widespread destruction.

[9]The dynamic magnification factor depends on the natural and forcing frequencies, the mass in motion, and the amount of damping (or, alternatively, on the damping ratio). Formulas for calculating the magnification factor for damped and undamped cases are given in virtually every textbook covering vibration theory.

Resonance is now known to be a prime factor in the collapse of the Oakland Interstate 880 Cypress structure during the October 17, 1989, Loma Prieta earthquake. The structure had a natural frequency of 2 Hz to 4 Hz, which coincided with the 3 Hz to 5 Hz natural period of the deep mud that underlaid piles that supported portions of the freeway that collapsed. The depth of the mud and the length of the piles varied between 20 ft and 80 ft. The natural period also varied. Portions of the freeway built on harder alluvial sediments remained standing.

Although the Cypress structure was built to the standards of its time, it was poorly designed, and it is now recognized that use of nonductile reinforced concrete joints and bents with only three hinges, and inadequate confinement of the structure, made its failure predictable.

15. IMPULSE RESPONSE—UNDAMPED SYSTEM

Seismic energy is applied to a structure in a nonregular manner. While a Fourier analysis can be used to analyze the structure response, it is also possible to break the irregular seismic loading into a series of short-duration rectangular impulses. An *impulse* is a force, F, that is applied over a duration, dt, that is much less than the natural period, T, of the structure. The product Fdt is 1.0 for a *unit impulse*. Therefore, a study of the response, $x(t)$, of a system to an impulse is of great interest.[10]

Equation 4.24 indicates that the same response will be achieved from all short-duration impulses (sine, rectangular, square, triangular, random, etc.) that have the same value of $\int F\,dt$. The response is sinusoidal even though the loading is not.

$$x(t) \approx \frac{g}{W\omega}\int (F\,dt)\sin\omega t\,dt \quad \text{[undamped]} \quad \text{[U.S.]} \qquad 4.24(a)$$

$$x(t) \approx \frac{1}{m\omega}\int (F\,dt)\sin\omega t\,dt \quad \text{[undamped]} \quad \text{[SI]} \qquad 4.24(b)$$

16. DUHAMEL'S INTEGRAL FOR AN UNDAMPED SYSTEM

If an undamped structure is acted upon by an irregular force of any duration, the loading can be treated as a series of impulses. The response in this case is given by Eq. 4.25, known as *Duhamel's integral*. Equation 4.25 is the application of superposition to a series of pulses, each ending at time τ.

$$x(t) = \frac{g}{W\omega}\int_0^t F(\tau)\sin\omega(t-\tau)\,d\tau$$

$$= \frac{g_c}{m\omega}\int_0^t F(\tau)\sin\omega(t-\tau)\,d\tau \quad \text{[U.S.]} \qquad 4.25(a)$$

[undamped]

$$x(t) = \frac{1}{m\omega}\int_0^t F(\tau)\sin\omega(t-\tau)\,d\tau \quad \text{[SI]} \qquad 4.25(b)$$

[undamped]

Several numerical methods can be used to evaluate the integral in Eq. 4.25. However, when the ground motion is not known in advance, such an integration is not possible. Since earthquake motions are both nonregular and generally unexpected, it is usually acceptable to work with a maximum value of acceleration (or velocity or displacement). This is the principle behind the spectral values discussed in Sec. 3.9. From Eq. 3.1 and Eq. 3.2, the total force (i.e., the *base shear*) on the structure is

$$F_{\max} = \frac{ma_{\max}}{g_c} \approx \frac{m\text{v}_{\max}\omega}{g_c} \quad \text{[U.S.]} \qquad 4.26(a)$$

$$F_{\max} = ma_{\max} \approx m\text{v}_{\max}\omega \quad \text{[SI]} \qquad 4.26(b)$$

Example 4.3

A mass of 2×10^6 lbm (9.1×10^5 kg) is supported on two vertical members with lateral stiffnesses of 25,000 lbf/in (4.4×10^6 N/m) each. The columns have no mass and are fixed at both ends. The lateral forcing function consists of a ramp up to 50,000 lbf (220 kN) taking 0.08 sec, a uniform loading for 0.08 sec, and a ramp down to zero taking 0.08 sec. Use Duhamel's integral to determine the response (displacement) as a function of time.

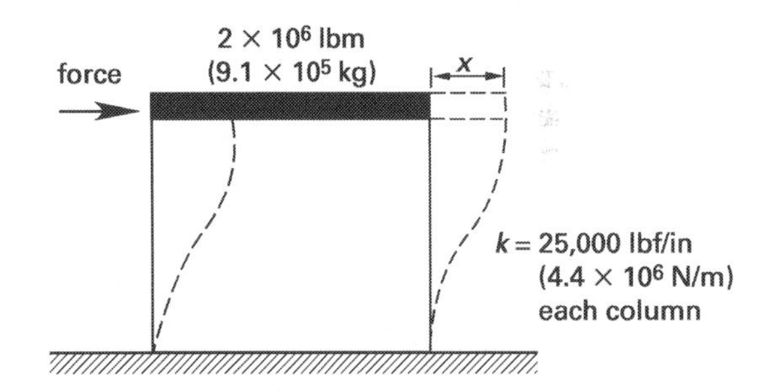

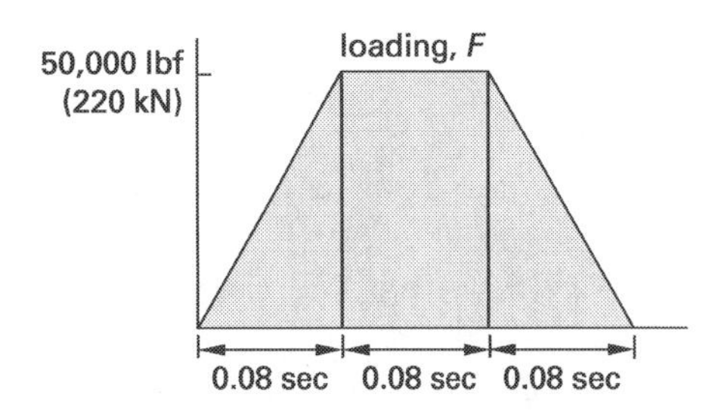

Customary U.S. Solution

The total combined stiffness, k_t, of the two columns is

$$k_t = (2)\left(25{,}000\ \frac{\text{lbf}}{\text{in}}\right) = 50{,}000\ \text{lbf/in}$$

[10]An impulse loading can occur from a projectile impact, bomb blast, sudden wind gust, or short-duration seismic tremor.

From Eq. 4.14 and Eq. 4.15, the natural period is

$$T = 2\pi\sqrt{\frac{m}{g_c k}}$$
$$= 2\pi\sqrt{\frac{2 \times 10^6 \text{ lbm}}{\left(32.2 \ \frac{\text{ft-lbm}}{\text{lbf-sec}^2}\right)\left(50{,}000 \ \frac{\text{lbf}}{\text{in}}\right)\left(12 \ \frac{\text{in}}{\text{ft}}\right)}}$$
$$= 2.02 \text{ sec}$$

Since the total period over which the loading is applied is much less than the period (3×0.08 sec < 2.02 sec), the loading can be considered an impulse.

The natural frequency is

$$\omega = \frac{2\pi}{T} = \frac{2\pi}{2.02 \text{ sec}} = 3.11 \text{ rad/sec}$$

The total impulse is

$$\int F\,dt = \left(\tfrac{1}{2}\right)(0.08 \text{ sec})(50{,}000 \text{ lbf}) + (0.08 \text{ sec})(50{,}000 \text{ lbf}) + \left(\tfrac{1}{2}\right)(0.08 \text{ sec})(50{,}000 \text{ lbf}) = 8000 \text{ lbf-sec}$$

Use Duhamel's integral, Eq. 4.25, to find the response. Since the mass is given in lbm (not lbf), make use of the unit equivalence, $W_{\text{lbf}}/g = m_{\text{lbm}}/g_c$.

$$x(t) = \int \frac{F\,dt}{m\omega} \sin \omega t$$
$$= \left(\frac{8000 \text{ lbf-sec}}{\frac{(2 \times 10^6 \text{ lbm})\left(3.11 \ \frac{\text{rad}}{\text{sec}}\right)}{\left(32.2 \ \frac{\text{ft-lbm}}{\text{lbf-sec}^2}\right)\left(12 \ \frac{\text{in}}{\text{ft}}\right)}}\right) \sin 3.11t$$
$$= (0.497 \text{ in}) \sin 3.11t$$

SI Solution

The total combined stiffness, k_t, of the two columns is

$$k_t = (2)\left(4.4 \times 10^6 \ \frac{\text{N}}{\text{m}}\right) = 8.8 \times 10^6 \text{ N/m}$$

From Eq. 4.14 and Eq. 4.15, the natural period is

$$T = 2\pi\sqrt{\frac{m}{k}} = 2\pi\sqrt{\frac{9.1 \times 10^5 \text{ kg}}{8.8 \times 10^6 \ \frac{\text{N}}{\text{m}}}} = 2.02 \text{ s}$$

Since the total period over which the loading is applied is much less than the period (3×0.08 s < 2.02 s), the loading can be considered an impulse.

The natural frequency is

$$\omega = \frac{2\pi}{T} = \frac{2\pi}{2.02 \text{ s}} = 3.11 \text{ rad/s}$$

The total impulse is

$$\int F\,dt = \left(\tfrac{1}{2}\right)(0.08 \text{ s})(220 \times 10^3 \text{ N}) + (0.08 \text{ s})(220 \times 10^3 \text{ N}) + \left(\tfrac{1}{2}\right)(0.08 \text{ s})(220 \times 10^3 \text{ N}) = 3.52 \times 10^4 \text{ N·s}$$

From Duhamel's integral, Eq. 4.25, the response is

$$x(t) = \int \frac{F\,dt}{m\omega} \sin \omega t$$
$$= \left(\frac{3.52 \times 10^4 \text{ N·s}}{(9.1 \times 10^5 \text{ kg})\left(3.11 \ \frac{\text{rad}}{\text{s}}\right)}\right) \sin 3.11t$$
$$= (1.24 \times 10^{-2} \text{ m}) \sin 3.11t$$

17. MULTIPLE-DEGREE-OF-FREEDOM SYSTEMS

A system with several lumped masses, such as a building with multiple concrete floors supported by steel columns, whose positions are independent of one another is a *multiple-degree-of-freedom* (MDOF) *system.*

An MDOF system has as many ways of oscillating as there are lumped masses. These "ways" are known as *modes.* Each mode has its own characteristic *mode shape* and natural frequency of vibration, each being some multiple of the previous mode's frequency. The mode with the longest period is known as the *first* or *fundamental mode.* Higher modes have higher frequencies (smaller periods), and the periods decrease rapidly from the fundamental mode.[11] Typical mode shapes of an MDOF system with three lumped masses are shown in Fig. 4.9.

18. RESPONSE OF MDOF SYSTEMS

Each modal frequency results in a specific mode shape, as illustrated in Fig. 4.9. However, an earthquake contains waveforms with varied frequency content; therefore, all of the modes may be present simultaneously in an earthquake. This makes it difficult to determine the building's response.

Since MDOF response can be determined as the superposition of many SDOF responses, matrix analysis (on a computer) can be used to evaluate MDOF systems based on the equivalent SDOF performance. As with SDOF

[11]For example, for a typical high-rise building with a uniform plan view and a moment-resisting frame, the decrease is in the order of 1, 1/3, 1/5, 1/7, 1/9, and so on.

Figure 4.9 *Typical Mode Shapes for a Three-Degree-of-Freedom System*

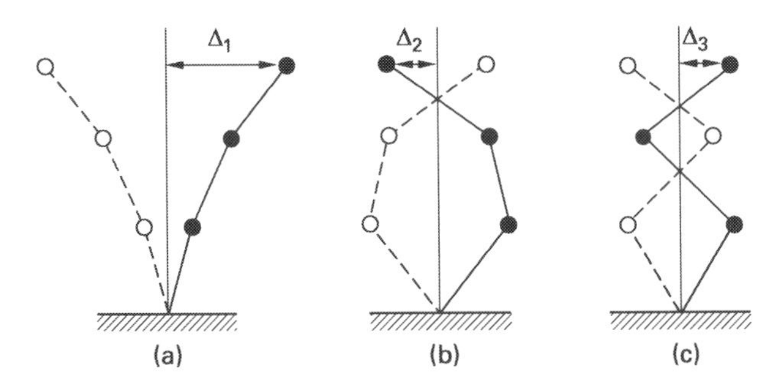

systems, considerable simplification can be achieved by limiting the analysis to the maximum deflections. However, even this simplification requires a probabilistic analysis because the modal maxima do not occur at the same time, nor do they necessarily have the same sign.

Various approximation formulas are used to combine the modal maxima. The sum-of-the-squares approximation is commonly quoted.[12] If the maximum displacements, Δ_i, are known for the first n modes for some particular point (e.g., the top story), Eq. 4.27 usually gives a conservative estimate of the total displacement.

$$\Delta_t = \sqrt{\sum \Delta_i^2} \qquad 4.27$$

The method of combining modal responses by taking the square root of the sum of the squares is referred to in the ASCE/SEI7 as the *SRSS method*. An alternative to SRSS is the *complete quadratic combination* (CQC) method required when modal periods are closely spaced [ASCE/SEI7 Sec. 12.9.3].

Theoretically, all mode shapes must be included in the summation, but, in practice, most of the vibration energy goes into the first three to six modes, and higher modes can be disregarded. (With the use of a computer, however, there is no need to stop with such a small number of modes.) Since the lower modes dominate, the response spectra for MDOF systems are similar to those of SDOF systems. (See Sec. 5.1.) For short periods (e.g., less than 1 sec), the MDOF response is usually slightly less than for first-mode SDOF systems. For periods exceeding 1 sec, the response usually slightly exceeds SDOF response.[13]

ASCE/SEI7 Sec. 12.9.1 requires that all significant modes be included. This can be accomplished by making sure that, for all modes considered, at least 90% of the mass of the structure is included in the calculation of response for the horizontal direction being investigated (i.e., the participation factor from Sec. 4.21 is at least 0.90).

[12]This is an easy computational approximation. Whether or not it is an accurate approximation is beyond the scope of this book.

[13]This generalization is highly dependent on the response spectrum and the soil type at the site.

Example 4.4

Determine the three modal frequencies for the MDOF system shown.

$m_3 = 0.5$, $k_3 = 100$, $m_2 = 1$, $k_2 = 100$, $m_1 = 1$, $k_1 = 100$

Solution

Let x_1, x_2, and x_3 be the displacements—measured with respect to the equilibrium position—of masses 1, 2, and 3, respectively. Then, neglecting the inertial (ma) force, the spring forces on each mass are

(3): $k_3(x_3-x_2)$

(2): $k_2(x_2-x_1)$, $k_3(x_3-x_2)$

(1): k_1x_1, $k_2(x_2-x_1)$

The free bodies shown are not in equilibrium. (This is particularly evident for mass 3.) According to D'Alembert's principle of dynamic equilibrium, an inertial force resisting motion must be added. This inertial force is

$$F_{\text{inertial}} = ma$$

However, from Eq. 3.2, the acceleration is approximately $\omega^2 x$.

$$F_{\text{inertial}} \approx m\omega^2 x$$

Therefore, the equilibrium equations for the three masses are found by adding the inertial force to the spring forces and then combining coefficients for the three displacements.

$$\text{mass 1:} \quad \left(m_1\omega^2 - (k_1 + k_2)\right)x_1 + k_2x_2 = 0$$

$$\text{mass 2:} \quad k_2x_1 + \left(m_2\omega^2 - (k_2 + k_3)\right)x_2 + k_3x_3 = 0$$

$$\text{mass 3:} \quad k_3x_2 + (m_3\omega^2 - k_3)x_3 = 0$$

The masses and stiffnesses are known. Writing the three equilibrium equations in matrix form,

$$\begin{bmatrix} \omega^2 - 200 & 100 & 0 \\ 100 & \omega^2 - 200 & 100 \\ 0 & 100 & 0.5\omega^2 - 100 \end{bmatrix} \begin{bmatrix} x_1 \\ x_2 \\ x_3 \end{bmatrix} = \begin{bmatrix} 0 \\ 0 \\ 0 \end{bmatrix}$$

Disregarding the trivial solution, the coefficient matrix must have a determinant of zero. Setting the determinant equation to zero results in the following equation.

$$\omega^6 - 600\omega^4 + 90{,}000\omega^2 - 2{,}000{,}000 = 0$$

Being a cubic, this equation has three roots. Each root is a modal frequency.

$$\omega_1 = 5.18 \text{ rad/sec}$$
$$\omega_2 = 14.14 \text{ rad/sec}$$
$$\omega_3 = 19.32 \text{ rad/sec}$$

19. MODE SHAPE FACTORS

The *mode shape factors*, ϕ, are relative numbers that represent the ratios of each of the story deflections (from the equilibrium position) to some common basis, usually the deflection of the first or last story. (See Fig. 4.10.) Since mode shape factors are relative, they can usually be determined by initially assuming a value of one of the deflections.

$$\phi_i = \frac{x_i}{x_1} \quad \text{4.28}$$

In some cases, the mode shape factors are normalized by dividing by $\sqrt{\Sigma m_i \phi_i^2}$. Then, Eq. 4.29 will be valid.

$$\sum m_i \phi_i^2 = 1 \quad \text{4.29}$$

Figure 4.10 *Mode Shape Factors*

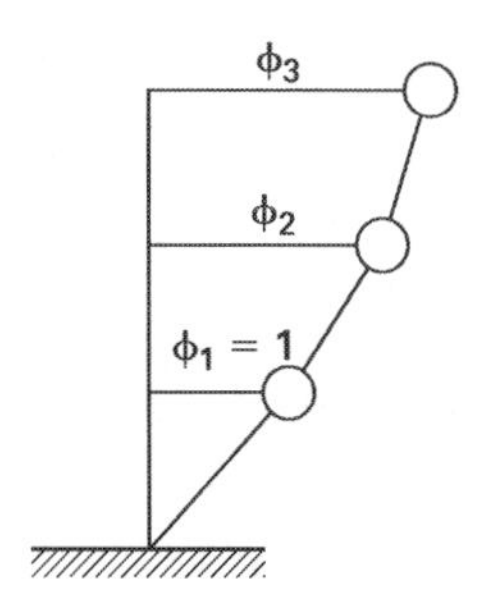

Example 4.5

Find the normalized first mode shape for the system in Ex. 4.4.

Solution

The first equilibrium equation is

$$\left(m_1\omega^2 - (k_1 + k_2)\right)x_1 + k_2 x_2 = 0$$
$$\left((1)(5.18)^2 - (100 + 100)\right)x_1 + 100x_2 = 0$$
$$-173.2x_1 + 100x_2 = 0$$

Since the mode shape factors are relative, let $x_1 = 1$. (This will result in an unnormalized mode shape.) Then, $x_2 = 1.732$.

Similarly, the equilibrium equation for mass 3 is

$$k_3 x_2 + (m_3\omega^2 - k_3)x_3 = 0$$
$$(100)(1.732) + \left((0.5)(5.18)^2 - 100\right)x_3 = 0$$
$$x_3 = 2$$

The unnormalized mode shape is

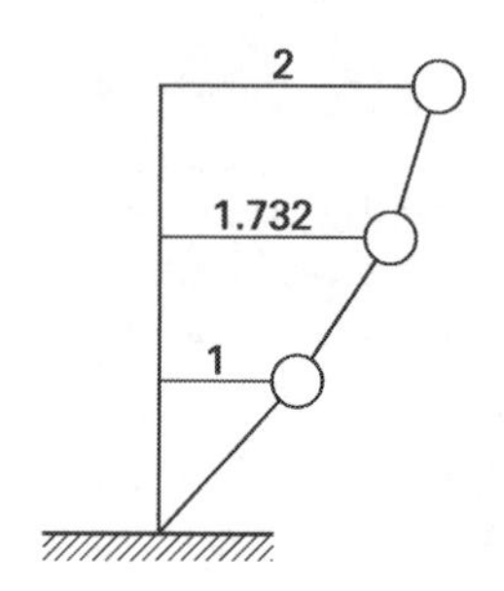

$$\sqrt{\sum m_i \phi_i^2} = \sqrt{(1)(1)^2 + (1)(1.732)^2 + (0.5)(2)^2}$$
$$= \sqrt{6}$$

Dividing each of the unnormalized mode shape factors by $\sqrt{6}$ results in the following mode shape.

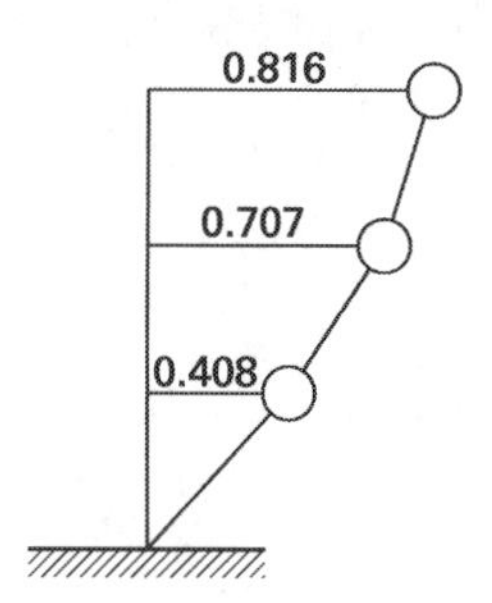

20. RAYLEIGH METHOD

Example 4.4 and Ex. 4.5 show the significant computational burden of performing a full dynamic analysis for even a simple MDOF system. While the computer is an ideal tool for doing this, there may be some situations in which such an analysis is unnecessary or inappropriate. (See Sec. 6.33.)

For such situations, it may be possible to use one of several iterative procedures, most of which are variations of the *Rayleigh method.* This method starts by assuming a mode shape. (Even poor initial assumptions converge rapidly to the correct answer.) Then, the maximum kinetic energy is set equal to the maximum potential energy. Eventually, the mode shape is calculated and used as the starting point for the subsequent iteration.

The *Stodola method*, consisting of the following steps, is one such iterative process.[14]

step 1: Assume a mode shape. That is, assume a deflection, x, for each mass. (A good starting point is the shape of static deflection taken by the structure when it is turned 90° and acted upon by gravity.)

step 2: Compute the inertial forces for each mass from Eq. 4.30.

$$F_{\text{inertial}} \approx m\omega^2 x \quad \text{4.30}$$

step 3: Compute the spring forces on each mass as the sum of the inertial forces acting on the springs.

step 4: Compute the spring deflections.

step 5: Calculate the mode deflections from the spring deflections. Repeat from step 1 as required.

Example 4.6

Use one iteration of the Stodola method to determine the mode shape of the system in Ex. 4.4.

Solution

step 1: Assume the following mode shape.

$$x_3 = 2.0$$
$$x_2 = 1.5$$
$$x_1 = 1$$
$$x_0 = 0 \text{ (ground)}$$

step 2: The inertial forces are given by Eq. 4.30.

$$F_{i3} = 0.5\omega^2(2) = \omega^2$$
$$F_{i2} = 1\omega^2(1.5) = 1.5\omega^2$$
$$F_{i1} = 1\omega^2(1) = \omega^2$$

step 3: The spring forces are

$$F_{s3} = F_{i3} = \omega^2$$
$$F_{s2} = F_{i2} + F_{i3} = 2.5\omega^2$$
$$F_{s1} = F_{i1} + F_{i2} + F_{i3} = 3.5\omega^2$$

step 4: The spring deflections are

$$x_{s3} = \frac{F_{s3}}{k_3} = 0.01\omega^2$$
$$x_{s2} = \frac{F_{s2}}{k_2} = 0.025\omega^2$$
$$x_{s1} = \frac{F_{s1}}{k_1} = 0.035\omega^2$$

step 5: Dividing by ω^2, the new relative mode deflections are

$$x_1 = x_{s1} = 0.035$$
$$x_2 = x_{s1} + x_{s2} = 0.060$$
$$x_3 = x_{s1} + x_{s2} + x_{s3} = 0.070$$

Dividing by 0.035, the mode shape is

$$x_3 = 2.00$$
$$x_2 = 1.71$$
$$x_1 = 1$$

These values can be used to repeat the procedure. Eventually, the values from steps 1 and 5 will agree.

21. PARTICIPATION FACTOR

The *participation factor*, Γ_j, is the fraction of the total building mass that acts in any particular mode, j. It can be used to calculate the *story drift*, x. (See also Sec. 4.20.) The denominator of Eq. 4.31 is the same as Eq. 4.29 and will be equal to 1.0 if normalized mode shape factors are used. (If the mode shape factors are normalized, the denominator is not needed.) Weight, W, can be substituted for mass, m, in Eq. 4.31.

$$\Gamma_j = \frac{\sum m_i \phi_{ij}}{\sqrt{\sum m_i \phi_{ij}^2}} \quad \text{4.31}$$

$$x = \Gamma S_d \phi \quad \text{4.32}$$

22. STORY SHEARS

The participation can also be used to calculate the *floor force*, F_x, that acts at story x (i.e., the force that acts at that level) and the cumulative *story shear*, V_x, that acts at that level and above. This can be done in two ways, one method derived from Hooke's law and using the spring constant, and the other method derived from Newton's law and using the mass. (Section 6.35 describes the ASCE/SEI7 method of distributing the base shear to the stories.)

$$F_x = \frac{\Gamma W S_a \phi}{g} = \Gamma k S_d \phi \quad \text{[U.S.]} \quad \text{4.33(a)}$$

$$F_x = \Gamma m S_a \phi = \Gamma k S_d \phi \quad \text{[SI]} \quad \text{4.33(b)}$$

Example 4.7

Determine the drifts, story shears, and total base shear for the structure in Ex. 4.5. Assume the spectral displacement and acceleration are 4 (arbitrary units) and 0.28 g (108 in/sec^2), respectively. Assume consistent units are used.

[14]The *Holzer method* is another iterative procedure; it is not discussed in this book. See Wakabayashi (1986) and other structural engineering analysis books.

Solution

First, calculate the participation factor from the normalized mode shape factors determined in Ex. 4.5. Since normalized values are used, the denominator has a value of 1.0 and is not needed.

$$\begin{aligned}\Gamma &= \sum m_i\phi_i \\ &= (1)(0.408) + (1)(0.707) + (0.5)(0.816) \\ &= 1.523\end{aligned}$$

Use Eq. 4.32 to calculate the total drifts.

$$\begin{aligned}x_1 &= \Gamma S_d\phi = (1.523)(4)(0.408) = 2.49 \\ x_2 &= (1.523)(4)(0.707) = 4.31 \\ x_3 &= (1.523)(4)(0.816) = 4.97\end{aligned}$$

The story drifts are relative to the floors below.

$$\begin{aligned}x_{3-2} &= 4.97 - 4.31 = 0.66 \\ x_{2-1} &= 4.31 - 2.49 = 1.82 \\ x_{1-\text{ground}} &= 2.49 - 0 = 2.49\end{aligned}$$

Calculate the story shears from the story drifts. Each of the lateral stiffnesses was 100.

$$\begin{aligned}V_1 &= kx_1 = (100)(2.49) = 249 \\ V_2 &= (100)(1.82) = 182 \\ V_3 &= (100)(0.66) = 66\end{aligned}$$

The floor forces can be calculated from the story shears or the participation factors.

$$\begin{aligned}F_1 &= V_1 - V_2 = 249 - 182 = 67 \\ F_2 &= V_2 - V_3 = 182 - 66 = 116 \\ F_3 &= V_3 = 66\end{aligned}$$

Alternatively,

$$\begin{aligned}F_1 &= \Gamma m S_a\phi = (1.523)(1)(108)(0.408) = 67.1 \\ F_2 &= (1.523)(1)(108)(0.707) = 116.3 \\ F_3 &= (1.523)(0.5)(108)(0.816) = 67.1\end{aligned}$$

The base shear is the sum of the floor forces.

$$\begin{aligned}V &= F_1 + F_2 + F_3 \\ &= 67.1 + 116.3 + 67.1 \\ &= 250.5\end{aligned}$$

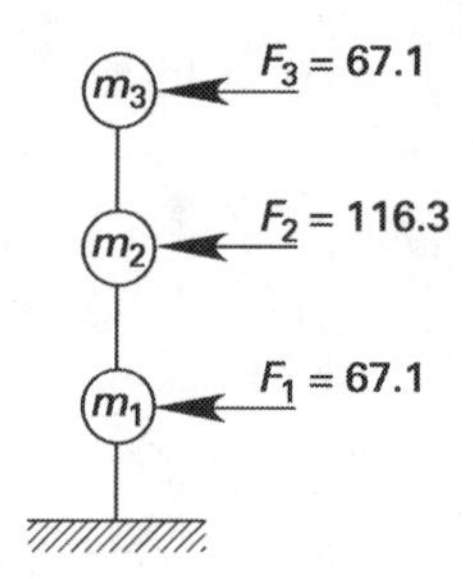

concentrated floor forces

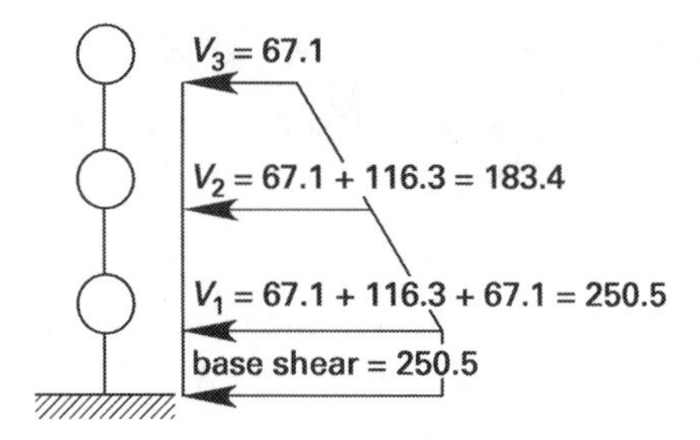

cumulative loading: story shears and base shear

5 Response of Structures

1. ELASTIC RESPONSE SPECTRA

The response of a building to earthquake ground motion depends on the dynamic characteristics of the building. Specifically, the natural period (see Sec. 3.8) and the damping ratio (see Sec. 4.8) affect building response more than do other factors. For a given damping ratio, ζ, and for a given ground motion, a curve known as a *response spectrum of spectral acceleration*, S_a, can be drawn that plots the maximum acceleration response of an elastic single-degree-of-freedom system against the natural period of the system. The response spectrum for a particular earthquake can be used to determine the theoretical maximum acceleration response of the building.

A family of curves (i.e., *response spectra*) for an actual earthquake for various damping ratios is illustrated in Fig. 5.1. Similar response spectra can be developed for *spectral velocity* and *spectral displacement.*

The spectra shown in Fig. 5.1 are for *elastic response* to an earthquake. That is, the structures used to develop the curves moved and swayed during the earthquake, but there was no yielding. For that reason, the curves are known as *elastic response spectra.*

There will always be a region on the response spectrum where the acceleration is highest. This occurs where the natural building period coincides with the predominant earthquake period—when the building is in resonance

Figure 5.1 *Typical Elastic Response Spectra (1940 El Centro earthquake in N-S direction)*

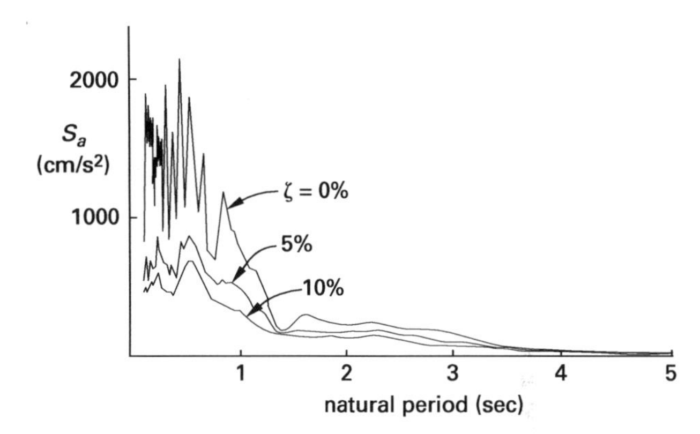

with the earthquake. For California earthquakes, the peak usually occurs in the 0.2 sec to 0.5 sec period range.[1] Theoretically, infinite resonant response (i.e., an infinite magnification factor) is possible, though it is highly unlikely since all real structures are damped.[2]

It seems intuitively logical that a building with large amounts of internal damping will resist acceleration (i.e., motion) to a greater extent than will a similar building with no damping. Such behavior is actually observed as spectral acceleration decreases because damping increases, although the effect of damping at lower periods is slight (since the natural periods of undamped and lightly damped structures are essentially the same).

2. IDEALIZED RESPONSE SPECTRA

The response spectra derived from the behavior of one SDOF system in one particular earthquake are usually quite jagged, as shown in Fig. 5.1. It is not possible to use such a historical record for design, since it is unlikely that an earthquake matching the original earthquake in duration, magnitude, or time history will occur. Also, even if the design earthquake was completely specified, the significant variation in spectral values over small period ranges would require an unreasonable accuracy in the determination of the building period. To get around these problems, a smoothed average *design*

[1]This is not always the case, as shown by the Loma Prieta earthquake.
[2]A properly designed and constructed building seldom experiences true resonance. Planned or unplanned yielding occurs before true resonant response is achieved, and this yielding damps out the resonance.

response spectrum based on the envelopes of performance of several earthquakes is developed, as illustrated in Fig. 5.2.

Figure 5.2 *Average Elastic Design Response Spectra (based on the 1940 El Centro earthquake) [multiplier = 1]*

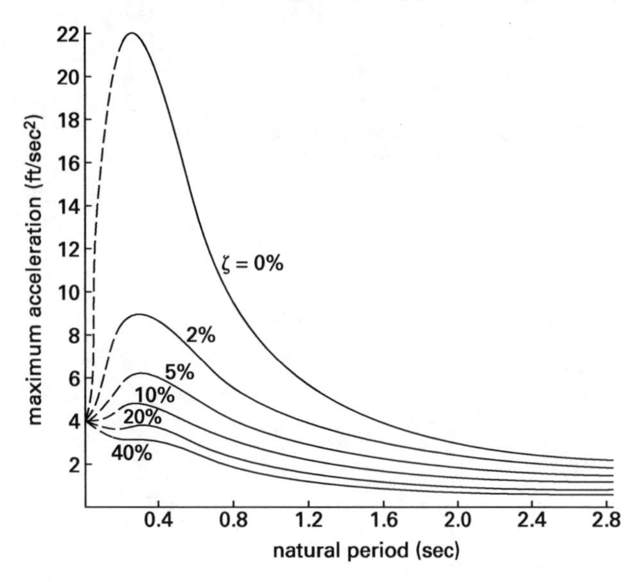

Example 5.1

The primary support for an industrial drill press with a mass of 100,000 lbm (45 000 kg) is the structural steel bent shown. The beam-column and base connections are rigid. The horizontal beam has a mass of 119 lbm/ft (160 kg/m), neglecting the weight of the vertical supports. The system has 5% damping. Determine the elastic response (i.e., base shear) for a 1940 El Centro earthquake in the north-south direction. Use average design spectra.

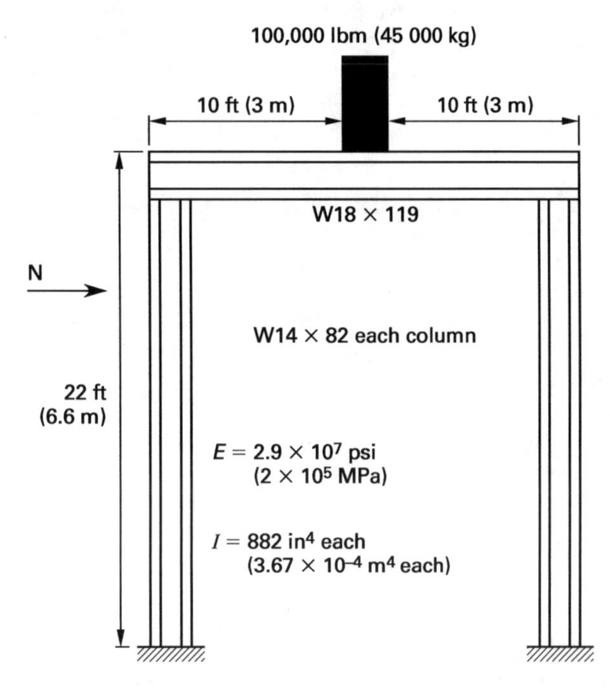

Customary U.S. Solution

The total mass, m, of the moving system is

$$m = 100{,}000 \text{ lbm} + (20 \text{ ft})\left(119 \ \frac{\text{lbm}}{\text{ft}}\right)$$
$$= 102{,}380 \text{ lbm}$$

From Table 4.1, the combined stiffness, k_t, of the two vertical supports is

$$k_t = (2)\left(\frac{12EI}{h^3}\right)$$
$$= \frac{(2)(12)\left(2.9 \times 10^7 \ \frac{\text{lbf}}{\text{in}^2}\right)(882 \text{ in}^4)}{(22 \text{ ft})^3\left(12 \ \frac{\text{in}}{\text{ft}}\right)^2}$$
$$= 4 \times 10^5 \text{ lbf/ft}$$

From Eq. 4.14 and Eq. 4.15, the natural period of vibration, T, is

$$T = 2\pi\sqrt{\frac{m}{g_c k}}$$
$$= 2\pi\sqrt{\frac{102{,}380 \text{ lbm}}{\left(32.2 \ \frac{\text{ft-lbm}}{\text{lbf-sec}^2}\right)\left(4 \times 10^5 \ \frac{\text{lbf}}{\text{ft}}\right)}}$$
$$= 0.56 \text{ sec}$$

From Fig. 5.2, the spectral acceleration for this period and 5% damping is $S_a = 5.5 \text{ ft/sec}^2$. From Eq. 3.1, the base shear is

$$V = \frac{mS_a}{g_c} = \frac{(102{,}380 \text{ lbm})\left(5.5 \ \frac{\text{ft}}{\text{sec}^2}\right)}{32.2 \ \frac{\text{ft-lbm}}{\text{lbf-sec}^2}}$$
$$= 1.75 \times 10^4 \text{ lbf}$$

SI Solution

The total mass, m, of the moving system is

$$m = 45\,000 \text{ kg} + (6 \text{ m})\left(160 \ \frac{\text{kg}}{\text{m}}\right) = 45\,960 \text{ kg}$$

From Table 4.1, the combined stiffness, k_t, of the two vertical supports is

$$k_t = (2)\left(\frac{12EI}{h^3}\right)$$
$$= \frac{(2)(12)(2 \times 10^5 \text{ MPa})\left(10^6 \ \frac{\text{Pa}}{\text{MPa}}\right)(3.67 \times 10^{-4} \text{ m}^4)}{(6.6 \text{ m})^3}$$
$$= 6.13 \times 10^6 \text{ N/m}$$

From Eq. 4.14 and Eq. 4.15, the natural period of vibration, T, is

$$T = 2\pi\sqrt{\frac{m}{k}} = 2\pi\sqrt{\frac{45\,960 \text{ kg}}{6.13 \times 10^6 \ \frac{\text{N}}{\text{m}}}}$$
$$= 0.54 \text{ s}$$

From Fig. 5.2, the spectral acceleration for this period and 5% damping is approximately $S_a = 1.68 \text{ m/s}^2$. From Eq. 3.1, the base shear is

$$V = mS_a = (45\,960 \text{ kg})\left(1.68 \ \frac{\text{m}}{\text{s}^2}\right)$$
$$= 7.72 \times 10^4 \text{ N}$$

3. RESPONSE SPECTRA FOR OTHER EARTHQUAKES

The design response spectra in Fig. 5.2, although normalized and averaged over several earthquakes, are adjusted for an earthquake of a specific magnitude and peak ground acceleration. Based on historical data and probability studies, the recurrence interval for an earthquake of that magnitude can be determined. For example, an earthquake of the 1940 El Centro magnitude is expected at that site, on the average, every 70 years. However, smaller earthquakes will be experienced more frequently than every 70 years, and larger earthquakes will be experienced less frequently than every 70 years.

In order to apply the average design response spectra to other earthquakes, they are simply scaled upward or downward for larger and smaller earthquakes, respectively. For example, Table 5.1 gives the scale factor (to be used to scale Fig. 5.2 downward) for other recurrence intervals at the El Centro site.

Table 5.1 *Scale Factors for Other Recurrence Intervals (based on elastic response to the 1940 El Centro earthquake)*

recurrence interval (yr)	scale factor
2	2.77
20	1.83
32	1.50
70	1.00

4. LOG TRIPARTITE GRAPH

Since spectral acceleration, velocity, and displacement for linear elastic response are all related (see Eq. 3.2), all three spectral quantities can be shown by a single curve on a graph with three different scales. Such a graph is done on a logarithmic scale and is known as a *log tripartite plot*. Both elastic and inelastic (see Sec. 5.9) tripartite plots are prepared. However, for inelastic response, the spectral acceleration, velocity, and displacement cannot be represented by a single curve on the tripartite plot.

Tripartite plots, both elastic and inelastic, can differ in how the axes are arranged. Figure 5.3 illustrates two common arrangements for presenting the information, while Fig. 5.4 gives an elastic log tripartite plot for the 1940 El Centro earthquake.

Figure 5.3 *Two Types of Log Tripartite Plots*

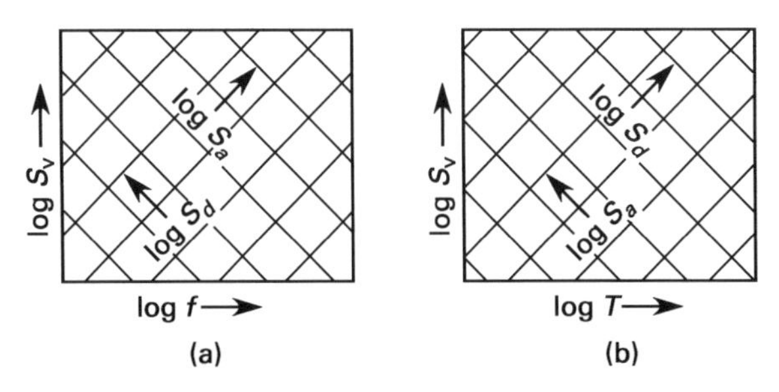

5. DUCTILITY

The expected magnitude of seismic loads and the nature of building codes make it necessary to accept some yielding during large earthquakes.[3] The design provisions in modern seismic codes could not create a purely elastic response during a large earthquake; in any case, building a structure with such a response would not be economical.

Displacement ductility (or just *ductility*) is the capability of a structural member or building to distort and yield without collapsing. During an earthquake, a ductile structure can dissipate large amounts of seismic energy after local yielding of connections, joints, and other members has begun.

The actual ductility of a joint or structural member is specified by its *ductility factor*, μ. There are a number of definitions of the ductility factor, all of which represent the ratio of some property at failure (i.e., fracture) to that same property at yielding. For example, the ductility factor may be specified in terms of energy absorption, as in Eq. 5.1.

$$\mu = \frac{U_{\text{fracture}}}{U_{\text{yield}}} \qquad 5.1$$

In addition to the definition based on the ratio of energies, there are definitions of the ductility factor based on ratios of linear strain and angular strain (rotation). These definitions are not interchangeable, although they are related.[4] Generally, however, the basic concept

[3]The high seismic loading expected in California and the high cost of a totally elastic design make it necessary to accept some yielding. Therefore, the building is designed to withstand a smaller effective peak acceleration (see Sec. 3.5) without yielding, thereby ensuring yielding when a larger ground acceleration is experienced.

[4]For ideal (linear) elastoplastic systems, the ductility based on energy absorption, μ_U, can be calculated from the ductility based on strain, μ_ϵ, as

$$\mu_U = 2\mu_\epsilon - 1$$

This means that if the ductility, as calculated from linear strain, is 4 to 6, the ductility will be 7 to 11 when calculated from Eq. 5.1.

Figure 5.4 *Elastic Log Tripartite Plot (1940 El Centro earthquake)*

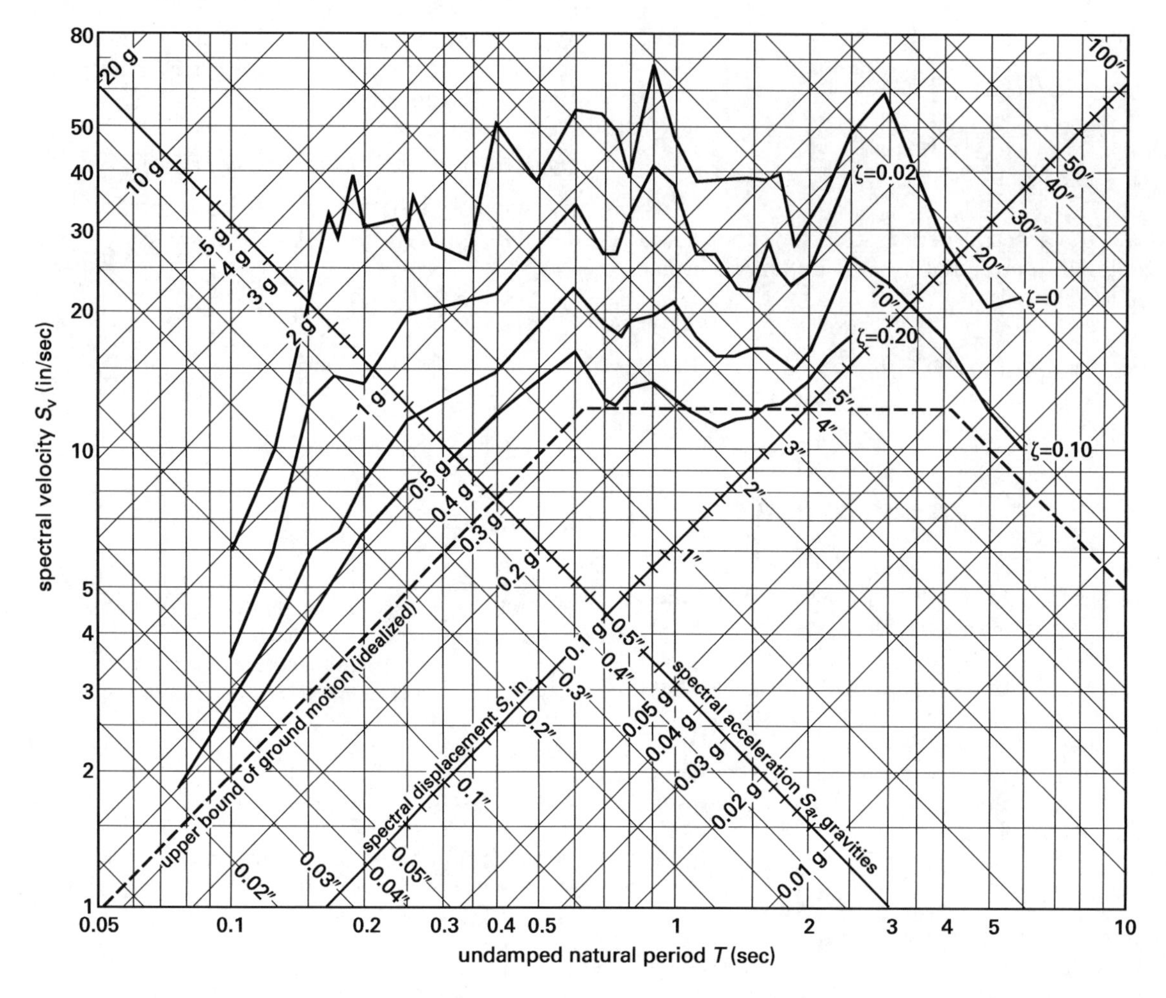

Reprinted from *Design of Multistory Reinforced Concrete Buildings for Earthquake Motions*, by John A. Blume, Nathan M. Newmark, and Leo H. Corning, 1961, with permission from the Portland Cement Association, Skokie, IL.

(i.e., the ratio of some failure property to the same yield property) is all that is needed to explain the significance of a ductile structure.

The minimum assumed ductility (based on strain or deformation) of building structures with good connections and good redundancy that are designed to modern seismic codes is 2.2. (Ductility of bridge structures is much less.) Desirable levels vary, although it is best to have large values of the ductility factor—4 to 6 for concrete frames and 6 to 8.5 for steel frames. In order to achieve these levels of ductility in the structure overall, the structural members themselves must have special detailing with inelastic deformation in mind.

6. STRAIN ENERGY AND DUCTILITY FACTOR

The area under the stress-strain curve represents the *strain energy* absorbed, U, as shown by Fig. 5.5. The maximum energy that can be absorbed without yielding

Figure 5.5 *Strain Energy*

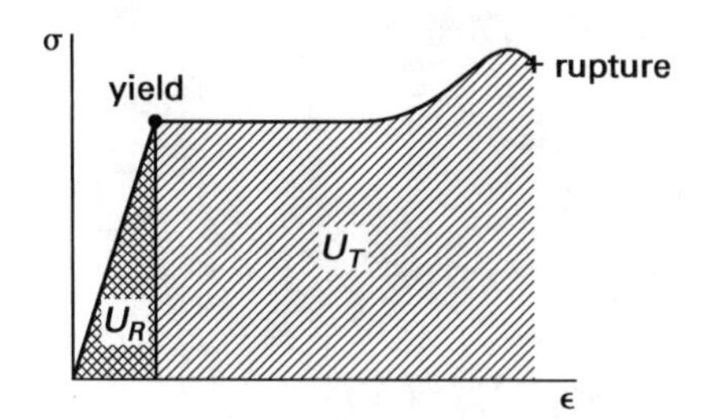

(i.e., the area under the curve up to the yield point) is known as the *modulus of resilience*, U_R. The maximum energy that can be absorbed without failure is the *modulus of toughness* (*rupture*), U_T. One definition of the ductility factor, μ, can be calculated from the ratio of these two quantities.

$$\mu = \frac{U_T}{U_R} \tag{5.2}$$

7. HYSTERESIS

Hysteresis (*hysteretic*) *damping* is the dissipation of part of the energy input when a structure is subjected to load reversals in the *inelastic* range. Such dissipation occurs in the structure itself as well as in the soil around the foundation and, therefore, depends on the nature of the building, foundation, and soil. The energy lost per cycle, U_H, is the area within the *hysteresis loop*, as shown in Fig. 5.6. Hysteresis losses are unaffected by the velocity of the structure.

Figure 5.6 *Hysteresis Loop*

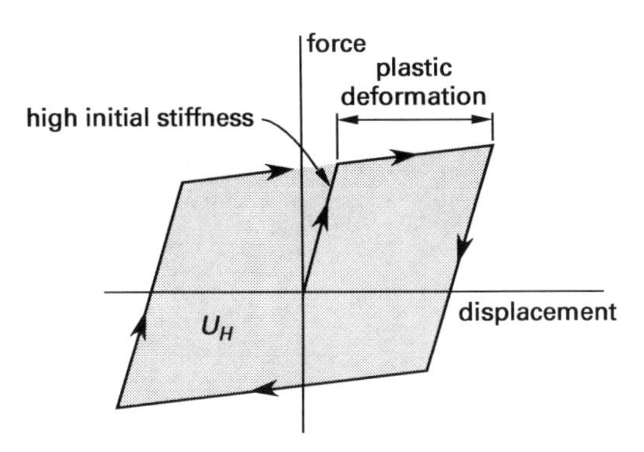

Inasmuch as it is difficult to evaluate hysteresis losses, hysteresis damping in seismic studies is sometimes accounted for by defining an equivalent internal viscous damping. Such an approximation works reasonably well in some cases, but the validity deteriorates as the deflection increases.

8. LARGE DUCTILITY SWINGS

The effect of *reversal deformation* after a few cycles of very high ductility is significant. Tests and actual experience indicate that even modern structures can fail after only a few deformation reversals if the strain is well into the inelastic region.[5] It is particularly easy to show that threaded rods are susceptible to such failure.

9. INELASTIC RESPONSE SPECTRA

The total seismic energy, U, received by a building structure in an earthquake is stored or dissipated in four primary ways. Some of the energy is stored as elastic strain energy, U_E; some is converted into kinetic energy, U_K; some is dissipated as hysteretic or plastic losses, U_H; and some is lost due to frictional and damping effects, U_ς. In the simplest models, the sum of these four terms equals the total energy input.

Particularly when the building is stressed in the elastic region, input energy is dissipated relatively slowly, primarily because of internal friction (i.e., damping) of the structure converting the kinetic energy into heat. However, it takes much more energy to plastically deform parts of the structure. Since the amount of input energy is limited to what was received and the frictional losses are approximately constant, an increase in this energy of deformation is accompanied by a decrease in the kinetic energy of oscillation. Thus, each yielding connection, every broken column, and every sheared pin dissipates a finite amount of kinetic energy. Therefore, a building's amplitude of oscillation and number of oscillation cycles decrease as major portions of the building yield.[6]

The *inelastic design response spectra* (IDRS) show what the acceleration will be when some of the seismic energy is removed inelastically. It is appropriate to consider the inelastic effects when the response of a building to a major earthquake is being determined. The inelastic response spectra are usually derived from the elastic response spectra.

There are several well-known methods of obtaining the inelastic response spectra from the elastic response spectra, but few of them are suitable for manual calculations. Perhaps the quickest and easiest, though not necessarily the most rigorous, method is simply to scale the elastic curves downward by some function of the ductility factor.

$$S_{a,\text{inelastic}} = \frac{S_{a,\text{elastic}}}{\text{factor}} \qquad 5.3$$

The "factor" in Eq. 5.3 depends on the period. For extremely small periods (i.e., frequencies greater than approximately 33 Hz), there is no reduction at all. For periods greater than approximately 0.5 sec to 1.5 sec (i.e., frequencies less than 2 Hz), the ductility factor (based on strain), μ_ϵ, itself can be used as the reduction factor. For intermediate periods (33 Hz > f > 2 Hz), the reduction factor is approximately $\sqrt{\mu_U} = \sqrt{2\mu_\epsilon - 1}$.

In converting an elastic response spectrum to an inelastic response spectrum, the ductility factor, μ_ϵ, used to calculate the reduction factor may be known as the *structure deflection ductility factor* or *design ductility factor*, μ_Δ. It is the ratio of the deflection at ultimate collapse to the deflection at first yield, measured at the roof of the structure. Estimates of this value are known to be unreliable at low natural periods (i.e., high frequencies), but division by μ_Δ or μ_ϵ is favored because of its simplicity.

Values of the design ductility factor in excess of 6 are not often used, as excessive damage (architectural as well as structural) would be experienced, even though larger values (up to 10 for ductile steel structures) are readily achievable.

At high periods (i.e., low frequencies), energy absorption effects dominate, and a ductility factor based on energy (rather than strain), μ_U, is more appropriate for use in determining the inelastic response spectrum. (See Ftn. 4.)

[5]This type of failure was first predicted in a controversial paper written by Vitelmo V. Bertero and Egor Popov in the 1960s. Such failures were actually observed in the 1964 Alaska and 1971 San Fernando earthquakes.

[6]A yielding structure experiences larger localized deformations than an elastic structure does. This is different, however, from the overall oscillation of the structure.

10. NORMALIZED DESIGN RESPONSE SPECTRA

For design purposes, the response spectrum should be representative of the characteristics of all seismic properties experienced at a specific site. The design response spectrum should be based on geologic, tectonic, seismological, and soil characteristics associated with that specific site if these are known. If not, it may be constructed according to the spectral shape presented by Fig. 5.7 (based on IBC and ASCE/SEI7). Figure 5.7 is normalized with respect to the peak ground acceleration (EPA) and uses the site-specific design spectral response acceleration values, S_{DS} and S_{D1} as defined by the IBC, spectral accelerations that define the shape of the response spectrum.

Figure 5.7 *Design Response Spectrum*

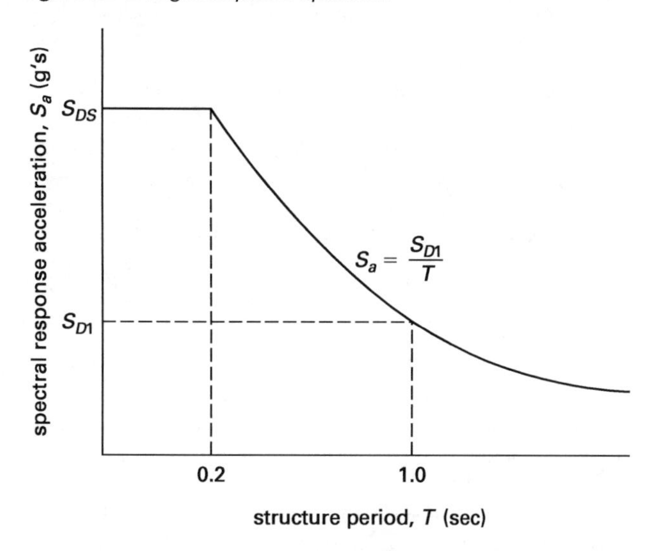

The design spectral response acceleration values S_{DS} and S_{D1} account for the potential amplification of the ground vibration generated at a specific site by an earthquake (seismic response throughout the spectral range). They are influenced by the regional seismicity and geology, the expected recurrence rates and maximum magnitudes of events on known faults and seismic zones, the proximity of the site to active seismic sources (i.e., faults), and the characteristics of the site soil profile.

The first term, S_{DS}, corresponds to the site-dependent effective peak acceleration and is controlled by the short period portion of the spectrum for structures having a fundamental period of 0.2 sec. S_{D1} corresponds to the site-dependent effective peak acceleration response at 1.0 sec period and is controlled by the longer period portion of the spectrum. The procedure for determining the numerical value for S_{DS} and S_{D1} is explained in detail in Chap. 6.

11. RESPONSE SPECTRUM FOR IBC-DEFINED SOIL PROFILES

Six different soil profile types are classified in the IBC based on average shear wave velocity in the top 100 ft (30 m) of the soil layer: site class A, B, C, D, E, and F (see Sec. 6.15). The IBC design response spectrum (as shown in Fig. 5.7) is site dependent by virtue of the soil-profile dependence of the design spectral response acceleration values S_{DS} and S_{D1}. The functions plotted in Fig. 5.8 are easily derived by computing quantities S_{DS} and S_{D1} from Eq. 6.8, Eq. 6.9, and T_s as shown in Fig. 5.7.

Figure 5.8 *Normalized Response Spectra Shapes*

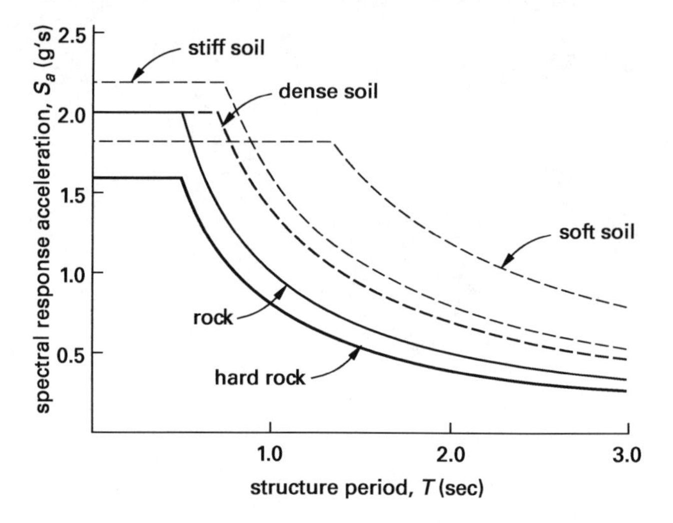

12. DRIFT

Drift is the lateral displacement (deflection) of one floor relative to some point below. *Total drift*, δ, is shown in Fig. 5.9 and is measured with respect to the ground. *Story drift*, Δ, is shown in Fig. 5.10 and is measured with respect to the story below. There are two main reasons to control drift. First, excessive movement in upper stories has strong adverse psychological and physical effects on occupants. Second, it is difficult to ensure structural and architectural integrity with large amounts of drift.[7] Excessive drift can be accompanied by large secondary bending moments and inelastic behavior. (See Sec. 5.13.) In a severe earthquake in which yielding is experienced, a modern high-rise building can be expected to experience a drift of approximately 2% of its total height at the roof level.[8]

There are three components of drift: (1) column and girder bending and shear, (2) joint rotation, and (3) frame bending. The first component is sometimes referred to as *bent action*. The first two components together are referred to as *shear drift*. The third component is known as *chord drift* and *cantilever displacement*.

[7] *Architectural failures* are such nonstructural damage as failure of partitions, windows, and hung ceilings. In low-rise construction, damage to stairwells and elevator shafts can also be considered nonstructural. However, in high-rise construction, stairwells and elevator shafts usually constitute the most critical structural elements in the structural core.

[8] ASCE/SEI7 Sec. 12.8 limits the drift under the code-specified design lateral forces based on the method of construction of the structure. (See Sec. 6.40.) Under larger forces, the drift will be larger.

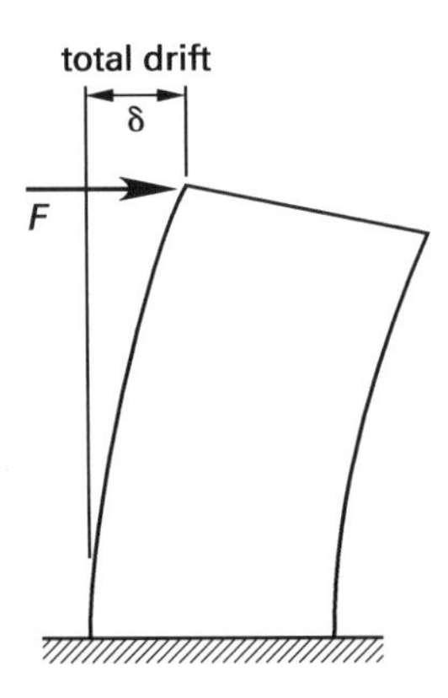

Figure 5.9 Total Drift

Table 5.2 contains generalizations about the effects of different variables on drift. The drift is proportional to the variables (raised to the powers indicated) defined in the table.

The *story drift ratio* is the story drift divided by the height (floor to floor) of the story.

The IBC requires computation of seismic building drifts based on the response that occurs during the design earthquake. Simply, the code's drift value is the drift that would occur when the structure responds inelastically to the design earthquake.

In the past, drifts, δ_w, were determined from "working stress level" lateral forces. Now, however, the IBC determines design seismic forces from "strength levels." The corresponding design level response displacements are denoted as δ_{xe}. Displacements are computed from static, elastic lateral analyses using the design seismic forces of ASCE/SEI7 Sec. 12.8.1, as specified in ASCE/SEI7 Sec. 12.8.6. The elastic, design-level deformations, δ_{xe}, are then converted to the actual drifts expected, δ_x, for the design earthquake by multiplying by the deflection amplification factor, C_d, and dividing by the importance factor, I, as specified in ASCE/SEI7 Sec. 12.8.6. The resulting drifts, δ_x, are called *maximum inelastic response displacements*, and they include an estimated

Table 5.2 Effect of Different Variables on Drift

variable	column drift	girder drift	joint drift	chord drift
story height, H	H^3	H^2	H^2	none
building height, H	none	none	none	H^3
girder length, L	none	L	none	L^3
column depth, D	D^2	none	D^{-1}	none
girder depth, D	none	D^2	D^{-1}	none
column height, H	H	none	H^2	none
shear load, F	none	none	none	F
frame length, L	none	none	none	L^{-2}

inelastic contribution to the total deformation. That is, the maximum actual drift expected at level x is

$$\delta_x = \frac{\delta_{xe} C_d}{I} \tag{5.4}$$

The maximum inelastic response displacement, δ_x, is the displacement of the center of mass of the story. In some cases, the actual displacement of portions of the story may be higher than the displacement at the center of mass, particularly when the building experiences torsional motion. The peak story displacement, Δ, is the largest displacement of any point of the story. This value must be within the code limits for allowable story drift, Δ_a. (See Table 6.14.)

With the design basic ground motion, structures experience forces larger than both the working stress and strength level design forces. The corresponding δ is several times larger than either δ_w or δ_{xe}. (See Sec. 6.40.)

13. P-δ EFFECT

The column members in a structure are loaded in compression by the vertical live and dead loads. Normally, these loads are concentric with the bases of the members. When the structure is acted upon by a lateral (horizontal) seismic load, the structure becomes laterally displaced and the applied vertical loads become eccentric with respect to

Figure 5.10 Story Drift

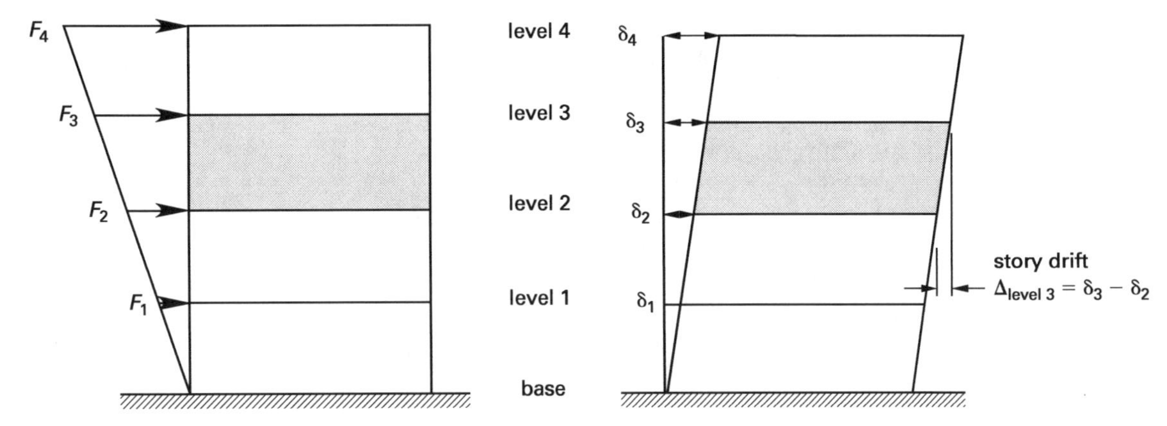

the bases. This results in additional forces and moments and increased story displacements. This *secondary effect* on shears, axial forces, moments, and displacements of frame members is referred to as the *P-δ*, *P-Δ*, or *P-delta effect.* (See Fig. 5.11.)

Figure 5.11 *P-δ Effects*

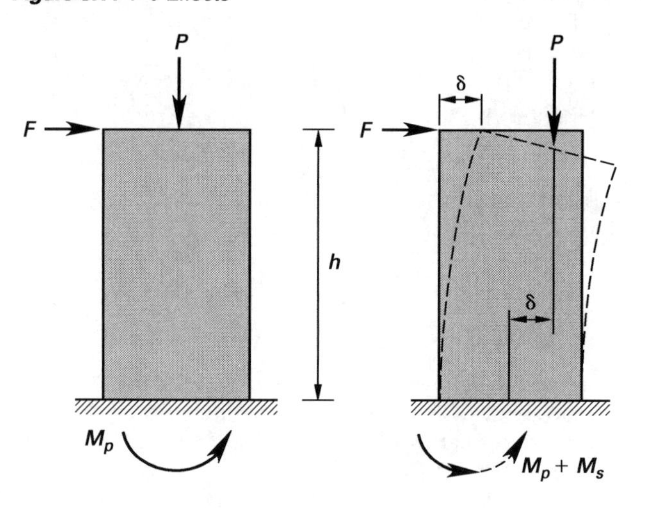

When the total vertical load is concentric with the base of the structure, the overturning moment is referred to as the *primary moment.* The magnitude of the primary moment is Fh, where F is the lateral seismic load and h is the height of the structure. When the vertical loads become eccentric with respect to the base, the overturning moment adds an eccentric bending stress to the columns. This additional column stress is referred to as the *secondary moment.* In the ground-level columns, the magnitude of the secondary moment is $P\delta$, where P is a function of the building weight (i.e., dead load, live load, and snow load) and δ is the drift.

According to ASCE/SEI7 Sec. 12.8.7, the member forces and moments and the story displacements generated by *P*-delta effects should be considered in the evaluation of overall structural frame stability. For this evaluation, the forces producing the story displacements of δ should be used. (See Sec. 6.40.)

If the *overturning moment* increases faster than the *restoring moment* from the frame stiffness, the frame will be unstable. Since the vertical load is constant (i.e., is not transient as is the seismic load that causes the initial eccentricity), the column members will eventually fail and the frame will buckle. Based on the 1991 Bernal research, *P*-delta has very little effect on structural response until dynamic instability is approached. Protection against instability failures is provided by wall X-bracing and thick shear walls.

Unstable frames can be inadvertently designed in nonseismic areas where only vertical loads are considered. Designing a frame to withstand large lateral (seismic) loads has the effect of limiting drift; such frames, therefore, are unlikely to experience a problem caused by *P*-delta instability.

ASCE/SEI7 Sec. 12.8.7 states that the *P*-delta effect need not be considered in the analysis of the entire structure when the ratio of secondary moment to primary moment (i.e., *stability coefficient,* see Eq. 6.34) in a story is less than or equal to 0.10.

14. TORSIONAL SHEAR STRESS

A building's *center of mass*, CM (shown on the plan view of Fig. 5.12), is a point through which the base shear (i.e., the total lateral seismic force) can be assumed to act. This base shear is resisted by the vertical members at the ground level. Each such member may have a different rigidity and, thus, provides a different lateral resisting force in the opposite direction of the base shear. The building's *center of rigidity*, CR, is a point through which the resultant of all the resisting forces acts.[9]

Figure 5.12 *Centers of Mass and Rigidity (building plan view)*

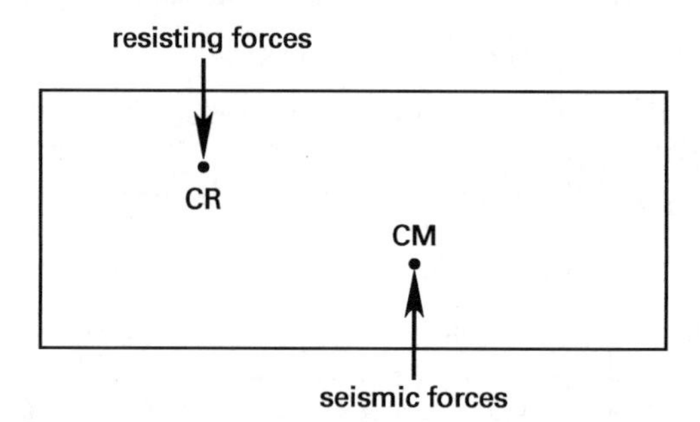

If the building's center of mass does not coincide with its center of rigidity, the building will tend to act as if it is "pinned" at its center of rigidity. It is said to be acted upon by a torsional moment, $M_{\text{torsional}}$, calculated as the product of the shear, V, and the eccentricity, e. This *eccentricity* is the distance (measured perpendicular to the direction of lateral load) between the centers of mass and rigidity. (See Fig. 5.13.)

$$M_{\text{torsional}} = Ve \qquad 5.5$$

Equation 5.6 calculates the maximum torsional shear stress in circular members. In Eq. 5.6, J is the polar moment of inertia which, for circular cross sections, can be calculated as the sum of the moments of inertia taken with respect to the x- and y-axes.

[9]The implication here is that the structure has rigid diaphragms (see Sec. 7.4) between the floors so that the torsional moment can be transferred to the various resisting members distributed at that level. Structures with flexible diaphragms (see Sec. 7.6) are incapable of distributing torsional moments to vertical resisting elements. ASCE/SEI7 Sec. 12.3.1 gives a method of determining whether or not a diaphragm can be considered flexible. Specifically, the diaphragm is flexible if the maximum lateral diaphragm deformation is more than twice the average story drift.

Figure 5.13 *Eccentricity*

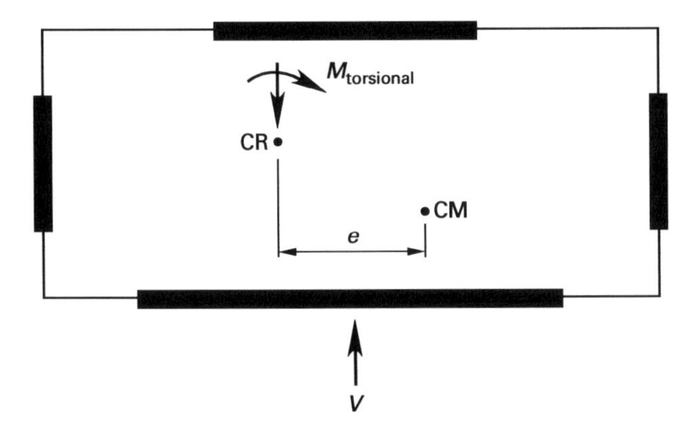

$$v = \frac{M_{\text{torsional}} r}{J} \quad 5.6$$

$$J = I_x + I_y \quad \left[\begin{array}{c}\text{circular cross}\\ \text{sections only}\end{array}\right] \quad 5.7$$

In buildings, the rotation about the center of rigidity is resisted by a torsional shear force (stress) in all members.[10] This torsional force (stress) is proportional to the distance, r, from the building's center of rigidity to the resisting member. A satisfactory substitute for the polar moment of inertia, J, can be calculated from the relative rigidities of the resisting elements. If $R_{\text{rel},i}$ is the relative rigidity of shear wall i, and r_i is the distance of wall i from the center of rigidity, then the polar moment of inertia is

$$J = \sum R_{\text{rel},i} r_i^2 \quad 5.8$$

The units of J are somewhat ambiguous since the units used to determine the relative rigidities are not necessarily known.

The shear force, $F_{i,\text{torsion}}$, in member i due to torsion is

$$F_{i,\text{torsion}} = \frac{R_{\text{rel},i} r_i M_{\text{torsional}}}{J} = \frac{R_{\text{rel},i} r_i M_{\text{torsional}}}{\sum R_{\text{rel},i} r_i^2} \quad 5.9$$

Then, the torsional shear stress in member i is found by dividing the shear force by the cross-sectional area of the member.

$$v = \frac{F_{i,\text{torsion}}}{A} \quad 5.10$$

[10]Unlike the base shear, which is resisted only by walls parallel to the seismic force, the torsional shear is resisted by all walls and columns.

Equation 5.8 shows that the contribution of a stiff element to torsional rigidity increases with the square of the distance of that element from the center of rigidity. If R_{rel} is the relative rigidity of a shear wall and r is the distance of the wall from the center of rigidity, then the contribution of the wall to the torsional moment of inertia is $J_w = R_{\text{rel}} r^2$. Therefore, shear walls should be located as near the building perimeter (and hence as far from the center of rigidity) as possible.

ASCE/SEI7 Sec. 12.8.4.2 requires that an *accidental eccentricity*, e_a, of $\pm 5\%$ (based on the maximum building dimension at that level perpendicular to the direction of the seismic load) be added to the actual eccentricity, if any, in the design of all buildings, even those that are symmetrical. (See Fig. 5.14. Also, see Sec. 6.39.) This eccentricity is included to account for accidental errors in workmanship, uncertainties in the actual location of the centers of mass and rigidity, nonuniform distribution of dead and live loads, nonuniformities that result from subsequent building modifications, and eccentricities that develop during an earthquake after the failure of certain structural elements.

Figure 5.14 *Accidental Eccentricity*

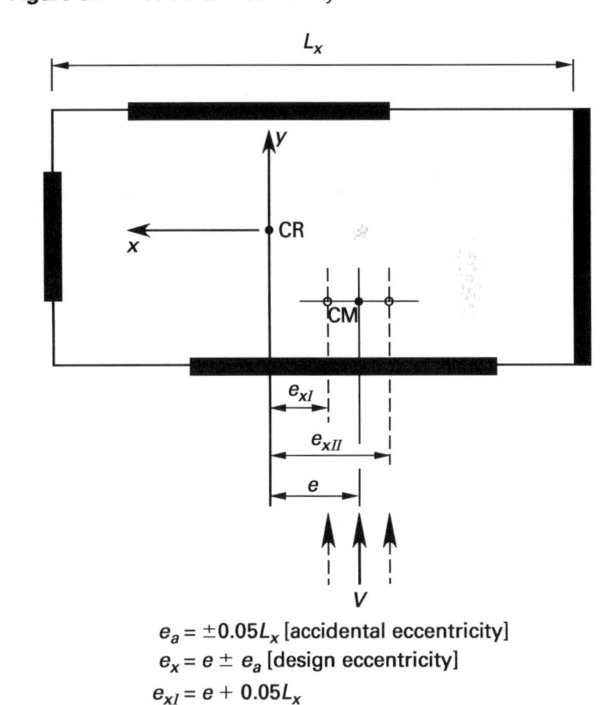

Example 5.2

A crane system is modeled as a 1000 lbm (455 kg) mass attached to the end of a 5 ft (1.5 m) cantilever beam supported by a 20 in (51 cm) diameter hollow tubular vertical column. Calculate the maximum torsional shear stress for a lateral acceleration in the x-direction of 0.3 g.

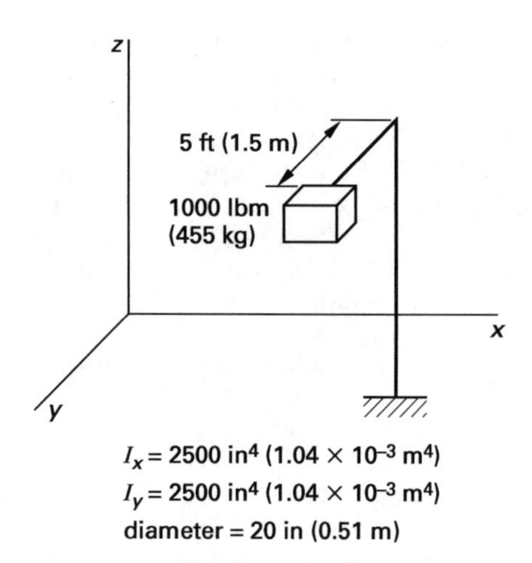

Customary U.S. Solution

The lateral (seismic) force is equal to the inertial force.

$$F_i = \frac{ma}{g_c} = \frac{(1000 \text{ lbm})(0.3 \text{ g})\left(32.2 \ \frac{\text{ft}}{\text{sec}^2\text{-g}}\right)}{32.2 \ \frac{\text{ft-lbm}}{\text{lbf-sec}^2}}$$
$$= 300 \text{ lbf}$$

From Eq. 5.5, the torsional moment is

$$M_{\text{torsional}} = Fe = (300 \text{ lbf})(5 \text{ ft})\left(12 \ \frac{\text{in}}{\text{ft}}\right)$$
$$= 18{,}000 \text{ in-lbf}$$

Equation 5.7 gives the polar moment of inertia.

$$J = I_x + I_y = 2500 \text{ in}^4 + 2500 \text{ in}^4$$
$$= 5000 \text{ in}^4$$

The distance from the center of rigidity (i.e., the center of the column) to the most exterior point on the column is 10 in. The maximum torsional shear stress is given by Eq. 5.6.

$$v = \frac{M_{\text{torsional}} r}{J} = \frac{(18{,}000 \text{ in-lbf})(10 \text{ in})}{5000 \text{ in}^4}$$
$$= 36 \text{ lbf/in}^2 \ (36 \text{ psi})$$

SI Solution

The lateral (seismic) force is equal to the inertial force.

$$F_i = ma = (455 \text{ kg})(0.3 \text{ g})\left(9.81 \ \frac{\text{m}}{\text{s}^2\text{·g}}\right)$$
$$= 1339 \text{ N}$$

The torsional moment is

$$M_{\text{torsional}} = Fe = (1339 \text{ N})(1.5 \text{ m})$$
$$= 2009 \text{ N·m}$$

Equation 5.7 gives the polar moment of inertia.

$$J = I_x + I_y$$
$$= 1.04 \times 10^{-3} \text{ m}^4 + 1.04 \times 10^{-3} \text{ m}^4$$
$$= 2.08 \times 10^{-3} \text{ m}^4$$

The distance from the center of rigidity (i.e., the center of the column) to the most exterior point on the column is 25.5 cm. The maximum torsional shear stress is given by Eq. 5.6.

$$v = \frac{M_{\text{torsional}} r}{J}$$
$$= \frac{(2009 \text{ N·m})(0.255 \text{ m})}{2.08 \times 10^{-3} \text{ m}^4}$$
$$= 2.46 \times 10^5 \text{ Pa} \quad (246 \text{ kPa})$$

15. NEGATIVE TORSIONAL SHEAR

The base shear causes a shear stress that acts in the same direction in all vertical base members. The torsional shear stress, however, has different signs on either side of the center of rigidity. (See Fig. 5.15.) On one side (i.e., where the resisting element is on the same side of the center of rigidity as the center of mass, wall 2 in the figure) the torsion increases the stress from the base shear; on the other side (wall 1 in the figure), the stress is decreased. The amount of decrease is known as *negative torsional shear.* The total lateral force is the sum of the shear force and the torsional force. Negative torsional shear should normally be neglected; that is, it should not be used to decrease the design capacity of a wall or member.

It is easy to make the error of reversing the signs of the induced stresses and adding the negative torsional stress

Figure 5.15 *Torsional Shear*

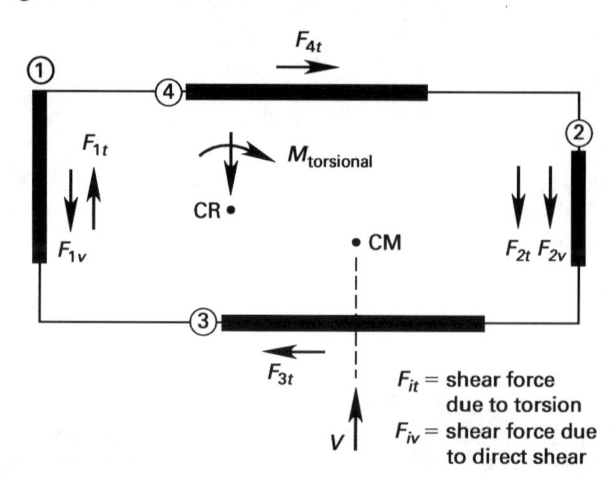

Figure 5.16 Overturning Moment

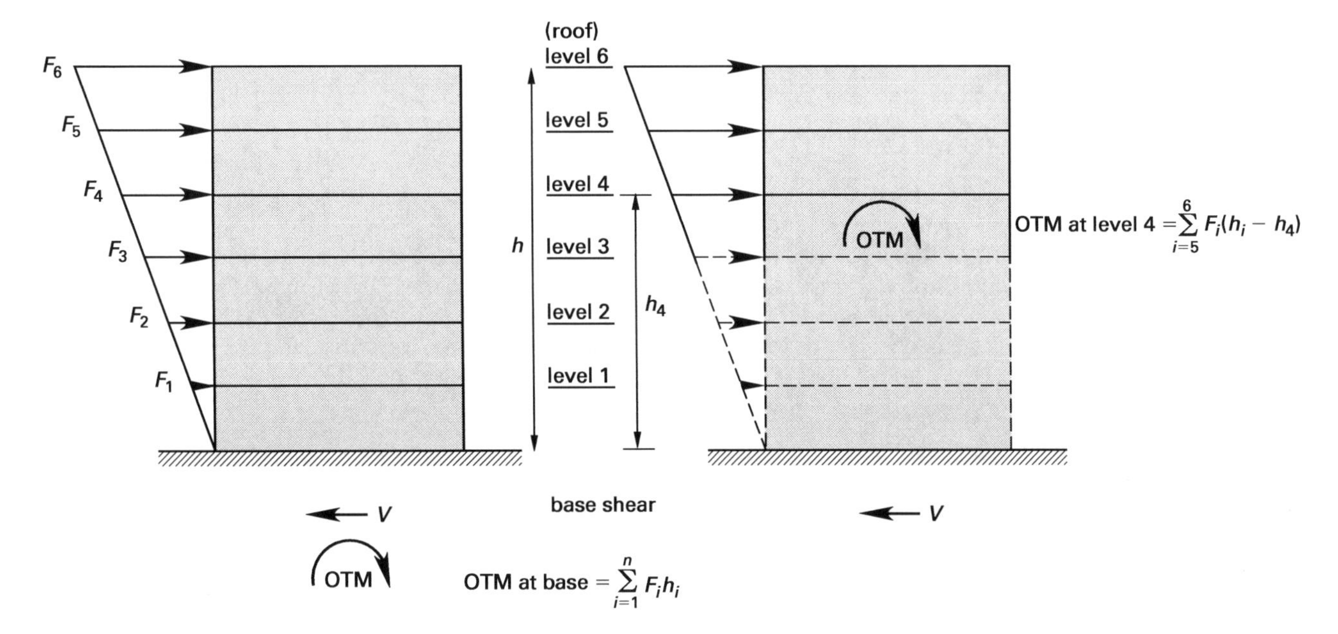

where it should be subtracted, and vice versa. The key to avoiding this error is to always work with the stresses that *resist* the forces and moments. Thus, the stress that resists the base shear acts in a direction opposite to the base shear (i.e., opposite to the direction of ground motion). Similarly, the torsional stresses that resist the torsional moment are in the direction opposite to the applied moment.

16. OVERTURNING MOMENT

The distribution of earthquake forces over the height of a structure causes the structure to experience overturning effects. According to ASCE/SEI7 Sec. 12.8.5, every structure is to be designed to resist the overturning effects caused by seismic forces. The design overturning moment is distributed to the various resisting elements. The intent is to transfer the overturning effects on all resisting elements to the foundation.

The summation of moments due to the distributed lateral forces (see Sec. 6.35) is the *overturning moment*, often given the symbol OTM. If the overturning moment is large enough, it can reverse the compression that normally exists in outer columns caused by the dead and live building loads. Because footings and concrete walls and columns can be placed in a state of tension, the overturning moment is more of a problem for concrete frame and shear wall construction (which cannot tolerate much tension) than it is for steel frame construction.

The overturning moment will increase the compressive stress in outer columns on the opposite side of the building. Such an increase must be countered by increasing the thickness of shear walls and using extra steel reinforcement in concrete columns.

Overturning moments should be calculated for each building level. (See Fig. 5.16.) The first overturning moment is the sum of all moments taken about the ground level. This moment should be used to size footings and to design the primary outer columns. The overturning moment for each subsequent floor considers only lateral forces above that floor. This moment is used to design the shear walls and other supporting structures at that floor.

17. RIGID FRAME BUILDINGS

Before 1965, when the design of structural systems was still in its infancy, most tall buildings were designed as *rigid frames*.[11] In a rigid-frame building, columns and beams were welded together to create a structural grid that resisted wind and earthquake forces elastically.[12]

Such buildings were expensive to construct because they used inordinately large amounts of material, usually steel, to keep the stresses in the elastic region.[13]

18. HIGH-RISE BUILDINGS

The optimum design for seismic loading often conflicts with that for wind loading, a significant factor for any

[11]The Empire State (completed 1931) and the Chase Manhattan Bank (completed 1961) buildings in New York City and the Tenneco Building in Houston are rigid-frame structures.

[12]Bracing in the core (see Sec. 5.18) was not used, although it was recognized that it contributed to structural performance. Using such bracing would have been prohibitively complicated because tools for the structural analysis, such as computers and software, had not been developed. The increasing cost of land after 1965 also made it worthwhile to use costlier designs.

[13]For smaller buildings, rigid frames may be more economical in some cases, particularly where wind forces prevail or when the cost of the additional material is less than the cost of increased design and testing.

tall building. For an earthquake, the building needs to be flexible, even though the full flexibility might be called upon only once in 500 years. However, the full flexibility might be experienced during large windstorms, say, once every 50 years. The greater flexibility required to resist large earthquakes makes for unpleasant motions in windstorms.[14]

The rigid-frame system relies on the bending of columns and beams for its lateral stiffness. However, bending is a poor way to tap a structural member's strength compared to axial loading.[15] A *tube building* resists lateral forces in a radically different way from a rigid-frame building. The tube is like a giant box beam cantilevering out of the ground. Axial forces in the columns mainly resist the tendency to move laterally.

In order to economically design for increasing numbers of stories, different flexible structural systems have been developed.[16] The most general systems are: (1) frames with bracing in the core, which creates a stiff vertical truss, good for buildings up to 30 or 40 stories; (2) framed tubes, good for up to 60 or 70 stories; and (3) diagonally braced tubes, good for up to 100 or 120 stories.

In a *pure tube* system, also known as a *framed tube*, all of the lateral resistance is in the structure's exterior tube, made up of closely spaced steel columns linked by stiff deep spandrel beams. This framing is usually (but not always) located on the perimeter of the structure. In the past, framed tubes were used for tall buildings in high-wind areas and are now finding some applicability in seismic design.

[14]This and the preceding sections are meant to document the trend in high-rise design, not to suggest using tube structures in designs for earthquake resistance.

[15]A measure of the "efficiency" of steel is the weight of steel per square foot of floor space for all stories. The 60-story rigid-frame Chase Manhattan Bank Building uses about 60 pounds of steel per square foot (290 kg/m^2). The 100-story John Hancock Center in Chicago uses a trussed tube structural system requiring half as much steel per unit area—about 30 pounds per square foot (145 kg/m^2).

[16]Dr. Fazlur Khan is acknowledged as being the first structural engineer to recognize the value of the alternate structural systems. One of the first (if not the first) flexible buildings was the 43-story concrete framed-tube system Chesternut Dewitt Apartment Building in Chicago (completed in 1965) that Khan designed. He also designed the One Shell Plaza Building (completed in 1971) in Houston (a 50-story structure using lightweight concrete, at the time, the tallest poured concrete building in the world).

6 Seismic Building Codes

1. HISTORICAL BASIS OF SEISMIC CODES IN CALIFORNIA

It is somewhat surprising that the formal study of earthquake-resistant design had to wait almost two decades after the 1906 San Francisco earthquake. After that earthquake, the only lateral force requirement placed on structures designed and constructed in San Francisco was a 30 lbf/ft^2 (1.44 kN/m^2) wind loading.

It was not until the 1925 Santa Barbara earthquake that the California legislature directed that significant effort be expended on the study of seismology.[1] Then, in 1933, the Long Beach earthquake resulted in 115 fatalities and widespread damage to schools in the area (caused by poor workmanship, design errors, and construction shortcuts). The *Field Act* was subsequently passed, giving the Division of Architecture, State Department of Public Works, responsibility for approving school designs.[2] Additionally, the *1933 Riley Act* was created to set minimum standards for lateral force resistance in all buildings (specifically, just 2% of the dead load). Early codes specified that a

[1]Based on damage reports, the 1925 Santa Barbara earthquake is estimated at 6.3 Richter magnitude, although Richter-style seismometers had not yet been developed. Fatalities were limited, primarily because the earthquake occurred in the early morning before people were in the business district and children were in school. Widespread damage similar to that which destroyed San Francisco was averted as city engineers detected the tremors by observing fluctuations in the water pressure gauges and shut off the gas valves and electrical mains. Despite this, there were 13 fatalities and significant building damage, particularly in brick, masonry, and tile construction. Steel, wood, and properly designed reinforced concrete construction sustained little or no damage, although damage to poorly designed concrete structures occurred.

[2]The 1933 Long Beach earthquake had a Richter magnitude of 6.3, according to the newly developed seismometer. (Although the seismometer's range was exceeded, there is essentially a full record of this earthquake.) As with the 1925 Santa Barbara earthquake, the 1933 Long Beach earthquake occurred in the early morning, before children were in school. The same types of structural failures were observed—that is, masonry and brick buildings, in particular, were damaged.

building had to be "strong" enough to resist a static lateral force, the *base shear*, V, of some fraction (e.g., 10% for masonry school buildings) of the weight, W.[3] The fraction was known as the *base shear coefficient*, C_s, with no units (as it is in fractions of gravities). Between 1943 and 1953, the base shear coefficient was modified several times based on the building period and/or the height of a building, but the *equivalent static force* concept remained and is in use to this day.[4,5] (See Eq. 6.23.)

$$V = C_s W \tag{6.1}$$

Nine lives were lost in the 1940 El Centro earthquake, a 7.1 magnitude event caused by the Imperial Fault. While only approximately 10 sec in duration, a relatively high ground acceleration, 0.33 g, was observed. This earthquake was significant, not because of the widespread damage, but because it was the first earthquake to occur in a heavily instrumented area and resulted in the first accelerometer yielding response data on building periods.

The 1966 Parkfield earthquake on the San Andreas Fault had a relatively low magnitude of 5.5 and a very short duration, but the ground acceleration of 0.5 g was the highest observed to that date. It became apparent that magnitude and acceleration are not necessarily correlated.

A great amount of accelerometer data was obtained from the 1971 San Fernando earthquake (6.6 magnitude, San Fernando Fault zone). This earthquake was significant for two reasons: First, an unbelievable ground acceleration of 1.24 g was experienced at the Pacoima Dam site. Second, there were failures of new buildings designed with the current seismic codes.[6]

The 1979 Imperial Valley earthquake (Richter magnitude 6.6) produced the first accelerometer data from a building with extensive damage. In a building that was partially supported at the ground level by concrete columns, the period and amplitude of oscillation decreased significantly each time one of the columns failed. This is consistent with the concept that seismic energy is removed from a yielding structure.[7]

The 1989 Loma Prieta earthquake (magnitude 7.1, 62 fatalities, San Andreas Fault) was significant because of the important lessons learned about soil and site conditions. The most significant damage outside the epicenter occurred in areas where soil resonance magnified the seismic energy. Although the overall seismic energy should have been (and probably was) greatly attenuated by the large distance between the epicenter and San Francisco, the site conditions under the San Francisco Marina district and Oakland's Interstate 880 Cypress structure magnified the remaining energy.

The 1994 Northridge earthquake (magnitude 6.7) was centered in the San Fernando Valley region of Southern California. The Northridge earthquake was significant in that it was centered in a major urban area and, although of short duration, produced spectral acceleration values at many sites exceeding the UBC design spectrum. Measured peak ground accelerations routinely exceeded 0.7 g. The duration of source rupture associated with the Northridge earthquake was 6 sec to 8 sec. The fault involved was the Northridge Thrust Fault (also known as the Pico Thrust Fault). Interestingly, before this earthquake, this fault was not considered a seismic danger. Because a thrust fault (blind thrust) was involved, the Northridge earthquake produced a significant vertical acceleration component, which contributed to economical losses exceeding $15 billion (the second most expensive U.S. earthquake).

One of the biggest surprises to come out of the Northridge earthquake was widespread damage to welded moment resisting steel structures, favored in earthquake country because of their supposed ductility. Review and analysis of the damage revealed no single cause, but rather a combination of factors, including the influence of vertical acceleration, connection ductility, and weld materials and quality. The majority of damage consisted of fractures in welded beam-column connections. These connections had cracks in either top or bottom flange welds, which occasionally propagated through the column flange and/or beam web. One prime reason for the Northridge failures has been identified as lack of ductility due to overconstrained joints. Appendix P presents more details regarding performance of the steel moment frame connections in this earthquake.

The 1995 Hyogoken-Nanbu (Kobe) earthquake, which struck Hyogo, located on the south-central part of Japan, had moderate to severe ground shaking with a magnitude of 6.9. The Kobe earthquake produced a fault rupture directly through the downtown section of a city, resulting in over 5400 deaths and injuries numbering in the tens of thousands, with estimated damage costs of $150 billion.

Kobe's setting is very similar to that of the San Francisco Bay Area. Both have large strike-slip faults adjacent to a bay and engineered buildings constructed on sedimentary deposits. The city of Oakland is particularly comparable to Kobe. Similarities include the proximity to a major active fault and considerable deposits of soft soils and bay mud in the downtown section, directly under many of the engineered buildings. These building structures are mostly constructed of nonductile concrete, steel, and masonry.

[3]This is simplifying the theory slightly, as the base shear coefficient was actually applied to the dead load and some part of the live load.

[4]The American Society of Civil Engineers (ASCE) and the Structural Engineers Association of Northern California (SEAONC) formed a committee in 1948 that recommended that the equivalent static force concept be used in San Francisco. In 1959, the Structural Engineers Association of California (SEAOC) code was expanded to a "uniform" code for all areas of the United States. This was the first *Blue Book*. At this time, the type of building—frame, box, and so on—was made significant in code specifications.

[5]There are cases in which ASCE/SEI7 Sec. 12.6 requires a dynamic analysis. (See Sec. 6.33.)

[6]In particular, the New Olive View Hospital and the San Fernando Veterans Administration Hospital were new structures that sustained major damage.

[7]That this event proves inelastic behavior removes seismic energy from a structure should not be used to legitimize intentional design for inelastic behavior. It would be a major flaw to design columns to lose their load-carrying ability in the way they did in this instance.

One major point driven home by the Kobe earthquake is the importance of characterizing near-source ground motion in the design of earthquake-resistant structures. A severe earthquake in a major urban area can create immense social and economic losses, and the impact on the economy will be significant. The main steps in mitigating this risk are to improve the model seismic design codes and to introduce mandatory requirements for seismic rehabilitation of vulnerable buildings.

2. SEISMIC CODES

A *code* is a set of rules adopted by an organization empowered to enforce the code. The mere publication of a set of guidelines such as those contained in ASCE/SEI7 does not constitute a governing law. The guidelines must be adopted by a law-making body to become legal documents.

Various regions of the United States have adopted different codes. Agencies whose codes are widely adopted are referred to as "model code agencies," such as the International Code Council (ICC), which publishes the *International Building Code* (IBC). ICC members are representatives of local, regional, and state governments who investigate and research principles concerning safety of life and property in the construction, function, and location of buildings and related structures. The IBC is dedicated to the development of better building construction and greater safety to the public. It is the most widely adopted model building code in the United States and contains extensive seismic provisions for structures.

Many of the early seismic provisions were influenced by the *SEAOC Blue Book: Seismic Design Recommendations* published by the Seismology Committee of the Structural Engineers Association of California (SEAOC).[8] The *Blue Book* included commentary that is invaluable in understanding the significance of the IBC provisions.

The *Blue Book* provisions were first incorporated into the *Uniform Building Code* (UBC) around 1960 (and following the UBC, into the IBC), but other building codes were slower to include more than limited seismic provisions, probably because the true seismic risk of the regions was not recognized.[9] However, following the 1971 San Fernando earthquake, when several buildings supposedly built according to current seismic provisions experienced substantial damage, other organizations received funding to develop seismic design recommendations.

- The Applied Technology Council (ATC) published ATC 3-06 in 1978. This was a massive 500-page document intended to serve as a reference for other code writers. It has now been superseded by the NEHRP provisions.
- The Building Seismic Safety Council (BSSC), with Federal Emergency Management Agency (FEMA) funding, first published the National Earthquake Hazards Reduction Program (NEHRP) provisions in 1985. These provisions have been updated regularly since then. They have now been superseded by ASCE/SEI7, which has adopted the provisions and incorporated them by reference.
- The American National Standards Institute (ANSI) published its A58.1 in 1982. This document dealt with determining seismic loading, but it did not address detailing.
- The American Concrete Institute (ACI) first included detailing to resist seismic loads in the 1983 edition of ACI 318. However, determination of seismic loads was not covered.
- The American Institute of Steel Construction (AISC) has developed detailing requirements for steel buildings. Specifically, AISC 358 is a compendium of prequalified connections and design processes suitable for use in special moment-resisting frames.
- The American Society of Civil Engineers (ASCE) first published *Minimum Design Loads for Buildings and Other Structures* (ASCE/SEI7) in 1993. ASCE/SEI7 has been updated regularly since then. It has become the preeminent standard affecting seismic design. Starting in 2006, with some modifications, ASCE/SEI7 was incorporated into the IBC by reference.

While the large number of seismic documents may seem confusing, it should be noted that all are used as "source documents" for the IBC.

Adoption of the IBC is up to each municipality. Most large cities have their own specific requirements that can supersede portions of the IBC or replace it entirely. Design of buildings located in Los Angeles, for example, is governed by a city code derived from the IBC.

General seismic provisions such as those in the IBC may be superseded by even more stringent statutory requirements. For example, Title 24 of the *California Administrative Code* requires that schools and hospitals be operational after an earthquake.

Bridges in California are designed according to CALTRANS seismic provisions. The CALTRANS

[8]SEAOC, 1400 K Street, Suite 212, Sacramento, CA 95814.

[9]Historically, the western United States has led in the requirements for seismic design, first using the *Uniform Building Code* (UBC) prepared by the International Conference of Building Officials (ICBO), and then using the *International Building Code* (IBC) prepared by the International Code Council. The nation's two other model building codes were written by the Building Officials and Code Administrators International (BOCA) in Country Club Hills, Illinois, and the Southern Building Code Congress International (SBCCI) in Birmingham, Alabama. Both codes significantly strengthened their seismic provisions in 1992. However, the two codes differ from the UBC in their methodology.

method evaluates each bridge two ways (the term "loading" refers to lateral loads, moments, and shears).

1. Transverse seismic loading plus 30% of the longitudinal seismic loading
2. Longitudinal seismic loading plus 30% of the transverse seismic loading

As with the IBC method, the CALTRANS method includes both static and dynamic analyses. The static method is used for well-balanced spans with supports that are all approximately equal in stiffness. Once determined, the maximum lateral load is applied uniformly along the bridge. The dynamic method is used when the bridge is irregular in configuration or strength.

With the static method, the earthquake design force, V, is calculated as

$$V = \frac{(\mathrm{ARS})W}{Z} \qquad 6.2$$

ARS is the *acceleration response spectrum* value obtained from one of four response spectra curves provided in the code or from site-specific curves, if available. Different curves are provided for different *alluvium* (i.e., soil) depths and peak rock accelerations for the *maximum credible earthquake* expected in that area. (The maximum credible earthquake is obtained from maps published by the California Division of Mines and Geology.)

Z is the *ductility risk reduction factor*, which depends on the type of structure and its period. Z varies from slightly less than 1.0 to slightly more than 8.0. A value of 5.0 is appropriate for new bridges. W is the dead load (i.e., the weight) of the bridge.

The bridge period, T (in seconds), is calculated from the following formula. P is the stiffness of the superstructure (i.e., the total force that, if applied uniformly, would cause the bridge to deflect one unit distance). Consistent units must be used.

$$T = 2\pi\sqrt{\frac{W}{P}} \qquad 6.3$$

3. INTERNATIONAL BUILDING CODE

In response to technical disparities among the three sets of model codes previously used in the United States, three model code agencies—the Building Officials and Code Administrators (BOCA), the International Conference of Building Officials (ICBO), and the Southern Building Code Congress International (SBCCI)—created the International Code Council (ICC) (December 9, 1994). The ICC was established as a nonprofit organization seeking to develop a single set of comprehensive and coordinated national codes. The ICC offers a single, complete set of construction codes without regional limitations—the International Codes. The *International Building Code* (IBC) was first published in 2000 and is revised every three years.

The IBC incorporates both the SEAOC Seismology Committee's recommendations for seismic design and the NEHRP-recommended seismic regulations for new buildings. The IBC contains many of the design requirements from the UBC-97, which had been acknowledged as the prominent code publication for earthquake design provisions for several decades.

The seismic provisions of the IBC code exemplify a number of notable lessons learned from earthquakes and recent advances in other seismic resource documents, including the NEHRP. Several provisions of the code include, but are not limited to

- new design response spectra based on detailed maps of expected future ground motion
- new devised base shear equations utilizing new soil-profile types, seismic response coefficients, and the maximum considered earthquake ground motion
- adoption of a revised structural response modification factor based on elastic ground motion response
- adoption of a strength-based design approach
- acceptance of more precise evaluation of the drift induced in structures in response to design ground motions
- acceptance of the effects of structural redundancy-reliability and overstrength factors
- nonstructural components provisions consistent with the 1997 NEHRP
- nonbuilding structures provisions consistent with the 1997 NEHRP

4. SURVIVABILITY DESIGN CRITERIA

While little effort is expended in trying to design buildings that will be totally elastic (i.e., experience no damage) during an earthquake, it is implicit in seismic codes that catastrophic collapse must be avoided. The purpose of the earthquake design provisions is primarily to safeguard against major structural failures and loss of life; these provisions are not intended to limit damage or maintain function. The following three design standards constitute the implied IBC seismic *survivability* (or *life-safety*) *design criteria* [1996 *Blue Book*, Sec. C101.1.1 and App. A]. It is notable, however, that these criteria are not actually specified in the IBC.

- *There should be no damage to buildings from a minor earthquake.*
- *There may be some architectural (nonstructural) damage, but no structural damage during a moderate earthquake.*

- *There may be possible structural and nonstructural damage but no collapse during a severe earthquake.*[10] Yielding is relied upon to dissipate the damaging seismic energy. Theoretically, all structural damage will be repairable when collapse is prevented, although some buildings may be condemned and replaced for reasons of economics or convenience.

The IBC provisions will not prevent structural and nonstructural damage from direct earth faulting, slides, or soil liquefaction.

5. EFFECTIVENESS OF SEISMIC PROVISIONS

The code provides "reasonable" but not complete assurance of the protection of life. Furthermore, the code does nothing to prevent construction on land that is subject to earth slides (of the type that occurred during the Alaskan earthquake) or liquefaction (as occurred in the Niigata, Japan, earthquake).

IBC seismic provisions are intended as minimum requirements. The level of protection can be increased by increasing the design lateral force, energy absorbing capacity, redundancy, and construction quality assurance.

Seismic design is both a science and an art that, unfortunately, must be verified in the field. Thus, the history of seismic codes has been to require design features or methods and then evaluate the effectiveness of those features in practice.

The seismic code used is not the only factor affecting the performance of a structure during an earthquake. In many cases of structural failure in modern buildings, earthquake severity, duration, soil conditions, inadequate design, poor control or material quality, and poor workmanship are found to be the major factors contributing to collapse.[11] Of course, modern seismic codes cannot be blamed when pre-1973 structures fail. (See Sec. 3.1.)

6. APPLICABLE SEISMIC SECTIONS IN THE IBC

General seismic provisions applicable to all structures are contained in IBC Chap. 16, Structural Design, Sec. 1613, "Earthquake Loads." However, the code provisions for sizing and detailing structures also appear in the following other chapters.

- Chapter 18, Soils and Foundations
- Chapter 19, Concrete (particularly Sec. 1908)
- Chapter 21, Masonry (particularly Sec. 2106)
- Chapter 22, Steel (particularly Sec. 2205 and Sec. 2206)
- Chapter 23, Wood (particularly Sec. 2305)

A section for nonbuilding structures (see Sec. 6.46) was first added in the 1980 UBC version and subsequently expanded in the 2003 edition of the IBC. This subject is now addressed by Chap. 15, Nonbuilding Structures, of the ASCE/SEI7 and covers structures such as tanks, towers, chimneys, signs, billboards, and storage racks, but not such structures as retaining walls, bridges, dams, docks, and offshore platforms.

7. NATURE OF IBC SEISMIC CODE PROVISIONS

Previous model building codes were often stand-alone documents providing all requirements from referenced documents. The IBC has specifically chosen a different format. It provides an organizational format, but refers the engineer to a wide variety of external documents for many of the most critical design aspects. However, the IBC contains many statements that overrule clauses of the external document (e.g., ASCE/SEI7), so design engineers must be familiar with all documents to perform their work.

There are two major categories of seismic provisions in the IBC: those that relate to proportioning the structure, and those that relate to detailing elements of the structure. The methods of proportioning structural elements are *allowable stress design* (ASD) and *load and resistance factor design* (LRFD). The proportions are chosen such that the structure's ability to absorb energy (i.e., its "strength") matches the application of energy, no matter how much yielding has occurred, and such that overall stability is maintained. This requires that the lateral force resisting elements be roughly distributed (in plan) throughout the structure. (Thus, arbitrarily increasing the strength of one element may actually have a negative effect on the overall seismic performance.) In equation form, the ratio of energy demand to energy capacity evaluation in plan should be roughly constant. The IBC uses the current ASCE/SEI7 as an external source document for many requirements related to proportioning the structure.

Design details prevent premature local failure by ensuring ductile behavior and preventing local instability and failure of elements that are cyclically stressed beyond their yield points. Unlike the IBC provisions for proportioning the structure, the design details can usually be determined without evaluating the stresses, drifts, or loads. The IBC uses a wide variety of external source documents for requirements related to detailing. These

[10]These italicized sentences are the standards of survivability as they are commonly stated, but it is understood the degree of damage is dependent on the severity of the ground shaking at the building site, not on the magnitude of the earthquake at some distant epicenter.

[11]In particular, the widespread structural failures that occurred in the 1985 Mexico City earthquake are examples of how even modern buildings can be "brought down" by these contributing factors. In fact, the failures that occurred seem to validate the need for the UBC provisions required in the 1985 code, as the very features required by that code were often not included in the design of buildings that collapsed.

source documents are usually developed by industry-specific organizations.

Controlled yielding in a major earthquake is implicitly anticipated by the IBC, and therefore, a code based on yield or ultimate strengths would be preferred. In early UBC versions, the seismic design forces were based on ASD (working stress or service level stress) and not on strength. This was primarily because the vertical load-carrying systems in the majority of steel highrise structures were, until recently, based on ASD.

The ASD method has essentially been replaced in structural steel work by the LRFD method, also known as the *ultimate strength design method.* In this method, the applied loads are multiplied by a load factor. The product (the "strength") must be less than the ultimate strength of the structural member multiplied by a resistance factor.

The term "LRFD" is used in the design of steel and wood structures, whereas in the design of concrete and masonry structures, the equivalent method is known as "strength design." The strength design inclusion in the IBC represents the most significant development toward consistency in national seismic requirements.

8. WIND LOADS

Wind loading is covered in ASCE/SEI7 Chap. 26 through Chap. 31. Wind loads depend on many factors, including location, height, exposure, topography, architectural features, and building dynamics. Wind load on a structure is defined as a pressure acting on the exterior surface. Equation 6.4 gives the wind pressure on a structure.

$$q_z = KV^2 \tag{6.4}$$

q_z is the wind pressure, K consists of one or more constants and adjustment coefficients, and V is the basic wind speed.

9. SNOW LOADS

In many locations, snow applies a significant load on structures that must be considered in design. Snow regularly causes the failure of roof systems and can cause progressive collapse of entire structures. Snow loads on roofs vary widely based on the geographic location, elevation, site exposure, and slope of the roof. Structural members must be capable of supporting snow loads, which in many cases constitute the largest design load for the roof system.

Snow on a structure's roof may result in a uniform loading condition (i.e., the same load over the entire roof) or a nonuniform loading condition (i.e., a varying load) caused by wind-induced drifting or melting and refreezing of snow. Conditions giving rise to uniform loading are the exception. Unbalanced accumulation of snow at valleys, parapets, and roof structures, and offsets in roofs of uneven configuration are typical. Compound roof systems may accumulate large unbalanced loads in valleys, particularly on the leeward side of roofs.

Snow loading is covered in ASCE/SEI7 Chap. 7. Specifically, the design snow load, p_f, for buildings and other structures is given by Eq. 6.5 [ASCE/SEI7 Eq. 7.3-1].

$$p_f = 0.7C_eC_tI_sp_g \tag{6.5}$$

C_e is the snow exposure factor from ASCE/SEI7 Table 7-2. C_t is the thermal factor from ASCE/SEI7 Table 7-3. I_s is the snow risk importance factor for snow loading from ASCE/SEI7 Table 1.5-2. The basic ground snow loads, p_g, are given in ASCE/SEI7 Fig. 7-1, except where controlled by the local building official. The roof snow load is assumed to act vertically upon the area projected upon a horizontal plane.

10. COMBINED SEISMIC, WIND, AND SNOW LOADING

Snow loads must be considered in seismic design (according to the load combinations specified in IBC Sec. 1605.2) because snow adds to the mass of the structure. However, the IBC permits a reduction in the design snow load when it is used in combination with earthquake loads [IBC Sec. 1605.3].

Although codes such as AISC, AITC, and ACI provide for the combination of various loadings (e.g., snow and wind, earthquake and wind), seismic and wind loads are distinctly different in origin. Wind loads are applied over an exterior surface of a structure, whereas seismic loads are inertial in nature.

IBC provisions for ductile seismic detailing must always be met, even if the wind load is greater than the seismic load. Seismic loads generally control for heavy structures and moderate-weight structures in seismic design categories D, E, and F, but wind loads often control the design in seismic design categories A, B, and C, and for lightweight construction in all design categories. However, buildings must be designed to withstand seismic detailing requirements and limitations, even when wind code prescribed load effects are greater than seismic load effects [IBC Sec. 1604.10].

There is no such thing as a "governing" load when a building is in a potential earthquake area. In some cases, the maximum expected lateral wind loading will result in larger drift (see Sec. 5.12) or larger lateral forces than will an earthquake. However, even in that instance, the design must include seismic detailing. The reason for this requirement is that the structure must be able to resist seismic loads in a ductile manner even when it can resist a larger design wind load elastically. Simply, the intent is to avoid catastrophic failure and to provide the necessary structural integrity to resist actual seismic forces, which are potentially much higher than the design seismic forces.

11. SEISMIC DESIGN CRITERIA SELECTION

As described in Sec. 3.10, the base shear, V, is the total design seismic force imposed by an earthquake on the structure at its base. The base shear is the sum of all the inertial story shears. The code calculates the base shear from the total structure weight and then apportions the base shear to the stories in accordance with dynamic theory. The IBC code refers to ASCE/SEI7 for most of the requirements for calculating the base shear. The design seismic forces can be determined based on the ASCE/SEI7 static lateral force procedure, referred to as the "equivalent lateral force procedure" [ASCE/SEI7 Sec. 12.8], and/or the dynamic lateral force procedure, referred to as "modal response spectrum analysis" [ASCE/SEI7 Sec. 12.9]. The dynamic force procedure is always acceptable for design of any structure. However, ASCE/SEI7 Sec. 12.9.4 specifies that the minimum design seismic force must be at least 85% of that prescribed by the static lateral force procedure.

The seismic design process involves consideration of a number of structural and site characteristics, including seismic zoning, risk, seismic importance factors, building fundamental period, site geology and soil characteristics and soil-profile types, maximum considered earthquake ground motion, response modification factor, configuration, structural system, and height. Furthermore, ASCE/SEI7 Sec. 12.12.5 requires that all parts of the structure be designed with adequate strength to withstand the lateral displacements induced by the design ground motion considering the inelastic response of the structure and the inherent redundancy, overstrength, and ductility of the lateral force resisting system. Table 6.1 furnishes the steps needed in seismic design of a structure.

Table 6.1 *Seismic Design Procedure*

step	description	this book's section reference
1.	Identify appropriate structural system.	6-21
2.	Classify risk category of the structure.	6-13
3.	Determine the components of seismic response coefficient.	6-30
4.	Identify structural system limitations and irregularities.	6-24, 6-25
5.	Select appropriate lateral force procedure.	6-31
6.	Determine the total design base shear.	6-32, 6-34, 6-50
7.	Distribute the design base shear over the structure's height.	6-35, 6-36
8.	Analyze P-delta effects for the structure.	6-43
9.	Examine overturning effects caused by earthquake forces.	6-42
10.	Evaluate torsional effects for the structure.	6-39
11.	Study story drift limitations.	6-40
12.	Consider redundancy of lateral force resisting system.	6-22
13.	Evaluate overstrength of lateral force resisting system.	6-20
14.	Design elements of the structure.	6-44
15.	Confirm seismic detailing requirements with the IBC.	6-6–6-12
16.	Verify structure's continuous load path completion.	

12. IBC PROCEDURE FOR DETERMINING DESIGN ACCELERATION

The design process is influenced predominantly by a building's anticipated maximum ground acceleration at the site. The values of the maximum considered acceleration are developed by considering historical records, geological data, and seismological information.

Determining the design spectra for static analysis has been dramatically changed with the IBC. The code provides maps for the United States detailing expected ground accelerations. From these maps, values are read and adjusted for the soil and a response spectrum of design spectral response accelerations is developed.

The first step in determining the design spectrum is to identify values for the mapped acceleration parameters. There are two different values required—one for structures with a fundamental period of vibration of 0.2 sec, S_S, and one for periods of 1.0 sec, S_1. Both values are needed for determining the base shear of a structure. These two values are read from U.S. Seismic Design Maps from the U.S. Geological Survey (USGS) provided in the IBC [IBC Fig. 1613.3.1(1) through Fig. 1613.3.1(6)]. The first two IBC figures show the entire conterminous United States, and the later figures show detailed regions. For each region, there are two maps that look similar; one for each of the two required values (S_S and S_1). The engineer must take care that the correct map is used for the appropriate value.

13. RISK (OCCUPANCY) CATEGORIES

When earthquake disaster strikes a community, any major structural and nonstructural damage to a facility that could threaten life safety demands complete closure until adequate repair measures are taken, and further evaluation must deem the facility safe for occupancy again. Certain facilities, such as hospitals and police and fire stations, cannot be shut down under these circumstances. Accordingly, law requires that these facilities be designed to remain operational after an earthquake.[12]

It is implicit in seismic codes that catastrophic collapse of structures must be avoided to safeguard lives, minimize economic losses, and avoid disruption in the event

[12]After an earthquake, based on Title 24 of the *California Administrative Code*, hospitals and schools must be operational. Also, the California Hospital Act requires that hospitals be fully functional and operational.

of an earthquake. For purposes of earthquake-resistant design, each structure is specified in one of the risk categories (previously known as "occupancy categories") listed in ASCE/SEI7 Table 1.5-1, which consists of four risk categories with their functions defined. Buildings in some of these categories require special review, inspection, and construction observation.

Essential facilities are emergency structures that must remain operational after an earthquake. They include hospitals with surgery and emergency treatment facilities, fire and police stations, emergency preparedness structures (including structures housing emergency vehicles), and government communication centers required for emergency response. *Hazardous facilities* are used to store or support dangerous toxic or explosive chemicals or substances, or are designed to house large numbers of people—for example, places of public assembly (300 or more people), schools (250 or more students), colleges and adult education centers (500 or more students), nursing homes, daycare centers, nurseries, and jails. All other structures that house occupancies or have functions not listed are considered *standard occupancy structures.* Examples are apartment buildings, hotels, office buildings, and wholesale or retail structures. *Miscellaneous structures* are buildings or parts of buildings. They include private garages, carports, sheds, factories, and agricultural buildings.

IBC Chap. 3, Use and Occupancy Classification, defines the occupancy categories in greater detail. Design and construction review requirements are specified in IBC Chap. 17.

14. SEISMIC IMPORTANCE FACTOR: I_e

Table 6.2 [ASCE/SEI7 Table 1.5-2] specifies the importance factor, I_e, that increases seismic design forces for critical structures based on the structure's risk category.[13,14] Details for classifying each risk category are given in ASCE/SEI7 Table 1.5-1.

The seismic importance factor is between 1.0 and 1.50, depending on how critical it is for the structure to survive a major earthquake with minimal damage. From Table 6.2, it is obvious that a higher importance factor, I_e, of 1.50 is designated for essential structures in order to ensure that they remain functional and operational after a severe earthquake. For these structures, the design base shear is increased by 50% compared to other structures. Increasing design base shear increases the seismic safety of a structure.

Table 6.2 *Risk Categories and Importance Factors for Earthquakes [ASCE/SEI7 Table 1.5-2]*

occupancy or functions of structure	risk category	seismic importance factor, I_e
miscellaneous structures	I	1.00
standard occupancy structures	II	1.00
hazardous structures	III	1.25
essential structures	IV	1.50

15. SOIL-PROFILE TYPES: SITE CLASS A THROUGH F

Soft soil may amplify earthquake ground motion. Amplification of vibrations due to unfavorable soil conditions has been strikingly illustrated in many earthquakes, such as the 1985 Mexico City earthquake and the 1989 Loma Prieta earthquake. To that effect, IBC Sec. 1613.3.2 specifies that each site be assigned a soil-profile type found by a properly substantiated geotechnical investigation.

A *soil profile* is a significant layer of soil with distinct characteristics extending from the surface into relatively unaltered material. IBC Sec. 1613.3.2 soil profiles are classified into six different soil types ranging from A to F in accordance with ASCE7/SEI7 Chap. 20. Type F is assigned to those sites that require specific evaluation. New soil profile designations are found in the 1997 NEHRP publication that are based on quantitative geological parameters and recent data.

Table 6.3 [ASCE/SEI7 Table 20.3-1] is based on the site categorization procedure of ASCE/SEI7 Chap. 20 and provides a detailed description of the IBC soil profiles. This table furnishes average shear wave velocity, $\overline{v}_s$, average field standard penetration resistance, $\overline{N}$, and average undrained shear strength, $\overline{s}_u$, values associated with soil profiles A through F, where applicable.

The average shear wave velocity, $\overline{v}_s$, may be measured on site or calculated using ASCE/SEI7 Eq. 20.4-1.[15] The average field standard penetration resistance, $\overline{N}$, and average standard penetration resistance for cohesionless soil layers, $\overline{N}_{ch}$, can be determined from ASCE/SEI7 Eq. 20.4-2 and Eq. 20.4-3.[16] The average

[13]Seismic importance factors for wind design and for nonstructural component design are discussed in Sec. 6.8 and Sec. 6.46, respectively.

[14]Increasing base shear in design and construction also increases costs. An earthquake risk management study may be used to evaluate the seismic hazard and vulnerability of a structure. The study may rationalize amplifying seismic loads to achieve greater performance goals, such as immediate occupancy in the event of a major earthquake. James L. Witt, previous director of the Federal Emergency Management Agency, said, "Mitigation saved the Anheuser-Busch facility in Los Angeles after Northridge. The Anheuser-Busch Engineering Department retrofitted the plant to conform to the L.A. seismic code—and the plant was functioning within days of the earthquake. Without those revisions—they would have sustained more than $300 million in direct and interruption losses."

[15]For these soil types, the average shear wave velocities, $\overline{v}_s$, and the average field standard penetration resistance, $\overline{N}$, are determined based on the 100 ft (30 480 mm) of soil profile.

[16]The standard penetration resistance, $\overline{N}$, of the soil layer should be obtained in accordance with approved, nationally recognized standards.

***Table 6.3** Site Class Definitions [ASCE/SEI7 Table 20.3-1]*

site class	soil profile	soil profile description: shear wave velocity ft/sec (m/s)	standard penetration blows/ft (blows/m)	undrained shear strength lbf/ft² (kPa)
A	hard rock	$\overline{v}_s > 5000$ ($\overline{v}_s > 1500$)	–	–
B	rock	$2500 < \overline{v}_s \le 5000$ ($760 < \overline{v}_s \le 1500$)	–	–
C	very dense soil and soft rock	$1200 < \overline{v}_s \le 2500$ ($360 < \overline{v}_s \le 760$)	$\overline{N} > 50$	$\overline{s}_u \ge 2000$ ($\overline{s}_u \ge 100$)
D	stiff soil	$600 < \overline{v}_s \le 1200$ ($180 < \overline{v}_s \le 360$)	$15 \le \overline{N} \le 50$	$1000 \le \overline{s}_u \le 2000$ ($50 \le \overline{s}_u \le 60$)
E	soft clay soil	$\overline{v}_s < 600$ ($\overline{v}_s < 180$)	$\overline{N} < 15$	$\overline{s}_u < 1000$ ($\overline{s}_u < 50$)
	Or, any profile > 10 ft (3048 mm) of PI > 20, $\overline{s}_u < 500$ ($\overline{s}_u < 24$) and $w \ge 40\%$			
F	Any profile containing soils having one or more of the following characteristics: 1. soils vulnerable to potential failure or collapse under seismic loading such as liquefiable soils, quick and highly organic clays, collapsible weakly cemented soils 2. peats and/or highly organic clays ($H >$ 10 ft (3 m) of peat and/or highly organic clay where H = thickness of soil) 3. very high plasticity clays ($H >$ 25 ft (7.6 m) with plasticity index PI > 75) 4. very thick soft/medium stiff clays ($H >$ 120 ft (37 m)), with $s_v < 1000$ lbf/ft² (50 kPa)			

undrained shear strength, $\overline{s}_u$, can be obtained from ASCE/SEI7 Eq. 20.4-4.[17]

Where soil properties are not known in sufficient detail, site class D can be used, not E or F. Site class E or F can only be used when a geotechnical study shows it to be valid, or when required by the local building official.[18]

16. NEAR SOURCE EFFECTS

In seismic areas where large-magnitude earthquakes are expected, severe damage to structures built near or directly on top of active faults is likely to occur. The ground acceleration that these structures experience may be up to twice the acceleration that more distant structures experience.

In the 1997 UBC, *near source factors* were used to adjust design parameters for sites located near active faults. The IBC has eliminated these factors by providing the detailed maps of Fig. 1613.3.1(1) through Fig. 1613.3.1(6).

17. SEISMIC RESPONSE COEFFICIENTS

The IBC maps provide numerical values of estimated ground motion (i.e., acceleration) based on geographic location. For any location, two mapped acceleration parameter values are required: S_S, the mapped spectral acceleration for short periods (see Fig. 3.1 [IBC Fig. 1613.3.1(1)]) and S_1, the mapped spectral acceleration for a 1 sec period (see Fig. 3.2 [IBC Fig. 1613.3.1(2)]). Throughout the IBC procedure, these two period values are quantified and used to determine the design loads for structures of any fundamental period of vibration. For any structure, a value for each parameter is necessary, requiring the use of two of the maps. In addition, the values from the maps are given as percentage values, and calculations need to be performed on decimal values of acceleration.

After determining S_S and S_1 from the IBC maps, the next step is to adjust them to appropriate seismic design parameters. IBC Sec. 1613.3 adjusts the values based on the soil profile of the site. Values of site coefficients F_a

[17]The undrained shear strength, $\overline{s}_u$, should be determined in accordance with approved, nationally recognized standards, not to exceed 5000 psf (250 kPa).

[18]The E soil profile is appropriate in areas with large deposits of very soft clay that are subject to large amplifications in seismic ground motion. Buildings constructed on San Francisco Bay mud or on the Mexico City lake bed are likely candidates for the E soil profile.

and F_v are given in Table 6.4 and Table 6.5 [IBC Table 1613.3.3(1) and Table 1613.3.3(2)], and used in Eq. 6.6 and Eq. 6.7 [IBC Eq. 16-37 and Eq. 16-38].

$$S_{MS} = F_a S_S \qquad 6.6$$

$$S_{M1} = F_v S_1 \qquad 6.7$$

Table 6.4 *Seismic Coefficient as a Function of Site Class [IBC Table 1613.3.3(1)]*

values of site coefficient F_a[a]

site class	mapped spectral response acceleration at short period				
	$S_S \le 0.25$	$S_S = 0.50$	$S_S = 0.75$	$S_S = 1.00$	$S_S \ge 1.25$
A	0.8	0.8	0.8	0.8	0.8
B	1.0	1.0	1.0	1.0	1.0
C	1.2	1.2	1.1	1.0	1.0
D	1.6	1.4	1.2	1.1	1.0
E	2.5	1.7	1.2	0.9	0.9
F	b	b	b	b	b

[a]Use straight-line interpolation for intermediate values of mapped spectral response acceleration at short period, S_S.
[b]Values shall be determined in accordance with Sec. 11.4.7 of ASCE/SEI7.

Table 6.5 *Seismic Coefficient as a Function of Site Class [IBC Table 1613.3.3(2)]*

values of site coefficient F_v[a]

site class	mapped spectral response acceleration at 1 sec period				
	$S_1 \le 0.1$	$S_1 = 0.2$	$S_1 = 0.3$	$S_1 = 0.4$	$S_1 \ge 0.5$
A	0.8	0.8	0.8	0.8	0.8
B	1.0	1.0	1.0	1.0	1.0
C	1.7	1.6	1.5	1.4	1.3
D	2.4	2.0	1.8	1.6	1.5
E	3.5	3.2	2.8	2.4	2.4
F	b	b	b	b	b

[a]Use straight-line interpolation for intermediate values of mapped spectral response acceleration at 1 sec period, S_1.
[b]Values shall be determined in accordance with Sec. 11.4.7 of ASCE/SEI7.

When reduced by 33%, Eq. 6.6 and Eq. 6.7 become Eq. 6.8 and Eq. 6.9 [IBC Sec. 1613.3.4, Eq. 16-39 and Eq. 16-40].

$$S_{DS} = \tfrac{2}{3} S_{MS} \qquad 6.8$$

$$S_{D1} = \tfrac{2}{3} S_{M1} \qquad 6.9$$

S_{MS} is the maximum considered earthquake response accelerations for short periods. S_{M1} is the maximum considered earthquake spectral response accelerations for a 1 sec period. Figure 6.1 shows a complete design response spectrum for a site. The x-axis has two critical periods of vibration: T_S and 1.0, where T_S is the ratio of S_{D1} to S_{DS}. These periods separate the spectrum into two different formulas for determining S_a, the design spectral response acceleration (the y-axis of the graph).

Figure 6.1 *Response Spectra*

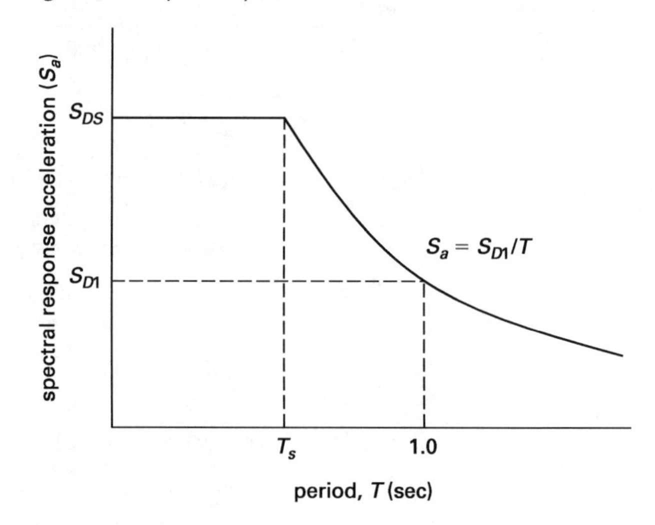

Structures with short fundamental periods of vibration ($T < T_S$) have a constant value for S_a, equal to the value of S_{DS}. Structures with long periods of vibration ($T > T_S$) use Eq. 6.10.

$$S_a = \frac{S_{D1}}{T} \qquad 6.10$$

When a simplified lateral force analysis is used (see Sec. 6.34), S_S need not be greater than 1.5.

Example 6.1

A public auditorium with a 6000 person capacity is being designed in northern California, near the San Andreas Fault. The average shear wave velocity of the supporting soil is 3000 ft/sec (914 m/s).

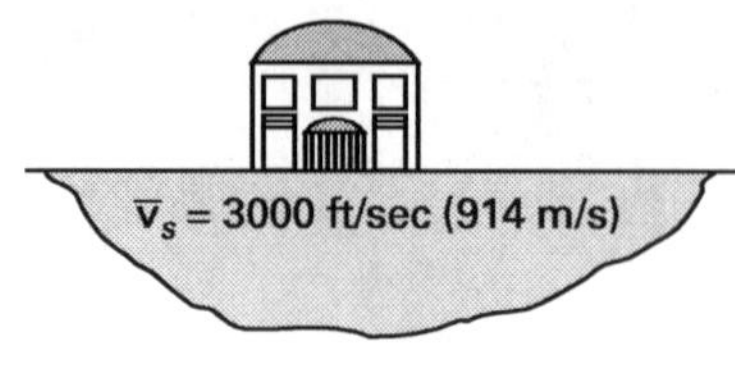

(a) What is the value of the importance factor?

(b) What is the soil-profile type?

Solution

(a) Because of the auditorium's 6000 person capacity (more than 300 persons) and social function of this building, this building is classified as a hazardous structure according to ASCE/SEI7 Table 1.5-1. According to Table 6.2, the value of the seismic importance factor, I_e, for this type of occupancy is 1.25.

(b) This building is located on a soil profile that has an average shear wave velocity ($\overline{v}_s$) of 3000 ft/sec (914 m/s). Based on the site categorization procedure of IBC Sec. 1613.3.2 and Table 6.3 [ASCE/SEI7 Table 20.3-1], the soil profile is type B. Type B is described as a rock soil profile with an average shear wave velocity of 2500 ft/sec $< \overline{v}_s \leq$ 5000 ft/sec (760 m/s $< \overline{v}_s \leq$ 1500 m/s).

18. SEISMIC DESIGN CATEGORY

The seismic design category (SDC) is determined based on a structure's risk category and the design spectral response acceleration parameters, S_{DS} and S_{D1} [IBC Sec. 1613.3.5]. The seismic design category can be A, B, C, D, E, or F, with F representing the most severe seismic design situations. (See Table 6.6.) Normally, Table 6.7 and Table 6.8 [ASCE/SEI7 Table 11.6-1 and Table 11.6-2 (corresponding to IBC Table 1613.3.5(1) and Table 1613.3.5(2))] are both used to determine the seismic design category. If the two determinations are different, the more severe of the two is taken as the seismic design category. Use of Table 6.7 alone is permitted when (a) $S_1 < 0.75$, (b) $T_a < 0.8T_s$ in both orthogonal directions, (c) $T < T_s$ in both orthogonal directions, and (d) other procedural provisions of ASCE/SEI7 Sec. 11.6 are followed [ASCE/SEI7 Sec. 11.6].

Table 6.7 and Table 6.8 assume $S_1 < 0.75$, which is intended to account for the structure not being immediately adjacent to an active fault. When $S_1 \geq 0.75$, ASCE/SEI7 Sec. 11.6 requires risk category I, II, and III structures to be assigned seismic design category E, and risk category IV structures to be assigned seismic design category F.

The seismic design category allows the code to adjust seismic design provisions for different regions of the country. For structures with low seismic design values (seismic design category A) the code allows a much wider range of design and construction practices. As a structure is placed in a higher seismic design category, the code begins to restrict more of the design practices to ensure better seismic performance.

Two design factors depend on the seismic design category. The first is the type of structural system that may be built or used. The second is the required amount of ductile detailing of various elements of the structure. Both of these parameters are discussed in more detail in this book.

There are very few locations in California where at least seismic design category D does not apply, although geotechnical report information has occasionally supported category C.

Table 6.6 *Qualitative Seismic Design Categories*

seismic design category A	buildings on good soils in areas where expected ground shaking will be minor
seismic design category B	buildings of risk categories I, II, and III, where expected ground shaking will be moderate, such as would be expected with stratified soils made up of good and poor soils
seismic design category C	buildings of risk categories I, II, and III, where more severe ground shaking will occur, and to buildings of risk category IV (hospitals, police stations, emergency control centers, etc.), where expected ground shaking will be moderate
seismic design category D	buildings and structures in areas expected to experience severe and destructive ground shaking, such as is expected with poor soils, but not located close to a major fault
seismic design category E	buildings of risk categories I, II, and III in areas near active major faults, regardless of soil or rock classification
seismic design category F	buildings of risk category IV (hospitals, police stations, emergency control centers, etc.), areas near active major faults, regardless of soil or rock classification

Table 6.7 *Seismic Design Category Based on S_{DS} (short-period response) Acceleration Parameter [ASCE/SEI7 Table 11.6-1 (IBC Table 1613.3.5(1))]*

	risk category	
value of S_{DS}*	I, II, or III	IV
$S_{DS} < 0.167$	A	A
$0.167 \leq S_{DS} < 0.33$	B	C
$0.33 \leq S_{DS} < 0.50$	C	D
$0.50 \leq S_{DS}$	D	D

*Values of S_{DS} are in gravities (g's).

Reproduced from the 2010 edition of *Minimum Design Loads for Buildings and Other Structures* by the American Society of Civil Engineers (ASCE), copyright © 2010. Used with permission from ASCE.

19. RESPONSE MODIFICATION FACTOR: *R*

Building a structure to respond 100% elastically in a large-magnitude earthquake is not economical. Therefore, the prescribed design lateral strengths are considerably lower than needed to maintain a structure in the

Table 6.8 *Seismic Design Category Based on S_{D1} (1 sec period response) Acceleration Parameter [ASCE/SEI7 Table 11.6-2 (IBC Table 1613.3.5(2))]*

	risk category	
value of S_{D1}*	I, II, or III	IV
$S_{D1} < 0.067$	A	A
$0.067 \leq S_{D1} < 0.133$	B	C
$0.133 \leq S_{D1} < 0.20$	C	D
$0.20 \leq S_{D1}$	D	D

*Values of S_{D1} are in gravities (g's).

Reproduced from the 2010 edition of *Minimum Design Loads for Buildings and Other Structures* by the American Society of Civil Engineers (ASCE), copyright © 2010. Used with permission from ASCE.

elastic range. This reduced design strength level results in nonlinear behavior and energy absorption at displacements in excess of initial yield. Strength reductions due to nonlinear behavior are influenced by the maximum allowable displacement ductility demand, the fundamental period of the system, and the soil-profile type. Strength reductions from the elastic strength are accomplished by using a response modification factor.

The structure's *response modification factor*, R, represents the inherent overstrength and global ductility capacity of structural components.[19] *Ductility* can be defined as a measure of the ability of a structural system to deform in the plastic range prior to failure. Ductile performance is important because seismic energy is dissipated through yielding of the structural components, and because it permits considerable displacements during intense earthquakes without risk to the structure's integrity and the occupants' life safety. In an earthquake, a structure's inherent overstrength allows the structure's members to form an initial plastic hinge. But it is the structure's global ductility that allows the structure to withstand additional seismic forces, moving from the formation of the initial plastic hinge to the formation of mechanisms.

Because all structures are designed for strengths less than would be needed in a completely elastic structure, the value of the response modification factor, R, always exceeds 1.0. Lightly damped structures constructed of brittle materials are assigned low values of R because they cannot support deformation in excess of initial yield. Highly damped structures constructed of ductile materials are assigned higher values of R.

The level of reduction specified in the ASCE/SEI7 seismic provisions is essentially based on the analysis of the historical performance of various structural systems in strong earthquakes. The structure response modification factor is determined from the type of structural system used in design of structures, as defined for buildings in App. H [ASCE/SEI7 Table 12.2-1]. Systems with higher ductility (e.g., steel moment-resisting frames) have higher R values associated with better seismic performance expectations.

20. SEISMIC SYSTEM OVERSTRENGTH FACTOR: Ω_O

The type of structural system and natural period of a structure significantly influence the structure's response to ground shaking. In the event of a severe earthquake, code-compliant structures are expected to deform beyond their elastic load-carrying capacities due to effects of system overstrength. Overloading of nonductile elements of the structure can occur if the effects of over-strength are not accounted for in design. *Overstrength* is defined as a characteristic of structures where the actual strength is greater than the design strength. The degree of overstrength depends on material type and structural system type.

A *seismic system overstrength factor*, Ω_O, also known as the *seismic force amplification factor*, has been assigned to each identified structural system. This factor accounts for overstrength of the structure in the inelastic range.

21. STRUCTURAL SYSTEMS

ASCE/SEI7 Sec. 12.2 recognizes eight major types of structural systems (bearing wall, building frame, moment-resisting frame, dual systems with special moment-resisting frames, dual systems with intermediate moment-resisting frames, shear wall-frame, cantilevered column, and steel systems without seismic detailing) capable of resisting lateral forces. In determining the base shear and design story drift for the structural systems mentioned, App. H [ASCE/SEI7 Table 12.2-1] provides corresponding height limitations, the appropriate response modification coefficient, R, the seismic force amplification factor, Ω_O, and the deflection amplification factor, C_d. Nonbuilding structures are defined in ASCE/SEI7 Chap. 15. (See Sec. 6.46.)

A. Bearing Wall Systems

A *bearing wall system*, shown in Fig. 6.2, is a structural system that relies on the same elements to resist both gravity and lateral loads. By itself, the word "wall" is ambiguous because there are two main types of structural walls. (Partition walls and curtain walls are not structural walls.) A *bearing wall* is designed and constructed to resist vertical (i.e., gravity) loads. A *shear wall* is designed and constructed to resist lateral loads. A wall can be used to resist both vertical and lateral loads. A bearing wall system does not have a complete vertical load-carrying space frame. Bearing walls or bracing systems support all of the gravity loads. Lateral forces are resisted by shear walls, light bracing in bearing walls, or braced frames (where the bracing also carries lateral load).

[19]The 1994 UBC R_w value is approximately equal to 1.4 times the IBC R value. This selection is for transition from service-level to strength-level design forces. The R factor is a partly empirical, partly judgmental factor that reduces the base shear to a predetermined value. This is the only judgment factor in the base shear equation.

Figure 6.2 *Bearing Wall System*

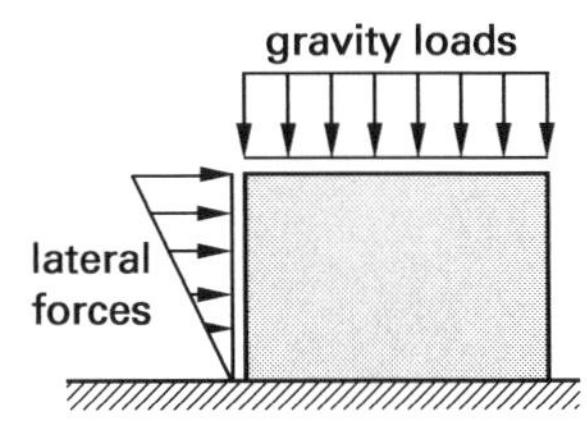

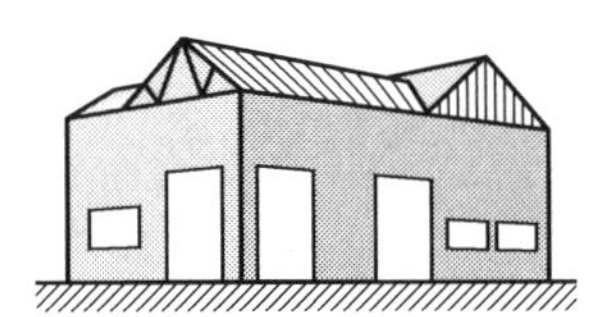

A bearing wall system lacks redundancy and has an inadequate inelastic response capacity. Such systems do not possess a complete vertical load-carrying frame and rely on walls or braced frames to carry the vertical (gravity) and lateral (seismic) loads.[20] The distinguishing factor of these systems is that the failure of the primary seismic system also compromises the ability of the structure to support its dead and live loads.

Typical bearing wall systems are: light-framed walls with shear panels, concrete or masonry shear walls, light steel-framed bearing walls with tension-only braces, and braced frames where the bracing carries gravity loads.

B. Building Frame Systems

A *building frame* (*vertical load-carrying frame*) is a complete, self-contained, three-dimensional unit composed of interconnected members. (See Fig. 6.3.) *Building frame systems* use a complete space frame to carry the vertical (gravity) loads and a separate system of non-bearing shear walls or braced frames to resist the lateral (seismic) load.[21] Unlike the bearing wall system, failure of the primary lateral support system does not compromise the ability of the structure to support gravity loads.

Figure 6.3 *Building Frame System*

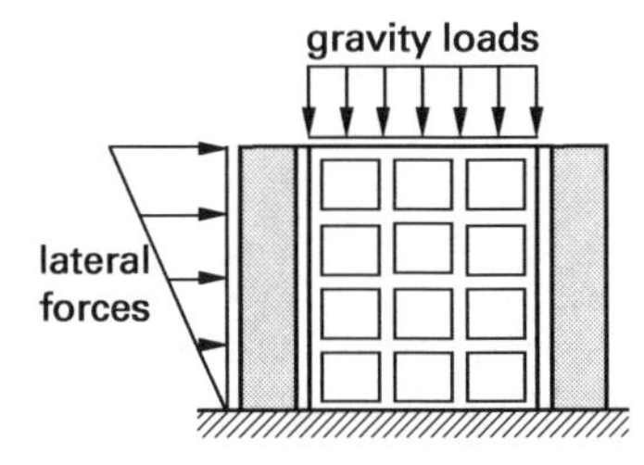

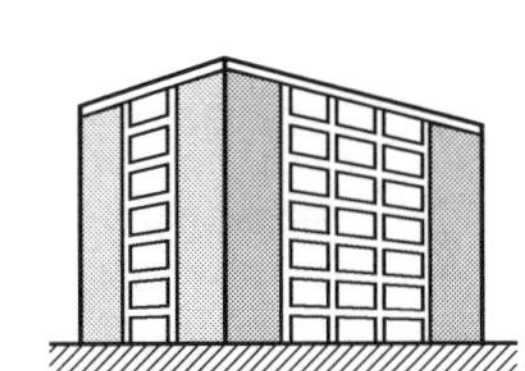

The requirements of the "deformation compatibility" of ASCE/SEI7 Sec. 12.1.3 provision should be satisfied for these systems. A frame may or may not have bracing. If it does, it is known as a *braced frame*. A braced frame is a vertical truss system of interconnected members designed to resist lateral loads through the development of axial loads in the members. Braced frames can be of the concentric or eccentric types that are relatively rigid and require special detailing to ensure adequate ductile performance. Typical building frame systems are: steel eccentrically braced frames; light-framed walls with shear panels; concrete or masonry shear walls; steel, concrete, or heavy ordinary braced frames; and special steel concentrically braced frames.

C. Moment-Resisting Frames

Moment-resisting frames, shown in Fig. 6.4, resist forces in members and joints primarily by flexure and rely on a frame to carry both vertical and lateral loads. Lateral loads are carried primarily by flexure in the members and joints. Theoretically, joints are completely rigid. Moment-resisting frames can be constructed of concrete, masonry, or steel, although masonry frame use is limited.

Figure 6.4 *Moment-Resisting Frame System*

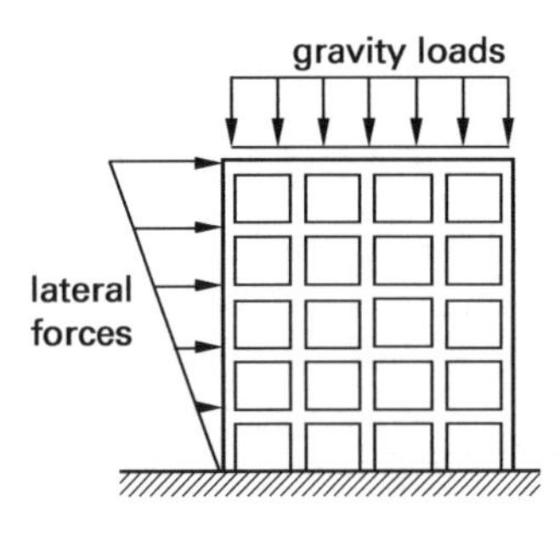

There are five types of moment-resisting frames: steel and concrete special moment-resisting frames (SMRF), masonry moment-resisting wall frames (MMRWF), concrete intermediate moment-resisting frames (IMRF), steel or concrete ordinary moment-resisting frames (OMRF), and special steel truss moment frames (STMF). These systems provide a sufficient degree of redundancy and have excellent inelastic response capacities.

Special moment-resisting frames are specially detailed to ensure ductile behavior and comply with ACI 318 (concrete) or AISC 341 (steel). *Intermediate moment-resisting frames*, concrete frames with less stringent requirements designed in accordance with ACI 318, cannot be used in seismic design categories D, E, or F. *Ordinary moment-resisting frames* are steel or concrete moment-resisting frames that do not meet the special detailing requirements for ductile behavior. Ordinary moment-resisting frames constructed of steel and concrete are restricted in use by App. H [ASCE/SEI7 Table 12.2-1].

D. Dual Systems

Dual systems, shown in Fig. 6.5, have essentially complete space frames that provide support for all vertical

[20]It is common to refer to this type of design as a *box system*.

[21]A *space frame* is a three-dimensional structural system, without bearing walls, consisting of interconnected members that operate as a single unit. A *frame* is a truss-like two-dimensional system with concentric or eccentric connection points in which the lateral forces are resisted by axial stresses in the members. A space frame functions with or without the aid of horizontal diaphragms or floor-bracing systems.

(gravity) loads and combine two of the previously mentioned systems (moment-resisting frames and shear wall-braced frames) to resist lateral loads. ASCE/SEI7 distinguishes between two types of dual systems: those with special moment-resisting frames and those with intermediate moment-resisting frames. All moment-resisting frames (SMRF, IMRF, MMRWF, or steel OMRF) in dual systems must be able to resist at least 25% of the design base shear independently. The two systems are designed to resist the total design base shear in proportion to their relative rigidities [ASCE/SEI7 Sec. 12.2.5.1].[22]

Figure 6.5 *Dual System*

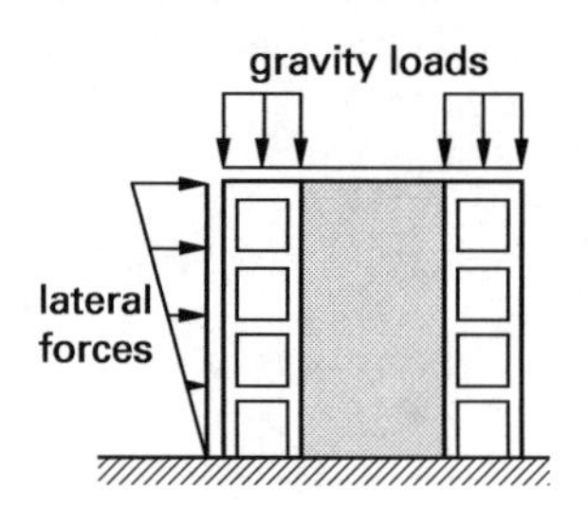

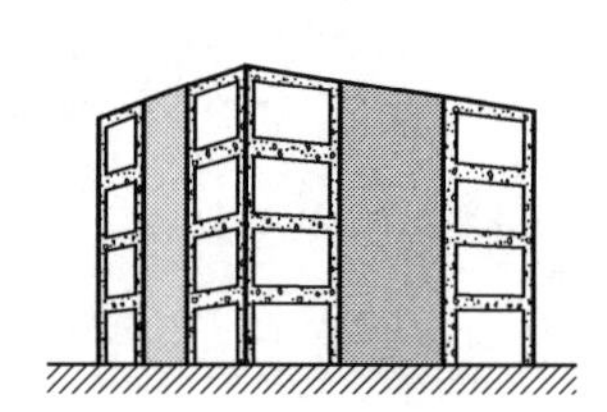

E. Shear Wall-Frame Interaction Systems

To resist lateral forces, *shear wall-frame interaction systems* primarily use a combination of shear walls and moment frames. (See Fig. 6.6.) Building frames that are part of the lateral force resisting systems are required to be concrete frames. These systems are restricted to locations of low seismicity.

Figure 6.6 *Shear Wall-Frame Interaction System*

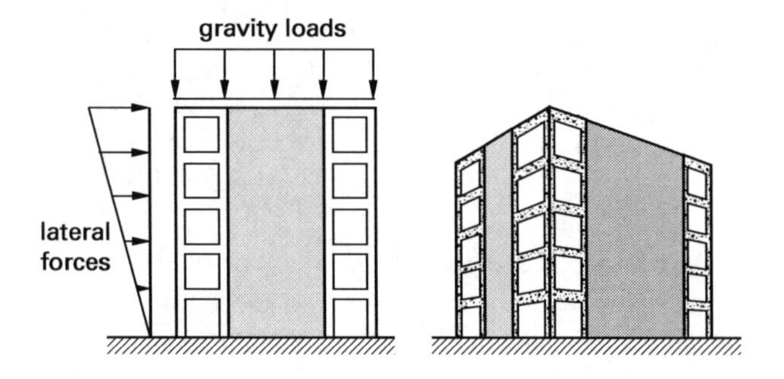

F. Cantilevered Column Systems

Cantilevered column (inverted pendulum) building systems have single cantilevered column elements supporting beams or framing at the top. (See Fig. 6.7.) These systems have a large portion of their mass concentrated near or at the top and are fixed at their bases. Design base shear is essentially applied at the top of the vertical base member. They are regarded as inverted pendulum-type structures since they extend from a fixed base and have zero moment restraint at the top. These systems have essentially one degree of freedom in horizontal translation.

Figure 6.7 *Cantilevered Column Building System*

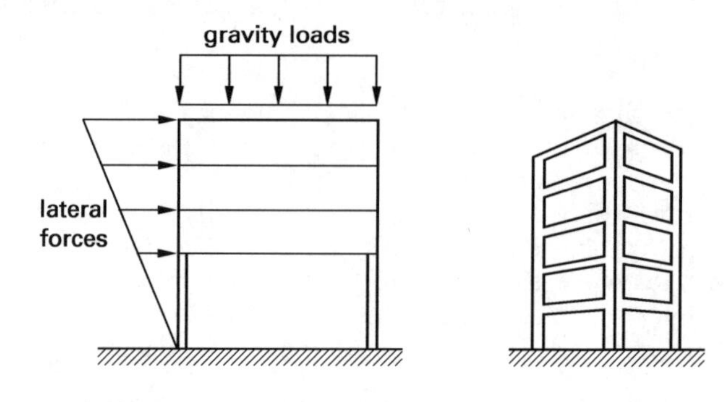

Supporting columns or piers of inverted pendulum-type structures should be designed from the bending moment determined at the base according to procedures of ASCE/SEI7 Sec. 12.2.5.3.

The cantilevered columns in this building system provide both lateral load resistance and gravity load resistance. These column elements have low redundancy and limited inelastic response capacity (energy dissipation). The potentially adverse effects that failure of the columns due to lateral forces will have on the gravity load-carrying capacity must be evaluated to determine stability. In seismic design categories D, E, or F, the maximum height for these structures is 35 ft (10.7 m).

G. Non-Seismic Steel Systems

Steel systems not specifically designed for seismic loads must, nevertheless, be designed and detailed in accordance with AISC 360. However, they do not need to be designed and detailed in accordance with AISC 341.

H. Undefined Systems

Undefined structural systems do not fit into any of these categories. The designer of such systems must submit a rational basis for the design force level used.

22. REDUNDANCY FACTOR: ρ

Redundancy is an important characteristic of a structure that provides multiple paths of resistance (i.e., load paths). Higher redundancy in a structure implies better reliability. Inelastic action of a structure during a major seismic event can cause part of the structure to fail. For structures expected to experience severe inelastic demands, the lateral load-resisting system of the structure should be made as redundant as possible so that loads can be distributed to other lateral force resisting elements.

[22]Moment-resisting frames can be steel, concrete, or masonry, but concrete intermediate moment-resisting frames are prohibited except as permitted in the IBC.

The *redundancy factor*, ρ, is applied to increase the horizontal seismic forces associated with the design force for elements of the structure [ASCE/SEI7 Sec. 12.3.4]. This factor effectively increases the design force based on the extent of structural redundancy inherent in the design configuration of the structure and its lateral force resisting system. The redundancy factor value varies between 1.0 and 1.3.

The value of ρ depends on the type of calculation that is being completed, the level of seismicity at the location, and the design of the structure. According to ASCE/SEI7 Sec. 12.3.4.1, ρ is 1.0 for

- calculations of drift and P-delta effects
- the design of nonstructural elements
- the design of nonbuilding structures that are not similar to buildings
- the design of systems where overstrength load combinations are used
- diaphragms and certain structures with damping systems
- all structures in seismic design categories B or C
- the design of structural walls for out-of-plane forces

For structures in seismic design categories D, E, or F, ρ is 1.3 unless either of the following conditions is met, in which case, ρ is 1.0 [ASCE/SEI7 12.3.4.2].

1. Each story resisting more than 35% of the base shear meets the requirements of Table 6.9 [ASCE/SEI7 Table 12.3-3].
2. Structures are regular in plan at all levels, provided that the seismic force-resisting systems consist of at least two bays of seismic-resisting perimeter framing on each side of the structure in each orthogonal direction at each story resisting more than 35% of the base shear.

For condition 2, the number of bays of a shear wall is determined by dividing the total length of the shear wall by the story height, or doubling this value for light-frame construction (e.g., timber and light-gauge steel structures).

Observation of structural performance in Northridge, Kobe, and other large earthquakes has shown that structures with adequately redundant systems perform better than do structures with few lateral load-resisting elements. For this reason, the redundancy factor has been introduced to persuade engineers to design more highly redundant structures. In certain situations, it may be difficult to achieve a redundant design. In those cases, the magnitude of the inelastic response and the ductility demand should be reduced by increasing earthquake design loads by way of the ρ factor.

Table 6.9 *Redundancy Requirements [ASCE/SEI7 Table 12.3-3]*

lateral force-resisting element	requirement
braced frames	Removal of an individual brace, or connection thereto, would not result in more than a 33% reduction in story strength, nor does the resulting system have an extreme torsional irregularity (horizontal structural irregularity type 1b).
moment frames	Loss of moment resistance at the beam-to-column connections at both ends of a single beam would not result in more than a 33% reduction in story strength, nor does the resulting system have an extreme torsional iregularity (horizontal structural irregularity type 1b).
shear walls or wall pier with a height-to-length ratio of greater than 1.0	Removal of a shear wall or wall pier with a height-to-length ratio greater than 1.0 within any story, or collector connections thereto, would not result in more than a 33% reduction in story strength, nor does the resulting system have an extreme torsional irregularity (horizontal structural irregularity type 1b).
cantilever columns	Loss of moment resistance at the base connections of any single cantilever column would not result in more than a 33% reduction in story strength, nor does the resulting system have an extreme torsional irregularity (horizontal structural irregularity type 1b).
other	no requirements

The redundancy factor is applied in the load combination equations rather than in the base shear equation because stiffness and drift control requirements are not directly influenced.

Example 6.2

The plan view of a one-story office building in San Francisco is shown. This structure has a one-story, light, wood-framed bearing wall. The building has a flexible wood structural panel roof diaphragm. The wood walls are sheathed with wood panels rated for shear resistance. The building is 14 ft (4.3 m) tall. The shear walls equally resist the seismic shear and have lengths equal to 75% of the total length of the building in each direction. (a) What is the value of the response modification factor? (b) What is the value of the seismic force amplification factor? (c) What is the value of the redundancy factor?

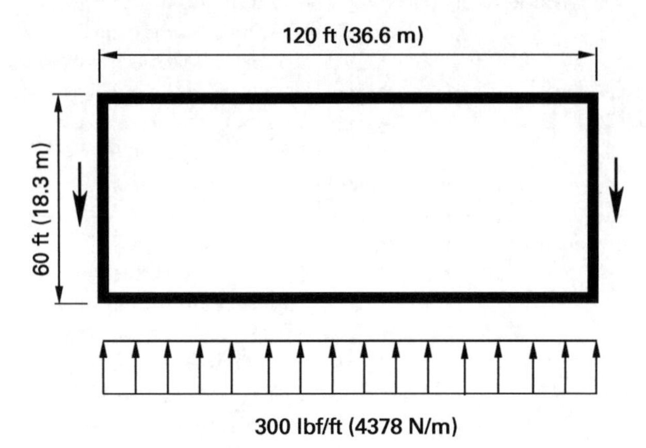

Customary U.S. Solution

(a) Based on App. H [ASCE/SEI7 Table 12.2-1], for bearing wall systems with lateral force resisting elements that have wood structural panel walls and are three stories or fewer in height, the response modification factor, R, is 6.5.

(b) Overstrength as a characteristic of a structure occurs when the actual strength is greater than the design strength. The degree of overstrength greatly depends on structural material and system. According to App. H [ASCE/SEI7 Table 12.2-1], the value of the seismic force amplification (overstrength) factor, Ω_O, is 3.0. From footnote g, this may be reduced by 0.5 as long as Ω_O is 2.5 or greater. Therefore, $\Omega_O = 3.0 - 0.5 = 2.5$.

(c) The ASCE/SEI7 requires that all structures be assigned a redundancy factor that accounts for the design of the lateral force resisting elements.

Since the building is in San Francisco, the seismic design category will be high due to the level of seismicity of the region. The value of ρ may be 1.0 if the structure meets the requirements of ASCE/SEI7 Sec. 12.3.4.2. The building has a regular plan and perimeter framing. The fewest number of bays on an exterior face will be in the shortest direction. For light-framed construction, the number of bays is

$$\frac{(2)(0.75)(60 \text{ ft})}{14 \text{ ft}} = 6.4$$

Since this is greater than 2.0, the code allows the use of ρ to be 1.0.

SI Solution

(a) Based on App. H [ASCE/SEI7 Table 12.2-1], for bearing wall systems with lateral force resisting elements that have wood structural panel walls and are three stories or fewer in height, the value of the response modification factor, R, is 6.5.

(b) Overstrength as a characteristic of a structure occurs when the actual strength is greater than the design strength. The degree of overstrength greatly depends on structural material and system. According to App. H [ASCE/SEI7 Table 12.2-1], the value of the seismic force amplification (overstrength) factor, Ω_O, is 3.0. From footnote g, this may be reduced by 0.5 as long as Ω_O is 2.5 or greater. Therefore, $\Omega_O = 3.0 - 0.5 = 2.5$.

(c) The ASCE/SEI7 requires that all structures be assigned a redundancy factor that accounts for the design of the lateral force resisting elements.

Since the building is in San Francisco, the seismic design category will be high due to the level of seismicity of the region. The value of ρ may be 1.0 if the structure meets the requirements of ASCE/SEI7 Sec. 12.3.4.2. The building has a regular plan and perimeter framing. The fewest number of bays on an exterior face will be in the shortest direction. For light-framed construction, the number of bays is

$$\frac{(2)(0.75)(18.3 \text{ m})}{4.3 \text{ m}} = 6.4$$

Since this is greater than 2.0, the code allows the value of ρ to be 1.0.

23. REGULAR/IRREGULAR STRUCTURES

In designing a structure, selection of the structure's basic plan, shape, and configuration is a critical step. The decision will influence the ability of the structure to withstand earthquake ground shaking. While configuration cannot be presumed to be the sole reason for building inadequacy, it is usually a significant contributor. Previous earthquake performance of buildings clearly illustrates that all other parameters being identical, the simpler the building is, the better its seismic performance will be.

A structure is an assemblage of framing members designed to support vertical (gravity) loads and resist lateral forces. For building structures, there are many types and configurations, and a dominant ideal configuration for any particular type of building structure does not exist. Despite the relationship between configuration and seismic performance, it was not until the 1973 edition of the UBC that building *configuration* was addressed in a specific provision.

ASCE/SEI7 Sec. 12.3.2 provides guidelines for classifying building configuration for a structure. Structures are classified as either *structurally regular* or *structurally irregular*, as shown in Fig. 6.8, and are based on horizontal and vertical configurations given in ASCE/SEI7 Table 12.3-1 and Table 12.3-2.

Regular structures, according to the ASCE/SEI7, are structures having no significant physical discontinuities in plan or vertical configuration or in their force-resisting systems. *Irregular structures* are defined as

Figure 6.8 *Structural Configuration*

(a) vertically regular structure

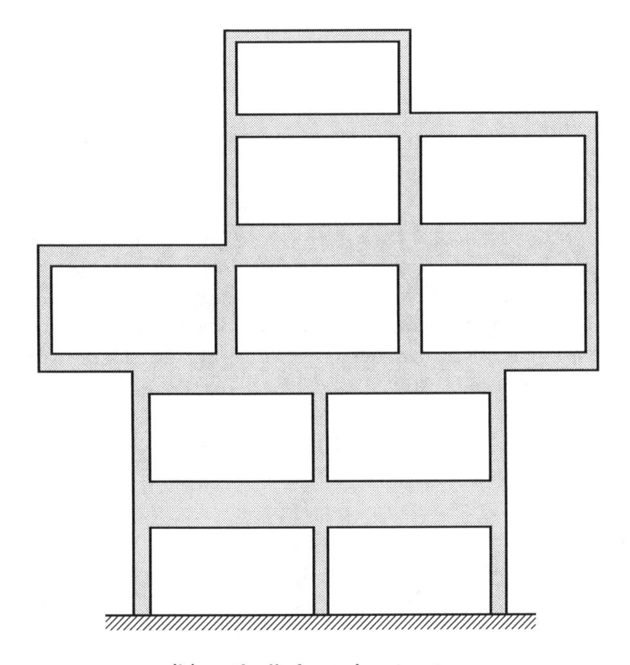

(b) vertically irregular structure

structures having significant physical discontinuities in configuration or in their lateral force resisting systems.

Regular structures have a uniform and continuous distribution of mass, stiffness, strength, and ductility with no significant torsional forces or large height-width ratio or large changes in plan area from floor to floor. They have relatively shorter spans than irregular structures, simple structural subsystems, and balanced stiffness and strength between members, connections, and supports. The code permits static analysis for a regular structure with few exceptions. (See Sec. 6.33.)

Structures with irregular shapes, changes in mass from floor to floor, variable stiffness with height, and unusual setbacks historically have not performed well during earthquakes. This is unfortunate from aesthetic and creative perspectives, because such buildings are generally among the most pleasing in appearance and interesting to design. With few exceptions (see Sec. 6.33), the code specifies that a dynamic analysis is required for an irregular structure.[23]

To identify requirements affecting irregular structures, Table 6.10 (similar to ASCE/SEI7 Table 12.3-1 and Table 12.3-2, combined) lists major physical discontinuity types along with corresponding reference sections.

Table 6.10 *Plan and Vertical Structural Irregularities [ASCE/SEI7 Table 12.3-1 and Table 12.3-2]*

structural irregularity	physical irregularity type	ASCE/SEI7 table
vertical	stiffness soft story	12.3-2, Item 1
	weight (mass)	12.3-2, Item 2
	vertical geometric irregularity	12.3-2, Item 3
	in-plane discontinuity	12.3-2, Item 4
	discontinuity in lateral strength—weak story	12.3-2, Item 5
horizontal	torsional irregularity	12.3-1, Item 1
	reentrant corners	12.3-1, Item 2
	diaphragm discontinuity	12.3-1, Item 3
	out-of-plane offsets	12.3-1, Item 4
	nonparallel systems	12.3-1, Item 5

24. VERTICAL STRUCTURAL IRREGULARITY

ASCE/SEI7 Table 12.3-2 lists and defines the following five types of *vertical structural irregularities.* (See Fig. 6.9.) None of the irregularities need to be considered for structures in seismic design category A. Depending on the type of irregularity and the seismic design category, the ASCE/SEI7 refers to Sec. 12.3.3 or Table 12.6-1 for restrictions or remediation.

1. A *soft story* has a stiffness less than 70% of the story immediately above, or less than 80% of the average stiffness of the three stories above. An *extreme soft story* has a stiffness less than 60% of the story immediately above, or less than 70% of the average stiffness of the three stories above.

[23]Since the code is specific, some of the design criteria were set primarily by judgment, as neither historical nor empirical data were available. Thus, the criteria are somewhat "loose," and the intent is to penalize only the worst cases of structural irregularity and to allow less severe structural configurations.

2. A story has a *mass (weight) irregularity* when its mass is more than 150% of the effective mass of a story above or below. (Roofs lighter than the floor immediately below are excluded.)
3. A story has *vertical geometric irregularity* when the horizontal dimension of a story's lateral force-resisting system is more than 130% of that in an adjacent story. (One-story penthouses are excluded.)
4. An *in-plane discontinuity* exists at a story when there is an in-plane offset of the lateral load-resisting elements greater than the length of those elements or there exists a reduction in stiffness of the resisting element in the story below.
5. A *weak story* has a story lateral strength less than 80% of the story immediately above. An *extreme weak story* has a story lateral strength less than 65% of the story immediately above.

Figure 6.9 *Vertical Irregularities*

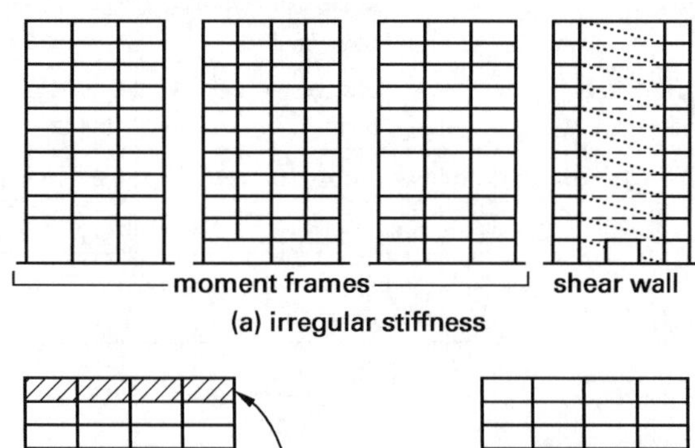

(a) irregular stiffness

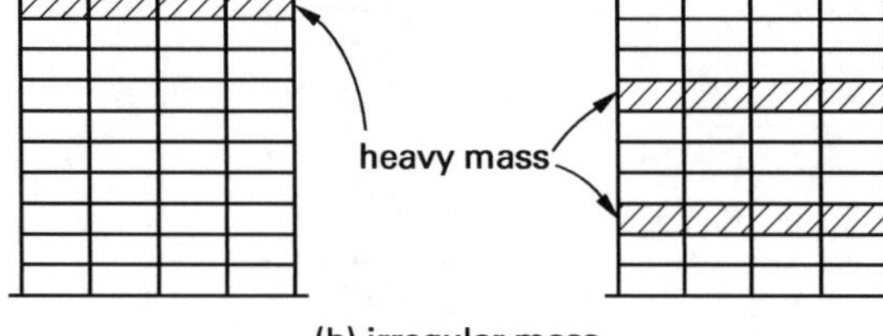

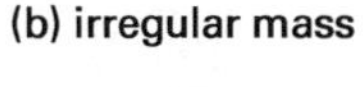

(b) irregular mass

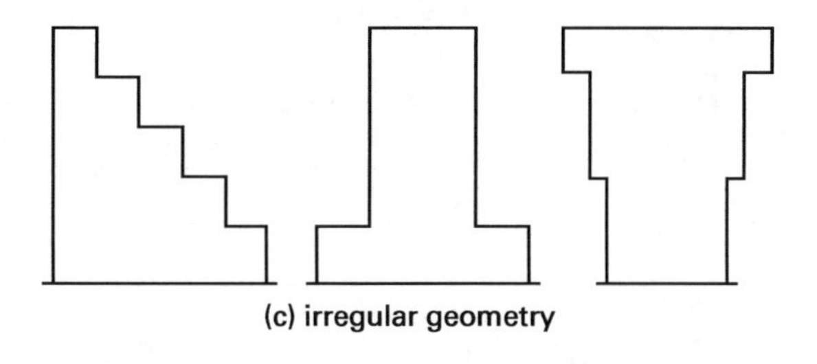

(c) irregular geometry

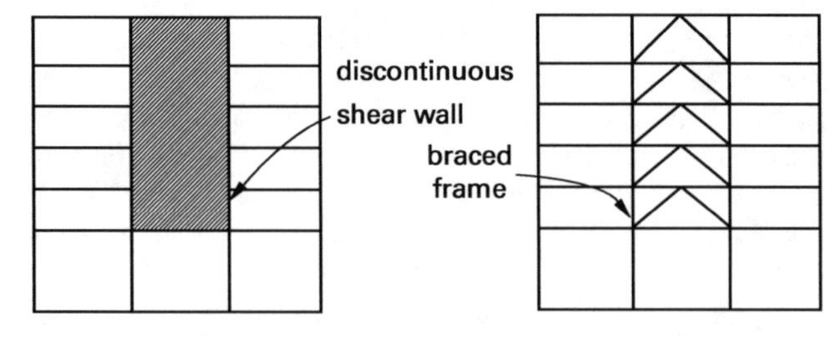

(d) in-plane discontinuity

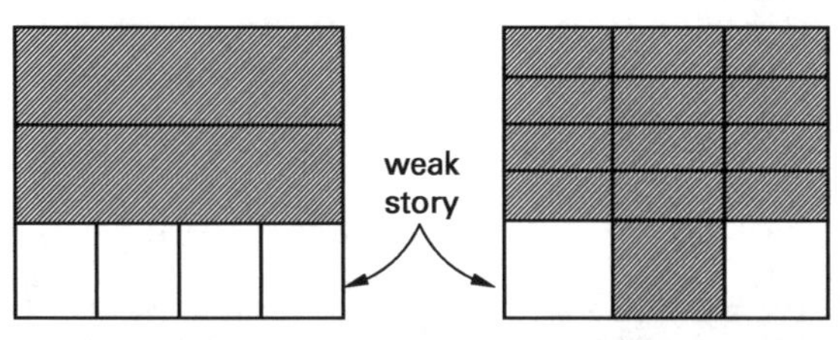

(e) discontinuity in lateral strength

A structure that meets one of the five conditions of ASCE/SEI7 Table 12.3-2 (corresponding to the five conditions listed previously) would normally be considered irregular. However, a structure that would otherwise be considered irregular under the first two conditions listed can be considered regular if the story drift ratio (as calculated from the design lateral forces and neglecting torsional effects) for each floor is less than 1.3 times the story drift ratio for the floor above. (The story drift ratio is the ratio of the actual drift relative to the floor below and the floor-to-floor height.) Furthermore, this condition does not have to be satisfied by the top two stories, as long as all stories below the top two satisfy it.

Example 6.3

Two 150 ft (45.7 m) building structures in California are shown in elevation. Structure I has a special reinforced concrete shear wall for lateral loads in one direction and a special steel moment-resisting frame system in the orthogonal direction. Structure II has a combined special steel moment-resisting frame with reinforced concrete shear wall in each direction, but the shear wall is discontinuous at the second floor. (a) What is the applicable response modification factor, R, for structure I? (b) What is the applicable response modification factor, R, for structure II?

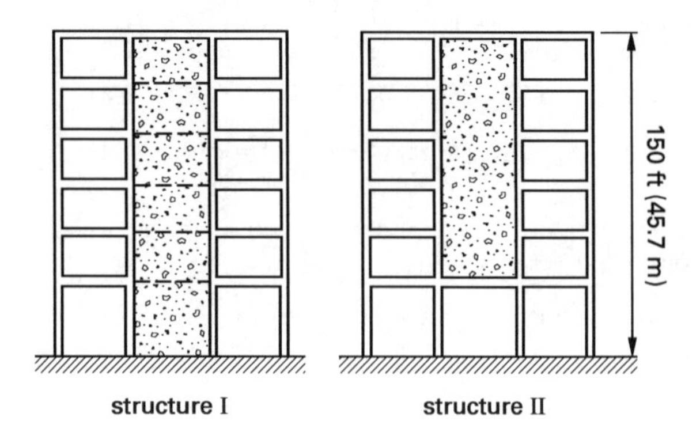

Solution

(a) Structure I is a building frame system with special reinforced concrete shear walls in one direction and a special steel moment-resisting frame in the orthogonal direction. Where combinations along different axes exist, ASCE/SEI7 Sec. 12.2.2 permits any combination of lateral systems for buildings less than 160 ft (49 m) in height, and restricts R only for systems in combination with a bearing wall system in one direction. Based on App. H [ASCE/SEI7 Table 12.2-1],

For a building frame system with special reinforced concrete shear wall, $R = 6.0$.
For special steel moment-resisting frame, $R = 8.0$.

Thus, R should be 6.0 for the elevation shown, and $R = 8.0$ should be used in the orthogonal direction.

(b) Structure II is a dual system in both directions at the upper stories, but not at the first level. ASCE/SEI7 Sec. 12.2.3.1 requires the value of R at any level be equal to or smaller than the value of R at a higher level. Based on App. H [ASCE/SEI7 Table 12.2-1],

For dual systems combining a special steel moment-resisting frame and special reinforced concrete shear walls, $R = 7.0$.

For special steel moment-resisting frames alone, the value of R is 8.0. Since this value is higher than the dual system, the value of R for both directions in structure II is 7.0.

25. HORIZONTAL STRUCTURAL IRREGULARITY

ASCE/SEI7 Table 12.3-1 lists and defines the following five types of *horizontal structural irregularities.* (See Fig. 6.10.) None of the irregularities need to be considered for structures in seismic design category A. Depending upon the type of irregularity and the seismic design category, the ASCE/SEI7 refers to Secs. 12.3.3, 12.5.3, 12.7.3, 12.8.4, 12.12.1, 16.2.2, or Table 12.6-1 for restrictions or remediation.

1. *Torsional irregularity* exists when the maximum story drift (caused by the lateral load *and* the accidental torsion) at one end of the structure transverse to its axis is more than 1.2 times the average story drift calculated from both ends. Only buildings with rigid diaphragms are affected by this type of irregularity.[24]

$$\Delta_2 > 1.2\left(\frac{\Delta_1 + \Delta_2}{2}\right)$$

   Extreme torsional irregularity exists when Δ_2 is more than 1.4 times the average story drift.

   ASCE/SEI7 Sec. 12.7.3 specifies that earthquake analysis should be conducted on a 3-D structural model when torsional irregularity exists.

2. A building has *reentrant corner irregularity* when one or more parts of the structure project beyond a reentrant corner a distance greater than 15% of the plan dimension in the given direction.

$$\text{projecting wing } D_2 > 0.15D_1$$

$$\text{projecting wing } L_2 > 0.15L_1$$

3. *Diaphragm discontinuity* occurs with diaphragms having abrupt discontinuities or variations in stiffness, including when there are cutout, or open, areas greater than 50% of the gross diaphragm area, or when the stiffness of the diaphragm changes more than 50% from story to adjacent story.

$$\text{opening area} > 0.50LD$$

4. An *out-of-plane offset* is a discontinuity in the lateral force path—an out-of-plane offset of the vertical elements.

5. A *nonparallel system* is one for which the vertical load-carrying elements are not parallel to or symmetrical about the major orthogonal axes of the lateral force-resisting system.

26. BUILDING PERIOD: *T*

The ASCE/SEI7 gives two methods for determining the *building period, T.* The first is an approximate method that implies that the natural period increases as the height of the structure increases (see Sec. 3.8). Equation 6.11 [ASCE/SEI7 Eq. 12.8-7] in accordance with ASCE/SEI7 Sec. 12.8.2.1 can be used for all buildings.[25] In Eq. 6.11 (traditionally known as *Method A*), h_n is the actual height (feet or meters) of the building above the base to the nth level. (The maximum allowable heights of the various types of structural systems are summarized in App. H [ASCE/SEI7 Table 12.2-1].[26])

$$T_a = C_t h_n^x \qquad 6.11$$

C_t and x are approximate period parameters determined from Table 6.11 [ASCE/SEIS7 Table 12.8-2]. Metric units are in parentheses, so that C_t is 0.028 (0.0724) for steel moment-resisting frames, 0.016 (0.0466) for reinforced concrete moment-resisting frames, 0.03 (0.0731) for eccentrically braced frames (see Sec. 10.2) and buckling-restrained steel frames, and 0.02 (0.0488) for all other buildings. x is 0.8 for steel moment-resisting frames, 0.9 for concrete moment-resisting frames, and 0.75 for both eccentrically braced frames and all other buildings.

[24]As defined in ASCE/SEI7 Sec. 12.3.1.3, a *flexible diaphragm* is one that has a maximum lateral deflection at a story more than two times the average story drift at that story.

[25]This first method will probably be used for almost all preliminary designs and many final designs.

[26]The 1988 *Blue Book* recommended that masonry bearing wall systems in regions of high seismicity be limited to 120 ft (36.6 m). The 1988 UBC set the limit at 160 ft (49 m). Thus, it is apparent that height limits are somewhat subjective. The discrepancy was eliminated in the 1990 *Blue Book.*

Figure 6.10 *Horizontal (Plan) Irregularities*

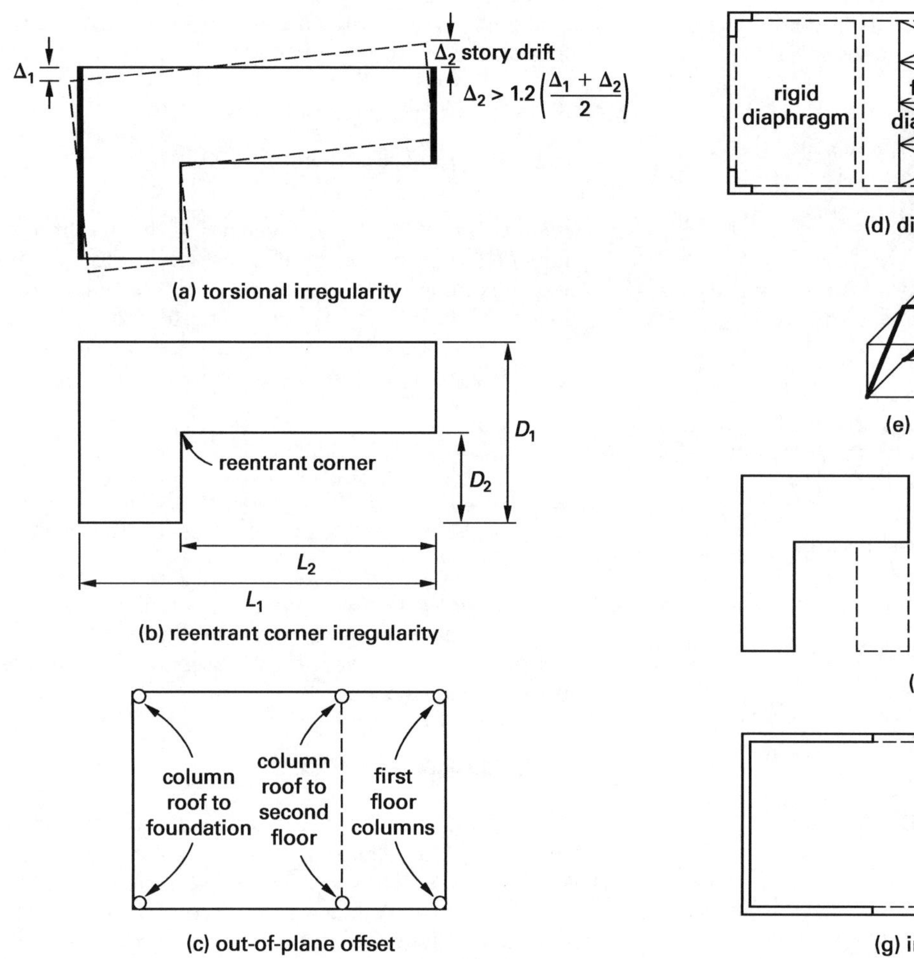

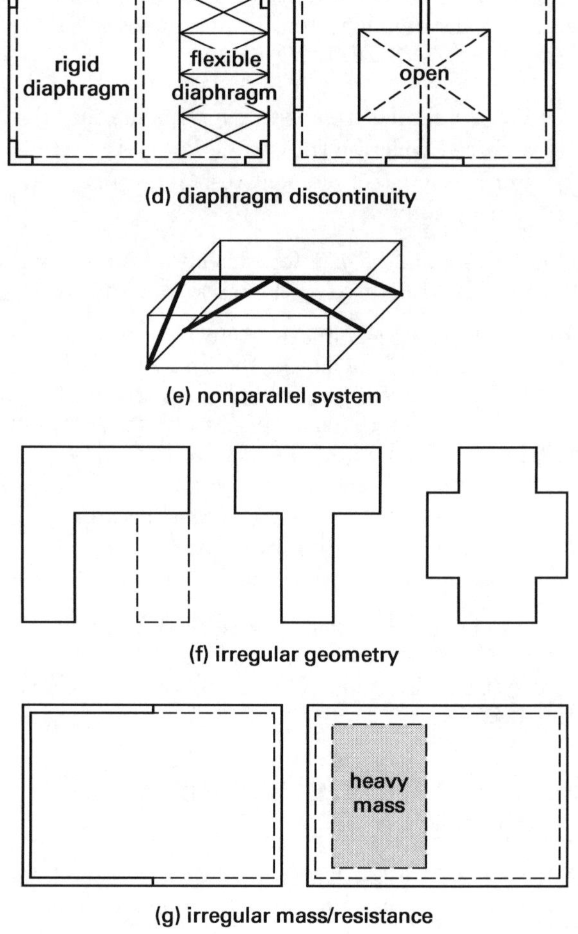

ASCE/SEI7 Sec. 12.8.2.1 contains an alternate method to be used in finding T_a for structures with concrete or masonry shear walls.[27]

$$T_a = \left(\frac{0.0019}{\sqrt{C_w}}\right)h_n \qquad 6.12$$

$$C_w = \frac{100}{A_B}\sum_{i=1}^{x}\left(\frac{h_n}{h_i}\right)^2\left(\frac{A_i}{1+0.83\left(\frac{h_i}{D_i}\right)^2}\right) \qquad 6.13$$

A_B is the base area of the structure, A_i is the web area of the shear wall, D_i is the length of the shear wall, h_i is the height of the shear wall, h_n is the building height, and x is the number of shear walls in the direction of loading.

Another method for finding the period, T, known as *Method B*, is based on the deformation characteristics of the resisting elements and is a more rational determination. ASCE/SEI7 Sec. 12.8.2 is sometimes referred to as the *Rayleigh method.*

If Method B is used to find the period, T, ASCE/SEI7 Sec. 12.8.2 requires that the value of T cannot be more than C_u greater than the value of T_a, determined from the empirical period given by Eq. 6.11 (i.e., Method A). The value of C_u varies depending upon the value of S_{D1} for the building site [ASCE/SEI7 Table 12.8-1]. For sites where S_{D1} is

- ≥ 0.4, C_u is 1.4
- 0.3, C_u is 1.4
- 0.2, C_u is 1.5
- 0.15, C_u is 1.6
- ≤ 0.1, C_u is 1.7

In addition to the ASCE/SEI7 method of determining the period, any other substantiated analysis method can be used. If a dynamic analysis is performed, the first modal period should be used for T.

[27]When Eq. 6.11 is used with shear wall structures, the base shear formula may be overly conservative in the lower seismic design categories.

Table 6.11 *Values of Approximate Period Parameters C_t and x [ASCE/SEI7 Table 12.8-2]*

structure type	C_t^*	x
moment-resisting frame systems in which the frames resist 100% of the required seismic force and are not enclosed or adjoined by components that are more rigid and will prevent the frames from deflecting where subjected to seismic forces		
steel moment-resisting frames	0.028 (0.0724)	0.8
concrete moment-resisting frames	0.016 (0.0466)	0.9
eccentrically braced steel frames	0.03 (0.0731)	0.75
buckling-restrained steel frames	0.03 (0.0731)	0.75
all other structural systems	0.02 (0.0488)	0.75

*Metric equivalents are shown in parentheses.

Example 6.4

In designing a 72 ft (21.95 m) steel moment-resisting frame structure with cast-in-place concrete shear walls, the fundamental (natural) period is calculated to be 0.8 sec using ASCE/SEI7's rational analysis. Utilizing ASCE/SEI7's approximate method, what is the fundamental period for this building?

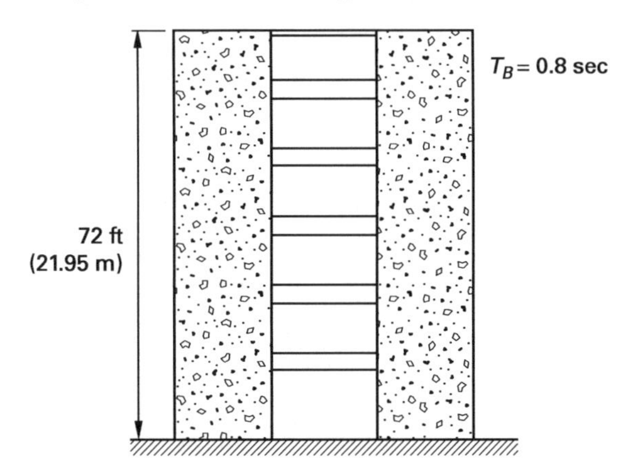

Customary U.S. Solution

The approximate formula (Method A) is given in ASCE/SEI7 Sec. 12.8.2.1. Since the concrete shear walls are more rigid than the moment-resisting frame, this is an "all other structural system." From Table 6.11 [ASCE/SEI7 Table 12.8-2], for all other systems, C_t is 0.02, x is 0.75. h_n is the building height. From Eq. 6.11,

$$T_a = C_t h_n^x = (0.02)(72\text{ ft})^{0.75} = 0.49\text{ sec}$$

SI Solution

The approximate formula (Method A) is given in ASCE/SEI7 Sec. 12.8.2.1. Since the concrete shear walls are more rigid than the moment-resisting frame, this is an "all other structural system." From Table 6.11 [ASCE/SEI7 Table 12.8-2], for all other systems, C_t is 0.0488, x is 0.75. h_n is the building height. From Eq. 6.11,

$$T_a = C_t h_n^x = (0.0488)(21.95\text{ m})^{0.75} = 0.49\text{ s}$$

27. SEISMIC LOAD EFFECT: *E*

In designing structures, the load effects of the horizontal and vertical components of the earthquake ground motion should be considered. The seismic load effect is a function of horizontal and vertical seismic-induced forces. This load on an element of the structure is denoted as E and is determined from Eq. 6.14 [ASCE/SEI7 Sec. 12.4.2].

$$E = E_h + E_v = \rho Q_E \pm 0.2 S_{DS} D \qquad 6.14$$

Q_E represents the forces associated with the horizontal component of the seismic load effect. Horizontal earthquake forces are due to the base shear, V, or the design lateral force (F_p or F_x). $0.2S_{DS}D$ represents loads resulting from the vertical component of the earthquake ground motion, E_v. The vertical earthquake effect is equal to an increase of $0.2S_{DS}D$ over the dead load effect D, where S_{DS} is the design spectral response acceleration at short periods, and ρ is the redundancy factor.

Based on ASCE/SEI7 Sec. 12.4.3.1, the estimated seismic load effect including the overstrength factor, E_m, that can be developed in a structure is determined from Eq. 6.15 [ASCE/SEI7 Sec. 12.4.3]. The value of E_m from Eq. 6.15 is only used when specifically required by the code.

$$E_m = E_{mh} \pm E_v = \Omega_O Q_E \pm 0.2 S_{DS} D \qquad 6.15$$

Ω_O, as provided in App. H [ASCE/SEI7 Table 12.2-1], represents the seismic force amplification factor that is required to account for structural overstrength. The seismic force amplification factor provides an upper bound approximation of actual seismic load acting on a structure with inelastic response capacity.

28. LOAD COMBINATIONS

Load combinations appear in all building codes and design standards, usually explicitly, but sometimes by reference or adoption. In the case of earthquake loading, load combinations prescribed by the IBC [IBC Sec. 1605] are the same as those in ASCE/SEI7 Sec. 2.3.2. However, ASCE/SEI7 contains additional combinations (such as those in ASCE/SEI7 Sec. 12.4, with *overstrength factors*) that may or may not have been adopted by the IBC. Clearly, it is important to know which authority to use.

For a structure, the basic contributing design loads are *floor live load* (L), *roof live load* (L_r), *dead load* (D), *seismic load effect* (E), *snow load* (S), and *wind load* (W).[28] These loads are referred to as the *design loads.* The IBC distinguishes between the seismic load effect, E, that combines horizontal and vertical loading, as described in ASCE/SEI7 Sec. 12.4.2, and the seismic load effect including the overstrength factor, E_m, from either horizontal or vertical loading, as described in ASCE/SEI7 Sec. 12.4.3. The IBC requires members to be designed for the most critical and unfavorable combination of loads [IBC Sec. 1605]. Thus, all combinations must be checked to find the controlling load combination.

Table 6.12 [IBC Sec. 1605] mentions a 1.33 allowance for ASD design using alternative basic load combinations. (See Sec. 6.50.)

The IBC presents two sets of load combinations. The primary set, from IBC Sec. 1605.2, is for use with strength design or load and resistance factor design (SD or LRFD). The second set, from IBC Sec. 1605.3, is for use with allowable strength design (ASD). This set is further divided into basic (IBC Sec. 1605.3.1) and alternative basic (IBC Sec. 1605.3.2) load combinations. A third set of load combinations (previously known as *special seismic load combinations*, a term that may still be encountered) incorporating overstrength factors is present only in ASCE/SEI7, but is required by the IBC when the simplified procedure of ASCE/SEI7 Sec. 12.14 is used and in cases of vertical seismic acceleration (vertical forces) [IBC Sec 1605.1.3]. Incorporating overstrength factors is required by ASCE/SEI7 for both ASD and strength design methods.

Example 6.5

A special steel moment-resisting frame is located in California at a site where $S_S = 1.4$. The soil profile is rock. The seismic design category is E. Assume that the importance factor, I_e, and redundancy factor, ρ, are both equal to 1.0. (a) What is the effect of horizontal seismic force, Q_E, on column A in the axial direction due to design base shear (i.e., resulting from the horizontal component of the earthquake ground motion)? (b) What is the load effect of the vertical component of the earthquake ground motion (i.e., vertical acceleration effect, E_v) on the column? (c) What is the seismic load effect, E, on one column in the axial direction resulting from the vertical and horizontal components of the earthquake ground motion? (d) What is the estimated seismic load effect including the overstrength factor that can be developed in the structure at the column?

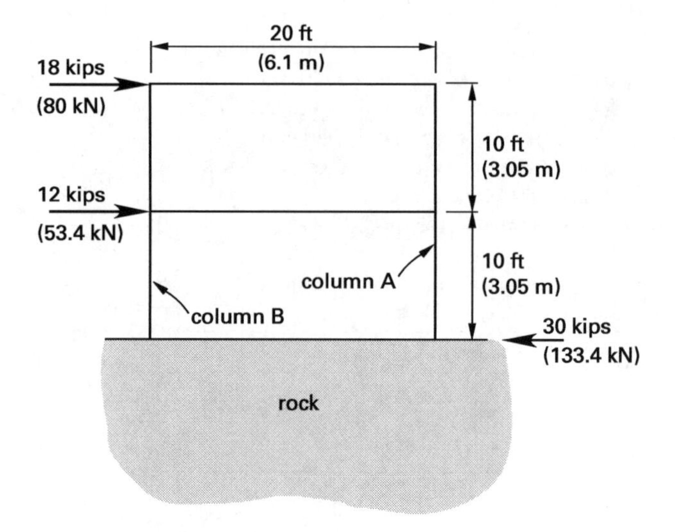

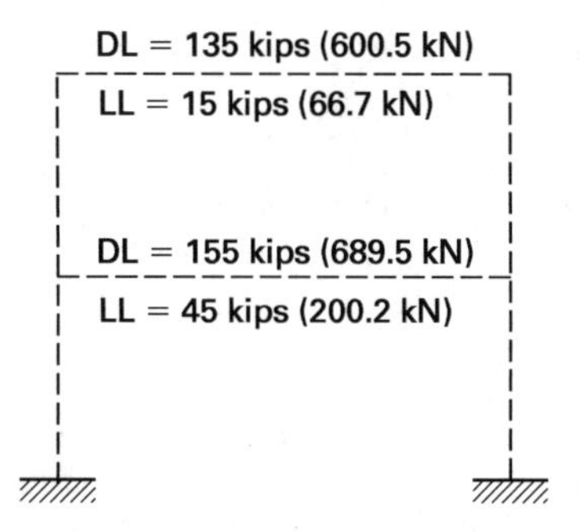

Customary U.S. Solution

(a) The effect of horizontal seismic force (design base shear) is found by summing moments about column B's footing.

$$\begin{aligned} Q_E &= \frac{\sum F_x h_x}{b} \\ &= \frac{(18 \text{ kips})(20 \text{ ft}) + (12 \text{ kips})(10 \text{ ft})}{20 \text{ ft}} \\ &= 24 \text{ kips} \end{aligned}$$

Accounting for structural redundancy, ρQ_E will be

$$\begin{aligned} \rho Q_E &= (1.0)(24 \text{ kips}) \\ &= 24 \text{ kips} \end{aligned}$$

(b) The seismic load effect of the vertical acceleration uses the equation

$$E_v = 0.2 S_{DS} D$$

[28]The other loads are: fluids (F); lateral pressure of soil and water in soil (H); ponding load (P); self-straining force and effects arising from contraction or expansion resulting from temperature change, shrinkage, or moisture change; creep in component materials; movement due to different settlement; or combinations thereof (T). The applicable load factors are given in IBC Sec. 1605.

Table 6.12 Combinations of Loads [IBC Sec. 1605]

design method	combination of loads	IBC section	IBC equation
basic load combinations			
	$1.4(D+F)$		16-1
	$1.2(D+F)+1.6(L+H)+0.5(L_r \text{ or } S \text{ or } R)$		16-2
	$1.2(D+F)+1.6(L_r \text{ or } S \text{ or } R)+1.6H+(f_1L \text{ or } 0.5W)$	1605.2	16-3
LRFD[a,b]	$1.2(D+F)+1.0W+f_1L+1.6H+0.5(L_r \text{ or } S \text{ or } R)$		16-4
	$1.2(D+F)+1.0E+f_1L+1.6H+f_2S$		16-5
	$0.9D+1.0W+1.6H$		16-6
	$0.9(D+F)+1.0E+1.6H$		16-7
	$D+F$		16-8
	$D+H+F+L$		16-9
	$D+H+F+(L_r \text{ or } S \text{ or } R)$	1605.3.1	16-10
	$D+H+F+0.75L+0.75(L_r \text{ or } S \text{ or } R)$		16-11
ASD (no 1.33 allowance for stress increase)[c,d,e,f,g]	$D+H+F+(0.6W \text{ or } 0.7E)$		16-12
	$D+H+F+0.75(0.6W)+0.75L+0.75(L_r \text{ or } S \text{ or } R)$		16-13
	$D+H+F+0.75(0.7E)+0.75L+0.75S$		16-14
	$0.6D+0.6W+H$		16-15
	$0.6(D+F)+0.7E+H$		16-16
alternate basic load combinations for ASD			
	$D+L+(L_r \text{ or } S \text{ or } R)$		16-17
	$D+L+0.6\omega W$		16-18
ASD (1.33 allowance for stress increase permitted)[c,d]	$D+L+0.6\omega W+S/2$	1605.3.2	16-19
	$D+L+S+0.6\omega W/2$		16-20
	$D+L+S+E/1.4$		16-21
	$0.9D+E/1.4$		16-22

f_1 = 1 for floors in places of public assembly, live loads greater than 100 lbf/ft^2 (4.79 kN/m^2), and parking garage live loads
= 0.5 for all other live loads

f_2 = 0.7 for roof configurations that do not shed snow
= 0.2 for all other roof configurations

[a]Where other factored load combinations are specifically required by other provisions of this code, such combinations shall take precedence.

[b]Where the effect of H resists the primary variable load effect, a load factor of 0.9 shall be included with H, where H is permanent. H shall be set to zero for all other conditions.

[c]Crane hook loads need not be combined with roof live loads or with more than $3/4S$, or $1/2W$.

[d]Flat roof snow loads and roof live loads of 30 psf (1.44 kN/m^2) or less need not be combined with seismic loads. Where flat roof snow loads exceed 30 psf (1.44 kN/m^2), 20% must be combined with seismic loads.

[e]Where the effect of H resists the primary variable load effect, a load factor of 0.6 shall be included with H, where H is permanent. H shall be set to zero for all other conditions.

[f]In Eq. 16-15, the wind load, W, is permitted to be reduced in accordance with exception 2 of ASCE/SEI7 Sec. 2.4.1.

[g]In Eq. 16-16, $0.6D$ may be increased to $0.9D$ for the design of special reinforced masonry shear walls complying with IBC Chap. 21.

Rock profiles are given site class B [ASCE/SEI7 Table 20.3-1]. With $S_S = 1.4$ and site class B, from Table 6.4, the value F_a is 1.0 [IBC Table 1613.3.3(1)]. Using Eq. 6.6 and Eq. 6.8,

$$S_{MS} = F_a S_S = (1.0)(1.4) = 1.4$$

$$S_{DS} = \tfrac{2}{3} S_{MS} = \left(\tfrac{2}{3}\right)(1.4) = 0.933$$

$$D = \frac{135 \text{ kips} + 155 \text{ kips}}{2} = 145 \text{ kips}$$

For strength design, E_v is

$$E_v = 0.2 S_{DS} D = (0.2)(0.933)(145 \text{ kips}) = 27.06 \text{ kips}$$

(c) The seismic load effect, E, is a function of horizontal and vertical seismic-induced forces and can be determined from Eq. 6.14 [ASCE/SEI7 Eq. 12.4-1].

$$E = E_h + E_v = \rho Q_E + 0.2 S_{DS} D = 24 \text{ kips} + 27.06 \text{ kips} = 51.06 \text{ kips}$$

(d) The estimated seismic load effect including the overstrength factor can be computed from Eq. 6.15. The seismic system overstrength factor, Ω_O, can be obtained from App. H [ASCE/SEI7 Table 12.2-1]. For a special steel moment-resisting frame, Ω_O is equal to 3.0.

$$\begin{aligned} E_m = E_{mh} + E_v &= \Omega_O Q_E + 0.2 S_{DS} D \\ &= (3.0)(24 \text{ kips}) + 27.06 \text{ kips} \\ &= 99.06 \text{ kips} \end{aligned}$$

SI Solution

(a) The effect of horizontal seismic force (design base shear) is found by summing moments about column B's footing.

$$\begin{aligned} Q_E &= \frac{\sum F_x h_x}{b} \\ &= \frac{(80 \text{ kN})(6.1 \text{ m}) + (53.4 \text{ kN})(3.05 \text{ m})}{6.1 \text{ m}} \\ &= 106.7 \text{ kN} \end{aligned}$$

Accounting for structural redundancy, ρQ_E will be

$$\begin{aligned} \rho Q_E &= (1.0)(106.70 \text{ kN}) \\ &= 106.7 \text{ kN} \end{aligned}$$

(b) The seismic load effect of the vertical acceleration uses the equation

$$E_v = 0.2 S_{DS} D$$

Rock profiles are given site class B [ASCE/SEI7 Table 20.3-1]. With $S_S = 1.4$ and site class B, from Table 6.4, the value F_a is 1.0 [IBC Table 1613.3.3(1)]. Using Eq. 6.6 and Eq. 6.8,

$$\begin{aligned} S_{MS} &= F_a S_S \\ &= (1.0)(1.4) \\ &= 1.4 \\ S_{DS} &= \tfrac{2}{3} S_{MS} \\ &= \left(\tfrac{2}{3}\right)(1.4) \\ &= 0.933 \\ D &= \frac{600.5 \text{ kN} + 689.5 \text{ kN}}{2} \\ &= 645 \text{ kN} \end{aligned}$$

For strength design, E_v is

$$\begin{aligned} E_v &= 0.2 S_{DS} D \\ &= (0.2)(0.933)(645 \text{ kN}) \\ &= 120.4 \text{ kN} \end{aligned}$$

(c) The seismic load effect, E, is a function of horizontal and vertical seismic-induced forces and can be determined from Eq. 6.14 [ASCE/SEI7 Eq. 12.4-1].

$$\begin{aligned} E = E_h + E_v &= \rho Q_E + 0.2 S_{DS} D \\ &= 106.7 \text{ kN} + 120.4 \text{ kN} \\ &= 227.1 \text{ kN} \end{aligned}$$

(d) The estimated seismic load effect including the overstrength factor can be computed from Eq. 6.15. The seismic system overstrength factor, Ω_O, can be obtained from App. H [ASCE/SEI7 Table 12.2-1]. For a special steel moment-resisting frame, Ω_O is equal to 3.0.

$$\begin{aligned} E_m = E_{mh} + E_v &= \Omega_O Q_E + 0.2 S_{DS} D \\ &= (3.0)(106.7 \text{ kN}) + 120.4 \text{ kips} \\ &= 440.5 \text{ kN} \end{aligned}$$

29. EFFECTIVE SEISMIC WEIGHT: *W*

The *effective seismic weight*, W (in pounds or newtons), used to calculate base shears and building periods is normally the total seismic dead load of the structure.[29] This includes the weight of the ceiling, partitions, pipes, ducts, and equipment that are normally attached. W does not include full design roof and live loads. The objective in determining seismic weight W is to include all contributions to mass likely to be present at the time of an earthquake. However, applicable portions of other loads should be included as follows [ASCE/SEI7 Sec. 12.7.2].

- A minimum of 25% of the floor live load (i.e., storage) is added in warehouses and storage buildings [ASCE/SEI7 Sec. 12.7.2, Item 1].
- No less than 10 lbf/ft^2 (0.48 kN/m^2) must be added when partition loads are used in the design of the floor [ASCE/SEI7 Sec. 12.7.2, Item 2].
- The total operating weight of permanent equipment must be included [ASCE/SEI7 Sec. 12.7.2, Item 3].
- 20% of flat roof snow load where flat roof snow load exceeds 30 lbf/ft^2 (1.44 kN/m^2) [ASCE/SEI7 Sec. 12.7.2, Item 4].
- The weight of landscaping and other materials at roof gardens or other similar areas must be included [ASCE/SEI7 Sec. 12.7.2, Item 5].

The units of effective seismic weight, W, will determine the units of base shear, V. Thus, if weight is expressed in kips, the base shear will be in kips. Mass is seldom, if

[29]As part of normal practice, all of the foundation weight and half of the first-story wall weight are commonly omitted in the analysis of seismic diaphragm loads. (See Sec. 6.34.) However, such a provision is not explicitly defined in the IBC for calculating base shear.

ever, used in practice, as all weights are given in pounds or kips. However, care must be taken in the unlikely event that the structure mass, as opposed to the structure weight, is specified. Equation 6.16 shows that the weight in pounds-force (lbf) is numerically the same as the mass in pounds-mass (lbm), although this is not true if other units of mass are used. g is the gravitational acceleration, 32.2 ft/sec^2 (9.81 m/s^2), and g_c is the gravitational constant, 32.2 ft-lbm/lbf-sec^2.

$$W_{\text{lbf}} = m_{\text{lbm}}\left(\frac{g}{g_c}\right) \quad (6.16)$$

$$W_{\text{lbf}} = m_{\text{slug}}g \quad (6.17)$$

$$W_{\text{newtons}} = m_{\text{kg}}g \quad (6.18)$$

For design of a multilevel structure, the effective seismic weight, W, is the sum of the effective seismic weights of all levels.

$$W = \sum_{x=1}^{n} W_x \quad (6.19)$$

30. SEISMIC RESPONSE COEFFICIENT

The seismic base shear is a function of the *seismic response coefficient*, C_s, and the *effective seismic weight*, W. The seismic response coefficient can be determined using Eq. 6.20 [ASCE/SEI7 Eq. 12.8-2], and is illustrated in Fig. 6.11 and Fig. 6.12.

$$C_s = \frac{S_{DS}}{\frac{R}{I_e}} \quad (6.20)$$

S_{DS} is the site-dependent spectral response acceleration parameter as determined from ASCE/SEI7 Sec. 11.4.4. The importance factor, I_e, and the response modification factor, R, influence the value of C_s. C_s need not exceed the following. (T is the fundamental period of the structure, and T_L is the long-period transition period, determined from ASCE/SEI7 Fig. 22-12.)

When $T \leq T_L$, use Eq. 6.21(a) [ASCE/SEI7 Eq. 12.8-3]. When $T > T_L$, use Eq. 6.21(b) [ASCE/SEI7 Eq. 12.8-4].

$$C_s = \frac{S_{D1}}{T\left(\frac{R}{I_e}\right)} \quad (6.21(a))$$

$$C_s = \frac{S_{D1}T_L}{T^2\left(\frac{R}{I_e}\right)} \quad (6.21(b))$$

Figure 6.11 *Minimum Seismic Response Coefficient*

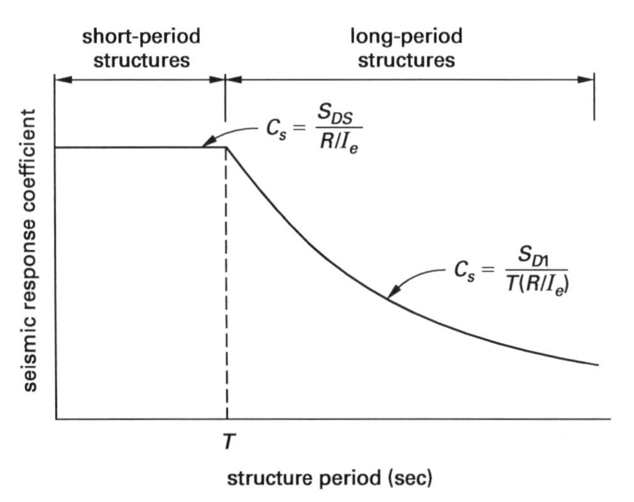

Figure 6.12 *Minimum Seismic Response Coefficient, High Seismic Region Only*

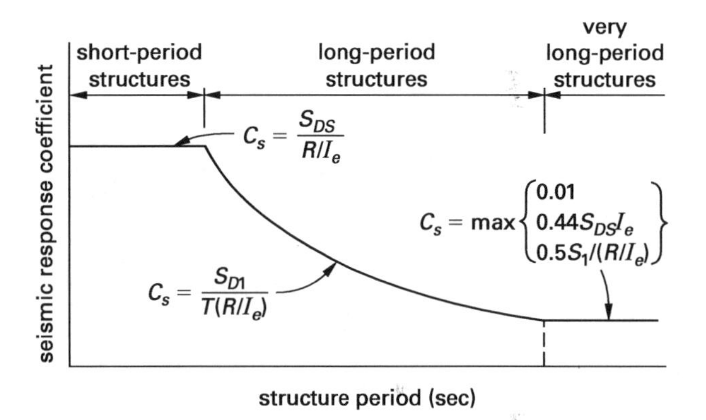

When $S_1 < 0.6$, C_s cannot be less than Eq. 6.22(a) [ASCE/SEI7 Eq. 12.8-5]. When $S_1 \geq 0.6$, C_s cannot be less than Eq. 6.22(b) [ASCE/SEI7 Eq. 12.8-6].

$$C_s = 0.044S_{DS}I_e \geq 0.01 \quad (6.22(a))$$

$$C_s = \frac{0.5S_1}{\frac{R}{I_e}} \quad (6.22(b))$$

The seismic response coefficient, and thus the base shear values, will be greater for structures having natural (fundamental) periods of less than $T = S_{D1}/S_{DS}$, whereas the natural period greater than T will result in a lower value. Coefficients S_{DS} and S_{D1}, which are site-dependent seismic response coefficients, are calculated using Eq. 6.8 and Eq. 6.9 [IBC Eq. 16-39 and Eq. 16-40]. For shorter (acceleration response) period structures, Eq. 6.21(a) can be used. Equation 6.21(b) can be used for structures having a longer (velocity response) period.

An increase in the value of I_e increases the seismic base shear coefficient, whereas an increase in the value of R decreases it.

Example 6.6

A ten-story, steel special concentrically braced frame for an emergency-preparedness center is being designed on a stiff soil profile. The value of S_S is 1.70 and S_1 is 0.85. The value of T_L is 12 sec. Determine the seismic response coefficient.

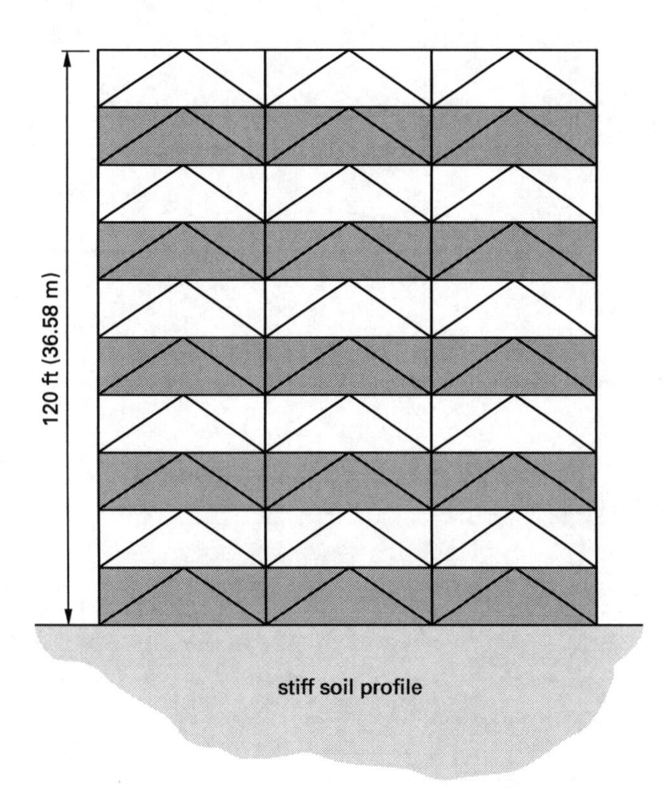

Solution

From Table 6.2 [ASCE/SEI7 Table 1.5-1], emergency preparedness buildings have a risk category of IV.

From Table 6.3 [ASCE/SEI7 Table 20.3-1], a stiff soil profile is categorized as site class D. From Table 6.4 [IBC Table 1613.3.3(1)], for an S_S of 1.70 and a site class D, F_a is 1.0.

From Table 6.5 [IBC Table 1613.3.3(2)], for an S_1 of 0.85 and a site class D, F_v is 1.5. Calculate the site class coefficients.

From Eq. 6.6,

$$\begin{aligned} S_{MS} &= F_a S_s = (1.0)(1.7) \\ &= 1.7 \end{aligned}$$

From Eq. 6.7,

$$\begin{aligned} S_{M1} &= F_v S_1 = (1.5)(0.85) \\ &= 1.275 \end{aligned}$$

From Eq. 6.8,

$$\begin{aligned} S_{DS} &= \tfrac{2}{3} S_{MS} = \left(\tfrac{2}{3}\right)(1.7) \\ &= 1.133 \end{aligned}$$

From Eq. 6.9,

$$\begin{aligned} S_{D1} &= \tfrac{2}{3} S_{M1} = \left(\tfrac{2}{3}\right)(1.275) \\ &= 0.85 \end{aligned}$$

Since $S_{D1} > 0.75$, the seismic design category is F.

Customary U.S. Solution

The natural period, T, can be determined from Eq. 6.11 [ASCE/SEI7 Eq. 12.8-7] (i.e., Method A). From Table 6.11 [ASCE/SEI7 Table 12.8-2], for a steel special concentrically braced frame, $C_t = 0.02$ and $x = 0.75$. h_n is 120 ft.

$$\begin{aligned} T_a &= C_t h_n^x = (0.02)(120\text{ ft})^{0.75} \\ &= 0.725\text{ sec} \end{aligned}$$

From App. H, for steel special concentrically braced frames, $R = 6.0$. From Table 6.2, the importance factor for structures in emergency-preparedness centers, I_e, is 1.5.

The seismic response coefficient, C_s, can be determined from Eq. 6.20.

$$\begin{aligned} C_s &= \frac{S_{DS}}{\frac{R}{I_e}} = \frac{1.133}{\frac{6}{1.5}} \\ &= 0.283 \end{aligned}$$

Based on Eq. 6.22(a) [ASCE/SEI7 Eq. 12.8-5], the minimum seismic base shear coefficient is 0.01. However, since S_1 is 0.85 and larger than 0.6, Eq. 6.22(b) [ASCE/SEI7 Eq. 12.8-6] must also be considered.

$$\begin{aligned} C_{s,\min} &= \frac{0.5 S_1}{\frac{R}{I_e}} = \frac{(0.5)(0.85)}{\frac{6}{1.5}} \\ &= 0.106 \end{aligned}$$

The larger of these two values must be used, so the minimum value of C_s is 0.106.

Since $T < T_L$, based on Eq. 6.21(a) [ASCE/SEI7 Eq. 12.8-3], the maximum seismic base shear coefficient is

$$\begin{aligned} C_{s,\max} &= \frac{S_{D1}}{T\left(\frac{R}{I_e}\right)} = \frac{0.85}{(0.725\text{ sec})\left(\frac{6}{1.5}\right)} \\ &= 0.293 \end{aligned}$$

Since $0.01 < 0.106 < 0.283 < 0.293$, $C_s = 0.283$ is the governing seismic base shear coefficient for this structure.

SI Solution

The natural period, T, can be determined from Eq. 6.11 [ASCE/SEI7 Eq. 12.8-7] (i.e., Method A). From Table 6.11 [ASCE/SEI7 Table 12.8-2], for a steel special concentrically braced frame, $C_t = 0.0488$ and $x = 0.75$. h_n is 36.58 m.

$$\begin{aligned} T_a &= C_t h_n^x \\ &= (0.0488)(36.58\ \text{m})^{0.75} \\ &= 0.726\ \text{s} \end{aligned}$$

From App. H, for steel special concentrically braced frames, $R = 6.0$. From Table 6.2, the importance factor for structures in emergency-preparedness centers, I_e, is 1.50.

The seismic base shear coefficient, C_s, can be determined from Eq. 6.20.

$$\begin{aligned} C_s &= \frac{S_{DS}}{\frac{R}{I_e}} = \frac{1.133}{\frac{6}{1.5}} \\ &= 0.283 \end{aligned}$$

Based on Eq. 6.22(a) [ASCE/SEI7 Eq. 12.8-5], the minimum seismic base shear coefficient is 0.01. However, since S_1 is 0.85 and larger than 0.6, Eq. 6.22(b) [ASCE/SEI7 Eq. 12.8-6] must also be considered.

$$\begin{aligned} C_{s,\text{min}} &= \frac{0.5 S_1}{\frac{R}{I_e}} = \frac{(0.5)(0.85)}{\frac{6}{1.5}} \\ &= 0.106 \end{aligned}$$

The larger of these two values must be used, so the minimum value of C_s is 0.106.

Since $T < T_L$, based on Eq. 6.21(a) [ASCE/SEI7 Eq. 12.8-3], the maximum seismic base shear coefficient is

$$\begin{aligned} C_{s,\text{max}} &= \frac{S_{D1}}{T\left(\frac{R}{I_e}\right)} = \frac{0.85}{(0.726\ \text{s})\left(\frac{6}{1.5}\right)} \\ &= 0.293 \end{aligned}$$

Since $0.01 < 0.106 < 0.283 < 0.293$, $C_s = 0.283$ is the governing seismic base shear coefficient for this structure.

31. LATERAL FORCE PROCEDURES

There are four alternate methods in ASCE/SEI7 for determining the lateral forces. The selection of the appropriate lateral force procedure primarily depends on the type of structure (i.e., regular versus irregular), number of stories, and height, among other factors. These procedures are: (1) simplified static, (2) static (equivalent), (3) modal response spectrum lateral force procedures, and (4) seismic response history procedures.

Modal response and response history are two types of dynamic procedures.

Any structure may be designed using the dynamic lateral force procedures of ASCE/SEI7 Sec. 12.9 or Chap. 16. Certain structures, such as those defined in Table 6.13 [ASCE/SEI7 Table 12.6-1], should be designed using only the dynamic lateral force procedures. These structures include all structures, regular or irregular, with fundamental periods greater than $3.5T_s$. The dynamic lateral force procedure should also be used for those structures that have horizontal structural irregularities 1 or vertical structural irregularities 1, 2, or 3 [ASCE/SEI7 Tables 12.3-1 and 12.3-2].

32. STATIC LATERAL FORCE PROCEDURE

The *static lateral force procedure* is also referred to as the *equivalent lateral force procedure*. The structures considered for this procedure are mainly regular structures. ASCE/SEI7 Sec. 12.8 provides the provisions for determining the seismic base shear by the static lateral force procedure. The total design *base shear* in a given direction should be determined in accordance with Eq. 6.23 [ASCE/SEI7 Eq. 12.8-1].

$$V = C_s W \qquad 6.23$$

C_s is the seismic base shear coefficient determined in Sec. 6.30.

Example 6.7

A 100 ft (30.48 m), ten-story office building has a total effective seismic weight of 15,000 kips (66 723 kN). The building is in risk category II and has values of $S_S = 1.75$, $S_1 = 0.74$, and $T_L = 8$ sec. It is designed with a special moment-resisting steel frame system and is constructed on rock (site class B). Use the ASCE/SEI7 static lateral force procedure to calculate the total design base shear.

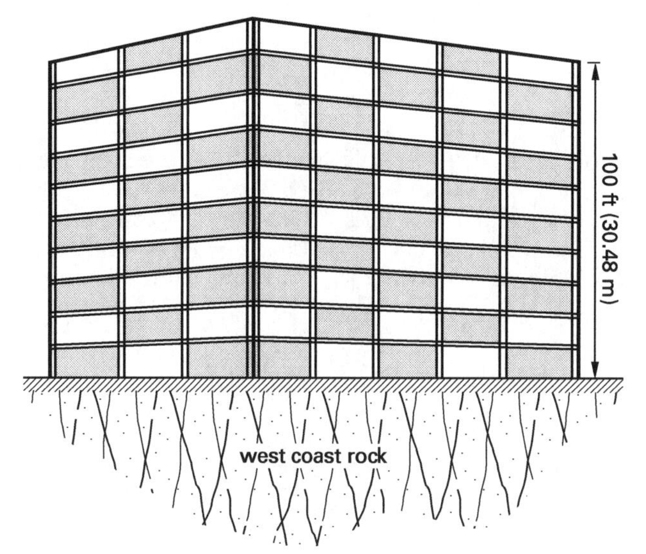

Table 6.13 *Permitted Analytical Procedures [ASCE/SEI7 Table 12.6-1]**

seismic design category	structural characteristics	equivalent lateral force analysis ASCE/SEI7 Sec. 12.8	modal response spectrum analysis ASCE/SEI7 Sec. 12.9	seismic response history procedures ASCE/SEI7 Chap. 16
B, C	all structures	P	P	P
D, E, F	risk category I or II buildings not exceeding 2 stories above the base	P	P	P
	structures of light frame construction	P	P	P
	structures with no structural irregularities and not exceeding 160 ft in structural height	P	P	P
	structures exceeding 160 ft in structural height with no structural irregularities and $T < 3.5T_s$	P	P	P
	structures not exceeding 160 ft in structural height and having only horizontal irregularities of Type 2, 3, 4, or 5 in Table 12.3-1 or vertical irregularities of Type 4, 5a, or 5b in Table 12.3-2	P	P	P
	all other structures	NP	P	P

*P: Permitted; NP: Not Permitted; $T_s = S_{D1}/S_{DS}$.

Customary U.S. Solution

To compute the total design base shear, V, the natural period, T, the seismic response modification factor, R, the importance factor, I_e, the seismic response coefficients, S_{DS} and S_{D1}, and the effective seismic weight, W, for this structure must be determined. The natural period, T, can be determined from Method A, Eq. 6.11 [ASCE/SEI7 Eq. 12.8-7]. From Table 6.11, for special moment-resisting steel frame systems, $C_t = 0.028$ and $x = 0.80$. h_n is 100 ft.

$$\begin{aligned} T_a &= C_t h_n^x \\ &= (0.028)(100\text{ ft})^{0.8} \\ &= 1.11\text{ sec} \end{aligned}$$

From App. H, for special moment-resisting steel frame systems, $R = 8.0$. From Table 6.2, the importance factor for risk category II is $I_e = 1.00$. The effective seismic weight, W, is 15,000 kips. From Table 6.4, with site class B and $S_S = 1.75$, F_a is 1.0. From Table 6.5, for site class B and $S_1 = 0.74$, F_v is 1.0.

$$\begin{aligned} S_{MS} &= F_a S_S = (1)(1.75) \\ &= 1.75 \\ S_{M1} &= F_v S_1 = (1)(0.74) \\ &= 0.74 \end{aligned}$$

From Eq. 6.8,

$$\begin{aligned} S_{DS} &= \tfrac{2}{3} S_{MS} = \left(\tfrac{2}{3}\right)(1.75) \\ &= 1.167 \end{aligned}$$

From Eq. 6.9,

$$\begin{aligned} S_{D1} &= \tfrac{2}{3} S_{M1} = \left(\tfrac{2}{3}\right)(0.74) \\ &= 0.493 \end{aligned}$$

From Eq. 6.20 [ASCE/SEIS7 12.8-2],

$$\begin{aligned} C_s &= \frac{S_{DS}}{\frac{R}{I_e}} = \frac{1.167}{\frac{8}{1.0}} \\ &= 0.146 \end{aligned}$$

Determine the minimum value of the seismic coefficient. From Eq. 6.22(a) [ASCE/SEI7 Eq. 12.8-5], the building must have a minimum coefficient of at least 0.01. In addition, since the value of S_1 is greater than 0.6, Eq. 6.22(b) [ASCE/SEI7 Eq. 12.8-6] must be considered.

$$\begin{aligned} C_{s,\min} &= \frac{0.5S_1}{\frac{R}{I_e}} = \frac{(0.5)(0.74)}{\frac{8}{1.0}} \\ &= 0.046 \end{aligned}$$

Since $T < T_L$, based on Eq. 6.21(a) [ASCE/SEI7 Eq. 12.8-3], the value of the seismic coefficient is

$$C_{s,\max} = \frac{S_{D1}}{T\left(\frac{R}{I_e}\right)} = \frac{0.493}{(1.11 \text{ sec})\left(\frac{8}{1.0}\right)} = 0.0555$$

Since $0.01 < 0.046 < 0.0555 < 0.146$, $C_s = 0.0555$. From Eq. 6.23,

$$V = C_s W = (0.0555)(15{,}000 \text{ kips}) = 833 \text{ kips}$$

SI Solution

To compute the total design base shear, V, the natural period, T, seismic response modification factor, R, importance factor, I_e, seismic response coefficients, S_{DS} and S_{D1}, and effective seismic weight, W, for this structure must be determined. The natural period, T, can be determined from Method A, Eq. 6.11 [ASCE/SEI7 Eq. 12.8-7]. From Table 6.11, for special moment-resisting steel frame systems, $C_t = 0.0724$ and $x = 0.80$. $h_n = 30.48$ m.

$$T_a = C_t h_n^x = (0.0724)(30.48 \text{ m})^{0.8} = 1.11 \text{ s}$$

From App. H, for special moment-resisting steel frame systems, $R = 8.0$. From Table 6.2, the importance factor for risk category II is $I_e = 1.00$. The effective seismic weight, W, is 66 723 kN. From Table 6.4, for site class B and $S_S = 1.75$, F_a is 1.0. From Table 6.5, for site class B and $S_1 = 0.74$, F_v is 1.0.

$$S_{MS} = F_a S_S = (1)(1.75) = 1.75$$

$$S_{M1} = F_v S_1 = (1)(0.74) = 0.74$$

From Eq. 6.8,

$$S_{DS} = \tfrac{2}{3} S_{MS} = \left(\tfrac{2}{3}\right)(1.75) = 1.167$$

From Eq. 6.9,

$$S_{D1} = \tfrac{2}{3} S_{M1} = \left(\tfrac{2}{3}\right)(0.74) = 0.493$$

From Eq. 6.20 [ASCE/SEIS7 12.8-2],

$$C_s = \frac{S_{DS}}{\frac{R}{I_e}} = \frac{1.167}{\frac{8}{1.0}} = 0.146$$

Determine the minimum value of the seismic coefficient. From Eq. 6.22(a) [ASCE/SEI7 Eq. 12.8-5], the building must have a minimum coefficient of at least 0.01. In addition, since the value of S_1 is greater than 0.6, Eq. 6.22(b) [ASCE/SEI7 Eq. 12.8-6] must also be considered.

$$C_{s,\min} = \frac{0.5 S_1}{\frac{R}{I_e}} = \frac{(0.5)(0.74)}{\frac{8}{1.0}} = 0.046$$

Since $T < T_L$, based on Eq. 6.21(a) [ASCE/SEI7 Eq. 12.8-3], the maximum value of the seismic coefficient is

$$C_{s,\max} = \frac{S_{D1}}{T\left(\frac{R}{I_e}\right)} = \frac{0.493}{(1.11 \text{ sec})\left(\frac{8}{1.0}\right)} = 0.0555$$

Since $0.01 < 0.046 < 0.0555 < 0.146$, $C_s = 0.0555$. From Eq. 6.23,

$$V = C_s W = (0.0555)(66\,723 \text{ kN}) = 3703 \text{ kN}$$

33. WHEN THE IBC PERMITS A STATIC ANALYSIS

The IBC refers to ASCE/SEI7 Sec. 12.6 to determine which type of analysis is used for a given structure. The ASCE/SEI7 is very specific about when the static method can be used. In general, any structure *may* be designed using the dynamic method at the option of the structural engineer, and some structures *must* use the dynamic method [ASCE/SEI7 Sec. 12.6].

Table 6.13 is based on ASCE/SEI7 Table 12.6-1 and identifies when certain types of analysis are permitted (P) and not permitted (NP). In general, irregular structures more than three stories tall in seismic design categories D, E, or F require dynamic analysis.

34. SIMPLIFIED LATERAL FORCE PROCEDURE

As an alternative to using the ASCE/SEI7 static lateral force procedure, the total design base shear may be determined by a simplified lateral force procedure, which is given in ASCE/SEI7 Sec. 12.14. The simplified lateral force procedure may be used in place of the methods listed in Table 6.13 if the structure meets the definitions of ASCE/SEI7 Sec. 12.14.1.1. Use is primarily restricted to one- or two-story bearing wall structures with limited irregularity. In general, this procedure is not intended for routine use.

The total design base shear in a given direction should be determined in accordance with Eq. 6.24 [ASCE/SEI7 Eq. 12.14-11].

$$V = \left(\frac{FS_{DS}}{R}\right)W \tag{6.24}$$

F is 1.0 for one-story buildings, 1.1 for two-story buildings, and 1.2 for three-story buildings.

The simplified lateral force procedure is a simple way to compute base shear and provides conservative results when compared to other IBC lateral force procedures. Figure 6.13 illustrates this comparison.

Figure 6.13 *Simplified Design Base Shear*

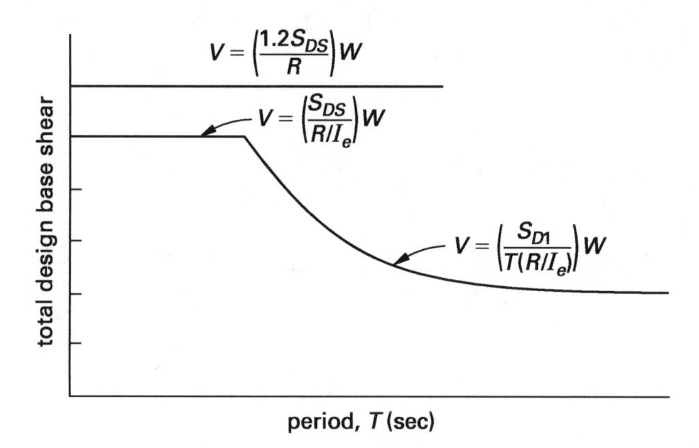

S_{DS} in Eq. 6.24 is the site-dependent design elastic response acceleration at short period, as determined in accordance with IBC Sec. 1613.3.4. When using Eq. 6.6 with the simplified lateral force procedure, S_S need not be taken larger than 1.5 [ASCE/SEI7 Sec. 12.14.8.1]. In earthquake-resistant design of a structure, if the soil profile is unknown, soil site class D can be presumed [IBC Sec. 1613.3.2].

Example 6.8

A one-story, single-family, light-framed wood residential dwelling is located on a site with very dense soil and a soft rock soil profile. The house has a total weight of 45 kips (200 kN). Using the ASCE/SEI7 simplified lateral force procedure, determine the total design base shear for this structure, assuming that (a) it is located at a location where $S_S = 0.40$ and $S_1 = 0.20$, (b) it is located at a location where $S_S = 2.95$ and $S_1 = 1.14$.

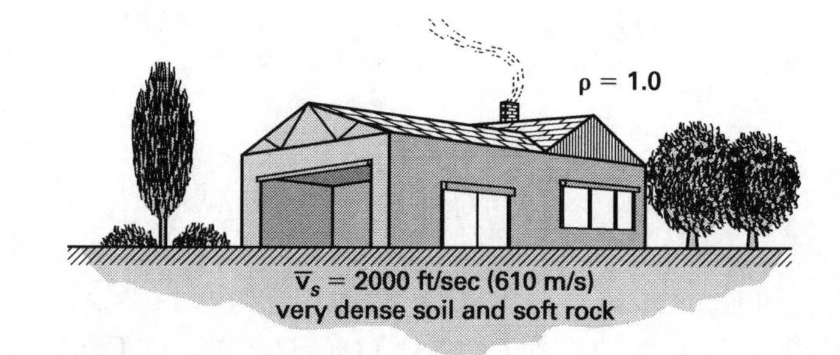

Customary U.S. Solution

The total design base shear, V, can be determined from Eq. 6.24 using the simplified lateral force procedure.

$$V = \left(\frac{FS_{DS}}{R}\right)W$$

For this one-story building, F is 1.0.

To compute V, the seismic response modification factor, R, seismic response coefficient, S_{DS}, and effective seismic weight, W, for this structure must be determined.

(a) From App. H, for structures three stories or fewer with light-framed wood structural bearing systems, $R = 6.5$. The effective seismic weight, W, is 45 kips. From Table 6.3, very dense soil and soft rock has a soil profile of site class C. From Table 6.4, for site class C and $S_S = 0.40$, F_a is 1.2. From Eq. 6.6,

$$\begin{aligned} S_{MS} &= F_a S_S = (1.2)(0.40) \\ &= 0.48 \end{aligned}$$

From Eq. 6.8,

$$\begin{aligned} S_{DS} &= \tfrac{2}{3} S_{MS} = \left(\tfrac{2}{3}\right)(0.48) \\ &= 0.32 \end{aligned}$$

Therefore, the total design base shear is

$$\begin{aligned} V &= \left(\frac{1.0S_{DS}}{R}\right)W = \left(\frac{(1.0)(0.32)}{6.5}\right)(45 \text{ kips}) \\ &= 2.22 \text{ kips} \end{aligned}$$

(b) Since a simplified procedure is to be used, S_S may be reduced to 1.5. $R = 6.5$ and $W = 45$ kips. From Table 6.4, for site class C and $S_S = 1.5$, F_a is 1.0. From Eq. 6.6,

$$\begin{aligned} S_{MS} &= F_a S_S = (1.0)(1.5) \\ &= 1.5 \end{aligned}$$

From Eq. 6.8,

$$\begin{aligned} S_{DS} &= \tfrac{2}{3} S_{MS} = \left(\tfrac{2}{3}\right)(1.5) \\ &= 1.0 \end{aligned}$$

Therefore, the total design base shear is

$$V = \left(\frac{1.0S_{DS}}{R}\right)W = \left(\frac{(1.0)(1.0)}{6.5}\right)(45 \text{ kips})$$
$$= 6.92 \text{ kips}$$

SI Solution

The total design base shear, V, can be determined from Eq. 6.24 using the simplified lateral force procedure.

$$V = \left(\frac{FS_{DS}}{R}\right)W$$

For this one-story building, F is 1.0.

To compute V, the seismic response modification factor, R, seismic response coefficient, S_{DS}, and effective seismic weight, W, for this structure must be determined.

(a) From App. H, for structures three stories or fewer with light-framed wood structural bearing systems, $R = 6.5$. The effective seismic weight, W, is 200 kN. From Table 6.3, very dense soil and soft rock has a soil profile of site class C. From Table 6.4, for site class C and $S_S = 0.40$, F_a is 1.2. From Eq. 6.6,

$$S_{MS} = F_a S_S = (1.2)(0.40)$$
$$= 0.48$$

From Eq. 6.8,

$$S_{DS} = \tfrac{2}{3}S_{MS} = \left(\tfrac{2}{3}\right)(0.48)$$
$$= 0.32$$

Therefore, the total design base shear is

$$V = \left(\frac{1.0S_{DS}}{R}\right)W = \left(\frac{(1.0)(0.32)}{6.5}\right)(200 \text{ kN})$$
$$= 9.85 \text{ kN}$$

(b) Since a simplified procedure is to be used, S_S may be reduced to 1.5. $R = 6.5$ and $W = 200$ kN. From Table 6.4, for site class C and $S_S = 1.5$, F_a is 1.0. From Eq. 6.6,

$$S_{MS} = F_a S_S = (1.0)(1.5)$$
$$= 1.5$$

From Eq. 6.8,

$$S_{DS} = \tfrac{2}{3}S_{MS} = \left(\tfrac{2}{3}\right)(1.5)$$
$$= 1.0$$

Therefore, the total design base shear is

$$V = \left(\frac{1.0S_{DS}}{R}\right)W = \left(\frac{(1.0)(1.0)}{6.5}\right)(200 \text{ kN})$$
$$= 30.8 \text{ kN}$$

35. VERTICAL DISTRIBUTION OF BASE SHEAR TO STORIES

In the absence of a justifiable, more rigorous approach, the base shear, V, is distributed to the n stories in accordance with Eq. 6.25 [ASCE/SEI7 Eq. 12.8-11 and Eq. 12.8-12]. The F_x forces increase with height above the base, as Fig. 6.14 illustrates. As building periods get longer, more force is applied toward the top to account for higher-mode effects. The exponent k varies related to the structure period. For structures having a period less than or equal to 0.5 sec, k is 1.0. For structures having a period greater than or equal to 2.5 sec, k is 2.0. For structures with periods between 0.5 sec and 2.5 sec, k should be determined using linear interpolation [ASCE/SEI7 Sec. 12.8.3]. The story shear, V_x, is the sum of the forces, F_x, above that story.

$$F_x = C_{vx}V \quad \text{[story force]} \qquad 6.25$$

$$C_{vx} = \frac{w_x h_x^k}{\sum_{i=1}^{n} w_i h_i^k} \qquad 6.26$$

$$V_x = \sum_{i=x}^{n} F_x \quad \text{[story shear]} \qquad 6.27$$

Figure 6.14 *Vertical Distribution of Story Shears*

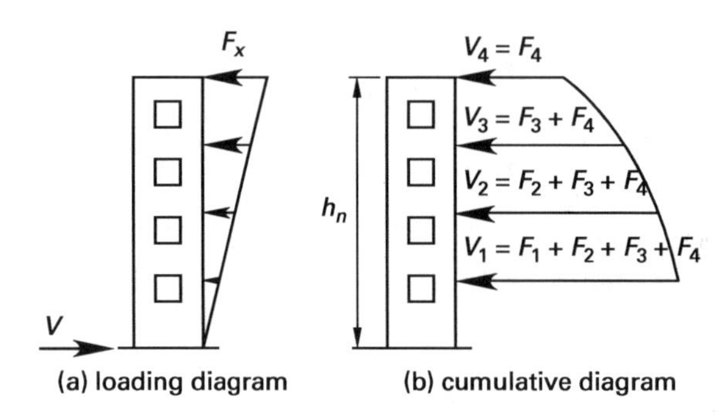

As Fig. 6.14 shows, the distribution of F_x is linear when k is 1.0. This distribution depends on the following requirements. Structures that do not meet these requirements must receive a dynamic analysis.

1. The building is regular and nearly symmetrical.
2. The lateral stiffnesses of each floor are approximately the same.
3. The lateral stiffnesses of each floor are uniformly distributed in plan.
4. The weight of each floor is approximately the same and is uniformly distributed in plan.
5. There is a continuous load path between members.
6. Torsional components of drift are small compared to translational components.
7. The first deflection mode (see Sec. 4.17) can be approximately represented as a straight line.

Example 6.9

A five-story building is constructed with 12 ft (3.7 m) story heights. The base shear has been calculated as 160 kips (0.71 MN). Each story floor has a weight of 800 kips (3.6 MN), and the roof has a weight of 700 kips (3.1 MN). The natural period of oscillation is 0.5 sec. What are the story forces?

Customary U.S. Solution

According to ASCE/SEI7 Sec. 12.8.3, the value of k in Eq. 6.26 is 1.0.

A table is the easiest way to set up the data for calculating and recording the story shears.

level x	h_x (ft)	w_x (kips)	$h_x w_x$ (ft-kips)	$\frac{h_x w_x}{\sum h_x w_x}$	F_x (kips)
5 (roof)	60	700	42,000	0.304	48.6
4	48	800	38,400	0.278	44.5
3	36	800	28,800	0.209	33.4
2	24	800	19,200	0.139	22.2
1	12	800	9600	0.070	11.2
totals		3900	138,000	1.000	160.0

The value of F_5 is given by Eq. 6.25 and Eq. 6.26.

$$F_5 = C_{vx}V = \left(\frac{w_5 h_5}{\sum w_i h_i}\right)V$$
$$= (0.304)(160 \text{ kips})$$
$$= 48.6 \text{ kips}$$

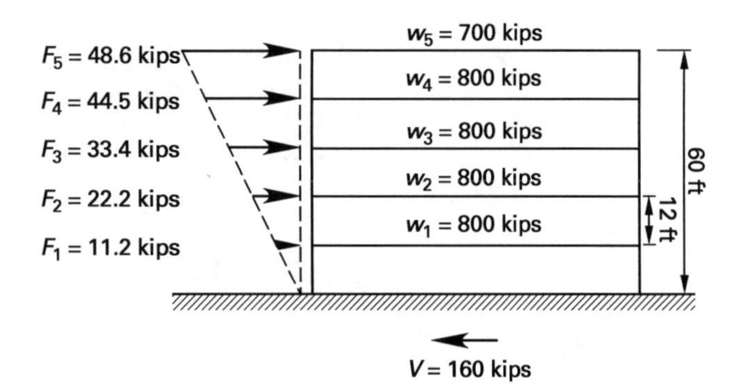

SI Solution

According to ASCE/SEI7 Sec. 12.8.3, the value of k in Eq. 6.26 is 1.0.

A table is the easiest way to set up the data for calculating and recording the story shears.

level x	h_x (m)	w_x (MN)	$h_x w_x$ (m·MN)	$\frac{h_x w_x}{\sum h_x w_x}$	F_x (MN)
5 (roof)	18.3	3.1	56.73	0.301	0.21
4	14.6	3.6	52.56	0.279	0.20
3	11.0	3.6	39.60	0.210	0.15
2	7.3	3.6	26.28	0.139	0.10
1	3.7	3.6	13.32	0.071	0.05
totals		17.5	188.49	1.000	0.71

The value of F_5 is given by Eq. 6.25 and Eq. 6.26.

$$F_5 = C_{vx}V = \left(\frac{w_5 h_5}{\sum w_i h_i}\right)V$$
$$= (0.304)(0.71 \text{ MN})$$
$$= 0.22 \text{ MN}$$

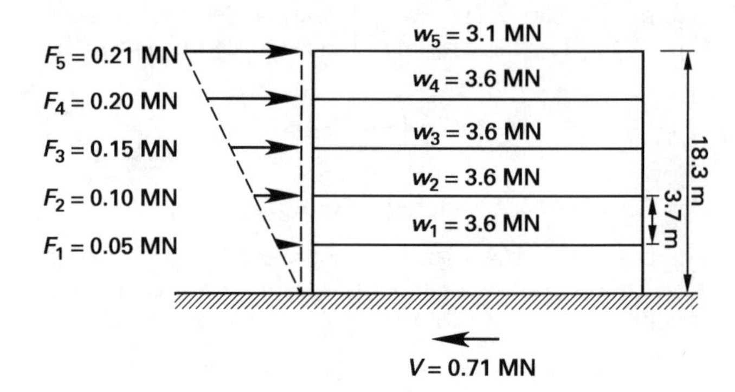

36. SIMPLIFIED VERTICAL DISTRIBUTION OF BASE SHEAR TO STORIES

Once the total design base shear, V, for simple structures (see Sec. 6.34) is determined by the simplified lateral force procedure, the lateral forces at each level may be determined from Eq. 6.28 [ASCE/SEI7 Eq. 12.14-12].

$$F_x = \left(\frac{w_x}{W}\right)V \qquad 6.28$$

w_x is the portion of W (effective seismic weight) at level x. In the simplified static procedure, F_x forces do not increase linearly with height above the base as they do in the standard static procedure. For the simplified vertical base shear distribution, natural period, T, P-delta effects, and story drift consideration may not be required.

Example 6.10

A two-story office building in seismic design category B, with an ordinary reinforced masonry bearing shear wall system, is located on a site with a stiff soil profile. The value of S_S is 2.05. Determine (a) the total design base shear using the simplified static lateral force procedure, and (b) the vertical distribution of base shear to stories.

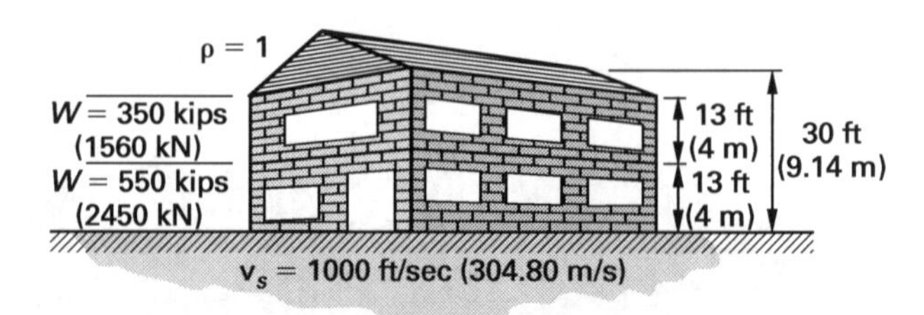

Customary U.S. Solution

(a) The total design base shear, V, can be determined from Eq. 6.24 using the simplified lateral force procedure.

$$V = \left(\frac{FS_{DS}}{R}\right)W$$

The value of F is 1.1 for two-story buildings.

To compute V, the seismic response modification factor, R, and effective seismic weight, W, for this structure must be determined.

From App. H, for a masonry bearing shear wall system, $R = 2.0$. From Table 6.3, stiff soil has a site class D soil profile. The effective seismic weight, W, for this building is

$$\begin{aligned} W &= 550 \text{ kips} + 350 \text{ kips} \\ &= 900 \text{ kips} \end{aligned}$$

From IBC Table 1613.3.3(1), for soil-profile site class D, the value of F_a is 1.0. Since the simplified static lateral force method is used, S_S does not need to be greater than 1.5. The applicable acceleration-controlled seismic response coefficient is

$$\begin{aligned} S_{DS} &= \tfrac{2}{3} S_{MS} = \tfrac{2}{3} F_a S_S \\ &= \left(\tfrac{2}{3}\right)(1.0)(1.5) \\ &= 1.00 \end{aligned}$$

Therefore, the total design base shear is

$$\begin{aligned} V &= \left(\frac{1.1 S_{DS}}{R}\right)W \\ &= \left(\frac{(1.1)(1.00)}{2.0}\right)(900 \text{ kips}) \\ &= 495 \text{ kips} \end{aligned}$$

(b) Because the simplified static lateral force procedure is used to determine the design base shear, the vertical distribution of base shear to stories is also based on the simplified vertical force distribution procedure of Eq. 6.28 [ASCE/SEI7 Eq. 12.14-12].

$$F_x = \left(\frac{w_x}{W}\right)V$$

$$\begin{aligned} F_1 &= \left(\frac{550 \text{ kips}}{900 \text{ kips}}\right)(495 \text{ kips}) \\ &= 302.5 \text{ kips} \end{aligned}$$

$$\begin{aligned} F_2 &= \left(\frac{350 \text{ kips}}{900 \text{ kips}}\right)(495 \text{ kips}) \\ &= 192.5 \text{ kips} \end{aligned}$$

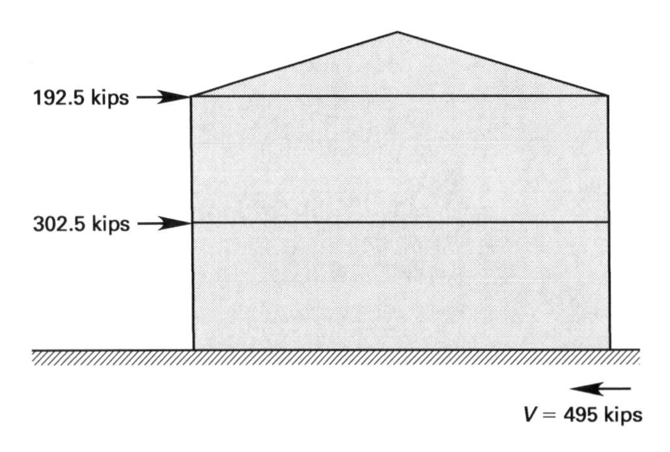

SI Solution

(a) The total design base shear, V, can be determined from Eq. 6.24 using the simplified lateral force procedure.

$$V = \left(\frac{FS_{DS}}{R}\right)W$$

The value of F is 1.1 for two-story buildings.

To compute V, the seismic response modification factor, R, and effective seismic weight, W, for this structure must be determined.

From App. H, for a masonry bearing shear wall system, $R = 2.0$. From Table 6.3, stiff soil has a site class D soil profile. The effective seismic weight, W, for this building is

$$\begin{aligned} W &= 2450 \text{ kN} + 1560 \text{ kN} \\ &= 4010 \text{ kN} \end{aligned}$$

From IBC Table 1613.3.3(1), for soil-profile site class D, the value of F_a is 1.0. Since the simplified static lateral force method is used, S_S does not need to be greater than 1.5. The applicable acceleration-controlled seismic response coefficient is

$$\begin{aligned} S_{DS} &= \tfrac{2}{3} S_{MS} = \tfrac{2}{3} F_a S_S \\ &= \left(\tfrac{2}{3}\right)(1.0)(1.5) \\ &= 1.00 \end{aligned}$$

Therefore, the total design base shear is

$$\begin{aligned} V &= \left(\frac{1.1 S_{DS}}{R}\right)W \\ &= \left(\frac{(1.1)(1.00)}{2.0}\right)(4010 \text{ kN}) \\ &= 2210 \text{ kN} \end{aligned}$$

(b) Because the simplified static lateral force procedure is used to determine the design base shear, the vertical distribution of base shear to stories is also based on the simplified vertical force distribution procedure of Eq. 6.28

[ASCE/SEI7 Eq. 12.14-12]. The forces at each level can be obtained from Eq. 6.28.

$$F_x = \left(\frac{w_x}{W}\right)V$$

$$F_1 = \left(\frac{2450 \text{ kN}}{4000 \text{ kN}}\right)(2210 \text{ kN}) = 1350 \text{ kN}$$

$$F_2 = \left(\frac{1560 \text{ kN}}{4000 \text{ kN}}\right)(2210 \text{ kN}) = 862 \text{ kN}$$

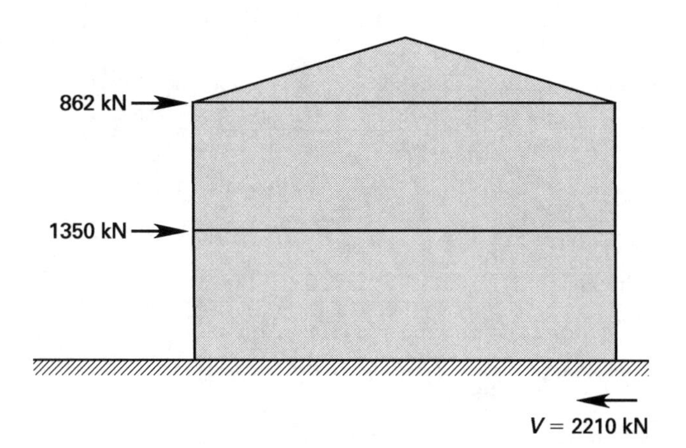

37. OTHER PROVISIONS FOR VERTICAL GROUND ACCELERATION

In the absence of specific information, it is commonly assumed that, for distant earthquakes, the vertical component of ground motion may be taken as two-thirds of the corresponding horizontal acceleration. This is not a specific code provision, but the approximation has been validated by numerous studies (with an approximate confidence level of 75%) for earthquakes more distant than 10 mi (15 km). The studies also noted that the vertical ground acceleration increases closer to the epicenter. At the epicenter, the magnitudes of vertical and horizontal accelerations can be the same. Other factors affecting the vertical ground acceleration are earthquake magnitude, frequency of shaking, and type of soil.

ASCE/SEI7 Sec. 12.4.2.2 provides methods for determining the effects of vertical acceleration on all structural elements (20% of the horizontal acceleration is based on ASCE/SEI7 Eq. 12.4-4), but this vertical acceleration is usually only critical for cantilever components and long-span prestressed components in high seismic regions. When considering seismic loads that are of the opposite direction of the dead load (such as upward vertical acceleration), the load cases from ASCE/SEI7 Sec. 2.3.2 and Sec. 2.4.1 reduce the amount of dead load available to counteract the seismic load.

38. PROVISIONS FOR DIRECTION OF THE EARTHQUAKE

ASCE/SEI7 requires that the ground motion producing structural response and seismic forces in any horizontal direction (i.e., x- and y-direction) be considered [ASCE/SEI7 Sec. 12.5]. Usually directions parallel and perpendicular to the major structure faces, regardless of the structure's orientation with respect to a major fault, will be used. It is assumed that a structure will have enough lateral strength to resist an *oblique earthquake* when requirements for these two orthogonal directions have been met.

ASCE/SEI7 seems to require an infinite number of analyses by requiring the structure to be designed for forces coming from "...any horizontal direction" [ASCE/SEI7 Sec. 12.5]. However, it is expected that structures designed using the longitudinal and transverse directions of the building will be adequate. Occasionally, individual elements that are part of the lateral force resisting system in both the longitudinal and transverse direction will be verified by combining the two directions vectorially.

A simultaneous application of earthquake forces from two directions could overstress parts of the lateral force-resisting system in certain cases, such as when there is torsional irregularity or a nonparallel structural system, or when a given member—usually a corner column—is part of two intersecting lateral force-resisting systems. ASCE/SEI7 Sec. 12.5 requires an analysis of such directional effects for seismic design categories C, D, E, and F.

39. PROVISIONS FOR HORIZONTAL TORSIONAL MOMENT

The ASCE/SEI7 provisions for torsion apply to rigid diaphragms only (see Sec. 7.4), where increased shears result from horizontal torsion. When the center of mass (CM) does not coincide with the center of rigidity (CR) of the vertical resisting elements in a story, a torsional moment, as shown in Fig. 6.15, is induced. For nonflexible diaphragms, ASCE/SEI7 Sec. 12.8.4.2 requires that a certain amount of accidental torsion (see Sec. 5.14) be planned for, even in regular buildings. Specifically, the center of mass at each level is assumed to be displaced from the calculated center of mass in each direction a distance of 5% of the building dimension at that level perpendicular to the direction of the seismic force. Thus, the accidental eccentricity will be different for the two orthogonal directions.

The accidental torsion is not a minimum value as it was in earlier UBCs. It is a value to be either added to or subtracted from the inherent calculated eccentricity. The term "design eccentricity" is used to represent the algebraic sum of the inherent and accidental (5%) eccentricities ($e_x + e_a$).

The accidental torsion may have to be increased above the 5% level for torsionally irregular structures by use of

Figure 6.15 *Torsional Moments*

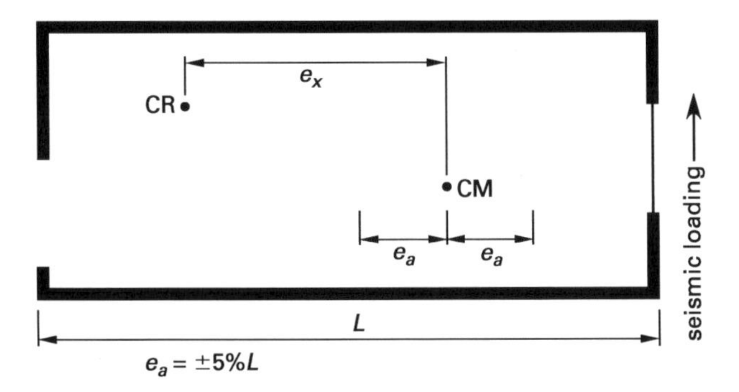

an amplification factor, A_x, determined from Eq. 6.29 [ASCE/SEI7 Eq. 12.8-14].

$$A_x = \left(\frac{\delta_{\text{max}}}{1.2\delta_{\text{avg}}}\right)^2 \qquad 6.29$$

$$A_x \leq 3.0$$

δ_{max} is the maximum displacement at level x, and δ_{avg} is the average of the displacements at the extreme points of the structure at level x.

40. STORY DRIFT DETERMINATION

Story drift is the lateral displacement of one level of a structure relative to the level above or below. In the IBC, drift requirements are based on the strength design method to conform with newly developed seismic base shear forces. In that regard, complete inelastic response drifts rather than force level drifts are used. Based on ASCE/SEI7 Sec. 12.8.6, story drifts should be determined using the maximum inelastic response displacement, δ_x, which is defined as the maximum total drift or total story drift caused by the design-level earthquake. Displacement includes both elastic and inelastic contributions to the total deformation. The maximum inelastic response displacement, δ_x, should be computed from Eq. 6.30 [ASCE/SEI7 Eq. 12.8-15].

$$\delta_x = \frac{C_d \delta_{xe}}{I_e} \qquad 6.30$$

δ_{xe} is a design level elastic response displacement found from the elastic analysis. The resulting deformations, δ_{xe}, should be determined at all critical locations in the structure under consideration. In calculation of δ_{xe}, translational and torsional deflections should be included. Figure 6.16 illustrates the concept behind determination of the elastic and inelastic response deformation δ_x from elastic deformation δ_{xe}.

Table 6.14 [ASCE/SEI7 Table 12.12-1] provides ASCE/SEI7 limitations imposed on the calculated story drift using δ_x. These limitations are intended to control inelastic deformations and potential instabilities in both structural and nonstructural elements that could affect life safety.

Furthermore, based on ASCE/SEI7 Sec. 12.14.8.5, when the simplified static lateral force procedure is used to determine the total design base shear, the drift, δ_x, can be assumed to be 1%.

Figure 6.16 *Elastic and Inelastic Response Deformation to Design Earthquakes*

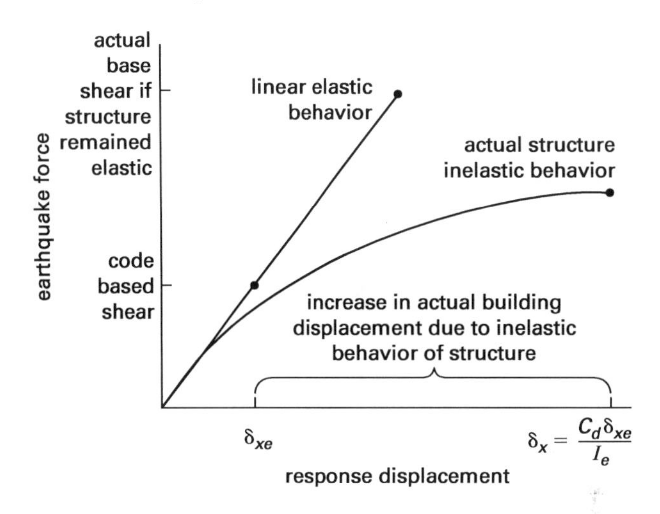

Table 6.14 *Story Drift Limitations [ASCE/SEI7 Table 12.12-1]*

allowable story drift, Δ_a[a,b]

structure	risk category I or II	risk category III	risk category IV
structures, other than masonry shear wall structures, 4 stories or less above the base as defined in ASCE/SEI7 Sec. 11.2, with interior walls, partitions, ceilings, and exterior wall systems designed to accommodate story drifts	$0.025h_{sx}$[c]	$0.020h_{sx}$	$0.015h_{sx}$
masonry cantilever shear wall structures[d]	$0.010h_{sx}$	$0.010h_{sx}$	$0.010h_{sx}$
other masonry shear wall structures	$0.007h_{sx}$	$0.007h_{sx}$	$0.007h_{sx}$
all other structures	$0.020h_{sx}$	$0.015h_{sx}$	$0.010h_{sx}$

[a] h_{sx} is the story height below level x.
[b] For seismic force-resisting systems comprised solely of moment frames in seismic design categories D, E, and F, the allowable story drift must comply with the requirements of ASCE/SEI7 Sec. 12.12.1.1.
[c] There is no drift limit for single-story structures with interior walls, partitions, ceilings, and exterior wall systems that have been designed to accommodate the story drifts. The structure separation requirement of ASCE/SEI7 Sec. 12.12.3 is not waived.
[d] Structures in which the basic structural system consists of masonry shear walls designed as vertical elements cantilevered from their base or foundation support which are so constructed that moment transfer between shear walls (coupling) is negligible.

Example 6.11

A four-story special moment-resisting steel structure is located on a site with rock-like soil in a region with frequent earthquakes. This structure has equal story heights of 12 ft (3.66 m) at each story, a natural period, T, of 0.70 sec, and an importance factor, I_e, of 1.0. The architectural components of the building are designed to accommodate story drift. The elastic displacements at each story are indicated in the following illustration.

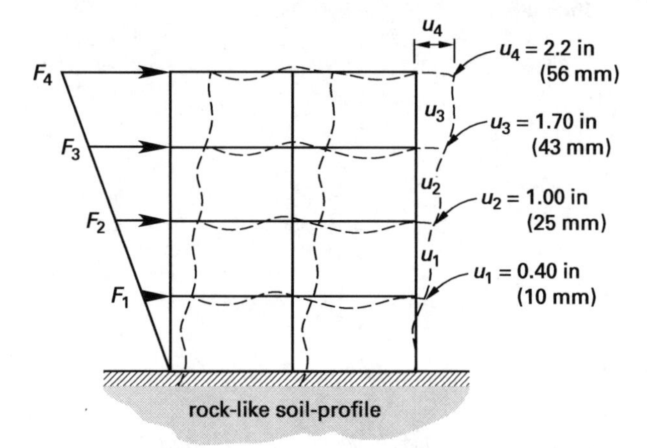

Determine (a) the design level response displacement in the third story, (b) the maximum inelastic response displacement in the first story, (c) the design level response story drift ratio in the top story, (d) the maximum allowable design level response story drift ratio, and (e) the design level response story drift that the building must be designed to tolerate if architectural cladding panels are to be installed at the second story, which can accommodate only 0.71 in (18 mm) of movement.

Customary U.S. Solution

(a) The design level response displacement, δ_{xe}, in the third story is

$$\begin{aligned}\delta_{3e} &= u_3 - u_2 \\ &= 1.7 \text{ in} - 1.0 \text{ in} \\ &= 0.7 \text{ in}\end{aligned}$$

(b) The maximum inelastic response displacement, δ_x, in the first story can be obtained from Eq. 6.30.

$$\delta_x = \frac{C_d \delta_{xe}}{I_e}$$

From App. H [ASCE/SEI7 Table 12.2-1], for special moment-resisting steel frame structures, $C_d = 5.5$. The design level response displacement, δ_{1e}, in the first story is

$$\delta_{1e} = u_1 = 0.4 \text{ in}$$

$$\delta_1 = \frac{(5.5)(0.4 \text{ in})}{1.0} = 2.20 \text{ in}$$

Based on Table 6.14 [ASCE/SEI7 Table 12.12-1], for nonmasonry buildings with four stories, the calculated story drift should not be greater than 0.025 times the story height.

$$\begin{aligned}\Delta_{1,\text{max}} &= 0.025 h_s \\ &= (0.025)(12 \text{ ft})\left(12 \ \frac{\text{in}}{\text{ft}}\right) \\ &= 3.60 \text{ in}\end{aligned}$$

Because $\delta_1 = 2.20$ in does not exceed 3.60 in, it is considered permissible.

(c) The design level response story drift in the top story is

$$\begin{aligned}\delta_{4e} &= u_4 - u_3 \\ &= 2.2 \text{ in} - 1.7 \text{ in} \\ &= 0.5 \text{ in}\end{aligned}$$

The story drift ratio is the story drift divided by the story height.

$$\begin{aligned}\alpha_{4,SR} &= \frac{0.5 \text{ in}}{(12 \text{ ft})\left(12 \ \frac{\text{in}}{\text{ft}}\right)} \\ &= 0.0035\end{aligned}$$

(d) The maximum allowable design level response story drift ratio for each story is

$$\begin{aligned}\alpha_{\text{allow}} &= \frac{0.025 h_s}{h_s} \\ &= 0.025\end{aligned}$$

(e) The external architectural cladding panels and their connections for this structure are designed to accommodate a 0.71 in movement ($\delta_x = 0.71$ in). The corresponding design level response story drift in the second story is

$$\delta_{ex} = \delta_x = \frac{C_d \delta_{2e}}{I_e}$$

$$\begin{aligned}\delta_{2e} &= \frac{\delta_x I_e}{C_d} \\ &= \frac{(0.71 \text{ in})(1.0)}{5.5} \\ &= 0.13 \text{ in}\end{aligned}$$

$$u_2 - u_1 = 0.6 \text{ in} > 0.13 \text{ in} \quad \text{[NG]}$$

The building as designed allows for lateral displacements that are too large.

SI Solution

(a) The design level response displacement, δ_{xe}, in the third story is

$$\begin{aligned}\delta_{3e} &= \Delta_3 - \Delta_2 \\ &= 43 \text{ mm} - 25 \text{ mm} \\ &= 18 \text{ mm}\end{aligned}$$

(b) The maximum inelastic response displacement, δ_x, in the first story can be obtained from Eq. 6.30.

$$\delta_x = \frac{C_d \delta_{xe}}{I_e}$$

From App. H [ASCE/SEI7 Table 12.2-1], for special moment-resisting steel frame structures, $C_d = 5.5$. The design level response displacement, δ_{1e}, in the first story is

$$\begin{aligned}\delta_{1e} &= \Delta_1 \\ &= 10 \text{ mm} \\ \delta_1 &= \frac{(5.5)(10 \text{ mm})}{1.0} \\ &= 55 \text{ mm}\end{aligned}$$

Based on Table 6.12 [ASCE/SEI7 Table 12.12-1], for nonmasonry buildings with four stories, the calculated story drift using δ_1 should not be greater than 0.025 times the story height.

$$\begin{aligned}\delta_{1,\max} &= 0.025 h_s \\ &= (0.025)(3.66 \text{ m})\left(1000 \ \frac{\text{mm}}{\text{m}}\right) \\ &= 92 \text{ mm}\end{aligned}$$

Because $\delta_1 = 55$ mm does not exceed 92 mm, it is considered permissible.

(c) The design level response story drift in the top story is

$$\begin{aligned}\delta_{4e} &= \Delta_4 - \Delta_3 \\ &= 56 \text{ mm} - 43 \text{ mm} \\ &= 13 \text{ mm}\end{aligned}$$

The story drift ratio is the story drift divided by the story height.

$$\begin{aligned}\Delta_{SR} &= \frac{13 \text{ mm}}{(3.66 \text{ m})\left(1000 \ \frac{\text{mm}}{\text{m}}\right)} \\ &= 0.0036\end{aligned}$$

(d) The maximum allowable design level response story drift ratio for each story is

$$\begin{aligned}\Delta_{SR} &= \frac{0.025 h_s}{h_s} \\ &= 0.025\end{aligned}$$

(e) The external architectural cladding panels and their connections for this structure are designed to accommodate a 0.71 in movement ($\delta_x = 18$ mm). The corresponding design level response story drift in the second story is

$$\delta_{ex} = \delta_x = \frac{C_d \delta_{2e}}{I_e}$$

$$\begin{aligned}\delta_{2e} &= \frac{\delta_x I_e}{C_d} = \frac{(18 \text{ mm})(1.0)}{5.5} \\ &= 3 \text{ mm}\end{aligned}$$

$$\Delta_2 - \Delta_1 = 15 \text{ mm} > 3 \text{ mm} \quad \text{[NG]}$$

The building as designed allows for lateral displacements that are too large.

41. PROVISIONS FOR BUILDING SEPARATIONS

The ASCE/SEI7 requires all structures to be separated from adjoining structures. (See Fig. 6.17 and Fig. 6.18.) *Separations* should permit structures to react to seismic forces independently, so that damaging impact between adjacent structures' elements will be avoided. According to ASCE/SEI7 Sec. 12.12.3, the separation should allow for the maximum inelastic response displacement, δ_M, according to Eq. 6.31 [ASCE/SEI7 Eq. 12.12-1], calculated individually for each building. In Eq. 6.31, C_d is the *deflection amplification factor* that scales up elastic deflection to inelastic deflection. Values of C_d are found in App. H [ASCE/SEI7 Table 12.2-1]. $\delta_{\max}$ is the maximum elastic deflection at the critical location.

$$\delta_M = \frac{C_d \delta_{\max}}{I_e} \qquad 6.31$$

However, from Eq. 6.32 [ASCE/SEI7 Eq. 12.12-2], for adjoining structures located on the same property, the separation should be reduced to δ_{MT} to account for potential movement due to out-of-phase building vibration. Therefore, the minimum required separation between structures is

$$\delta_{MT} = \sqrt{\delta_{M1}^2 + \delta_{M2}^2} \qquad 6.32$$

Figure 6.17 Building Separation

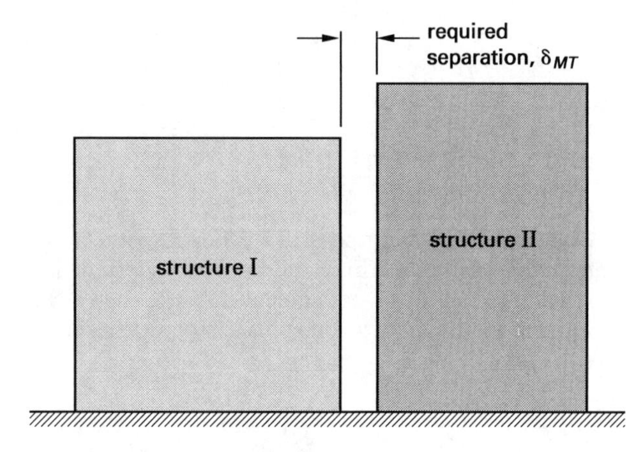

Figure 6.18 Inadequate Building Separation

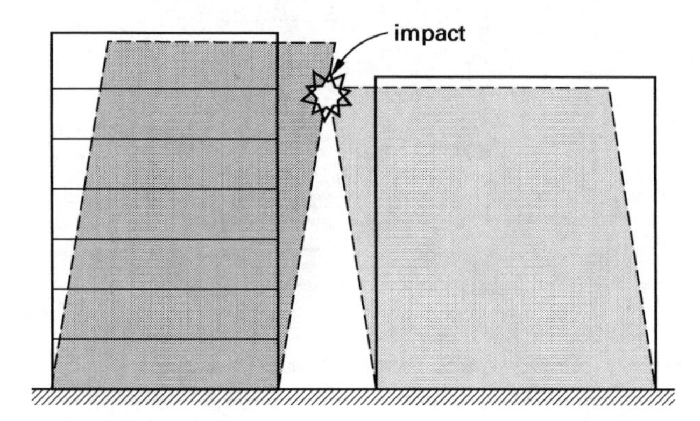

Example 6.12

Two office buildings will be built adjacent to each other on the same property in Northern California. A special steel concentrically braced frame as a dual system with a special moment-resisting steel frame with an importance factor, I_e, of 1.25 is proposed for structure I. Structure II will be a special concentrically steel braced frame with an importance factor, I_e, of 1.0. The elastic displacements for the two buildings are known. What should be the sufficient distance between all parts of these two buildings to avoid damaging impact while the buildings are responding to earthquake motion, independently?

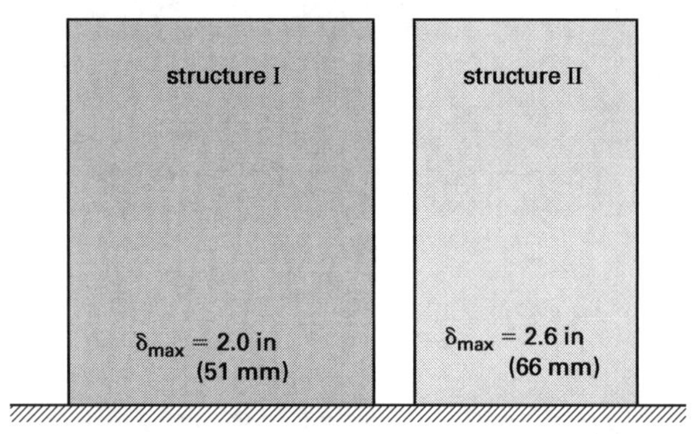

Customary U.S. Solution

Based on ASCE/SEI7 Sec. 12.12.3, all structures are required to be separated from adjoining structures on the same property, and the separation should be based on the maximum inelastic response displacement, δ_M, for each building.

The elastic displacements of these structures are 2.0 in and 2.6 in, respectively. From App. H [ASCE/SEI7 Table 12.2-1], for a special concentrically braced dual system consisting of steel bracing and special moment-resisting steel frames, $C_d = 5.5$. For a special concentrically steel braced frame, $C_d = 5.0$. Rearranging from Eq. 6.31, for structure I,

$$\begin{aligned}\delta_{M1} &= \frac{C_d \delta_{\max}}{I_e} \\ &= \frac{(5.5)(2.0 \text{ in})}{1.25} \\ &= 8.8 \text{ in}\end{aligned}$$

For structure II,

$$\begin{aligned}\delta_{M2} &= \frac{C_d \delta_{\max}}{I_e} \\ &= \frac{(5.0)(2.6 \text{ in})}{1.0} \\ &= 13.0 \text{ in}\end{aligned}$$

From Eq. 6.32, the minimum required separation is

$$\begin{aligned}\delta_{MT} &= \sqrt{\delta_{M1}^2 + \delta_{M2}^2} \\ &= \sqrt{(8.8 \text{ in})^2 + (13.0 \text{ in})^2} \\ &= 15.7 \text{ in}\end{aligned}$$

SI Solution

Based on ASCE/SEI7 Sec. 12.12.3, all structures are required to be separated from adjoining structures on the same property, and the separation should be based on the maximum inelastic response displacement, δ_M, for each building.

The elastic displacements of these structures are 51 mm and 66 mm, respectively. From App. H [ASCE/SEI7 Table 12.2-1], for a special concentrically braced dual system consisting of steel bracing and special moment-resisting steel frames, $C_d = 5.5$. For a special concentrically steel braced frame, $C_d = 5.0$. Rearranging from Eq. 6.31, for structure I,

$$\begin{aligned}\delta_{M1} &= \frac{C_d \delta_{\max}}{I_e} \\ &= \frac{(5.5)(51 \text{ mm})}{1.25} \\ &= 224 \text{ mm}\end{aligned}$$

For structure II,

$$\delta_{M2} = \frac{C_d\delta_{max}}{I_e}$$
$$= \frac{(5.0)(66 \text{ mm})}{1.0}$$
$$= 330 \text{ mm}$$

From Eq. 6.32, the minimum required separation is

$$\delta_{MT} = \sqrt{\delta_{M1}^2 + \delta_{M2}^2}$$
$$= \sqrt{(224 \text{ mm})^2 + (330 \text{ mm})^2}$$
$$= 399 \text{ mm}$$

42. PROVISIONS FOR OVERTURNING MOMENT

Building codes require that every designed structure be capable of resisting overturning effects induced by earthquake forces. At any level, the overturning moment (OTM) must be determined using the seismic forces, F_x, defined in Eq. 6.25, that act on all of the levels above the level under consideration.

$$\text{OTM} = \sum F_x h_x \quad 6.33$$

The overturning effects on every lateral force resisting element must be carried down to the foundation. The incremental increases of the design overturning moment at each higher level should be distributed to the various lateral force resisting elements. The distribution should be in the same proportion as the distribution of the horizontal shears to those resisting elements. The effects of uplift caused by seismic loads must also be analyzed. Any net tension must be resisted by interaction with the soil (e.g., by use of friction piles that resist uplift).

When calculating the overturning effects at the soil-foundation interface, including the calculation of soil pressure under typical footings and the soil-pile frictional forces during uplift, only 75% of the foundation overturning design moment, OTM_f, need be resisted [ASCE/SEI7 Sec. 12.13.4]. This 25% reduction is permitted because the overturning moment includes forces that represent higher mode forces, and the moments at the base associated with the higher modes are unlikely to occur simultaneously with mode 1 response. (See Fig. 6.19.)

43. PROVISIONS FOR *P*-DELTA EFFECTS

The *P*-delta effect is defined as the secondary effect on shears, axial forces, and moments of frame members induced by the gravity loads acting on the laterally displaced structure frame. For the overall structural frame stability evaluation, it is required that the forces producing the displacements of the design story drift, Δ_x, be used.

Figure 6.19 *Overturning Moment at the Base*

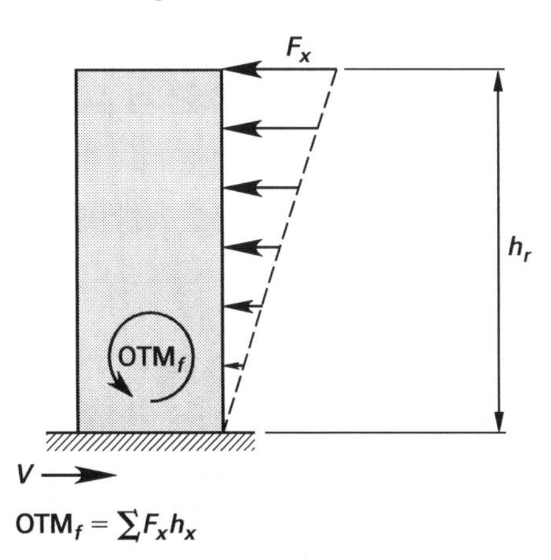

The ASCE/SEI7 uses the concept of a *stability coefficient*, θ, which is the ratio of secondary moments to primary moments.[30] If this stability coefficient is equal to or less than 0.10, the *P*-delta effects can be disregarded. Otherwise, a rational analysis must be used to evaluate the *P*-delta effects [ASCE/SEI7 Sec. 12.8.7]. Equation 6.34 is used to calculate θ at each level [ASCE/SEI7 Eq. 12.8-16]. Variables are illustrated in Fig. 6.20.

$$\theta = \frac{P_x \Delta I_e}{V_x h_{sx} C_d} \quad 6.34$$

Figure 6.20 *Stability Coefficient Variables*

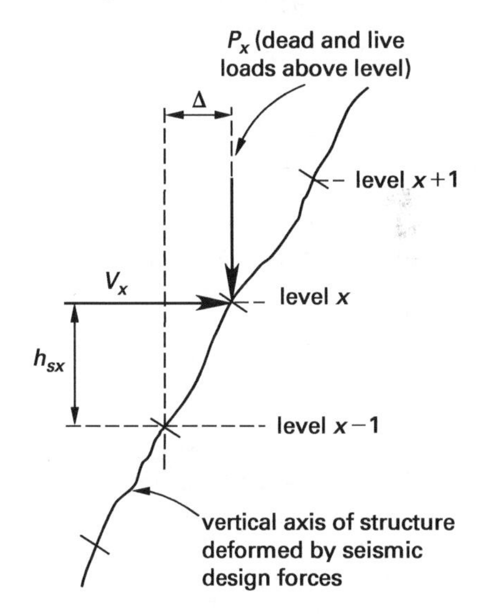

Example 6.13

Columns of an ordinary moment-resisting steel frame (OMRF) and a special moment-resisting steel frame (SMRF) are in seismic design categories B and D, respectively, and are loaded and respond as shown. Both buildings have an importance factor, I_e, of 1.0. Should the *P*-delta effects be considered for (a) the ordinary moment-resisting steel frame and/or (b) the special moment-resisting steel frame?

[30]The ratio may be evaluated for any story using the total dead load, floor live load, and snow load above the story.

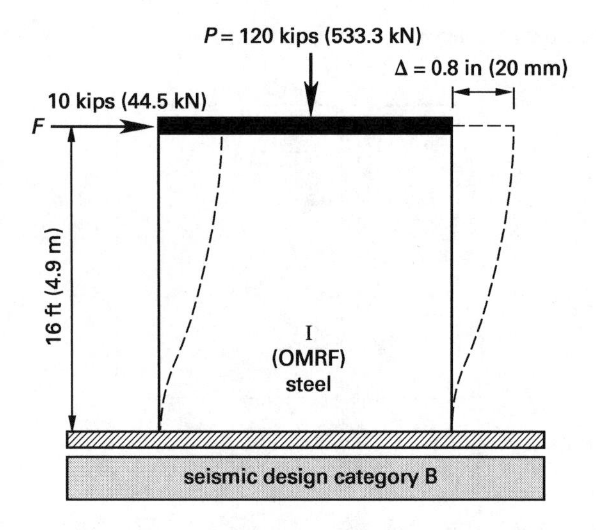

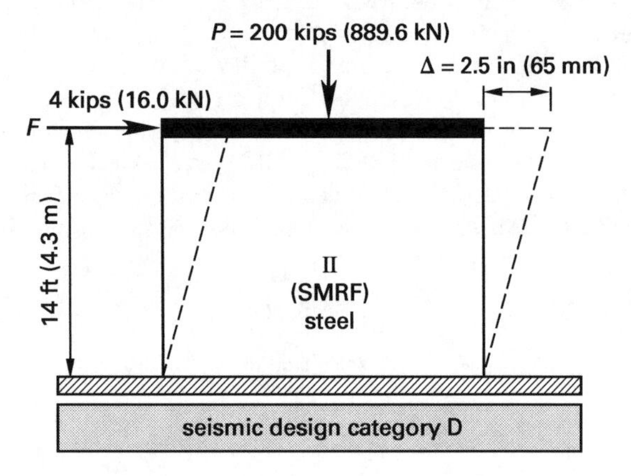

Customary U.S. Solution

(a) Structure I is located in seismic design category B. Based on ASCE/SEI7 Sec. 12.8.7, the *P*-delta effects (i.e., the effects of the secondary moments) should be considered in the evaluation of overall structural stability when the ratio of secondary moment to primary moment exceeds 0.1 for any story.

For structure I, from App. H [ASCE/SEI7 Table 12.2-1] for ordinary moment-resisting steel frames, $C_d = 3.0$. The design story drift, Δ, is 0.80 in, and $h_s = 16$ ft. The lateral seismic force, *F*, is 10 kips, and the axial load, *P*, is 120 kips. From Eq. 6.34 [ASCE/SEI7 Eq. 12.8-16],

$$\theta = \frac{P_x \Delta I_e}{V_x h_{sx} C_d}$$

$$= \frac{(120 \text{ kips})(0.8 \text{ in})(1.0)}{(10 \text{ kips})(16 \text{ ft})\left(12 \ \frac{\text{in}}{\text{ft}}\right)(3.0)}$$

$$= 0.017 \text{ radians}$$

Since $\theta = 0.017 < 0.1$, *P*-delta effects do not need to be considered for structure I.

(b) Based on ASCE/SEI7 Sec. 12.8.7, the *P*-delta effects need to be considered in the evaluation of overall structural stability if the value of θ exceeds 0.1 for any story.

From App. H [ASCE/SEI7 Table 12.2-1], for special moment-resisting steel frames, $C_d = 5.5$. The design story drift, Δ, is 2.5 in, and $h_s = 14$ ft. The lateral seismic force, *F*, is 4 kips, and the axial load, *P*, is 200 kips. From Eq. 6.34 [ASCE/SEI7 Eq. 12.8-16],

$$\theta = \frac{P_x \Delta I_e}{V_x h_{sx} C_d}$$

$$= \frac{(200 \text{ kips})(2.5 \text{ in})(1.0)}{(4 \text{ kips})(14 \text{ ft})\left(12 \ \frac{\text{in}}{\text{ft}}\right)(5.5)}$$

$$= 0.135 \text{ radians}$$

Since 0.135 > 0.10, *P*-delta effects should be considered for structure II.

SI Solution

(a) Structure I is located in seismic design category B. Based on ASCE/SEI7 Sec. 12.8.7, the *P*-delta effects (i.e., the effects of the secondary moments) should be considered in the evaluation of overall structural stability when the ratio of secondary moment to primary moment exceeds 0.1 for any story.

For structure I, from App. H [ASCE/SEI7 Table 12.2-1] for ordinary moment-resisting steel frames, $C_d = 3.0$. The design story drift, Δ, is 20 mm, and $h_s = 4.9$ m. The lateral seismic force, *F*, is 44.5 kN, and the axial load, *P*, is 533.3 kN. From Eq. 6.34 [ASCE/SEI7 Eq. 12.8-16],

$$\theta = \frac{P_x \Delta I_e}{V_x h_{sx} C_d}$$

$$= \frac{(533.3 \text{ kN})(20 \text{ mm})(1.0)}{(44.5 \text{ kN})(4.9 \text{ m})\left(1000 \ \frac{\text{mm}}{\text{m}}\right)(3.0)}$$

$$= 0.016 \text{ radians}$$

Since $\theta = 0.016 < 0.1$, *P*-delta effects do not need to be considered for structure I.

(b) Based on ASCE/SEI7 Sec. 12.8.7, the *P*-delta effects need to be considered in the evaluation of overall structural stability if the value of θ exceeds 0.1 for any story.

From App. H [ASCE/SEI7 Table 12.2-1], for special moment-resisting steel frames, $C_d = 5.5$. The design story drift, Δ, is 65 mm, and $h_s = 4.3$ m. The lateral seismic force, F, is 16 kN, and the axial load, P, is 889.6 kN. From Eq. 6.34 [ASCE/SEI7 Eq. 12.8-16],

$$\begin{aligned}\theta &= \frac{P_x \Delta I_e}{V_x h_{sx} C_d} \\ &= \frac{(889.6\text{ kN})(65\text{ mm})(1.0)}{(16\text{ kN})(4.3\text{ m})\left(1000\ \frac{\text{mm}}{\text{m}}\right)(5.5)} \\ &= 0.153\text{ radians}\end{aligned}$$

Since $0.153 > 0.10$, P-delta effects should be considered for structure II.

44. SEISMIC FORCE ON NONSTRUCTURAL COMPONENTS

ASCE/SEI7 Chap. 13 discusses nonstructural architectural and mechanical/electrical components. Nonstructural components are permanently installed building items that do not assist in the primary load carrying resistance of the building.

Examples of architectural components are unbraced parapet walls, interior bearing and nonbearing walls, retaining walls, partitions, and penthouses. Mechanical/electrical components include tanks and vessels with support systems; electrical, mechanical, and plumbing equipment; and emergency communication equipment. Table 6.15 [ASCE/SEI7 Table 13.5-1] provides a list of these categories.

To ensure life safety and preserve the functionality of essential facilities, these components should be designed to resist the total design seismic forces of ASCE/SEI7 Sec. 13.3. The seismic loading, F_p, on these elements and their attachments can be calculated from Eq. 6.35 [ASCE/SEI7 Eq. 13.3-1].

$$F_p = \left(\frac{0.4 a_p S_{DS} W_p}{\frac{R_p}{I_p}}\right)\left(1 + \frac{2z}{h_r}\right) \qquad 6.35$$

S_{DS} is the spectral acceleration for short period, I_p is the importance factor, a_p is the amplification factor of the component, R_p is the response modification factor associated with the component, and W_p is the *operating weight* of the component.

The minimum and maximum values of F_p are given by Eq. 6.36 [ASCE/SEI7 Eq. 13.3-2 and Eq. 13.3-3].

$$0.3 S_{DS} I_p W_p \le F_p \le 1.6 S_{DS} I_p W_p \qquad 6.36$$

The value of the spectral acceleration, S_{DS}, is the same as that used for the building, and depends on the location, proximity of the site to active seismic sources, and site soil-profile characteristics. The value of the seismic importance factor, I_p, is either 1.0 or 1.5, depending on the function of the component. ASCE/SEI7 Sec. 13.1.3 assigns a value of $I_p = 1.5$ for anchoring machinery and equipment required for life-safety systems, as well as for tanks containing toxic and explosive substances. The value of I_p is also 1.5 if the component is part of a risk category IV structure and is needed for the continued operation of the structure. The natural period of the elements, components, and equipment is not a factor in calculating the force on the elements, components, and equipment, except indirectly in the determination of a_p. All other components are given a value of 1.0.

The values of the *component amplification factor*, a_p, are given in Table 6.15 [ASCE/SEI7 Table 13.5-1]. The values of a_p vary from 1.0 to 2.5. Rigid components experience little amplification and are assigned an a_p of 1.0. Alternatively, dynamic properties or empirical data of the component and the structure that supports it may be used to determine this factor; however, its value may not be less than 1.0.

Values of the *component response modification factor*, R_p, are furnished in Table 6.15 [ASCE/SEI7 Table 13.5-1]. The maximum value of R_p is 3.5, and is reserved for penthouses and elements that allow high deformability. Most components and equipment are allowed an R_p of 2.5. A value of $R_p = 1.5$ is used for equipment that only allows low levels of deformability. For fasteners of exterior nonstructural wall elements, $R_p = 1.0$.

Equation 6.35 includes an amplification factor of $(1 + 2z/h_r)$ based on the relative height of the components in the building. z and h_r signify elevations with respect to grade of the component in the building. z denotes the height of an element or component attachment above base, and h_r represents the elevation of the structure roof above base. The resulting factor is 1.0 at the base and 3.0 at the roof. For example, for a retaining wall, from Eq. 6.35, at the top of the wall, $z = h$, $a_p = 1.0$, and $R_p = 2.5$. Therefore, $F_p/W_p = 0.48 S_{DS}$. At the base of the wall, $z = 0$ and $F_p/W_p = 0.16 S_{DS}$. However, per ASCE/SEI7 Eq. 13.3-3, F_p/W_p cannot be less than $0.3 S_{DS}$. Therefore, the average for design, assuming uniform over full height is $(0.48 + 0.30)/2 = 0.39 S_{DS}$.

The design seismic forces obtained from Eq. 6.35 should be distributed in proportion to the mass distribution of the element or component. The computed seismic forces, F_p, are to be used to design the members and connections that transfer these forces into the structure seismic-resisting systems.

Exterior cladding panels must resist a seismic force F_p calculated from Eq. 6.35, and they must accommodate relative seismic displacements of the structure, D_p, as calculated from ASCE/SEI7 Sec. 13.3.2, or 0.5 in (13 mm), whichever is greater. Panel connectors (e.g., bolts, inserts, welds, and dowels) and connection bodies (e.g., angles, bars, and plates) are designed for the seismic force F_p determined from Eq. 6.35, where for connectors, the values of R_p and a_p are from Table 6.15 [ASCE/SEI7 Sec. 13.5.3, Item d]. Drift, movement, and other standards also apply [ASCE/SEI7 Sec. 13.5.3].

Table 6.15 *Coefficients for Architectural Components [ASCE/SEI7 Table 13.5-1]*

architectural component or element	a_p[a]	R_p
interior nonstructural walls and partitions[b]		
plain (unreinforced) masonry walls	1.0	1.5
all other walls and partitions	1.0	2.5
cantilever elements (unbraced or braced to structural frame below its center of mass)		
parapets and cantilever interior nonstructural walls	2.5	2.5
chimneys and stacks where laterally braced or supported by the structural frame	2.5	2.5
cantilever elements (braced to structural frame above its center of mass)		
parapets	1.0	2.5
chimneys and stacks	1.0	2.5
exterior nonstructural walls[b]	1.0[b]	2.5
exterior nonstructural wall elements and connections[b]		
wall element	1.0	2.5
body of wall panel connections	1.0	2.5
fasteners of the connecting system	1.25	1.0
veneer		
limited deformability elements and attachments	1.0	2.5
low deformability elements and attachments	1.0	1.5
penthouses (except where framed by an extension of the building frame)	2.5	3.5
ceilings		
all	1.0	2.5
cabinets		
permanent floor-supported storage cabinets over 6 ft (1829 mm) tall, including contents	1.0	2.5
permanent floor-supported library shelving, book stacks, and bookshelves over 6 ft (1829 mm) tall, including contents	1.0	2.5
laboratory equipment	1.0	2.5
access floors		
special access floors (designed in accordance with ASCE/SEI7 Sec. 13.5.7.2)	1.0	2.5
all other	1.0	1.5
appendages and ornamentations	2.5	2.5
signs and billboards	2.5	3.0
other rigid components		
high deformability elements and attachments	1.0	3.5
limited deformability elements and attachments	1.0	2.5
low deformability materials and attachments	1.0	1.5
other flexible components		
high deformability elements and attachments	2.5	3.5
limited deformability elements and attachments	2.5	2.5
low deformability materials and attachments	2.5	1.5
egress stairways not part of the building structure	1.0	2.5

[a]A lower value for a_p must not be used unless justified by detailed dynamic analysis. The value for a_p must not be less than 1.00. The value of $a_p = 1$ is for rigid components and rigidly attached components. The value of $a_p = 2.5$ is for flexible components and flexibly attached components. See ASCE/SEI7 Sec. 11.2 for definitions of rigid and flexible.

[b]Where flexible diaphragms provide lateral support for concrete or masonry walls and partitions, the design forces for anchorage to the diaphragm shall be as specified in ASCE/SEI7 Sec. 12.11.2.

Example 6.14

A parapet wall extends 4 ft (1.22 m) above the roof line of a one-story reinforced concrete-walled commercial building, as shown. The walls are 8 in (203 mm) thick. Consider all connections between the roof, walls, and footings to be pinned (free to rotate). The building is located on a site with a stiff soil profile where $S_S = 1.87$ and $S_1 = 0.89$. Consider earthquake forces perpendicular to the wall face. Determine (a) the location and magnitude of the design force that the wall must be designed for based on out-of-plane forces defined in ASCE/SEI7 Sec. 12.11.1 and (b) the moment at the base of the parapet wall based on nonstructural wall element requirements given in ASCE/SEI7 Chap. 13.

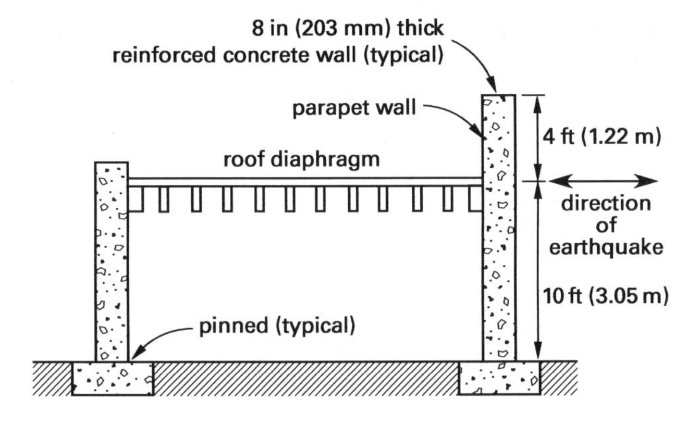

Customary U.S. Solution

(a) The wall length is not given, so work with a 1 ft strip of wall. Reinforced concrete has a mass of approximately 150 lbf/ft^3, so the distributed weight per foot of width is

$$\begin{aligned} w_p &= \gamma(\text{volume}) \\ &= \gamma(\text{width})(\text{height})(\text{thickness}) \\ &= \frac{\left(150\ \frac{\text{lbf}}{\text{ft}^3}\right)(1\ \text{ft})(14\ \text{ft})(8\ \text{in})}{12\ \frac{\text{in}}{\text{ft}}} \\ &= 1400\ \text{lbf}\quad [\text{per foot of wall}] \end{aligned}$$

The entire wall must first be analyzed. From Table 6.3, for a stiff soil profile, the site class is D. From Table 6.4, for $S_S = 1.87$ and site class D, F_a is 1.0. Using Eq. 6.6 [IBC Eq. 16-37],

$$\begin{aligned} S_{MS} &= F_a S_S = (1.0)(1.87) \\ &= 1.87 \end{aligned}$$

From Eq. 6.8 [IBC Eq. 16-39],

$$\begin{aligned} S_{DS} &= \tfrac{2}{3} S_{MS} = \left(\tfrac{2}{3}\right)(1.87) \\ &= 1.247 \end{aligned}$$

From ASCE/SEI7 Sec. 12.11.1 (also, see Sec. 7.2), the distributed seismic force per foot of wall width need not exceed

$$\begin{aligned} F_p &= 0.4 S_{DS} I_E w_p \\ &= (0.4)(1.247)(1.0)(1400\ \text{lbf}) \\ &= 698\ \text{lbf}\quad [\text{per foot of wall}] \end{aligned}$$

F_p must be at least 10% of the wall weight.

$$\begin{aligned} F_{p,\text{min}} &= (0.10)(1400\ \text{lbf}) \\ &= 140\ \text{lbf} \end{aligned}$$

Since $F_p = 698$ lbf is greater than $F_{p,\text{min}} = 140$ lbf, the value of 698 lbf governs. Use this as the design force.

For the purpose of finding reactions, this force can be assumed to act at the mid-height of the wall, at

$$\left(\tfrac{1}{2}\right)(14\ \text{ft}) = 7\ \text{ft}$$

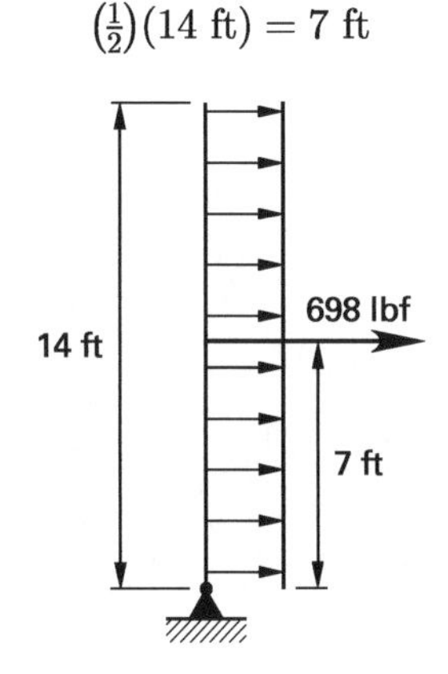

(b) The weight of the parapet alone is

$$\begin{aligned} W_p &= \gamma(\text{volume}) \\ &= \gamma(\text{width})(\text{height})(\text{thickness}) \\ &= \frac{\left(150\ \frac{\text{lbf}}{\text{ft}^3}\right)(1\ \text{ft})(4\ \text{ft})(8\ \text{in})}{12\ \frac{\text{in}}{\text{ft}}} \\ &= 400\ \text{lbf}\quad [\text{per foot of wall}] \end{aligned}$$

From Table 6.15, the values of a_p and R_p are both 2.5 for cantilever parapets. From ASCE/SEI7 Sec. 13.1.3, the seismic importance factor, I_p, is 1.0 because the parapet is not part of the life-safety protection system for the building.

From Eq. 6.35, the distributed seismic force per foot of parapet wall width is

$$F_p = \left(\frac{0.4 a_p S_{DS} W_p}{\frac{R_p}{I_p}} \right) \left(1 + \frac{2z}{h_r} \right)$$

$$= \left(\frac{(0.4)(2.5)(1.247)(400 \text{ lbf})}{\frac{2.5}{1.0}} \right) \left(1 + \frac{(2)(10 \text{ ft})}{10 \text{ ft}} \right)$$

$$= 599 \text{ lbf} \quad \text{[per foot of wall]}$$

The maximum design force is given by ASCE/SEI7 Eq. 13.3-2.

$$\begin{aligned} F_{p,\max} &= 1.6 S_{DS} I_p W_p \\ &= (1.6)(1.247)(1.0)(400 \text{ lbf}) \\ &= 798 \text{ lbf} \end{aligned}$$

The minimum design force is given by ASCE/SEI7 Eq. 13.3-3.

$$\begin{aligned} F_{p,\min} &= 0.3 S_{DS} I_p W_p \\ &= (0.3)(1.247)(1.0)(400 \text{ lbf}) \\ &= 150 \text{ lbf} \end{aligned}$$

Since $F_p = 599$ lbf is between $F_{p,\max} = 798$ lbf and $F_{p,\min} = 150$ lbf, the parapet is to be designed for $F_p = 599$ lbf.

For the purpose of determining the moment at the base of the parapet, this force can be assumed to act at the mid-height, 2 ft up from the base. In effect, the parapet acts as a vertical cantilever wall. The net moment at the parapet base (i.e., where it joins the roof) is

$$\begin{aligned} M &= (2 \text{ ft})(599 \text{ lbf}) \\ &= 1198 \text{ ft-lbf} \quad \text{[per foot of wall]} \end{aligned}$$

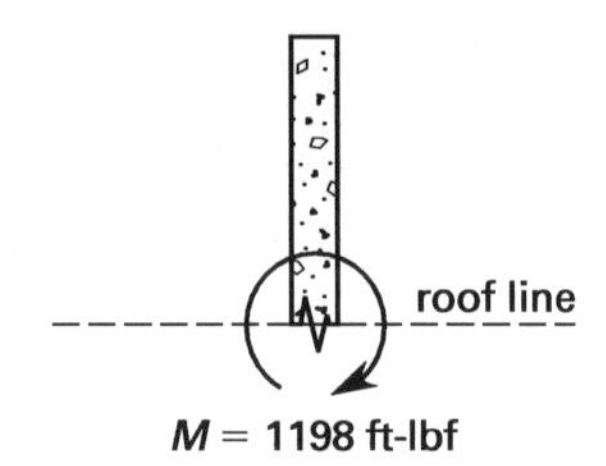

SI Solution

(a) The wall length is not given, so work with a 1 m strip of wall. Reinforced concrete has a mass of approximately 2400 kg/m^3, so the distributed weight per meter of width is

$$\begin{aligned} w_p &= \gamma(\text{volume}) \\ &= \gamma(\text{width})(\text{height})(\text{thickness}) \\ &= \frac{\left(2400 \ \frac{\text{kg}}{\text{m}^3}\right)\left(9.81 \ \frac{\text{m}}{\text{s}^2}\right)(1 \text{ m})(4.27 \text{ m})(203 \text{ mm})}{\left(1000 \ \frac{\text{mm}}{\text{m}}\right)\left(1000 \ \frac{\text{N}}{\text{kN}}\right)} \\ &= 20.4 \text{ kN} \quad \text{[per meter of wall]} \end{aligned}$$

The entire wall must first be analyzed. From Table 6.3, for stiff soil-profile, the site class is D. From Table 6.4, for $S_S = 1.87$ and site class D, F_a is 1.0. From Eq. 6.6 [IBC Eq. 16-37],

$$\begin{aligned} S_{MS} &= F_a S_S = (1.0)(1.87) \\ &= 1.87 \end{aligned}$$

From Eq. 6.8 [IBC Eq. 16-39],

$$\begin{aligned} S_{DS} &= \tfrac{2}{3} S_{MS} = \left(\tfrac{2}{3}\right)(1.87) \\ &= 1.247 \end{aligned}$$

From ASCE/SEI7 Sec. 12.11.1 (also, see Sec. 7.2), the distributed seismic force per meter of wall width need not exceed

$$\begin{aligned} F_p &= 0.4 S_{DS} I_E w_p \\ &= (0.4)(1.247)(1.0)(20.4 \text{ kN}) \\ &= 10.2 \text{ kN} \quad \text{[per meter of wall]} \end{aligned}$$

F_p must be at least 10% of the wall weight.

$$\begin{aligned} F_{p,\min} &= (0.10)(20.4 \text{ kN}) \\ &= 2.04 \text{ kN} \end{aligned}$$

Since F_p = 10.2 kN is greater than $F_{p,\min} = 2.04$ kN, the value of 10.2 kN governs.

For the purpose of finding reactions, this force can be assumed to act at the mid-height of the wall, at

$$\left(\tfrac{1}{2}\right)(4.27 \text{ m}) = 2.14 \text{ m}$$

4.27 m
10.2 kN
2.14 m

(b) The weight of the parapet alone is

$$\begin{aligned} W_p &= \gamma(\text{volume}) \\ &= \gamma(\text{width})(\text{height})(\text{thickness}) \\ &= \frac{\left(2400\ \frac{\text{kg}}{\text{m}^3}\right)\left(9.81\ \frac{\text{m}}{\text{s}^2}\right)(1\ \text{m})(1.22\ \text{m})(203\ \text{mm})}{\left(1000\ \frac{\text{mm}}{\text{m}}\right)\left(1000\ \frac{\text{N}}{\text{kN}}\right)} \\ &= 5.83\ \text{kN} \quad \text{[per meter of wall]} \end{aligned}$$

From Table 6.15, the values of a_p and R_p are both 2.5 for cantilever parapets. From ASCE/SEI7 Sec. 13.1.3, the seismic importance factor, I_p, is 1.0 because the parapet is not part of the life-safety protection system for the building.

From Eq. 6.36, the distributed seismic force per meter of parapet wall width is

$$\begin{aligned} F_p &= \left(\frac{0.4a_pS_{DS}W_p}{\frac{R_p}{I_p}}\right)\left(1+\frac{2z}{h_r}\right) \\ &= \left(\frac{(0.4)(2.5)(1.247)(5.83\ \text{kN})}{\frac{2.5}{1.0}}\right)\left(1+\frac{(2)(3.05\ \text{m})}{3.05\ \text{m}}\right) \\ &= 8.72\ \text{kN} \quad \text{[per meter of wall]} \end{aligned}$$

The maximum design force is given by ASCE/SEI7 Eq. 13.3-2.

$$\begin{aligned} F_{p,\text{max}} &= 1.6S_{DS}I_pW_p \\ &= (1.6)(1.247)(1.0)(5.83\ \text{kN}) \\ &= 11.6\ \text{kN} \end{aligned}$$

The minimum design force is given by ASCE/SEI7 Eq. 13.3-3.

$$\begin{aligned} F_{p,\text{min}} &= 0.3S_{DS}I_pW_p \\ &= (0.3)(1.247)(1.0)(5.83\ \text{kN}) \\ &= 2.18\ \text{kN} \end{aligned}$$

Since $F_p = 8.72$ kN is between $F_{p,\text{max}} = 11.6$ kN and $F_{p,\text{min}} = 2.18$ kN, the parapet is to be designed for $F_p = 8.72$ kN.

For the purpose of determining the moment at the base of the parapet, this force can be assumed to act at the mid-height, 0.61 m up from the base. In effect, the parapet acts as a vertical cantilever wall. The net moment at the parapet base (i.e., where it joins the roof) is

$$\begin{aligned} M &= (0.61\ \text{m})(8.72\ \text{kN}) \\ &= 5.32\ \text{kN·m} \quad \text{[per meter of wall]} \end{aligned}$$

roof line

$M = 5.32$ kN·m

45. PROVISIONS FOR COLLECTORS AND STRUCTURAL CONTINUITY

ASCE/SEI7 Sec. 12.10.2 is concerned with ensuring continuity in designed structures. The provisions are intended to provide a minimum design force for connections tying portions of a structure together, for example, collectors at projecting wings, or beam splices. Often, such collector forces are difficult to calculate by rational method without separating the structure or system into two parts at the connection being analyzed. The ASCE/SEI7 specifies a minimum continuity force to use in designing such connections, and requires that the minimum design load include overstrength factors as designated in Sec. 12.4.3 of the ASCE/SEI7.

46. NONBUILDING STRUCTURES

Nonbuilding structures [ASCE/SEI7 Chap. 15] are self-supporting structures other than buildings that, nevertheless, come under the jurisdiction of the local building official.[31] (See Fig. 6.21.)

Most nonbuilding structures, even though they are not designed to accommodate people, are supported by structural systems traditionally found in occupied building structures. Nonbuilding structures carry gravity loads and resist the effects of earthquakes. Consequently, the strength required to resist the displacements caused by the minimum design seismic forces on nonbuilding structures is calculated in the same manner as those for building structures. For nonbuilding structures, as opposed to building structures, the standard building drift limitations need not be met. However, P-δ effects should be evaluated.

There are some differences, however, in the total design base shear equations for building and nonbuilding structures. The natural period, T, must be determined by rational methods.

[31]Other items specifically not included in the building code are nuclear power generation plants, dams, and highway and railroad bridges. These structures are not normally within the jurisdiction of the local building official.

Figure 6.21 *Typical Nonbuilding Structures (not similar to buildings)**

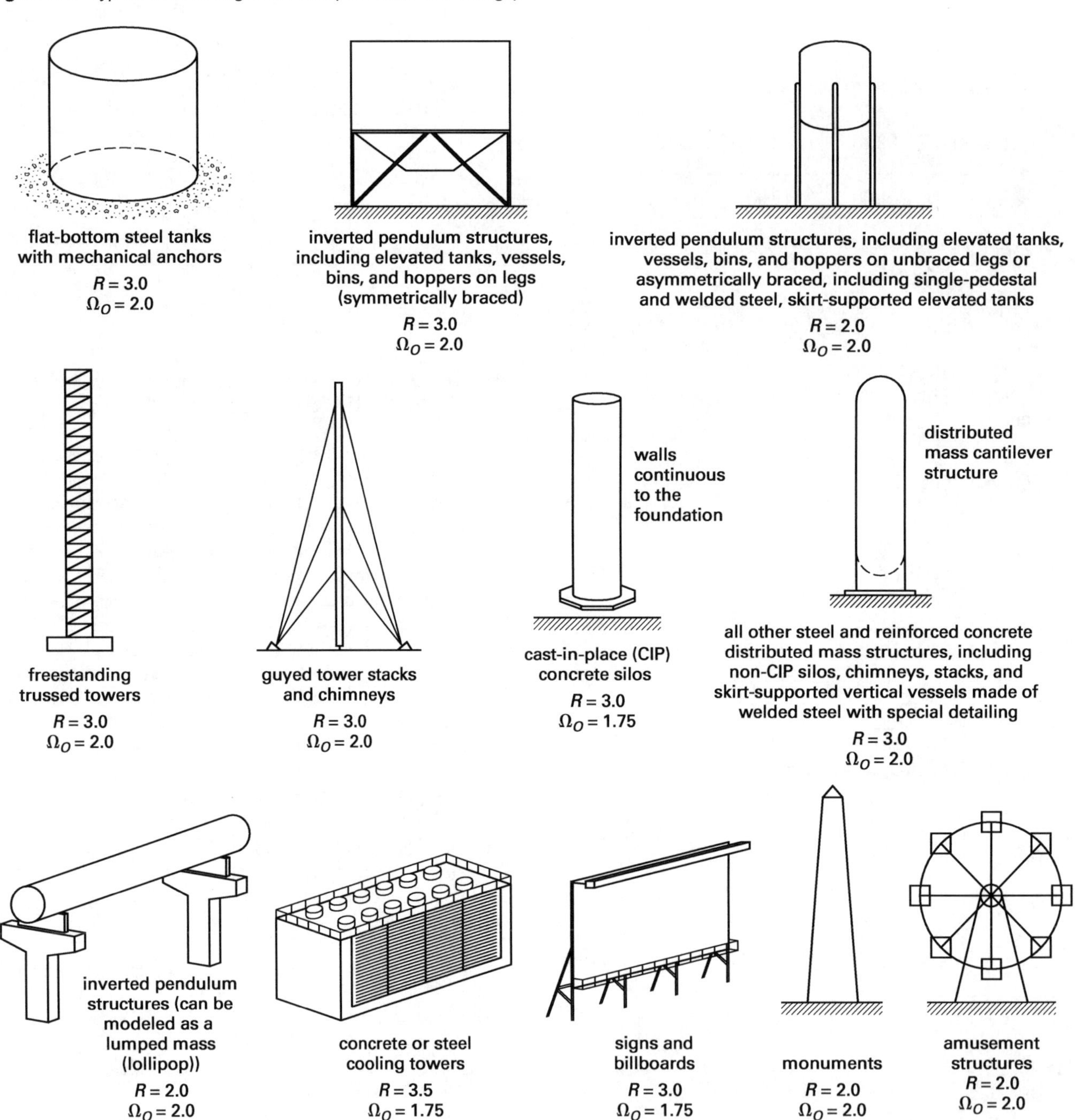

*Based on ASCE/SEI7 Table 15.4-2, which should be consulted in order to distinguish between structure categories.

The total design base shear given in Eq. 6.23 is used for nonbuilding structures. In locations where $S_1 > 0.6g$, C_s must also be equal or greater than the value of Eq. 6.37. For reduction in these forces, the values of the response modification factor, R, and seismic force amplification factor, Ω_O, are given in Table 6.16 [ASCE/SEI7 Table 15.4-2] for nonbuilding structures that are not similar to buildings. (ASCE/SEI7 Table 15.4-1 gives values for Ω_O and R for structures that are similar to buildings.) Generally speaking, the values of R assigned to nonbuilding structures are less than for building structures. This is considered justified because nonbuilding structures do not have structural redundancy of multiple bays and nonstructural panels that effectively give buildings greater strength and damping than is considered in the design process.

$$C_s = \frac{0.8S_1}{\frac{R}{I_e}} \qquad 6.37$$

Table 6.16 Selected Values of R and Ω_O for Nonbuilding Structures [ASCE/SEI7 Table 15.4-2]

structure type	R	Ω_O
1. Elevated tanks, vessels, bins, or hoppers on symmetrical braced legs	3.0	2.0
2. Elevated tanks, vessels, bins, and hoppers on asymmetrical or unbraced legs	2.0	2.0
3. Cast-in-place concrete silos having walls continuous to the foundation	3.0	1.75
4. Distributed mass cantilever structures such as stacks, chimneys, silos, and skirt-supported vertical vessels made from welded steel with special detailing	3.0	2.0
5. Trussed towers (freestanding or guyed), guyed stacks, and chimneys	3.0	2.0
6. Inverted pendulum type structures	2.0	2.0
7. Concrete or steel frame cooling towers	3.5	1.75
8. Signs and billboards	3.0	1.75
9. Amusement structures and monuments	2.0	2.0
10. Steel frame (truss) telecommunication towers	3.0	1.5

Rigid nonbuilding structures, however, are covered in ASCE/SEI7 Sec. 15.4.2 and are handled differently. Rigid nonbuilding structures are structures with a natural period, T, of less than 0.06 sec. The natural period is the determining factor of whether the structure is rigid. An example would be a concrete pedestal structure at grade level. The lateral force, V, on rigid structures and their anchorages is given by Eq. 6.38 [ASCE/SEI7 Eq. 15.4-5]. The weight, W, includes the weight of the full contents (if any) of the structure [ASCE/SEI7 Sec. 15.4.3].

$$V = 0.30 S_{DS} W I_e \qquad 6.38$$

The force, V, is distributed over the height according to the distribution of the mass. It is assumed to act in any horizontal direction.

The size of the supporting structural system for some short (i.e., less than 50 ft (15.24 m) in height) nonbuilding structures is determined by the footprint of the structure, vibration limitation, or other operational considerations rather than traditional lateral loadings. In such cases, the support can be much stronger than required for seismic resistance, and it can be expected to remain in the elastic range during a maximum earthquake. Therefore, ductility is not considered.

Certain concrete pedestal-type structures can also be expected to remain in the elastic range during a maximum earthquake. However, it is recommended that some ductility be included in the design. Such ductility is obtained (during the design stage) by providing sufficient transverse reinforcement to avoid brittle shear and/or development failures and by providing continuity and development of longitudinal reinforcement.

Tanks with ground support are provided with detailed specific provisions [ASCE/SEI7 Sec. 15.7.6]. Flat-bottom tanks and other tanks with supported bottoms at or below grade (as opposed to tanks on legs) are considered to be nonbuilding structures. Due primarily to the fact that liquid contents slosh around and add their own dynamic forces, the seismic performance of tanks is more complex than their simple appearance would suggest.[32]

ASCE/SEI7 allows the use of approved national standards developed for specific nonbuilding structures when such standards exist [ASCE/SEI7 Chap. 23].[33]

Particular attention must be paid to preventing uplift of tanks from their cradles or supports. Specifically, tanks must be anchored to their foundations. Furthermore, sloshing and freeboard should be considered in the design of the tank. Sloshing can be reduced by including baffles in the tank. Sufficient freeboard should be included in open tanks to prevent the contents from spilling out over the top.

Nonbuilding structures should be bolted to their foundations. In order to ensure ductile response, these bolts must stretch inelastically without failure and without being pulled from their concrete embedment. The bolt size (diameters and lengths) and placement patterns should be chosen so that the bolts achieve their full strength in a maximum earthquake without failure, using the loads specified in ASCE/SEI7.

For other nonbuilding structures, when applicable, the site design response acceleration, risk category, soil site class, seismic response coefficient, and response modification factor are the same as for building structures. The vertical distribution of design seismic forces may be determined either by Eq. 6.25 or from a dynamic analysis.

Example 6.15

A simple billboard-type sign is constructed on site class B, where S_1 is 1.35, S_S is 1.8, $S_{D1} = 0.9$, and $S_{DS} = 1.2$. It has a total weight of 3000 lbf (13.34 kN) distributed evenly across its width and 18 ft (5.5 m) of height. The total effective cross-sectional moment of inertia at the base of its supports is 0.05 ft^4 (4.3×10^{-4} m^4). The modulus of elasticity of the three support posts is 2×10^6 psi (1.38×10^4 MPa). The importance factor is 1.0. What is the ASCE/SEI7 total design base shear for an earthquake acting perpendicular to the sign face?

[32]Sloshing has very little damping (i.e., 0.1% or less).
[33]For example, American Petroleum Institute (API) Publication 650 is an approved standard.

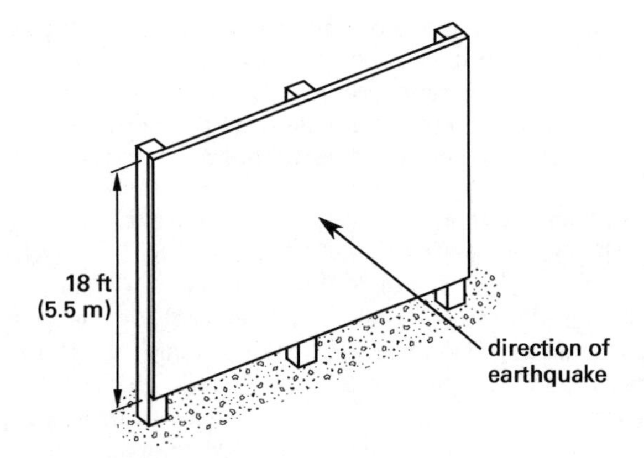

Customary U.S. Solution

One rational method of determining the sign's natural period is to consider the sign as an SDOF system. The supporting elements (i.e., the three posts) are essentially uniformly loaded cantilevers. From standard beam deflection tables, the tip deflection, x, for such a configuration with distributed load w per unit length is

$$x = \frac{wL^4}{8EI}$$

The stiffness, k, at the tip is

$$\begin{aligned} k &= \frac{\text{total load}}{x} = \frac{wL}{x} = \frac{8EI}{L^3} \\ &= \frac{(8)\left(2\times 10^6\ \frac{\text{lbf}}{\text{in}^2}\right)\left(12\ \frac{\text{in}}{\text{ft}}\right)^2(0.05\ \text{ft}^4)}{(18\ \text{ft})^3} \\ &= 19{,}750\ \text{lbf/ft} \end{aligned}$$

From Eq. 4.14 and Eq. 4.15, the natural period of oscillation is

$$\begin{aligned} T &= \frac{2\pi}{\omega} = 2\pi\sqrt{\frac{W}{kg}} \\ &= 2\pi\sqrt{\frac{3000\ \text{lbf}}{\left(19{,}750\ \frac{\text{lbf}}{\text{ft}}\right)\left(32.2\ \frac{\text{ft}}{\text{sec}^2}\right)}} \\ &= 0.43\ \text{sec} \end{aligned}$$

This is greater than 0.06 sec, so the billboard is not rigid as defined in ASCE/SEI7 Sec. 15.4.2.

From Table 6.16, the response modification factor, R, for a billboard is 3.0.

Determine the seismic response coefficient from Eq. 6.20.

$$\begin{aligned} C_s &= \frac{S_{DS}}{\frac{R}{I_e}} = \frac{1.2}{\frac{3.0}{1.0}} \\ &= 0.40 \end{aligned}$$

The minimum design seismic coefficient is determined from Eq. 6.37 since $S_1 > 0.6g$.

$$\begin{aligned} C_{s,\text{min}} &= \frac{0.8S_1}{\frac{R}{I_e}} = \frac{(0.8)(1.35)}{\frac{3.0}{1.0}} \\ &= 0.36 \end{aligned}$$

Since $0.40 > 0.36$, the total design base shear is

$$\begin{aligned} V &= C_s W \\ &= (0.40)(3000\ \text{lbf}) \\ &= 1200\ \text{lbf} \end{aligned}$$

SI Solution

One rational method of determining the sign's natural period is to consider the sign as an SDOF system. The supporting elements (i.e., the three posts) are essentially uniformly loaded cantilevers. From standard beam deflection tables, the tip deflection, x, for such a configuration with distributed load w per unit length is

$$x = \frac{wL^4}{8EI}$$

The stiffness, k, at the tip is

$$\begin{aligned} k &= \frac{\text{total load}}{x} = \frac{wL}{x} = \frac{8EI}{L^3} \\ &= \frac{(8)(1.38\times 10^4\ \text{MPa})\left(1000\ \frac{\text{kPa}}{\text{MPa}}\right)(4.3\times 10^{-4}\ \text{m}^4)}{(5.5\ \text{m})^3} \\ &= 285.3\ \text{kN/m} \end{aligned}$$

From Eq. 4.14 and Eq. 4.15, the natural period of oscillation is

$$\begin{aligned} T &= \frac{2\pi}{\omega} = 2\pi\sqrt{\frac{W}{kg}} \\ &= 2\pi\sqrt{\frac{13.34\ \text{kN}}{\left(285.3\ \frac{\text{kN}}{\text{m}}\right)\left(9.81\ \frac{\text{m}}{\text{s}^2}\right)}} \\ &= 0.434\ \text{s} \end{aligned}$$

This is greater than 0.06 s, so the billboard is not rigid as defined in ASCE/SEI7 Sec. 15.4.2.

From Table 6.16, the response modification factor, R, for a billboard is 3.0.

Determine the seismic response coefficient from Eq. 6.20.

$$\begin{aligned} C_s &= \frac{S_{DS}}{\frac{R}{I_e}} = \frac{1.2}{\frac{3.0}{1.0}} \\ &= 0.40 \end{aligned}$$

The minimum design seismic coefficient is determined from Eq. 6.37 since $S_1 > 0.6g$.

$$C_{s,\text{min}} = \frac{0.8S_1}{\frac{R}{I_e}} = \frac{(0.8)(1.35)}{\frac{3.0}{1.0}} = 0.36$$

Since 0.40 > 0.36, the total design base shear is

$$V = C_s W = (0.40)(13.34 \text{ kN}) = 5.34 \text{ kN}$$

47. PROVISIONS FOR DEFORMATION COMPATIBILITY

For structures in seismic design categories D, E, or F, any structural framing elements and their connections that are not part of the lateral force resisting system, but are nonetheless subjected to the deformations resulting from seismic forces, the ASCE/SEI7 requires design and detailing to be adequate to maintain support of design gravity (dead plus live) loads under expected seismic deformations. The ASCE/SEI7 Sec. 12.12.5 provisions are more rigorous than previous building code specifications. This is because the importance of *deformation compatibility* has been demonstrated in recent earthquakes in population regions. For example, many vertical load-carrying structural framing elements and their connections performed poorly in the 1994 Northridge earthquake. (See App. P.)

When computing expected deformations for the abovementioned elements, consider the following.

- P-δ effects on such elements should be considered.
- Expected deformations should be the greater of the story drift or the maximum inelastic response displacement, δ, considering P-δ.
- The stiffening effect of such elements should be considered when determining forces in the structure.
- Additional deformations that may result from foundation flexibility and diaphragm deflections should be considered.

48. PENALTIES FOR STRUCTURAL IRREGULARITY

The so-called "penalty" for some irregular structures is the requirement of a dynamic lateral force analysis. For example, buildings greater than three stories in height in seismic design categories D, E, or F with vertical stiffness, mass, or geometry irregularities (types 1, 2, and 3 in ASCE/SEI7 Tables 12.3-1 and 12.3-2) must receive a dynamic treatment.

However, adjustments for other types of irregularity are not adequately accomplished by a dynamic analysis. The ASCE/SEI7 penalizes irregular structures both by imposing additional requirements and by eliminating special allowances given to regular structures. For example, structures with in-plane discontinuity in seismic design categories D through F must increase the design force by 25% when designing the connection between diaphragms and vertical elements [ASCE/SEI7 Sec. 12.3.3.4]. When a structure is regular, this increased design force is not required.

Irregularity is also discouraged, for example, where weak stories are prohibited [ASCE/SEI7 Sec. 12.3.3.2] in buildings over two stories in height when the strength ratio based on the story above is below 65%.

Torsional irregularity is penalized by requiring the use of an increased accidental eccentricity [ASCE/SEI7 Sec. 12.8.4.3]. Seismic forces in two orthogonal directions must be evaluated when nonparallel systems exist [ASCE/SEI7 Sec. 12.5.3].

49. DYNAMIC ANALYSIS PROCEDURE

Although the details of how to perform an elastic dynamic analysis are beyond the scope of this book, the basic dynamic analysis procedure required by the building codes consist of three steps: (1) the static base shear is calculated; (2) a dynamic analysis using an elastic response spectrum is performed to determine the building period, base shear, story shears, and drifts; and (3) the force results of the analysis are adjusted based upon a comparison of the base shear from the dynamic analysis and that determined from the static methods [ASCE/SEI7 Sec. 12.9]. If the base shear of the required modal combination, V_t, is more than 85% of the static base shear, V, no adjustment to the results of the dynamic analysis is required. If the base shear is less than 85% of the static base shear, then all force values from the dynamic analysis must be scaled upward by the ratio $0.85V/V_t$ [ASCE/SEI7 Sec. 12.9.4].

The ASCE/SEI7 code provides much more detailed guidelines about the use of dynamic analysis than earlier buildings codes have provided. ASCE/SEI7 Sec. 12.9 provides guidelines for response spectra analysis. Chapter 16 of the ASCE/SEI7 discusses time history analysis, a less common dynamic analysis for building design.

Specific requirements about the modeling for response spectra are provided in the building code. In ASCE/SEI7 Sec. 12.9.1, the number of modes to be considered in the analysis must be large enough that the combined modal mass participation is at least 90%. The results of individual modes are to be combined using either a sum of the squares (SRSS) or complete quadratic combination (CQC) procedure [ASCE/SEI7 Sec. 12.9.3]. Once forces in individual elements are determined from the analysis, ASCE/SEI7 Sec. 12.9.2 allows for the design force to be obtained by dividing the element force parameters by the value of (R/I_e) and multiplying the element deformation values by (C_d/I_e).

Response spectra may be site-specific according to the requirements of ASCE/SEI7 Chap. 21. When preparing a site-specific response spectra, 5% damping is assumed and the earthquake should represent a 2% probability of exceedance within a 50-year period [ASCE/SEI7 Sec. 21.2.1]. Major earthquakes near 8.0 on the Richter scale (e.g., the 1906 San Francisco and the 1985 Mexico City earthquakes) are considered to be exceptional situations.

Dynamic analysis is usually performed on a computer. However, the following steps can be used to carry out a manual dynamic analysis on a simple multistory structure when desired. (It is not practical to perform a dynamic analysis on structures with irregularities by hand.)

step 1: Construct a lumped-mass, two-dimensional model of the structure. (i represents the mode index; x represents the floor index.)

step 2: Calculate the mode shape factors, $\phi_{i,m}$. (See Sec. 4.19.) Normalize the mode shape factors so that $\phi = 1$ at the highest level.

step 3: Calculate the period, T_m, for each mode.

step 4: For each mode shape, calculate

$$L_m = \sum_{i=1}^{n} \left(\frac{W_i}{g} \right) \phi_{i,m} \quad \text{6.39}$$

$$M_m = \sum_{i=1}^{n} \left(\frac{W_i}{g} \right) \phi_{i,m}^2 \quad \text{6.40}$$

step 5: Calculate the spectral acceleration ($S_{a,m}$) and seismic design coefficient for each mode from the response spectra.

$$C_m = \frac{S_{a,m} I}{R} = \frac{C_v I}{RT} \quad \text{6.41}$$

step 6: Calculate the base shear for each mode.

$$W_m = \frac{L_m^2 g}{M_m} \quad \text{[effective weight]} \quad \text{6.42}$$

$$V_m = C_m \left(\frac{W_m}{g} \right) \quad \text{6.43}$$

step 7: Calculate the participating mass (PM) fraction for each mode.

$$\text{PM}_m = \frac{L_m^2 g}{M_m W_t} \quad \text{6.44}$$

$$W_t = \sum W_x \quad \text{6.45}$$

step 8: Combine the base shears into the design dynamic lateral force, V_{dynamic}, using the SRSS (i.e., square root of the sum of the squares) method, with as many modes as are necessary to include at least 90% of the participating mass of the structural (i.e., until $\Sigma(\text{PM}) \geq 0.90$).

step 9: Calculate the lateral force, V_{static}, according to the static provisions of ASCE/SEI7 Sec. 12.8.

step 10: Determine the adjustment factor according to ASCE/SEI7 Sec. 12.9.4.

step 11: Adjust V_{dynamic} as is required by ASCE/SEI7 Sec. 12.9.4.

$$V_{\text{dynamic}} = \sqrt{(V_1^2 + V_2^2 + \cdots + V_n)^2} \quad \text{6.46}$$

step 12: Distribute the scaled-up base shear to each level.

$$F_{x,m} = V_{\text{dynamic}} \left(\frac{W_x \phi_{x,m}}{\sum (W_i \phi_{i,m})} \right) \quad \text{6.47}$$

step 13: Determine the raw deflections, moments, and shears for each mode.

step 14: Use SRSS to combine the raw deflections, moments, and shears into effective values [ASCE/SEI7 Sec. 12.9.3].

50. ALLOWABLE STRESS LEVELS

The allowable stress (working stress) method is a method of proportioning structural elements. The primary requirement of this design method is that calculated stresses produced in the elements by the allowable stress design load combinations (service level loads) do not exceed specified allowable stress limits. The calculated allowable stress is based on the yield stress and a reasonable factor of safety.

The IBC does not permit an increase in allowable stresses to be used with the basic ASD load combinations [IBC Sec. 1605.3.1].

The IBC provides alternate basic load combinations for all materials. IBC Sec. 1605.3.2 permits a one-third increase in allowable stresses for all combinations, including W (wind) or E (earthquake). In some cases, this one-third stress allowance has already been "built in" to tables provided by the IBC and vendors. For example, the wood structural panel diaphragm staple requirements (as in Table 12.1 [IBC Table 2306.2(1)]) have already considered this increase, as have the connector/connection/strap/tie recommendations published by certain vendors. The table footnotes may have to be read to determine if this increase has already been included in published data.

There are exceptions in considering the duration of load increase and the one-third stress increase permitted in allowable stresses for elements resisting earthquake forces, such as when certain structural irregularities exist.

The IBC suggests that allowable stresses may be increased by one-third for transitory wind, snow, and seismic loading when allowable stress design (ASD) is used. While this may still be appropriate for soil bearing strength, such an allowance for steel is not present in *AISC Specifications*, the IBC, or ASCE/SEI7.

7 Diaphragm Theory

1. DIAPHRAGM ACTION

The story shears calculated by Eq. 6.27 are assumed to be applied to a lumped mass representing the floor/ceiling layer in a building. The ceiling does not actually resist the story shear, but it does distribute the force among the resisting elements (e.g., shear walls, columns, moment-resisting frames, and other structural systems).

Ceilings and floors that transmit lateral forces to the resisting elements are known as *horizontal diaphragms.* The diaphragm's function of distributing the story shears is known as *diaphragm action.* It is common to refer to the story shear as the *diaphragm force.* However, it should be recognized that the diaphragm force may include some of the story shears for the level *and above* [ASCE/SEI7 Sec. 12.10.1].

Depending on the displacement of the diaphragm relative to the supporting shear walls, diaphragms can be considered either flexible or rigid.

Wood structural panel diaphragms are almost always considered flexible. Generally, concrete slab floors and diaphragms are considered to be rigid. Steel deck diaphragms and poured-in-place gypsum floors can be either rigid or flexible, depending on their design.

2. SEISMIC WALL AND DIAPHRAGM FORCES

Figure 7.1 shows a simple (regular) one-story box building with a flexible diaphragm roof. (While this discussion is applicable to larger buildings, Sec. 7.2 through Sec. 7.19 are primarily concerned with simple one-story buildings with masonry or reinforced concrete walls.) Both walls are identical. An earthquake acceleration occurs in the direction shown by the ground motion arrow.

When discussing seismic forces in structures with diaphragms (e.g., one- or two-story masonry buildings with wood structural panel diaphragm floors and ceilings), it is important to distinguish between forces in the parallel and perpendicular walls. (See Fig. 7.1.) The forces in the parallel walls are shear forces, while the forces in the perpendicular walls are normal forces (i.e., perpendicular to the face of the wall). This section is primarily concerned with the shear force in the parallel walls.

Figure 7.1 *Simplified Building and Roof Diaphragm*

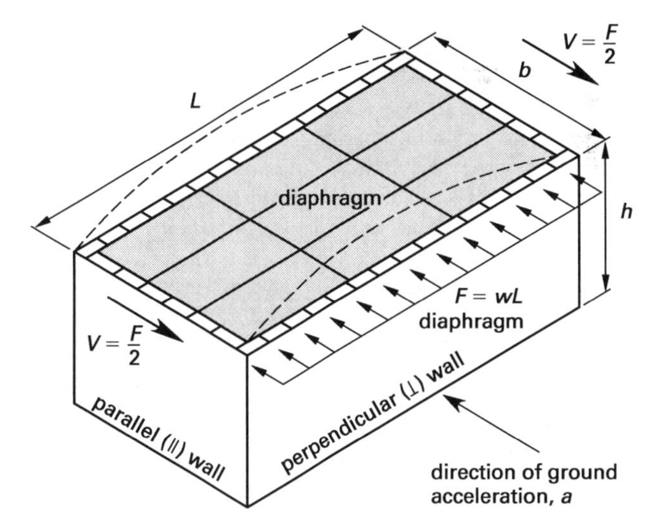

The seismic shear force acting on the parallel walls depends on the mass being accelerated, which consists of the diaphragm weight and some portion, usually assumed to be half, of the total wall weight. (It is assumed that the seismic shear force from the remaining half of the total wall weight passes directly to the foundation without stressing the wall.) The story weight includes the weight of any equipment mounted on the roof, anything suspended inside the building from the roof, and anything mounted on the upper half of the walls. The wall weight includes all weight of any parapet that projects above the roof line. Openings in the walls, such as windows and doors, that reduce the wall weight are usually disregarded when determining wall weight.

In rare cases, such as when the wall is a large proportion of the entire story weight, the diaphragm must also be able to resist the seismic force required by the appropriate IBC equations (see Eq. 6.35 and Eq. 6.36).

Wall forces are calculated from the IBC equations (see Eq. 6.35 and Eq. 6.36), which depend on the weight, W, of the wall. (See Sec. 6.29.) Openings in the walls, such as windows and doors, that reduce the wall weight are usually disregarded when determining wall weight.

The total seismic force resisted by the two parallel walls near the ground level is the sum of seismic forces resulting from the diaphragm and wall weights. In the simple illustration of Fig. 7.1, the force on one wall is half of the total force for both rigid and flexible diaphragms.

The portion of seismic load originating from the acceleration of the perpendicular walls is given by Eq. 7.1. The symbols ⊥ and || refer to "perpendicular" and "parallel," respectively.

$$F_{\perp\text{walls}} = \tfrac{1}{2}(\text{SC}_f)W_{\perp\text{walls}} \qquad 7.1$$

SC_f is the *seismic coefficient* of the floor. SC_f is the ratio of the story force from Eq. 6.23 to the story weight.

The portion of seismic load originating from the acceleration of the parallel walls is

$$F_{\|\text{walls}} = \tfrac{1}{2}(\text{SC}_f)W_{\|\text{walls}} \qquad 7.2$$

All of the inertial load from the accelerating walls and roof masses must be carried by the wall-roof connections. In calculating the total shear force on the parallel walls, the total walls and diaphragm weights should be used in design seismic base shear formulas.

$$W_{\text{total}} = W_{\text{diaphragm}} + W_{\perp\text{walls}} + W_{\|\text{walls}} \qquad 7.3$$

$$F_{\text{total}} = \tfrac{1}{2}(\text{SC}_f)W_{\text{total}} \qquad 7.4$$

There are two reasons for calculating the forces from the diaphragm and parallel walls separately. The first reason is to distinguish between the two for the purpose of subsequent calculations; that is, the parallel wall force does not contribute to chord loads and diaphragm shear where the diaphragm is flexible. The second reason is to emphasize the timing difference that occurs in a real earthquake.

The perpendicular and parallel walls experience an almost immediate force due to ground acceleration. However, the parallel walls receive the diaphragm force only after some delay. Unfortunately, an accurate analysis of this aspect of seismic behavior is almost impossible. For simple structures with three or fewer floors, the simple method of adding all forces together is used for convenience.

As with any seismic analysis, the diaphragm force must be evaluated in both orthogonal directions.

Based on the ASCE/SEI7 Sec. 12.10.1.1, floor and roof diaphragms in multi-story buildings should be designed to resist the forces determined from Eq. 7.5 [ASCE/SEI7 Eq. 12.10-1]. (See Sec. 6.44.)

$$F_{px} = \left(\frac{\sum_{i=x}^{n} F_i}{\sum_{i=x}^{n} w_i}\right) w_{px} \qquad 7.5$$

The force F_{px} should be equal to or greater than $0.2S_{DS}I_e w_{px}$, and need not exceed $0.4S_{DS}I_e w_{px}$.

3. WALL SHEAR STRESS

In the simple building shown in Fig. 7.1, the rigidities (for a rigid diaphragm) and tributary areas (for a flexible diaphragm) are identical for the two walls. Therefore, half of the total seismic force is carried by each parallel wall. The shear flow, q, in a parallel wall is

$$q = \frac{F_{\text{total}}}{2b} \quad \text{[per unit length]} \qquad 7.6$$

The shear stress, v, in a parallel wall of thickness t is

$$v_{\text{total}} = \frac{F_{\text{total}}}{2bt} \quad \text{[per unit area]} \qquad 7.7$$

Shear walls located on adjoining levels should be structurally continuous and should not be offset. There should be a complete transmission path from a shear wall on one level to another shear wall below.

Horizontally and vertically stacked openings in shear walls need special attention. (See Fig. 7.2.) Vertical shears need to be transferred to adjacent piers or boundary columns.

Figure 7.2 *Stacked Openings*

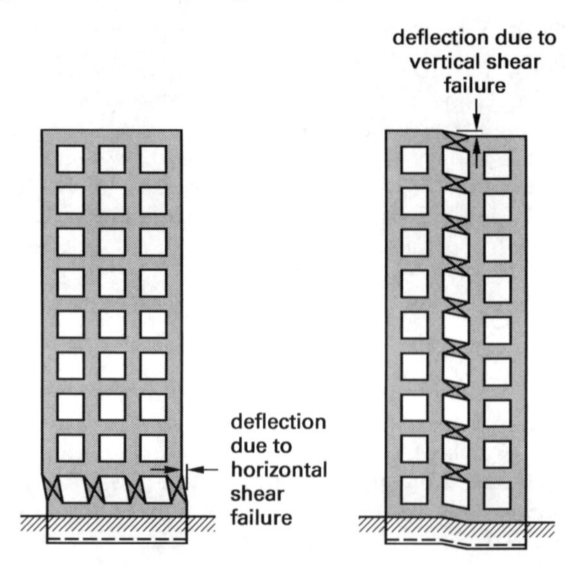

For masonry shear walls that are designed for in-plane shear forces by allowable stress design procedures, IBC Sec. 2106.1 (ACI 530 Sec. 1.18.3.2.6.1.2) requires a design shear stress that is 150% of the shear stress determined for the base shear force for buildings in seismic design categories D, E, or F.

4. RIGID DIAPHRAGM ACTION

A *rigid diaphragm* does not change its plan shape when subjected to lateral loads. It remains the same size, and square corners remain square. There is no flexure. Rigid diaphragms are capable of transmitting torsion to the major resisting elements (usually the outermost elements). The lateral story shear is distributed to the resisting elements in proportion to the rigidities of those elements.

Figure 7.3 illustrates a simple arrangement of a rigid diaphragm distributing the seismic load to two shear walls.

Figure 7.3 *Rigid Diaphragm Action*

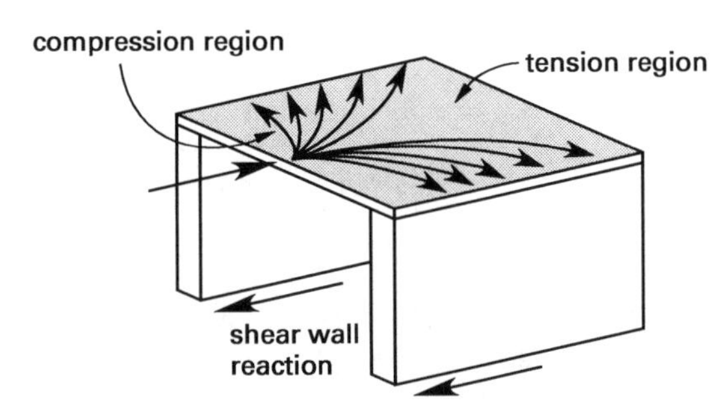

Since the diaphragm force is distributed to the resisting elements in proportion to the rigidities of those elements, the rigidities must be determined. In practice, a few guidelines are needed to do so.

1. The relative rigidities of masonry or concrete structures can be calculated using Eq. 4.11 and Eq. 4.12. Alternatively, App. D and App. E can be used. There is no need to use actual values of E and G, since only relative values are needed.

2. If a wall extends above roof level (i.e., has a *parapet*), the distance above the roof (i.e., the parapet height) should be disregarded when calculating the rigidity.

3. Shear walls with openings such as doors and windows require special attention. As a first approximation, such a wall can be treated as a solid shear wall. However, other methods (see Sec. 7.5) exist for evaluating the overall wall rigidity.

4. The rigidities of *transverse walls* (i.e., walls running perpendicular to the direction of the lateral force) are usually disregarded for calculating direct loads. This is called "omitting the *weak walls*." However, rigidities of all walls must be known in order to calculate torsional loads.

Example 7.1

Two walls—wall A with rigidity of 3.278 and wall B with rigidity of 7.895—support a rigid diaphragm roof in a one-story reinforced masonry building. A total seismic force of 120,000 lbf (533.8 kN) is applied parallel to the walls in such a way that there is no rotation. Determine the shear carried by each wall.

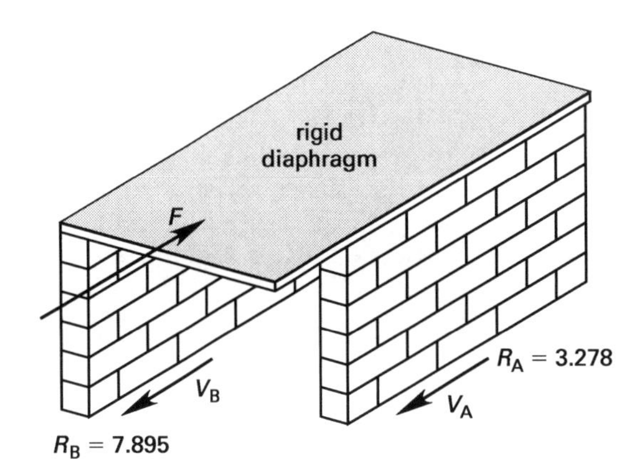

Customary U.S. Solution

The total seismic force is distributed to the walls in proportion to the relative rigidities.

$$V_A = R_{rel,A}F = \left(\frac{R_{tab,A}}{R_{tab,A}+R_{tab,B}}\right)F$$

$$= \left(\frac{3.278}{3.278+7.895}\right)(120{,}000 \text{ lbf})$$

$$= 35{,}206 \text{ lbf}$$

$$V_B = R_{rel,B}F = \left(\frac{R_{tab,B}}{R_{tab,A}+R_{tab,B}}\right)F$$

$$= \left(\frac{7.895}{3.278+7.895}\right)(120{,}000 \text{ lbf})$$

$$= 84{,}794 \text{ lbf}$$

SI Solution

The total seismic force is distributed to the walls in proportion to the rigidities.

$$V_A = R_{rel,A}F = \left(\frac{R_{tab,A}}{R_{tab,A}+R_{tab,B}}\right)F$$

$$= \left(\frac{3.278}{3.278+7.895}\right)(533.8 \text{ kN})$$

$$= 156.6 \text{ kN}$$

$$V_B = R_{rel,B}F = \left(\frac{R_{tab,B}}{R_{tab,A} + R_{tab,B}}\right)F$$

$$= \left(\frac{7.895}{3.278 + 7.895}\right)(533.8\ \text{kN})$$

$$= 377.2\ \text{kN}$$

5. CALCULATING WALL RIGIDITY

In order to determine the total rigidity (observed or tabulated) of a wall with openings, it is necessary to divide the wall into piers and beams. A *pier* is a vertical portion of the wall whose height is taken as the smaller of the heights of the openings on either side of it. A *beam* is a horizontal portion left after the piers have been located.

Figure 7.4 illustrates a wall with two windows. P_1, P_2, and P_3 are piers. B_1 and B_2 are beams.

Figure 7.4 *Wall with Openings*

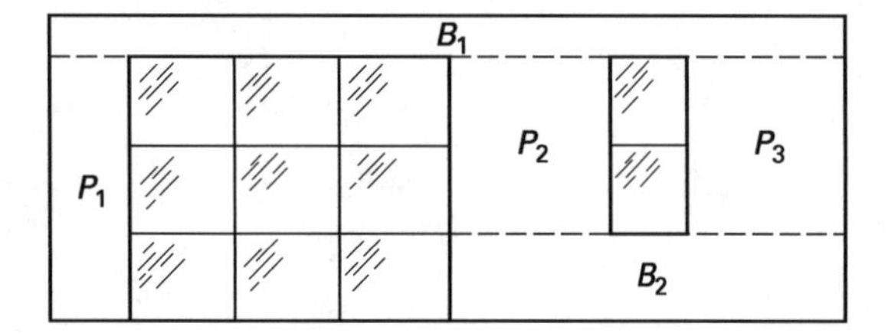

There are several methods of calculating wall rigidity from the characteristics of the wall's piers and beams, all of which yield slightly different answers.[2] The actual rigidity for any particular wall in a structure is unimportant. What is important is the *relative rigidity* of the wall compared to all other resisting walls in the structure. Thus, it is essential that the same method be used to calculate the rigidities of all walls and their piers. Two methods for calculating wall rigidity are discussed below.

Although it is generally assumed that cantilever conditions (see Sec. 4.5) prevail for the walls in one- and two-story buildings taken in their entireties, piers within the wall can be considered either fixed or cantilevered. For example, piers between openings may be considered to be fixed at their tops and bottoms although the wall taken as a whole is cantilevered.

The accuracy in wall rigidity calculations is not great. There are many assumptions made about material properties and wall performance, and the analysis procedure, though formalized, is less than rigorous. Therefore, values with more than three or four significant digits are unwarranted.

[2]These other methods are approximate and often do not agree. Also, it is not uncommon for the methods to determine the rigidity of a wall with openings—particularly a wall with fixed-pier assumptions—to be greater than a solid wall (an obvious impossibility). The rigidity of a wall with openings should be compared to the rigidity of a solid wall of the same dimensions.

Method A

With this fast and simple method (used only for preliminary analyses), the rigidity of a wall is calculated as the sum of the rigidities of the individual piers framed between openings in a wall. All piers are assumed to be fixed. The pier height is the height of the shortest adjacent opening.[3] Beams and wall portions above and below the openings are not considered.

Method B

By far, the most commonly used method of evaluating wall rigidities calculates "deflections" from standardized values of force, thickness, and modulus of elasticity. These deflections are recognized as being the reciprocals of rigidity.

To start, the gross deflection of the solid wall is calculated, ignoring all openings and assuming cantilever action. Then, the strip deflection of an interior strip having length equal to the wall length and height equal to the tallest opening is calculated, again assuming cantilever action. This strip deflection is subtracted from the solid wall's gross deflection.

Next, the rigidities (not the deflections) of all piers (assuming fixed ends) within the removed strip are summed, and the pier deflection correction is calculated as the reciprocal of the sum. The pier deflection correction is added to the difference of the gross and strip deflections to give the net deflection. The wall rigidity is the reciprocal of this net deflection.

The rigidity of the wall must be calculated by this method one opening at a time, considering the fixed pier adjacent to the opening and the wall section below the opening. The calculation of the pier deflection correction becomes recursive when openings in the wall are of different heights. For this reason, Method B can take a long time if the wall is relatively complex.

Example 7.2

The masonry wall shown in Fig. 7.4 and dimensioned as shown has a uniform thickness, and is part of a one-story building. Determine the rigidity using the two methods described in Sec. 7.5.

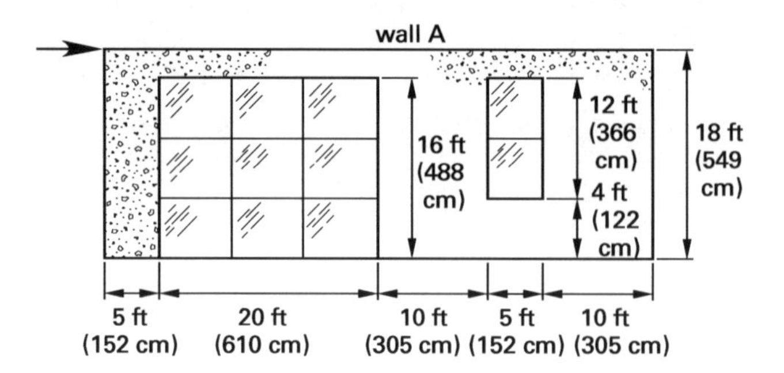

[3]A common error is to use the ground-to-ceiling distance.

Customary U.S. Solution

Method A

The total rigidity of the wall is the sum of all the pier rigidities. Beams are disregarded.

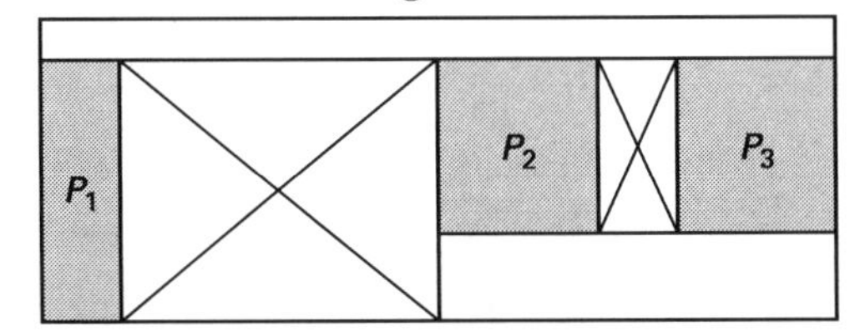

The rigidities, R, of all piers (assumed fixed) are obtained from Eq. 4.11 or App. D after calculating the height/depth ratio, h/d, of each.

element	h	d	h/d	R
P_1	16	5	3.2	0.236
P_2	12	10	1.2	1.877
P_3	12	10	1.2	1.877

In this method, the rigidity of the wall is the sum of the individual fixed-pier rigidities.

$$R = 0.236 + 1.877 + 1.877 = 3.99$$

Method B

First determine the deflection of a solid wall. The height/depth ratio of the entire wall is

$$\frac{h}{d} = \frac{18\text{ ft}}{50\text{ ft}} = 0.36$$

Since this is a one-story building, the wall taken as a whole is assumed to be cantilevered. From App. E, the rigidity is 7.895. The "deflection" of the solid wall is

$$\Delta_{\text{solid}} = \frac{1}{R} = \frac{1}{7.895} = 0.1267$$

The tallest opening has a height of 16 ft, so a "mid-strip" 16 ft high and 50 ft long is removed. Assuming a solid wall, the height/depth ratio is

$$\frac{h}{d} = \frac{16\text{ ft}}{50\text{ ft}} = 0.32$$

The entire wall was assumed to be a cantilever pier, and this mid-strip represents the majority of the wall. Therefore, it also is a cantilever member. Appendix E gives the rigidity as 9.165. The "deflection" of an assumed solid mid-strip is

$$\Delta_{\text{mid-strip}} = \frac{1}{9.165} = 0.109$$

The mid-strip, however, is not solid. It consists of two functional parts: a 5 ft by 16 ft solid pier, P_1, at the left and a larger section with a small window at the right. The large window section running floor-to-ceiling is assumed to contribute no rigidity to the wall. The rigidity of this mid-strip is the sum of the rigidities of pier P_1 and the larger section.

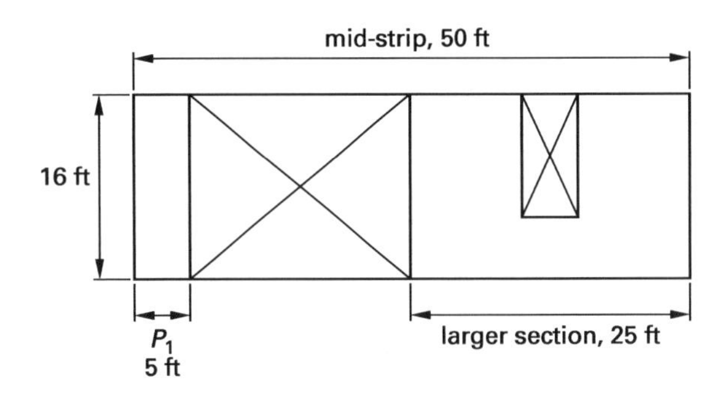

For the solid pier, P_1, at the left,

$$\frac{h}{d} = \frac{16\text{ ft}}{5\text{ ft}} = 3.2$$

For the same reason for considering the mid-strip to be a cantilever, this pier is assumed to act as a cantilever. From App. E, $R = 0.071$. (The "deflection" of pier P_1 is not needed.)

The larger section at the right of the wall has dimensions of 16 ft by 25 ft. Taken as a solid cantilever pier, the gross rigidity and deflection are

$$\frac{h}{d} = \frac{16\text{ ft}}{25\text{ ft}} = 0.64$$

$$R_{\text{larger section,gross}} = 3.37$$

$$\Delta_{\text{larger section,gross}} = \frac{1}{3.37} = 0.297$$

However, the larger section is not itself solid. There is a 12 ft high window. For a 12 ft by 25 ft solid window strip assumed to act as a cantilever,

$$\frac{h}{d} = \frac{12\text{ ft}}{25\text{ ft}} = 0.48$$

$$R_{\text{window strip}} = 5.31$$

$$\Delta_{\text{window strip}} = \frac{1}{5.31} = 0.188$$

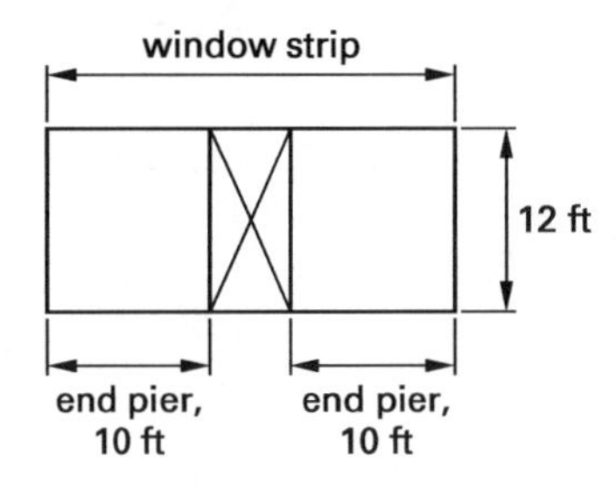

The window strip contributes stiffness, as there are two 12 ft by 10 ft end piers. Although cantilever performance could be argued as well, these two end piers are assumed to be fixed. Their combined rigidity and deflection are

$$\frac{h}{d} = \frac{12 \text{ ft}}{10 \text{ ft}} = 1.2$$

$$R_{\text{end piers}} = (2)(1.877) = 3.754$$

$$\Delta_{\text{end piers}} = \frac{1}{3.754} = 0.266$$

Now that all the pieces have been evaluated, the rigidity of the entire wall can be built up.

The net deflection and net rigidity of the larger section is

$$\begin{aligned}\Delta_{\text{larger section,net}} &= 0.297 - 0.188 + 0.266 \\ &= 0.375 \\ R_{\text{larger section,net}} &= \frac{1}{0.375} = 2.67\end{aligned}$$

(Check that 2.67 is less than the gross value of 3.37.)

Since the highest opening in the mid-strip extends the full height, the rigidity is merely the sum of the rigidities of pier P_1 and the larger section.

$$R_{\text{mid-strip,net}} = 0.071 + 2.67 = 2.74$$

$$\Delta_{\text{mid-strip,net}} = \frac{1}{2.74} = 0.365$$

(Check that 2.74 is less than the gross value of 9.165.)

The deflection and rigidity of the entire wall are

$$\Delta_{\text{entire wall}} = 0.127 - 0.109 + 0.365 = 0.383$$

$$R_{\text{entire wall}} = \frac{1}{0.383} = 2.61$$

(Check that 2.61 is less than the gross value of 7.895.)

SI Solution

Method A

The total rigidity of the wall is the sum of all the pier rigidities. Beams are disregarded.

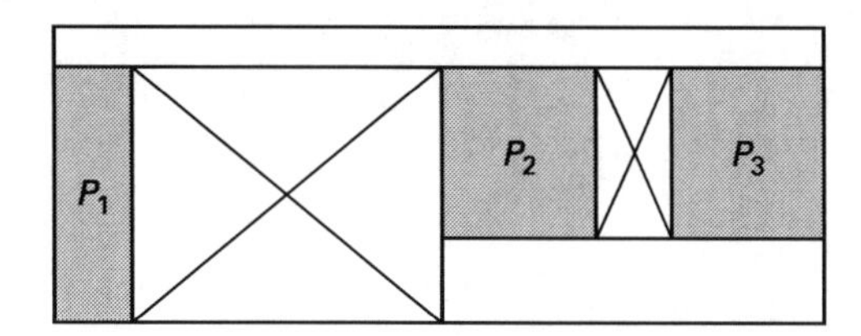

The rigidities, R, of all piers (assumed fixed) are obtained from Eq. 4.11 or App. D after calculating the height/depth ratio, h/d, of each.

element	h	d	h/d	R
P_1	488	152	3.2	0.236
P_2	366	305	1.2	1.877
P_3	366	305	1.2	1.877

In this method, the rigidity of the wall is the sum of the individual fixed-pier rigidities.

$$R = 0.236 + 1.877 + 1.877 = 3.99$$

Method B

First determine the deflection of a solid wall. The height/depth ratio of the entire wall is

$$\frac{h}{d} = \frac{549 \text{ cm}}{1524 \text{ cm}} = 0.36$$

Since this is a one-story building, the wall taken as a whole is assumed to be cantilevered. From App. E, the rigidity is 7.895. The "deflection" of the solid wall is

$$\Delta_{\text{solid}} = \frac{1}{R} = \frac{1}{7.895} = 0.1267$$

The tallest opening has a height of 488 cm, so a "mid-strip" 488 cm high and 1524 cm long is removed. Assuming a solid wall, the height/depth ratio is

$$\frac{h}{d} = \frac{488 \text{ cm}}{1524 \text{ cm}} = 0.32$$

The entire wall was assumed to be a cantilever pier, and this mid-strip represents the majority of the wall. Therefore, it also is a cantilever member. Appendix E gives the rigidity as 9.165. The "deflection" of an assumed solid mid-strip is

$$\Delta_{\text{mid-strip}} = \frac{1}{9.165} = 0.109$$

The mid-strip, however, is not solid. It consists of two functional parts: a 152 cm by 488 cm solid pier, P_1, at the left and a larger section with a small window at the right. The large window section running floor-to-ceiling is assumed to contribute no rigidity to the wall. The rigidity of this mid-strip is the sum of the rigidities of pier P_1 and the larger section.

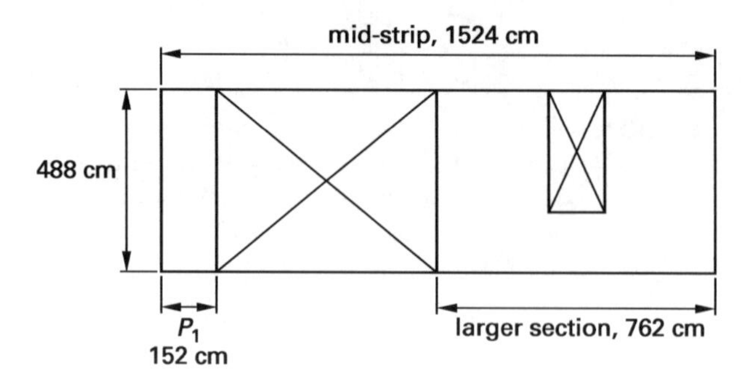

For the solid pier, P_1, at the left,

$$\frac{h}{d} = \frac{488 \text{ cm}}{152 \text{ cm}} = 3.2$$

For the same reason for considering the mid-strip to be a cantilever, this pier is assumed to act as a cantilever.

From App. E, $R = 0.071$. (The "deflection" of pier P_1 is not needed.)

The larger section at the right of the wall has dimensions of 488 cm by 762 cm. Taken as a solid cantilever pier, the gross rigidity and deflection are

$$\frac{h}{d} = \frac{488 \text{ cm}}{762 \text{ cm}} = 0.64$$

$$R_{\text{larger section,gross}} = 3.37$$

$$\Delta_{\text{larger section,gross}} = \frac{1}{3.37} = 0.297$$

However, the larger section is not itself solid. There is a 366 cm high window. For a 366 cm by 762 cm solid window strip assumed to act as a cantilever,

$$\frac{h}{d} = \frac{366 \text{ cm}}{762 \text{ cm}} = 0.48$$

$$R_{\text{window strip}} = 5.31$$

$$\Delta_{\text{window strip}} = \frac{1}{5.31} = 0.188$$

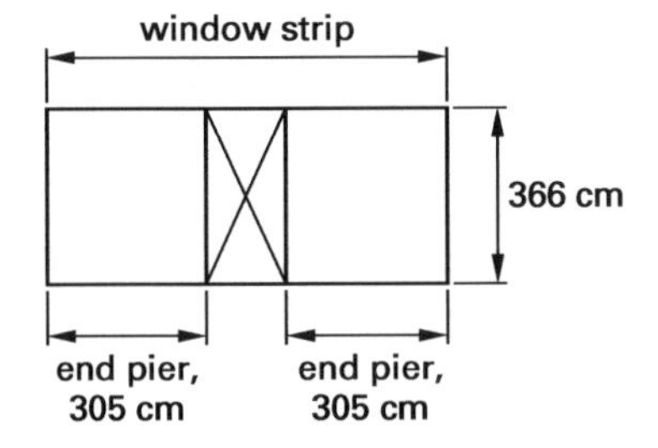

The window strip contributes stiffness, as there are two 366 cm by 305 cm end piers. Although cantilever performance could be argued as well, these two end piers are assumed to be fixed. Their combined rigidity and deflection are

$$\frac{h}{d} = \frac{366 \text{ cm}}{305 \text{ cm}} = 1.2$$

$$R_{\text{end piers}} = (2)(1.877) = 3.754$$

$$\Delta_{\text{end piers}} = \frac{1}{3.754} = 0.266$$

Now that all the pieces have been evaluated, the rigidity of the entire wall can be built up.

The net deflection and net rigidity of the larger section is

$$\Delta_{\text{larger section,net}} = 0.297 - 0.188 + 0.266 = 0.375$$

$$R_{\text{larger section,net}} = \frac{1}{0.375} = 2.67$$

(Check that 2.67 is less than the gross value of 3.37.)

Since the highest opening in the mid-strip extends the full height, the rigidity is merely the sum of the rigidities of pier P_1 and the larger section.

$$R_{\text{mid-strip,net}} = 0.071 + 2.67 = 2.74$$

$$\Delta_{\text{mid-strip,net}} = \frac{1}{2.74} = 0.365$$

(Check that 2.74 is less than the gross value of 9.165.)

The deflection and rigidity of the entire wall are

$$\Delta_{\text{entire wall}} = 0.127 - 0.109 + 0.365 = 0.383$$

$$R_{\text{entire wall}} = \frac{1}{0.383} = 2.61$$

(Check that 2.61 is less than the gross value of 7.895.)

6. FLEXIBLE DIAPHRAGMS

A flexible diaphragm changes shape when subjected to lateral loads. Its tension chord bends outward, and its compression chord bends inward, with a deflection shape similar to that of a simply supported beam loaded uniformly. Flexible diaphragms are assumed to be incapable of transmitting torsion to the resisting elements. (Also, see Chap. 5, Ftn. 9.)

As defined by ASCE/SEI7 Sec. 12.3.1.3, a *flexible diaphragm* is one that has a maximum lateral deflection more than two times the average story drift. To determine if a diaphragm is flexible, compare the in-plane deflection at the midpoint of the diaphragm to the story drift of the adjoining vertical resisting elements under equivalent tributary load.

A flexible diaphragm distributes the diaphragm force in proportion to the tributary areas of the diaphragm, as opposed to distributing it in proportion to the rigidities of the vertical resisting elements, as does a rigid diaphragm.

7. FRAMING TERMINOLOGY

Figure 7.5 illustrates such common wood framing terms as *sheathing*, *girder*, *beam*, *purlin*, and *joist* (or *sub-purlin*), as well as the *bridging* and *blocking* that are used to prevent lateral buckling. Blocking (e.g., often cut from the same material as the joists, although other blocking techniques are used) usually frames into joists or sub-purlins (two-by-sixes, two-by-eights, etc.); joists frame into purlins (e.g., four-by-eights); purlins frame into beams (e.g., four-by-fourteens); beams frame into girders (e.g., glulams); and girders frame into the walls.[4]

[4]These terms are not so rigidly defined that they preclude incorrect usage.

Figure 7.5 *Framing Members*

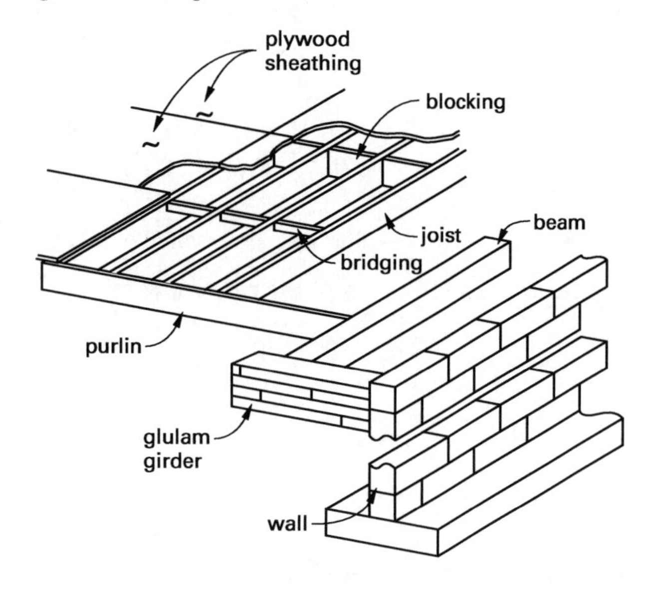

8. COLLECTORS

It is required that *collectors* (also known as *drag struts*, *braces*, or merely *struts* or *ties*) be used to transmit diaphragm reactions to shear walls at points of discontinuity (irregularity) in the plan [ASCE/SEI7 Sec. 12.10.2]. These collectors effectively separate the diaphragm into subdiaphragms that are analyzed independently. (See Fig. 7.6.)

Figure 7.6 *Use of a Collector*

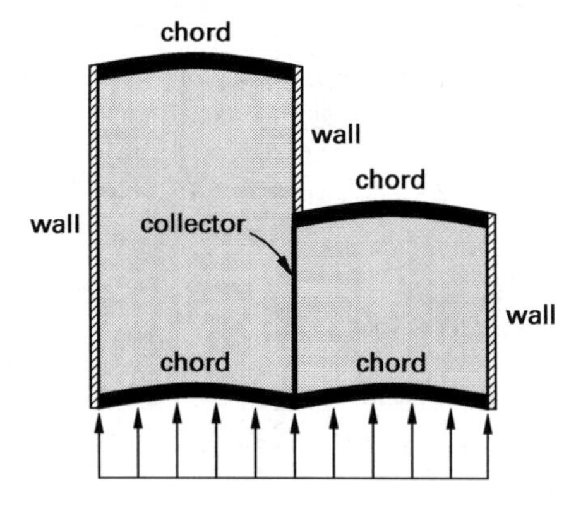

9. COLLECTOR FORCE

The collector force is the product of the diaphragm load (per unit area) and the areas tributary to the collector. Tributary areas are usually taken as some fraction of the subdiaphragm areas located on either side of the collector. The force is not the same along the length of the collector, but increases to a maximum at the point where the collector frames into a shear wall.

10. COLUMNS SUPPORTING PARTS OF FLEXIBLE DIAPHRAGMS

An interior or exterior column can be used to support the vertical roof or floor load, but such a column usually provides no lateral support when the diaphragm is flexible. (The seismic response coefficient, R, is at most 2.5 for systems where the column is presumed to be part of the lateral system of the building.) A girder that frames into such a column at one girder end and a shear wall at the other girder end almost always acts as a collector for seismic forces parallel to the girder direction.

Example 7.3

The plan view of an irregular building with a flexible diaphragm is shown. (a) Determine the tributary areas for an earthquake in the north-south direction. (b) Determine where the force in the collector is maximum.

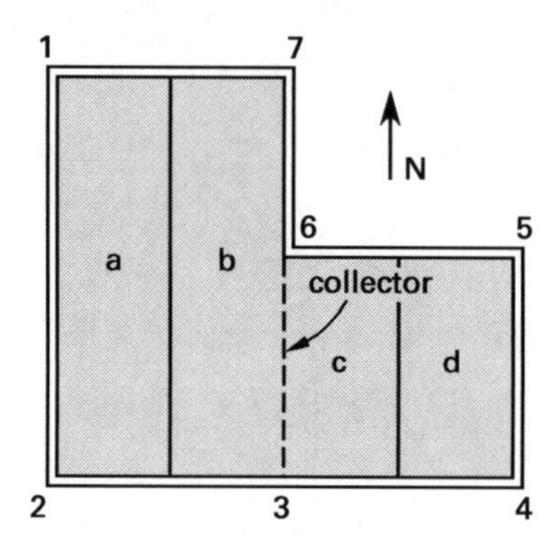

Solution

(a) The walls along sides 1-2, 6-7, and 4-5 will resist the seismic force. (Walls 1-7, 5-6, and 2-3-4 are *perpendicular walls*. See Sec. 7.2.) The collector between areas b and c splits the diaphragm into two rectangular diaphragms: a-b and c-d.

Area a is tributary to wall 1-2. Area b is tributary to wall 6-7. The forces transmitted to the collector are carried by the collector back to wall 6-7.

Area d is tributary to wall 4-5. Area c is tributary to the collector, which transmits all of the area c diaphragm force into wall 6-7.

(b) The collector carries half of area b's diaphragm force and all of area c's diaphragm force. The maximum value of this force occurs at point 6, where the collector frames into the shear wall 6-7.

11. FLEXIBLE DIAPHRAGM CONSTRUCTION

A flexible diaphragm is a relatively thin structural element such as a roof or floor attached to relatively rigid walls. It can be constructed as a braced frame with nonstructural covering, or as joists sheathed with wood structural panel, boards, or gypsum sheets.

Horizontal diaphragms are most common and are the simplest to analyze. However, diaphragms do not need to be horizontal in order to resist shear. A plywood roof can be inclined, peaked (i.e., folded-plate), or curved. In such cases, the roof trusses act as web stiffeners. A non-horizontal diaphragm is analyzed according to its footprint (plan) dimensions.

There are three structural requirements imposed on flexible diaphragms.

1. The diaphragm must be strong enough to remain intact under the action of wind and seismic loads.
2. The diaphragm must be securely attached to a wall in order to resist forces parallel to the wall.
3. The diaphragm must be securely attached to a wall in order to react to forces perpendicular to the wall.

12. FLEXIBLE DIAPHRAGM TORSION

There is no torsional shear stress (see Sec. 5.14) from eccentric mass placement in either the walls or diaphragm because flexible diaphragms are not considered capable of distributing torsional shear stresses.

13. DIAPHRAGM SHEAR FLOW

In most cases, the criterion by which diaphragm construction and connections are evaluated[5] is the *diaphragm shear flow*, q.[6] The shear flow is assumed to exist uniformly across the length b, known as the *diaphragm depth*. (In a building similar to Fig. 7.1, the total force is shared by the two parallel walls, each of length b. The perpendicular walls do not resist the applied seismic force.) The seismic load generated by parallel walls is not included in the diaphragm shear loading.

$$q = \frac{F_{\text{diaphragm}}}{2b} \quad \text{[per unit length]} \qquad 7.8$$

14. DIAPHRAGM FLEXURE

A flexible diaphragm is designed to withstand shear in its plane. It has no bending strength of its own. Rather, the diaphragm relies on the stiffness of its chords to limit overall diaphragm deflection. A common analogy is to assume the diaphragm acts like a girder, where the flanges (i.e., the chord members) resist the bending moment, and the web (i.e., the diaphragm) resists the shear. This is illustrated in Fig. 7.7.

[5]The larger of the seismic and wind shear flows will be used in the design, but not both.

[6]The maximum allowable shear flow on wood structural panel diaphragms depends on the staple size and spacing, panel thickness, and width of framing members, and on whether or not the panel edges are blocked, among other factors. Typical values range from 100 lbf/ft to 800 lbf/ft (1460 N/m to 11 675 N/m), with the lower values (i.e., 100 lbf/ft to 300 lbf/ft (1460 N/m to 4380 N/m)) applying to unblocked diaphragms. Refer to the code for exact limits and required staple spacing. (See Table 12.1 [IBC Table 2306.2(1)].)

Figure 7.7 Web and Flange of a Girder

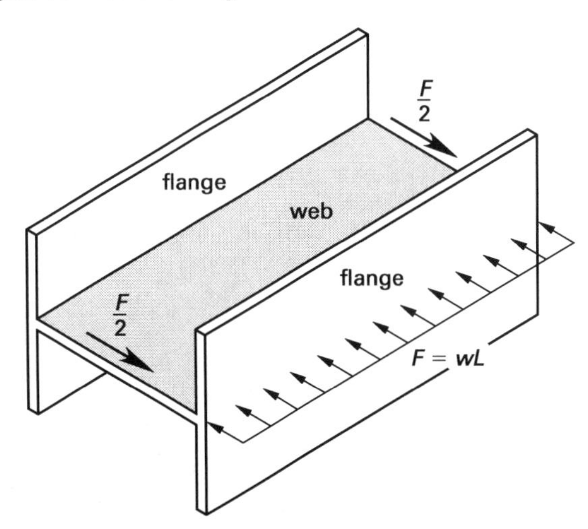

The diaphragm shear flow is assumed to be linearly distributed from zero at the midpoint (i.e., at $L/2$) to a maximum value at the parallel walls. (Distance L in Fig. 7.8 is known as the *diaphragm span*.) At any particular point, the shear flow is traditionally assumed to be uniform across the diaphragm depth between perpendicular walls. Edge fastening of the wood structural panel to the framing keeps the shear resistance continuous across the diaphragm.

Figure 7.8 Shear Flow Distribution on a Diaphragm

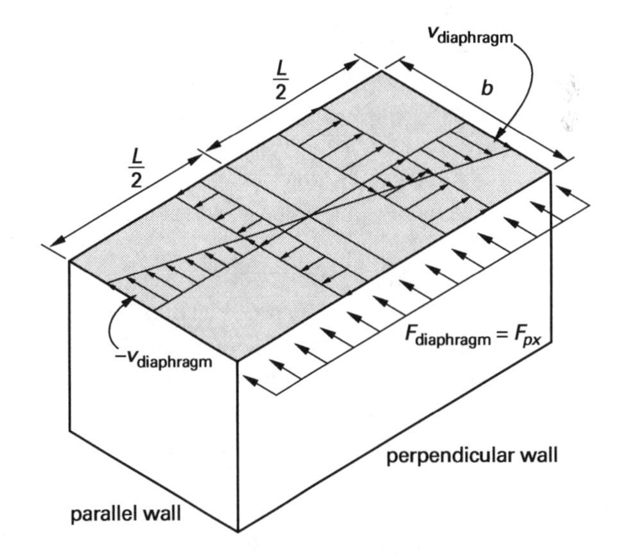

The assumption that the diaphragm shear flow is constant over the depth is inconsistent with the girder analogy. Diaphragms would typically be considered deep beams and have complex internal force distributions. However, assuming constant shear flow provides suitable accuracy for building design. This assumption has been verified by building performance in past earthquakes and experimental testing.

15. DEFLECTION OF FLEXIBLE DIAPHRAGMS

A flexible diaphragm will deflect. The beam analogy described in Sec. 7.14 is also valid here, as the diaphragm assumes the deflected shape of a simply supported beam loaded by a uniform load. The deflection is resisted by the perpendicular walls. Since the perpendicular wall deflection is the same as the diaphragm deflection, this deflection may be the factor that limits the force that can be safely applied to the diaphragm.

Actual determination of the deflection, Δ, in a wood structural panel diaphragm is complex, as it is for any wood/timber structural member, but procedures are available. Equation 7.9 [IBC Eq. 23-1] calculates the plywood diaphragm deflection as a sum of *flexural distortion* (the first term), *shear distortion* (the second term), and *staple distortion* (the third term). In some cases, such as in wood-framed perpendicular walls where a wood double-plate serves as the chord, the fourth term may be added to account for *chord-splice slip* values. Such slippage is neglected with masonry walls.

$$\Delta_{\text{in}} = \frac{5vL^3}{8EAb} + \frac{vL}{4Gt} + 0.122Le_n + \frac{\sum(\Delta_c X)}{2b} \quad \text{[U.S.]} \qquad 7.9(a)$$

$$\Delta_{\text{mm}} = \frac{0.052vL^3}{EAb} + \frac{vL}{4Gt} + \frac{Le_n}{1627} + \frac{\sum(\Delta_c X)}{2b} \quad \text{[SI]} \qquad 7.9(b)$$

A = section area of chord (in^2 or cm^2)
b = diaphragm width or depth (ft or m)
e_n = staple slip, at load per staple (in or cm)
E = modulus of elasticity of chord (psi or kPa)
G = wood structural panel modulus of rigidity (psi or kPa, typically taken as 90,000 psi (620 530 kPa) for wood structural panel; alternatively, $G = E/20$ for panels with exterior glue)
L = diaphragm length (ft or m)
t = wood structural panel thickness (in or cm)
v = maximum shear (lbf/ft or N/m)
$\sum(\Delta_c X)$ = sum of individual chord-splice slip values on both sides of diaphragm, each multiplied by its distance to the nearest support

The staple slip, e_n, depends on the staple size, wood structural panel thickness, load per staple, and type of lumber. Usually, worst-case green lumber is assumed. Staple slip is usually obtained graphically from appropriate sources. Values are typically less than 0.15 in (4 mm), with most values being half that amount.

The maximum deflection of the diaphragm is the acceptable limitation of deflection or drift for the perpendicular walls directly below the diaphragm. This limitation will depend on whether the walls are masonry or concrete and on which authority specifies the limitation. The IBC and ASCE/SEI7 do not specify actual limitations on diaphragm deflection, but the ASCE/SEI7 limits such deflection to amounts that maintain the structural integrity and protect occupants by supporting prescribed loads [ASCE/SEI7 Sec. 12.12.2].[7]

If the deflection is excessive, it can be reduced by increasing the wood structural panel thickness, decreasing the staple spacing, adding a collector strut, or placing an additional shear wall within the building to reduce the diaphragm span.

16. CHORDS

The elements—wall top plates or reinforcement—capable of supporting chord (i.e., compressive and tensile) forces at the edges of the diaphragm along the perpendicular walls are known as *chords*. Chords are generally considered to be tension and compression members, analogous to the flanges of the beam shown in Fig. 7.7. A diaphragm will be constructed with chord elements along all outer edges. The chords that run perpendicular to the applied force, that is, along the perpendicular walls, and are stressed during an earthquake are called the *active chords*. The chord elements in parallel walls are known as the *passive chords*. (See Fig. 7.9.)

Figure 7.9 Chords

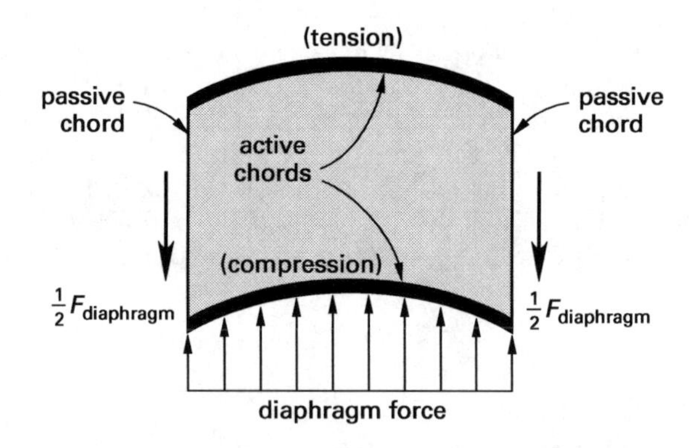

[7] In the 1980s, SEAOC developed the following formula for the maximum allowable deflection.

$$\Delta_{\text{in}} = \frac{75h^2F_b}{Et}$$

In the above equation, h is the wall height (ft), F_b is the allowable masonry flexural (compressible) stress (psi, increased by the 1/3 allowance for seismic loads), E is the masonry modulus of elasticity (1,500,000 psi for masonry, 2,000,000 psi for concrete), and t is the wall thickness (in).

A similar equation has been proposed by the American Institute of Timber Construction, in which the 75 is replaced by 96.

The *Reinforced Masonry Engineering Handbook* suggests the following equation for maximum deflection. This equation can be derived from the SEAOC equation if E=1,500,000 psi and F_b=(4/3)(900 psi) is used. 900 psi is appropriate for concrete walls but may be too high for masonry walls.

$$\Delta_{\text{in}} = \frac{2h^2}{45t}$$

In masonry buildings, reinforcement in the perpendicular walls themselves can (and is intended to) serve as chords if the force is transferred through solidly attached *ledgers* (see Sec. 12.9). This assumes that the masonry walls have adequate tensile reinforcement. Alternatively, a properly spliced wood ledger beam or a bond beam using an embedded reinforcing bar could serve as the chord with masonry walls, depending on the method of attachment. Chords can also be wood (e.g., the double top-plate in conventional wood stud walls), steel, or any other continuous material connected to the diaphragm edge.

The diaphragm may be functionally divided into independent parts, known as *subdiaphragms*. In this case, chords and struts will run through the diaphragm in addition to around it. One method of supporting internal chord members in masonry construction is with pilasters, as shown in Fig. 7.10. A *pilaster* is constructed as part of the masonry wall and is designed as a column.

Figure 7.10 *Pilaster*

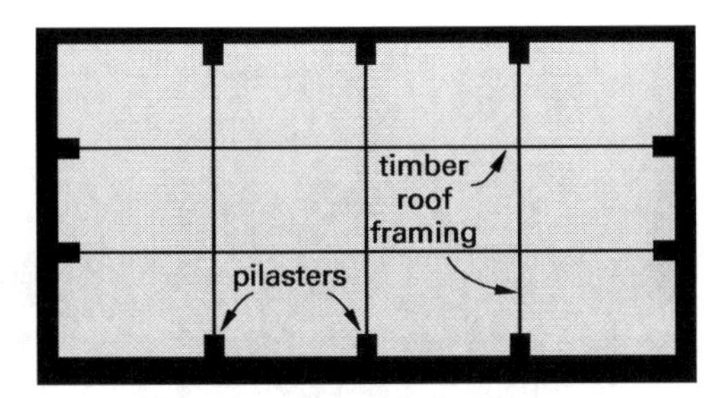

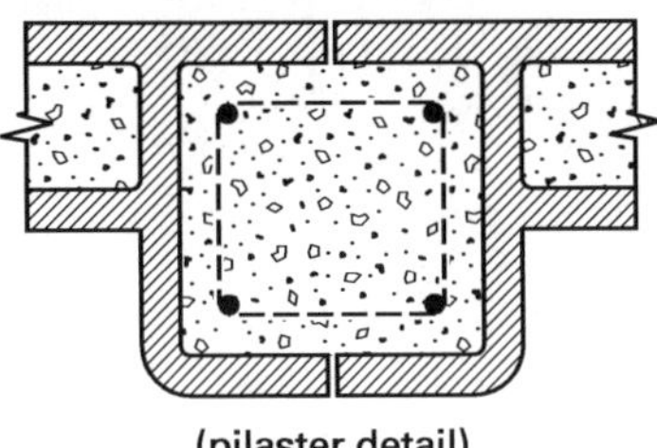

(pilaster detail)

17. CHORD FORCE

A diaphragm is assumed incapable of supporting a normal (bending) stress. Chords are designed to carry all tensile forces.

Many simple buildings have a single rectangular diaphragm with shear walls distributed along the perimeter. For such buildings the maximum chord force occurs at mid-span ($L/2$). The maximum chord force, C, is calculated as the bending moment of a simple beam under a distributed load (i.e., $wL^2/8$, with w in lbf/ft or N/m) divided by the depth, b, of the diaphragm. The distributed load is $w = F_{\text{diaphragm}}/L$, where $F_{\text{diaphragm}}$ is in pounds (N).

$$C = \frac{M}{b} = \frac{wL^2}{8b} = \frac{F_{\text{diaphragm}}L}{8b} \qquad 7.10$$

Equation 7.10 is derived by equating the applied moment to the resisting moment. The applied moment is $wL^2/8$. The resisting moment comes from two sources: the two chords. One chord is in compression; the other is in tension. Both act with a moment arm of $b/2$ (with respect to a neutral axis passing through the midpoint of the diaphragm).

$$\frac{wL^2}{8} = C\left(\frac{b}{2}\right) + C\left(\frac{b}{2}\right) = Cb \qquad 7.11$$

$$C = \frac{wL^2}{8b} \qquad 7.12$$

The minimum chord force is zero and occurs at the chord ends. At intermediate locations, the force follows the shape of the bending moment between the two parallel walls.

18. CHORD SIZE

The required chord size (cross-sectional area) can be determined from the allowable stress in tension or compression, whichever is less, for the chord material. A one-third increase in allowable stress may be permitted when using ASD with alternate load combinations because the loading is seismic. (See Sec. 6.50.)

$$A_{\text{chord}} = \frac{C}{\text{allowable stress}} \qquad 7.13$$

The chord area may be reduced near the ends of the chord, but the reduction must be in accordance with the actual moment distribution.

The allowable tensile stress for steel reinforcing bar depends on its grade (equivalent to its minimum yield strength in ksi (MPa)). Allowable tensile stresses are based on ACI 318 requirements.

19. OVERTURNING MOMENT

In addition to seismic shear loading, a shear wall will be subjected to overturning moments as well. (See Fig. 7.11.) Overturning will not be a problem, however, if there is a larger resisting moment.

Figure 7.11 *Overturning Forces on a Parallel Wall*

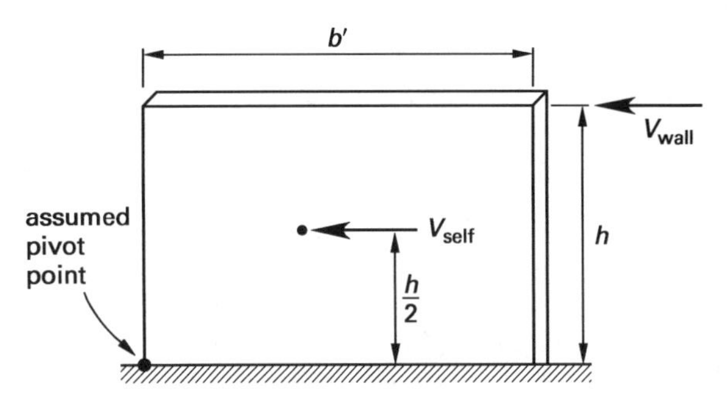

The force causing overturning is the seismic force of V_{wall} at the top of the shear wall. If the shear wall has a length of b', the total overturning force parallel to the ground is $V_{\text{wall}} = v_{\text{diaphragm}} b'$. ($b'$ and b may be the same if the parallel wall is one piece, or b' may correspond to the length of a tilt-up wall section.) This total roof load acts with a moment arm of h, the height of the force above the ground.

Also contributing to overturning is the seismic force due to the wall's self-weight, calculated as either $W_{\text{wall}} a/g$ or $C_s W_{\text{wall}}$. This seismic force acts halfway up the parallel wall, with a moment arm of $h/2$.

$$M_{\text{overturning}} = V_{\text{wall}} h + \frac{V_{\text{self}} h}{2} = V_{\text{wall}} h + \frac{W_{\text{wall}} a h}{2g}$$

$$= V_{\text{wall}} h + \frac{C_s W_{\text{wall}} h}{2} \qquad 7.14$$

The overturning moment is resisted by the weight of the parallel wall and the distributed roof dead load (calculated as the roof load tributary to that panel), both acting with a moment arm of $b'/2$. (See Fig. 7.12.)

$$M_{\text{resisting}} = (D_{\text{roof}} + W_{\text{wall}}) \left(\frac{b'}{2}\right) \qquad 7.15$$

Figure 7.12 *Resisting Forces on a Parallel Wall*

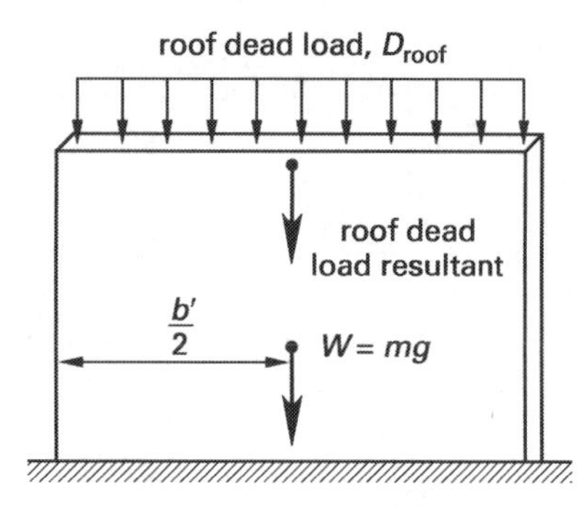

Measures to resist overturning are well known and include anchoring the parallel wall panels to the foundations and, in the case of tilt-up construction, interconnecting adjacent panels.

Example 7.4

A simple four-walled 40 ft by 50 ft (12.2 m by 15.2 m) building is part of a hazardous chemical chlorine storage facility located at a site where S_{D1} is $0.22g$ and S_{DS} is $0.45g$. It is constructed with a wood structural panel sheathed roof on 10 ft high (3 m) special reinforced concrete shear walls 8 in (203 mm) thick. All walls are reinforced vertically and horizontally. The average weight of the roof diaphragm and mounted equipment is 20 lbf/ft^2 (0.96 kN/m^2). All connections between walls, roof, and foundation are pinned. The building performance is being analyzed for an earthquake acting parallel to the short dimension. Disregard all openings in the walls.

(a) Find the shear flow in the diaphragm on line A-B.

(b) Find the required diaphragm edge fastener spacing on line A-B. (Assume $1^1/_2$ in 16-gage staples, case 1 plywood layout, 3/8 in (10 mm) DOC PS 1 sheathing grade on blocked 2 in (51 mm) frame members.)

(c) Find the maximum chord force on line B-B′.

(d) Find the horizontal shear stress (in psi or kPa) in wall A-B at a point 5 ft (1.5 m) above the foundation.

(e) Determine the design shear stress on the shear walls.

(f) If wall A-B was to be reduced to only 10 ft (3 m) long and the remaining 30 ft (9.1 m) of roof supported by a collector, find the collector force at the end of the wall.

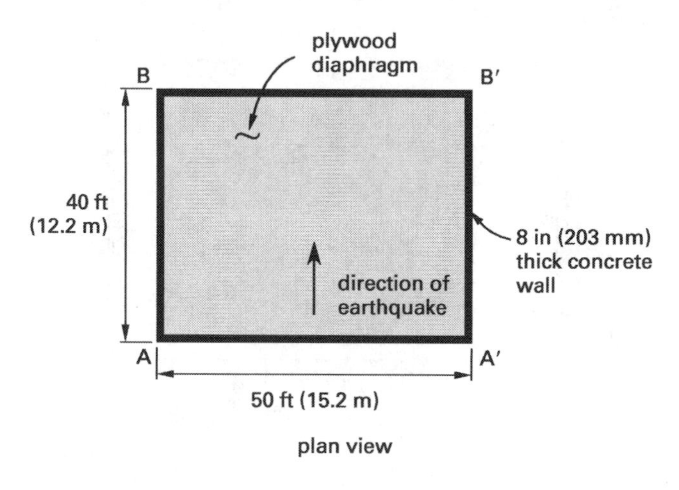

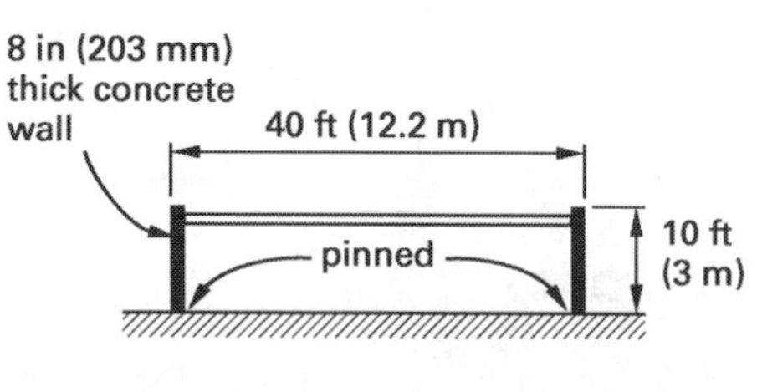

Solution

From Table 6.2 and Table 6.3, for a hazardous facility, the risk category is III and the importance factor, I_e, is 1.25. As the soil type is unknown, site class type D must be used. (See Sec. 6.15.)

From App. H [ASCE/SEI7 Table 12.2-1], for a bearing wall system consisting of special reinforced concrete shear walls, $R=5$. From IBC Table 1613.3.5(1), for risk category III and S_{DS} of $0.45g$, the seismic design category is C. From IBC Table 1613.3.5(2), for risk category III and S_{D1} of $0.22g$, the seismic design category is D. The higher seismic design category governs, so the seismic design category for the building is D.

One-story buildings have small periods of vibration, and in general the maximum limit on the base shear will always govern, so the seismic coefficient can be calculated using Eq. 6.20 [ASCE/SEI7 Eq. 12.8-2].

$$C_s = \frac{S_{DS}}{\frac{R}{I_e}} = \frac{0.45}{\frac{5}{1.25}} = 0.1125g$$

In calculating the seismic shear force as $V = C_s W$, where W is in pounds, C_s is needed in fractions of a gravity (i.e., as a/g or S_{DS}/g). Only if the building mass, in slugs, were given, would the actual value of g need to be retained, $V = C_s mg$, parallel in concept to the ubiquitous engineering equation $F = ma$.

Customary U.S. Solution

The weight being accelerated by the earthquake consists of the diaphragm weight and a portion of the wall weight. The weight of the diaphragm is

$$W_{\text{diaphragm}} = \left(20\ \frac{\text{lbf}}{\text{ft}^2}\right)(40\ \text{ft})(50\ \text{ft}) = 40{,}000\ \text{lbf}$$

Since reinforced concrete has a density of 150 lbf/ft^3, the weight of 1 ft^2 of an 8 in thick wall is

$$\gamma = \frac{(8\ \text{in})\left(150\ \frac{\text{lbf}}{\text{ft}^3}\right)}{12\ \frac{\text{in}}{\text{ft}}} = 100\ \text{lbf/ft}^2$$

For the purpose of determining the diaphragm force, the upper half (i.e., only the upper 5 ft) of both perpendicular walls (i.e., the chords) is used to calculate the wall weight. (See Sec. 6.29 and Sec. 7.2.) The remaining seismic force passes directly into the foundation without being carried by the wall-diaphragm connection. The weight of half the perpendicular walls (chords) is

$$W_{\perp\text{walls}} = \left(100\ \frac{\text{lbf}}{\text{ft}^2}\right)(5\ \text{ft})(2\ \text{walls})\left(50\ \frac{\text{ft}}{\text{wall}}\right) = 50{,}000\ \text{lbf}$$

(a) Equation 6.20 [ASCE/SEI7 Eq. 12.8-2] is used for calculating the base shear passing through to the foundation, but the code has additional requirements for calculating the force on certain elements in the building. For diaphragms, based on ASCE/SEI7 Sec. 12.10.1.1, the roof diaphragm is to be designed to resist a portion of the floor forces, as weighted by the floor weights, and the diaphragm force must be within $0.4 S_{DS} I_e w_{px}$ and $0.2 S_{DS} I_e w_{px}$. With the values of $S_{DS} = 0.45$ and $I_e = 1.25$, the limits on diaphragm force are $0.113 w_{px}$ and $0.225 w_{px}$. Since $0.1125 < 0.113$, use 0.113.

As this is a simple one-story building with only one diaphragm, all of the inertial load from the accelerating wall and roof masses must be carried by the diaphragm.

$$\begin{aligned} F_{\text{diaphragm}} &= 0.113W \\ &= (0.113)(40{,}000\ \text{lbf} + 50{,}000\ \text{lbf}) \\ &= 10{,}170\ \text{lbf} \end{aligned}$$

The shear per foot of diaphragm width, b, is given by Eq. 7.8.

$$q = \frac{F_{\text{diaphragm}}}{2b} = \frac{10{,}170\ \text{lbf}}{(2)(40\ \text{ft})} = 127\ \text{lbf/ft}$$

(b) Table 12.1 [IBC Table 2306.2(1)] gives the staple spacing directly. The allowable shear is 160 lbf/ft (> 127 lbf/ft), with a staple spacing of 6.0 in.

(c) The distributed seismic force, w, across the face of the diaphragm is

$$w = \frac{F_{\text{diaphragm}}}{L} = \frac{10{,}170\ \text{lbf}}{50\ \text{ft}} = 203\ \text{lbf/ft}$$

From Eq. 7.10, the chord force, C, is

$$C = \frac{wL^2}{8b} = \frac{\left(203\ \frac{\text{lbf}}{\text{ft}}\right)(50\ \text{ft})^2}{(8)(40\ \text{ft})} = 1586\ \text{lbf}$$

(d) The net effect on the shear walls is to include the inertial force for accelerating the diaphragm mass and half of all the walls. The diaphragm and perpendicular wall weights have already been determined. Half the weight of the parallel walls (i.e., the shear walls) is

$$W_{\parallel\text{walls}} = \left(100\ \frac{\text{lbf}}{\text{ft}^2}\right)(5\ \text{ft})(2\ \text{walls})\left(40\ \frac{\text{ft}}{\text{wall}}\right) = 40{,}000\ \text{lbf}$$

The total weight is

$$\begin{aligned} W_{\text{total}} &= W_{\text{diaphragm}} + W_{\perp\text{walls}} + W_{\parallel\text{walls}} \\ &= 40{,}000\ \text{lbf} + 50{,}000\ \text{lbf} + 40{,}000\ \text{lbf} \\ &= 130{,}000\ \text{lbf} \end{aligned}$$

The seismic force is

$$\begin{aligned} V &= 0.113W = (0.113)(130{,}000\ \text{lbf}) \\ &= 14{,}690\ \text{lbf} \end{aligned}$$

Since the perpendicular walls (chords) have no rigidity, all of the seismic force is resisted by the two parallel walls (shear walls). The shear stress is

$$v = \frac{V}{A} = \frac{14{,}690 \text{ lbf}}{(2 \text{ walls})(8 \text{ in})\left(40 \frac{\text{ft}}{\text{wall}}\right)\left(12 \frac{\text{in}}{\text{ft}}\right)}$$
$$= 1.91 \text{ psi}$$

(e) The stress on the shear walls was determined in part (d) to be 1.91 psi. For specially reinforced concrete shear walls, the stress on the shear walls is the design stress.

If the walls had been masonry, the design stress would include a 50% increase [ACI 530 Sec. 1.18.3.2.6.1.2]. (See Sec. 7.3.)

(f) Since the shear load along the wall was calculated in part (a) to be 127 lbf/ft, the connection between the 10 ft stub wall and the collector must carry 30 ft of compressive or tensile loading.

$$F_{\text{collector}} = (30 \text{ ft})\left(127 \frac{\text{lbf}}{\text{ft}}\right) = 3810 \text{ lbf}$$

Eliminating 30 ft of parallel wall (shear wall) reduces the accelerating mass but does not reduce the diaphragm force. Only the perpendicular walls affect the diaphragm force.

SI Solution

The weight being accelerated by the earthquake consists of the diaphragm weight and a portion of the wall weight. The weight of the diaphragm is

$$W_{\text{diaphragm}} = \left(0.96 \frac{\text{kN}}{\text{m}^2}\right)(12.2 \text{ m})(15.2 \text{ m})$$
$$= 178 \text{ kN}$$

Since reinforced concrete has a density of 2400 kg/m^3, the weight of 1 m^2 of a 203 mm thick wall is

$$\gamma = \frac{(203 \text{ mm})\left(2400 \frac{\text{kg}}{\text{m}^3}\right)\left(9.81 \frac{\text{m}}{\text{s}^2}\right)}{\left(1000 \frac{\text{mm}}{\text{m}}\right)\left(1000 \frac{\text{N}}{\text{kN}}\right)}$$
$$= 4.78 \text{ kN/m}^2$$

For the purpose of determining the diaphragm force, the upper half (i.e., only the upper 1.5 m) of both perpendicular walls (i.e., the chords) is used to calculate the wall weight. (See Sec. 6.29 and Sec. 7.2.) The remaining seismic force passes directly into the foundation without being carried by the wall-diaphragm connection. The weight of half the perpendicular walls (chords) is

$$W_{\perp\text{walls}} = \left(4.78 \frac{\text{kN}}{\text{m}^2}\right)(1.5 \text{ m})(2 \text{ walls})\left(15.2 \frac{\text{m}}{\text{wall}}\right)$$
$$= 218 \text{ kN}$$

(a) Equation 6.20 [ASCE/SEI7 Eq. 12.8-2] is used for calculating the base shear passing through to the foundation, but the code has additional requirements for calculating the force on certain elements in the building. For diaphragms, based on ASCE/SEI7 Sec. 12.10.1.1, the roof diaphragm is to be designed to resist a portion of the floor forces above it, as weighted by the floor weights, and the diaphragm force must be within $0.4S_{DS}I_e w_{px}$ and $0.2S_{DS}I_e w_{px}$. With the values of $S_{DS}=0.45$ and $I_e=1.25$, the limits on diaphragm force are $0.113w_{px}$ and $0.225w_{px}$. Since 0.1125 < 0.113, use 0.113.

As this is a simple one-story building with only one diaphragm, all of the inertial load from the accelerating wall and roof masses must be carried by the diaphragm.

$$F_{\text{diaphragm}} = 0.113W$$
$$= (0.113)(178 \text{ kN} + 218 \text{ kN})$$
$$= 44.75 \text{ kN}$$

The shear per meter of diaphragm width, b, is given by Eq. 7.8.

$$q = \frac{F_{\text{diaphragm}}}{2b} = \frac{(44.75 \text{ kN})\left(1000 \frac{\text{N}}{\text{kN}}\right)}{(2)(12.2 \text{ m})}$$
$$= 1834 \text{ N/m}$$

(b) Table 12.1 [IBC Table 2306.2(1)] gives the staple spacing directly. The allowable shear is 2335 N/m (>1834 N/m), with a staple spacing of 152 mm.

(c) The distributed seismic force, w, across the face of the diaphragm is

$$w = \frac{F_{\text{diaphragm}}}{L} = \frac{44.75 \text{ kN}}{15.2 \text{ m}}$$
$$= 2.94 \text{ kN/m}$$

From Eq. 7.10, the chord force, C, is

$$C = \frac{wL^2}{8b} = \frac{\left(2.94 \frac{\text{kN}}{\text{m}}\right)(15.2 \text{ m})^2}{(8)(12.2 \text{ m})}$$
$$= 6.96 \text{ kN}$$

(d) The net effect on the shear walls is to include the inertial force for accelerating the diaphragm mass and half of all the walls. The diaphragm and perpendicular wall weights have already been determined. Half the weight of the parallel walls (i.e., the shear walls) is

$$W_{\|\text{walls}} = \left(4.78 \frac{\text{kN}}{\text{m}^2}\right)(1.5 \text{ m})(2 \text{ walls})\left(12.2 \frac{\text{ft}}{\text{wall}}\right)$$
$$= 175 \text{ kN}$$

The total weight is

$$\begin{aligned} W_{\text{total}} &= W_{\text{diaphragm}} + W_{\perp\text{walls}} + W_{\|\text{walls}} \\ &= 178\ \text{kN} + 218\ \text{kN} + 175\ \text{kN} \\ &= 571\ \text{kN} \end{aligned}$$

The seismic force is

$$\begin{aligned} V &= 0.113W = (0.113)(571\ \text{kN}) \\ &= 64.5\ \text{kN} \end{aligned}$$

Since the perpendicular walls (chords) have no rigidity, all of the seismic force is resisted by the two parallel walls (shear walls). The shear stress is

$$\begin{aligned} v = \frac{V}{A} &= \frac{(64.5\ \text{kN})\left(1000\ \frac{\text{mm}}{\text{m}}\right)}{(2\ \text{walls})(203\ \text{mm})\left(12.2\ \frac{\text{m}}{\text{wall}}\right)} \\ &= 13.0\ \text{kPa} \end{aligned}$$

(e) The stress on the shear walls was determined in part (d) to be 13.0 kPa. For specially reinforced concrete shear walls, the stress on the shear walls is the design stress. If the walls had been masonry, the design stress would include a 50% increase [ACI 530 Sec. 1.18.3.2.6.1.2]. (See Sec. 7.3.)

(f) Since the shear load along the wall was calculated in part (a) to be 1834 N/m, the connection between the 3 m stub wall and the collector must carry 9.1 m of compressive or tensile loading.

$$F_{\text{collector}} = \frac{(9.1\ \text{m})\left(1834\ \frac{\text{N}}{\text{m}}\right)}{1000\ \frac{\text{N}}{\text{kN}}} = 16.7\ \text{kN}$$

Eliminating 9.1 m of parallel wall (shear wall) reduces the accelerating mass but does not reduce the diaphragm force. Only the perpendicular walls affect the diaphragm force.

8 General Structural Design

1. DISTRIBUTING STORY SHEARS TO MEMBERS OF UNKNOWN SIZE

Chapter 7 dealt with distributing seismic forces in frames whose resisting elements (fixed columns, shear walls, etc.) were already designed. (The seismic force to be resisted is the sum of story shears for a particular level and all levels above.) If the cross-sectional area and moment of inertia of these resisting elements are not known (as they will not be initially), assumptions must be made about the amount of lateral force each element carries.

There are several approximate methods (e.g., the portal, cantilever, and Spurr methods discussed in the next sections) that distribute the lateral forces to the resisting elements. These methods eliminate the need to use indeterminate solution methods such as moment distribution.[1]

2. PORTAL METHOD

The *portal method* is ideal for cases where the framing system is regularly spaced. It assumes that all interior columns carry the same shear, while exterior columns carry half the shear of interior columns.[2] Inflection points are assumed to occur at mid-span in each girder and column. (See Fig. 8.1.) Changes in length due to compression, tension, and deflection are disregarded. Each bay is treated independently of adjacent bays.

[1]Since the member geometries (areas and moments of inertia) are initially unknown, a rigorous method cannot be used.

[2]If the bay sizes (i.e., the distances between columns) are not the same, the shear may be distributed to the columns in proportion to the bay sizes.

Figure 8.1 *Portal Method Frame Deflection*

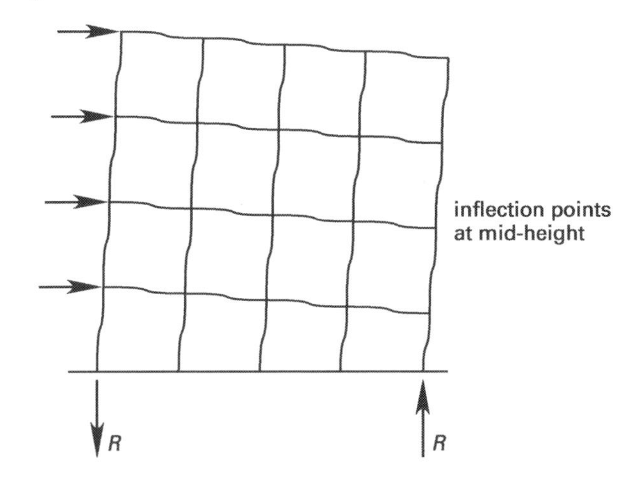

Example 8.1

The sum of story shears for a particular layer and those above it is 110 kips (490 kN). Use the portal method to find the shears in members 2-3 (i.e., the vertical shear on the horizontal members) and 2-7 (i.e., the horizontal shear on the vertical members).

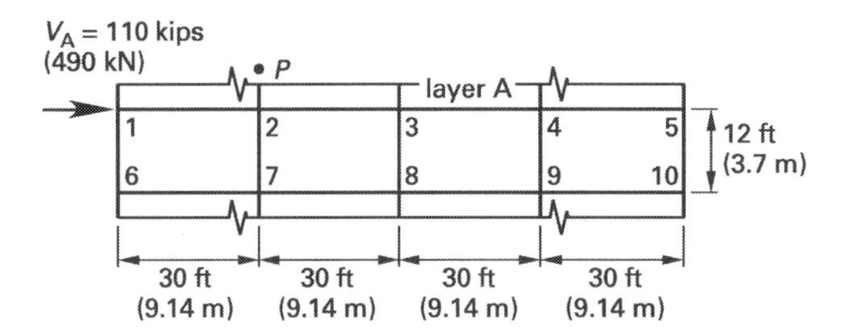

Customary U.S. Solution

Exterior columns 1-6 and 5-10 carry only half the loads of the interior columns. Therefore, there are a total of four full-strength columns (three interior full-strength columns plus two exterior half-strength columns). The horizontal shear carried by member 2-7 and all other interior columns is

$$V_{\text{interior}} = \frac{110 \text{ kips}}{4} = 27.5 \text{ kips}$$

Column 2-7 carries no axial load.

Forces are assumed to be applied at the inflection points of the columns and girders. The vertical girder shears, V_{girder}, are obtained by taking a free-body diagram of joint 2 and summing moments about point 2.

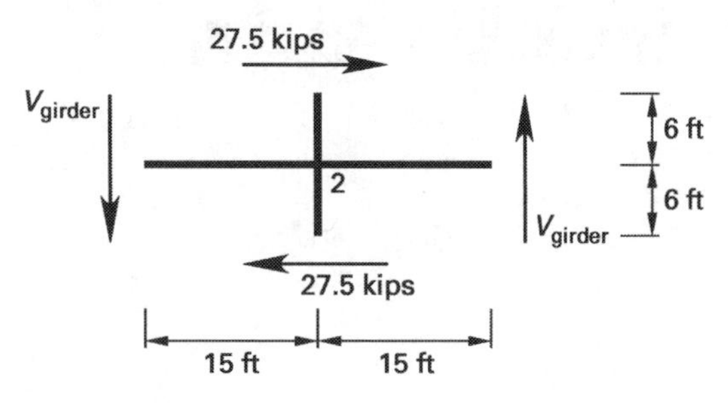

$$\sum M_2 = (2)(27.5 \text{ kips})(6 \text{ ft}) - 2V_{girder}(15 \text{ ft}) = 0$$

$$V_{girder} = 11 \text{ kips}$$

SI Solution

Exterior columns 1-6 and 5-10 carry only half the loads of the interior columns. Therefore, there are a total of four full-strength columns (three interior full-strength columns plus two exterior half-strength columns). The horizontal shear carried by member 2-7 and all other interior columns is

$$V_{interior} = \frac{490 \text{ kN}}{4} = 123 \text{ kN}$$

Column 2-7 carries no axial load.

Forces are assumed to be applied at the inflection points of the columns and girders. The vertical girder shears, V_{girder}, are obtained by taking a free-body diagram of joint 2 and summing moments about point 2.

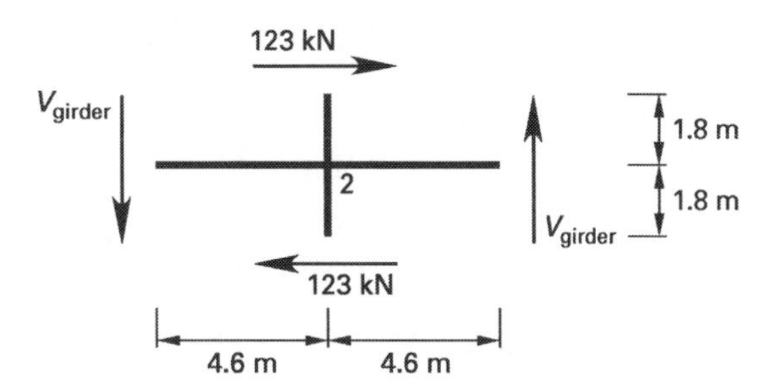

$$\sum M_2 = (2)(123 \text{ kN})(1.8 \text{ m}) - 2V_{girder}(4.6 \text{ m}) = 0$$

$$V_{girder} = 48.1 \text{ kN}$$

3. CANTILEVER METHOD

Unlike the portal method, which treats each bay independently of the others, the *cantilever method* assumes the entire floor works together as a unit. (See Fig. 8.2.) Although analysis of the cantilever method must begin at the roof level and work down, this method is preferred for buildings with more than 25 stories. It assumes that the floors remain plane (though not horizontal) and the force in a column is proportional to the distance of the column from the frame's center of gravity. As with the portal method, inflection points of columns and girders are assumed to occur at the mid-lengths.

Figure 8.2 *Cantilever Method Frame Deflection*

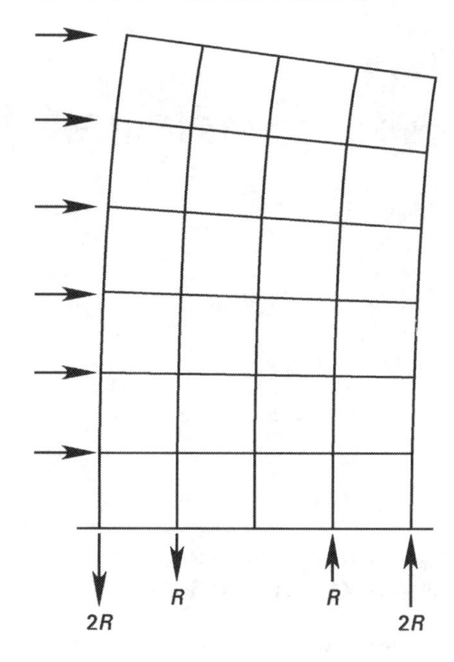

Example 8.2

The cumulative lateral force at the roof level of the frame shown is 110 kips (490 kN). Use the cantilever method to determine the shear in members 2-3 and 2-7.

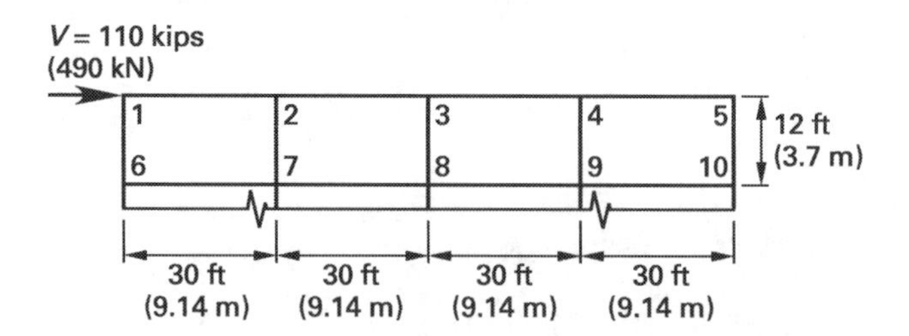

Customary U.S. Solution

The center of gravity of this level is located at point Q. Columns 1-6 and 2-7 are in tension. Columns 4-9 and 5-10 are in compression. Column 3-8 is along the neutral axis and is not stressed axially.

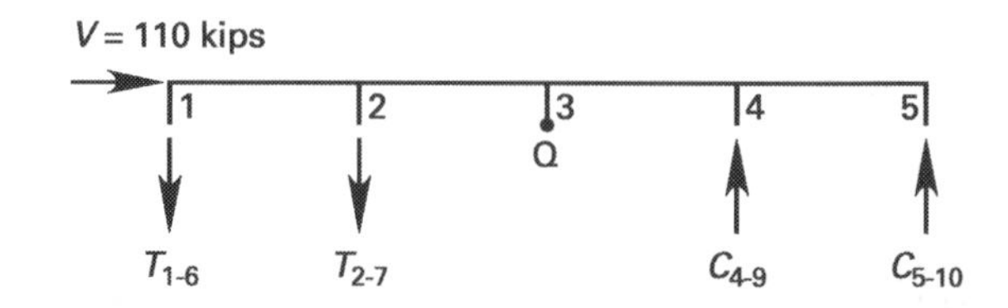

The sum of moments in a member at an inflection point is zero. Summing moments about point Q (halfway between points 3 and 8),

$$\begin{aligned}\sum M_Q &= (6\ \text{ft})(110\ \text{kips}) - (60\ \text{ft})T_{1\text{-}6} - (30\ \text{ft})T_{2\text{-}7} \\ &\quad - (30\ \text{ft})C_{4\text{-}9} - (60\ \text{ft})C_{5\text{-}10} \\ &= 0\end{aligned}$$

Since the column forces are proportional to the distance from column 3-8 and all columns are separated by the same distance, it is apparent that the relationships between the tension forces, T, and compressive forces, C, are

$$T_{1\text{-}6} = 2T_{2\text{-}7} = 2C_{4\text{-}9} = C_{5\text{-}10}$$

Making these substitutions,

$$T_{1\text{-}6} = C_{5\text{-}10} = 4.4\ \text{kips}$$
$$T_{2\text{-}7} = C_{4\text{-}9} = 2.2\ \text{kips}$$

The vertical shear in member 1-2 is equal to the sum of vertical loads to the left of its inflection point.

$$V_{1\text{-}2} = T_{1\text{-}6} = 4.4\ \text{kips}$$

Taking the free body of the inflection points and summing moments about point 1 gives the horizontal shear in member 1-6.

$$\sum M_1 = (6\ \text{ft})V_{1\text{-}6} - (15\ \text{ft})(4.4\ \text{kips}) = 0$$
$$V_{1\text{-}6} = 11\ \text{kips}$$

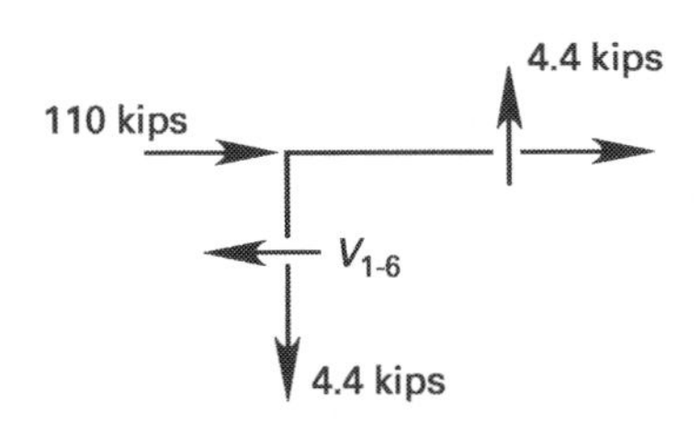

The vertical shear in member 2-3 is

$$V_{2\text{-}3} = 2.2\ \text{kips} + 4.4\ \text{kips} = 6.6\ \text{kips}$$

(This is approximately half the value calculated with the portal method in Ex. 8.1.)

Summing moments about point 2 gives the horizontal shear in member 2-7.

$$\begin{aligned}\sum M_2 &= (6\ \text{ft})V_{2\text{-}7} - (15\ \text{ft})(4.4\ \text{kips}) - (15\ \text{ft})(6.6\ \text{kips}) \\ &= 0\end{aligned}$$
$$V_{2\text{-}7} = 27.5\ \text{kips}$$

(This is the same as the value calculated with the portal method in Ex. 8.1.)

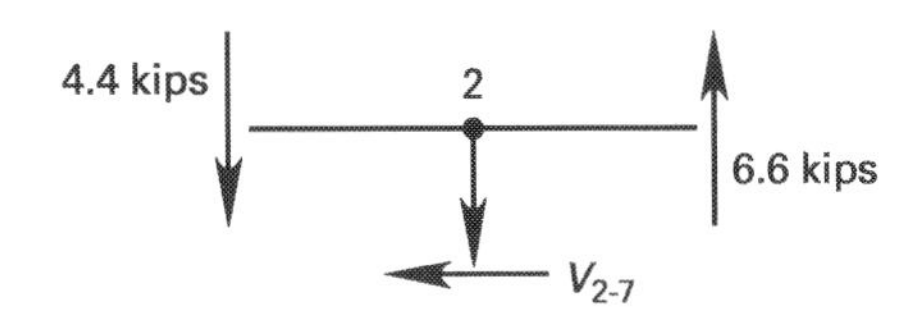

SI Solution

The center of gravity of this level is located at point Q. Columns 1-6 and 2-7 are in tension. Columns 4-9 and 5-10 are in compression. Column 3-8 is along the neutral axis and is not stressed axially.

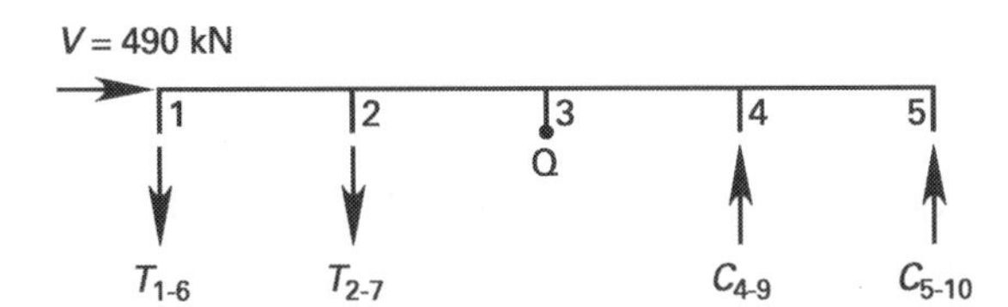

The sum of moments in a member at an inflection point is zero. Summing moments about point Q (halfway between points 3 and 8),

$$\begin{aligned}\sum M_Q &= \left(\frac{3.7\ \text{m}}{2}\right)(490\ \text{kN}) - (18.28\ \text{m})T_{1\text{-}6} \\ &\quad - (9.14\ \text{m})T_{2\text{-}7} - (9.14\ \text{m})C_{4\text{-}9} \\ &\quad - (18.28\ \text{m})C_{5\text{-}10} \\ &= 0\end{aligned}$$

Since the column forces are proportional to the distance from column 3-8 and all columns are separated by the same distance, it is apparent that the relationships between the tension forces, T, and compressive forces, C, are

$$T_{1\text{-}6} = 2T_{2\text{-}7} = 2C_{4\text{-}9} = C_{5\text{-}10}$$

Making these substitutions,

$$T_{1\text{-}6} = C_{5\text{-}10} = 20\ \text{kN}$$
$$T_{2\text{-}7} = C_{4\text{-}9} = 10\ \text{kN}$$

The vertical shear in member 1-2 is equal to the sum of vertical loads to the left of its inflection point.

$$V_{1\text{-}2} = T_{1\text{-}6} = 20\ \text{kN}$$

Taking the free body of the inflection points and summing moments about point 1 gives the horizontal shear in member 1-6.

$$\sum M_1 = (1.83\ \text{m})V_{1\text{-}6} - (4.57\ \text{m})(20\ \text{kN}) = 0$$
$$V_{1\text{-}6} = 50\ \text{kN}$$

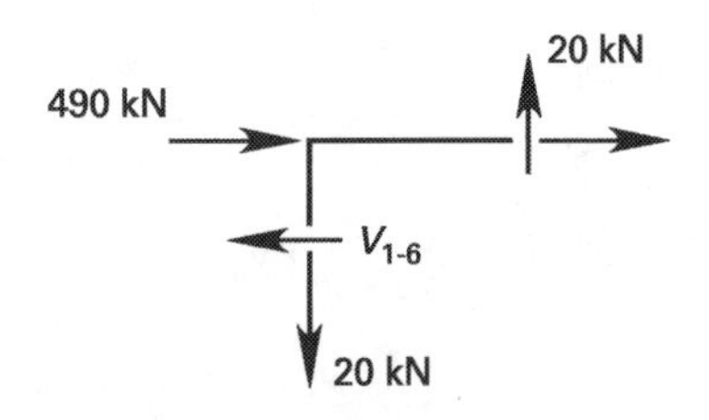

The vertical shear in member 2-3 is

$$\begin{aligned} V_{2\text{-}3} &= 10\ \text{kN} + 20\ \text{kN} \\ &= 30\ \text{kN} \end{aligned}$$

(This is approximately half the value calculated with the portal method in Ex. 8.1.)

Summing moments about point 2 gives the horizontal shear in member 2-7.

$$\begin{aligned} \sum M_2 = (1.83\ \text{m})V_{2\text{-}7} - (4.57\ \text{m})(20\ \text{kN}) \\ - (4.57\ \text{m})(30\ \text{kN}) = 0 \\ V_{2\text{-}7} = 125\ \text{kN} \end{aligned}$$

(This is essentially the same as the value calculated with the portal method in Ex. 8.1.)

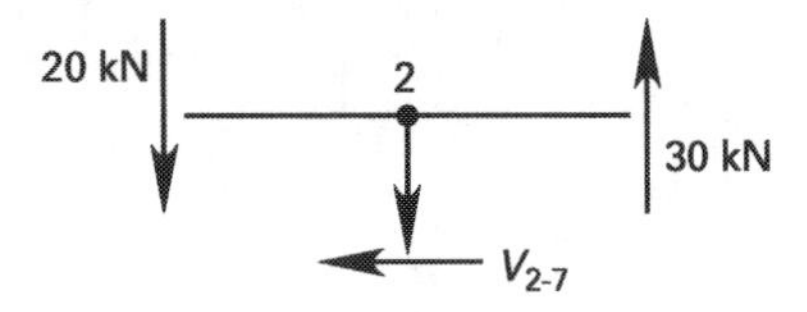

4. SPURR METHOD

Although the cantilever method assumes that the floor will remain plane, it does nothing to ensure a plane floor. Irregular lengthening and shortening of the columns produce secondary bending stresses that can become excessive in tall structures and disturb the planar nature of the floor.

If the height-to-bay length ratio is less than 4, the floor will indeed remain fairly plane. However, when the height-to-bay length ratio exceeds 4, interior column elongation and shortening will cause noticeable floor deflection.

The *Spurr method* uses the cantilever method to calculate girder shears, but girders are given specific strengths in order to justify the assumption that inflection points are located at mid-span.[3] Essentially, the girder moments of inertia are made proportional to the girder shear times the square of the span length.

5. FRAME MEMBER SIZING

Once the column and girder shears and axial loads are known, these members can be sized by traditional methods. For example, a girder can be considered to be loaded at its mid-points (between two sets of columns) by the vertical girder shear acting in opposite directions. (See Fig. 8.3.)

Figure 8.3 *Girder Shear and Moment Diagrams*

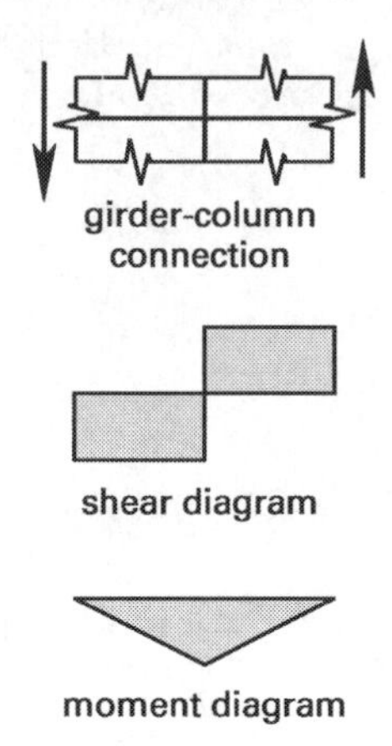

Another method for sizing frame members is to select the columns and girders so that the drift is limited to the ASCE/SEI7 Sec. 12.12 maximum under the action of the horizontal forces. This method, however, requires making assumptions about the joint rigidity. For example, assuming that a bend consisting of a girder of length L and two columns of height h must limit the drift to 2.5% of the story height ($T < 0.7$ sec) when a total shear force of V is experienced, and assuming all joints are rigid, the drift criterion is

$$\left(\frac{Vh^2}{12E}\right)\left(\frac{h}{I_{\text{column}}} + \frac{L}{I_{\text{girder}}}\right) = 0.025h \qquad 8.1$$

There are many possible combinations of column and girder strengths that will satisfy this criterion.

[3]This method is named for H. V. Spurr, who published the method in his book *Wind Bracing: The Importance of Rigidity in High Towers*, New York: McGraw-Hill, 1930.

9 Details of Seismic-Resistant Concrete Structures

1. CONCRETE CONSTRUCTION DETAILS

Concrete itself has poor ductility in shear and is, therefore, a brittle material. This limited ductility, combined with its higher mass compared to steel (higher mass increases the seismic force) and lower tolerance to errors in design and workmanship, usually makes concrete a second choice for high-rise construction. However, concrete ductile moment-resisting frames now make it possible to design structures with the ductility and energy dissipation capability of steel structures. In some cases (particularly those not requiring seismic provisions), a concrete building may be less expensive to construct.[1] The primary disadvantage of concrete is its weight, which increases the seismic forces. Some advantage can be gained, however, by using lightweight concrete.

Concrete structures can be constructed to behave in a ductile manner. Such structures are loosely referred to as having been constructed of "ductile concrete," "special concrete," or "California concrete." In order to make concrete into a ductile structure, much attention must be given to connection and confinement details.[2] (See Fig. 9.1.) "Ordinary concrete," or concrete construction that does not behave in a ductile way, cannot be used in California.

The two most important concepts in designing ductile concrete are (1) continuity and (2) confinement. Concrete members must not pull apart, and they must not disintegrate when the core becomes cracked or crushed.

The IBC discusses the required details of ductile concrete in Sec. 1905. The following lettered items list some of the more important provisions. Because all provisions have been greatly simplified in this book, and since the IBC modifies some detailing requirements, the IBC and ACI 318 references should be used for clarification. In almost all cases, other provisions also apply. (In some of the following cases, the accompanying figures contain additional information about special seismic provisions.)

[1]The assumed economic superiority of steel over concrete is traditional and is a very controversial issue; it is not, by any means, an absolute fact.

[2]It is not always easy, however, to get all the special steel reinforcing and confinement into the beam-column connection area. The joint congestion, called a "tangle" by some, can be considerable.

Figure 9.1 *Typical Ductile Frame Joint for Region of High Seismic Risk [ACI 318 Chap. 21]*

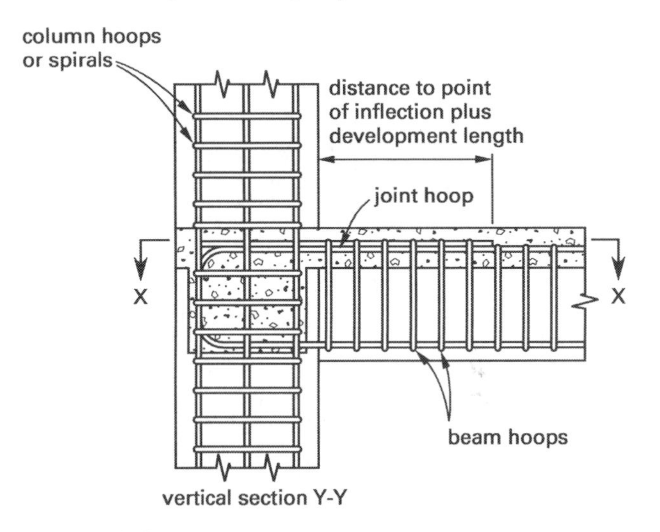

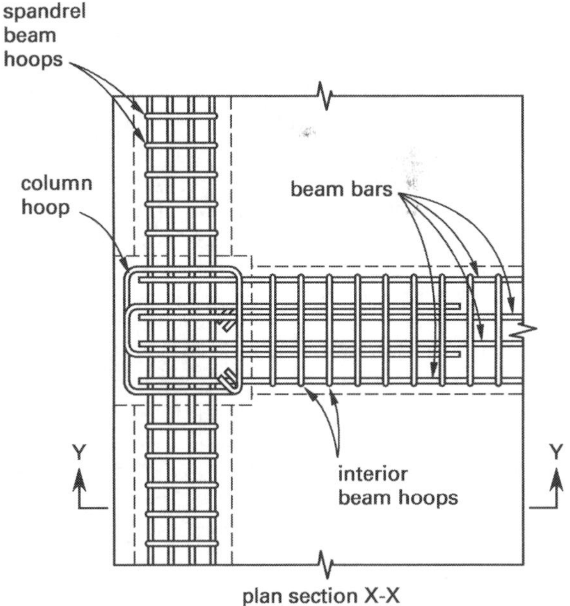

The special provisions of the IBC apply to reinforced concrete frames and shear wall structures that resist earthquake loading. ACI 318 requires members that are not part of the lateral force resisting system to satisfy minimum reinforcement requirements.

The design of concrete structures should be in accordance with the strength design (load and resistance factor design) or allowable stress design method.

A. Orthogonal Effects [ASCE/SEI7 Sec. 12.5]

Orthogonal effects should be investigated in cases of torsional irregularity, nonparallel structural systems, and where a member is part of two intersecting lateral force resisting systems (e.g., where the member is a corner column more than lightly loaded by seismic forces).[3]

B. Connections
[IBC Sec. 1908; ASCE/SEI7 Sec. 12.14.7.1; and ACI 318 Sec. 21.3, Sec. 21.6, and Sec. 21.8]

Connections are elements that interconnect two precast members or a precast member and a cast-in-place member. Connections that resist seismic forces must be designed and detailed by an engineer and shown on the drawings.

A strong connection is one that remains elastic while the specified nonlinear action regions (i.e., the member length over which nonlinear action occurs) endure inelastic response under the design basis ground motion. When a connection employs any of the code splicing methods to join precast members and utilizes cast-in-place concrete or grout to fill the splicing closure, it is referred to as a "wet connection." A connection used between precast members that does not qualify as a wet connection is called a "dry connection."

C. Deformation Compatibility
[ASCE/SEI7 Sec. 12.12.3 and Sec. 12.12.5]

Not all members in a structure are part of the lateral force resisting system. However, members that are not part of that system and their connections must be adequate to maintain support of design dead plus live loads when subjected to the expected deformations caused by seismic forces. For concrete elements that are not part of the lateral force resisting system, provisions of Sec. 21.13 of ACI 318 and ASCE/SEI7 Sec. 12.12.5 apply.

For concrete elements that are part of the lateral force resisting system, the accepted flexural and shear stiffness properties should not exceed one-half of the gross section properties, except on the condition that a rational cracked-section analysis is performed. Foundation flexibility and diaphragm deflections may create additional deformation that should also be considered.

D. Ties and Continuity [ASCE/SEI7 Sec. 12.10.2.1]

It is important that structural members do not pull apart. For buildings in seismic design categories C, D, E, or F, as a minimum, all smaller portions of a building must be tied to the rest of the building with elements having at least a strength with overstrength requirements to resist the special load combinations of ASCE/SEI7 Sec. 12.4.2.3.

[3]The column is more than lightly loaded if its factored axial load due to seismic forces is greater than 20% of the column axial strength.

E. Collector Elements [ASCE/SEI7 Sec. 12.10.2 and Sec. 12.14.7.3]

Collector elements transfer lateral forces from a portion of a structure to vertical elements of the lateral force resisting system. Concrete collector elements must rely on reinforcing steel to carry drag forces into shear walls.

For good engineering practice, when collector elements in topping slabs are placed over precast floor or roof elements, the slab thickness should not be less than 3 in (76 mm) or six times the diameter of the largest reinforcement ($6d_b$).

F. Concrete Frames [ASCE/SEI7 Sec. 12.2.1]

Concrete frames used in lateral force resisting systems in seismic design categories D, E, or F must be *special moment-resisting frames* (SMRF). In seismic design category C, as a minimum, these frames should be *intermediate moment-resisting frames* (IMRF).

G. Anchorage to Concrete and Masonry Walls [ASCE/SEI7 Sec. 12.11.2 and Sec. 13.4.2]

To provide out-of-plane support, the roof should be anchored to the walls and columns, and walls and columns should be anchored to the foundations. The provisions of ASCE/SEI7 Sec. 12.11.2 apply. Using the greater of the wind or earthquake loads in design, these anchorages should be capable of resisting load combinations using strength design or allowable stress design [ASCE/SEI7 Sec. 12.4.2.3]. As a minimum, these connections to and between walls must be capable of resisting the horizontal force of $0.4S_{DS}I_e$ times the weight of the wall, or 10% of the weight of the wall substituted for earthquake load, E [ASCE/SEI7 Sec. 12.11]. When anchor spacing exceeds 4 ft (1219 mm), walls must resist bending between the horizontal anchors.

In seismic design categories C, D, E, or F, when diaphragms are anchored to walls using embedded straps providing out-of-plane lateral support of the wall, the straps must either attach to or hook around the reinforcing steel or terminate so that forces can effectively be transferred to the reinforcing steel. Where flexible diaphragms are used to provide lateral support for the walls (out-of-plane wall anchorage to flexible diaphragms) in seismic design categories C, D, E, or F, the value of total design lateral seismic force, F_p, used for the design of the elements of the wall anchorage system should be determined by ASCE/SEI7 Eq. 12.11-1.

H. Boundary Elements of Shear Walls and Diaphragms [ACI 318 Sec. 21.9.6]

For special reinforced concrete structural shear walls, the reinforcement ratio cannot be less than 0.0025 along both the longitudinal and transverse axes of shear walls and diaphragms. Reinforcement spacing cannot exceed 18 in (457 mm). (See Fig. 9.2.)

Figure 9.2 Requirements for Shear Walls and Boundary Members

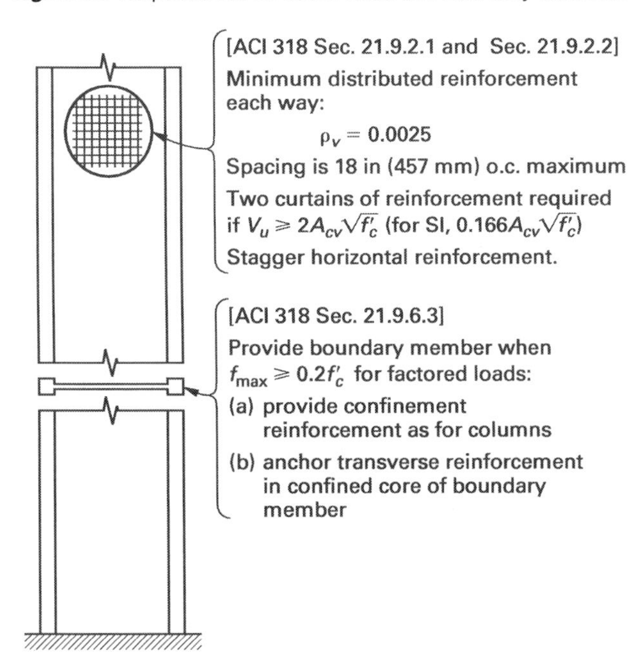

Figure 9.3 Seismic Hook, Crosstie, and Hoop [ACI 318 Sec. 2.2]

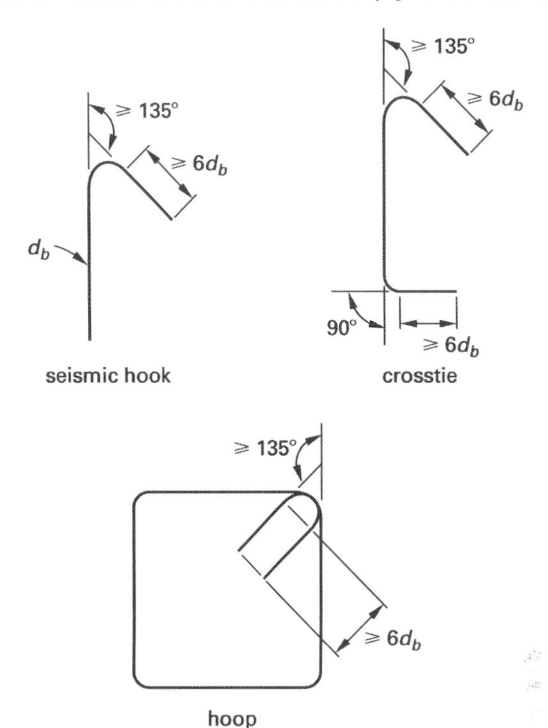

Boundary elements of shear walls must not fail due to deformations caused by a major earthquake, particularly when the design is governed by flexure. The design should specify special detailing to prevent tensile fracture as well as compressive crushing and buckling. This is accomplished by requiring two curtains of shear reinforcement for walls with factored shear in excess of $2A_{cv}\sqrt{f'_c}$ (in SI units, $0.166A_{cv}\sqrt{f'_c}$) [ACI 318 Sec. 21.9.2.2]. Additional reinforcement is required at the edges of all shear walls and diaphragms, as well as at the boundaries of all openings.

I. Seismic Hooks, Crossties, and Hoops [ACI 318 Sec. 2.2]

A *seismic hook* is a hook on a stirrup that engages the longitudinal reinforcement and projects into the interior of the stirrup or hoop. For a seismic hook, the bend must be at least 135° with an extension past the bend of at least 6 bar diameters (but not less than 3 in (76 mm)). A *crosstie* is a continuous reinforcing bar. It has a seismic hook at one end and a hook at least 90° with an extension past the bend of at least 6 bar diameters at the other end. A *hoop* is a closed tie made up of one or more reinforcing elements, each consisting of seismic hooks at both ends. The details of seismic hooks, crossties, and hoops are shown in Fig. 9.3.

J. Strength Reduction Factors [ACI 318 Sec. 9.3]

In general, the normal strength reduction factor, ϕ, defined in ACI 318 Sec. 9.3 is applicable. For structures that rely on special moment-resisting frames or special reinforced concrete structural walls to resist earthquake effects, provisions of ACI 318 Sec. 9.3.4 apply to strength reduction factors. The strength reduction factor is 0.60 for structural framing members (joints are excluded) when their nominal shear strength is less than the shear corresponding to development of their nominal flexural strength. The strength reduction factor for concrete diaphragms is the minimum of any strength reduction factor used for shear in any of the vertical components of the primary lateral force resisting system. The strength reduction factor is 0.85 for beam-column joints. The nominal flexural strength should concur with the most critical factored axial loads, including earthquake effects. The strength reduction factor is 0.60 for flexure, compression, shear, and bearing of structural plain concrete [ACI 318 Sec. 9.3.5].

K. Concrete Strength [ACI 318 Sec. 21.1.4]

The compressive strength, f'_c, of normal weight concrete may not be less than 3000 psi (20.69 MPa). The compressive strength of lightweight concrete may not exceed 5000 psi (34.48 MPa) unless experimental data are submitted to show that it has properties equivalent to normal weight concrete.[4]

L. Steel Reinforcement [ACI 318 Sec. 21.1.5]

All reinforcing steel must be of the deformed variety; plain bars are not permitted. Steel used must be low-alloy complying with ASTM A706. ASTM A615 grades 40 or 60 may be used if (1) the actual yield strength is no more than 18,000 psi (124.1 MPa) higher than the specified yield, and (2) the actual ultimate tensile stress is at least 125% of the actual yield strength. Restrictions on welding apply.

[4] ACI 318 uses a variable, λ, to adjust formulas for normal weight concrete to be appropriate for lightweight concrete. Due to the limited use of lightweight concrete for seismic-resistant structures, this book is written without using λ. See ACI 318 Sec. 8.6.1 for more information.

M. Special Moment-Resisting Frame Flexural Members [ACI 318 Sec. 21.5]

(This ACI section is specifically for flexural members used in lateral force resisting systems.) ACI 318 imposes a geometric restraint (i.e., a width/depth ratio greater than or equal to 0.3) and other constraints [ACI 318 Sec. 21.5.1.1 through Sec. 21.5.1.4] in order to reduce lateral instability and ensure a transfer of moment during inelastic response. (See Fig. 9.4 and Fig. 9.5.)

Seismic hoops (confinement) are required in all areas of flexural members where yielding is expected (e.g., at the ends of built-in beams and at other plastic hinge points). The first hoop must be within 2 in (51 mm) of the face of the supporting member. Maximum spacing must be less than (a) $d/4$, (b) 6 times the diameter of the smallest longitudinal bar, and (c) 6 in (151 mm) [ACI 318 Sec. 21.5.3.2]. Where hoops are not required, stirrup spacing with seismic hooks at both ends must be spaced at no more than $d/2$ throughout the member [ACI 318 Sec. 21.5.3.4].

N. Top and Bottom Reinforcement [ACI 318 Sec. 21.5.2]

At least two top and two bottom bars must run the entire length of the member. For both top and bottom steel, the amount of reinforcement steel must be greater than $200b_wd/f_y$ (for SI, $1.38b_wd/f_y$). The maximum steel is $0.025b_wd$. At the face of a joint, the positive moment strength cannot be less than 50% of the negative moment strength. At any section along the member length, neither the positive nor the negative moment strengths can be less than 25% of the corresponding strength at a joint.

O. Tension Lap Splices [ACI 318 Sec. 21.5.2.3]

Lap splices are classified as Class A (with a lap length of one development length) and Class B (with a lap length of 1.3 development lengths) [ACI 318 Sec. 12.15.1]. (The Class C splice (1.7 development lengths) was eliminated in 1989, and (in some cases) the basic development length was increased.)

Figure 9.4 *Special Flexural Member Reinforcement [ACI 318 Sec. 21.5]*

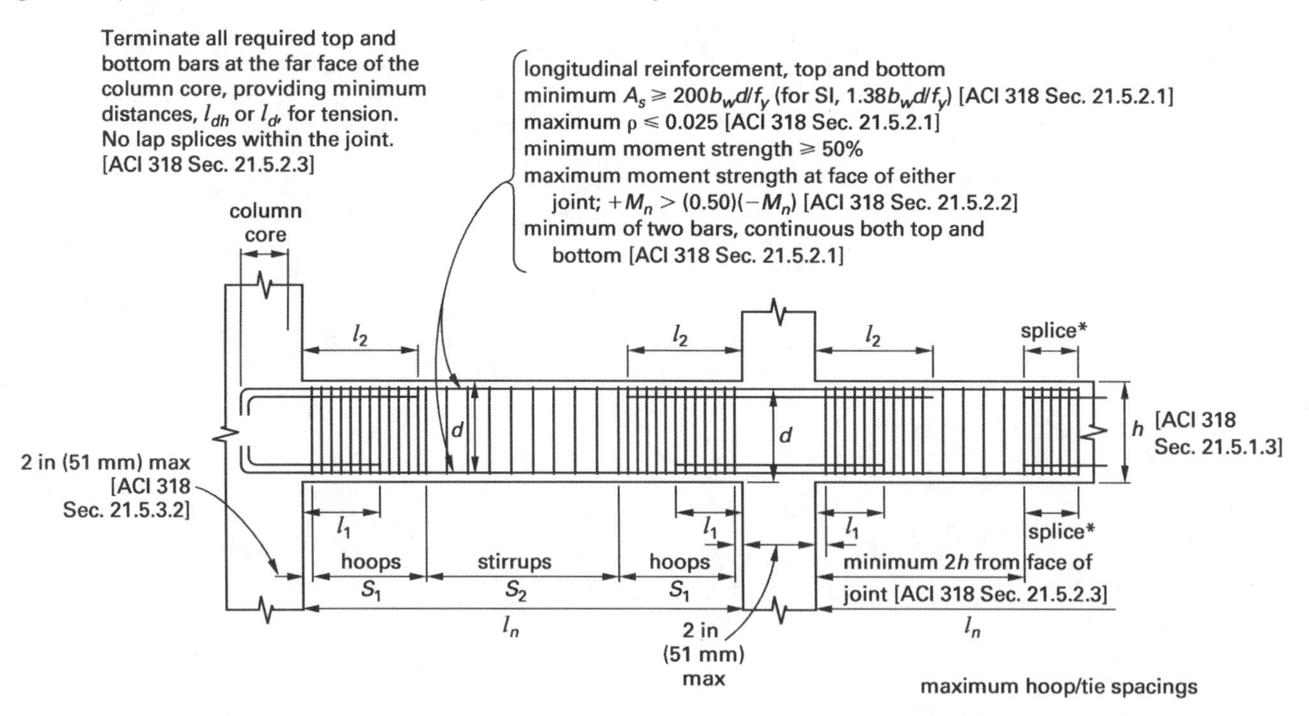

Dimensions l_1, l_2, S_1, S_2, hoop and stirrup spacing, anchorage length, cut-off points of discontinuous bars, l_d, or l_{dh} must be provided if less than across column core.

$l_n \geq 4d$ [ACI 318 Sec. 21.5.1.2]

l_1 = distance required by design for moment plus anchorage length

l_2 = distance to point of inflection plus anchorage length

d = design depth for $-M$ and $+M$

In length S_1, spacing for hoops $\leq d/4$; $8d_b$ of smallest bar; $24d_b$ of hoop; or 12 in (305 mm) [ACI 318 Sec. 21.5.3.2]

*At lap splices, spacing of hoops $\leq d/4$ but no greater than 4 in (102 mm) [ACI 318 Sec. 21.5.2.3]

In length S_2, spacing stirrups $\leq d/2$ [ACI 318 Sec. 21.5.3.4]

Figure 9.5 *Details of Transverse Confinement Reinforcement in Flexural Members [ACI 318 Sec. 21.5.3]*

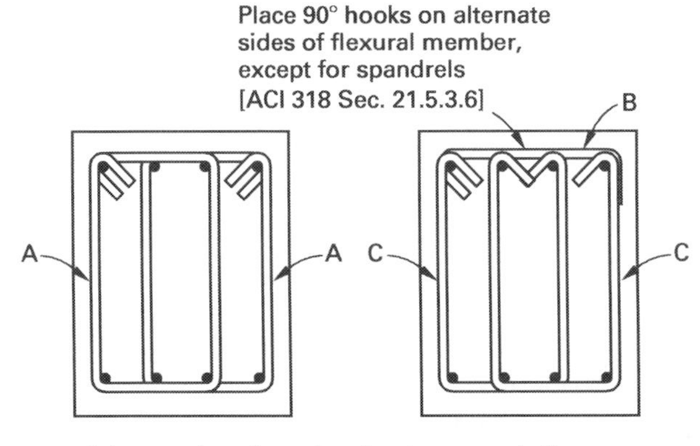

(a) examples of overlapping hoops and stirrups

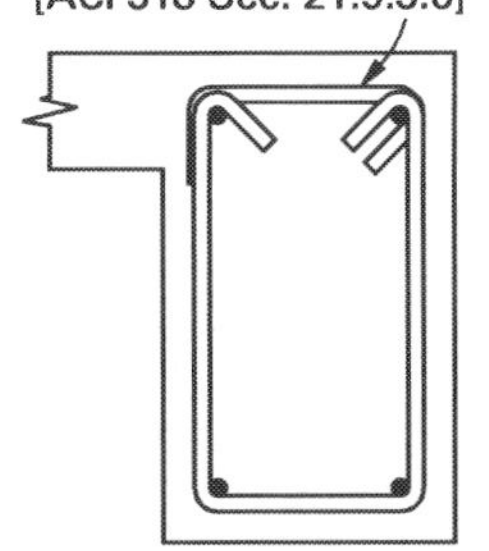

(b) spandrel beam

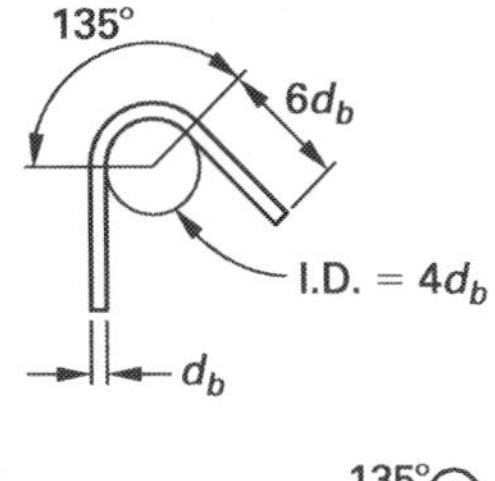

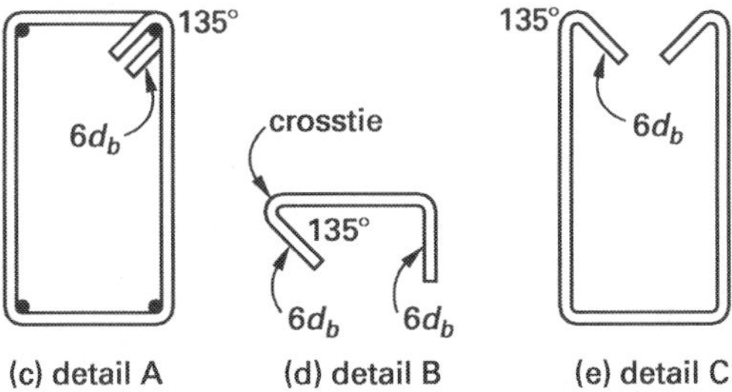

(c) detail A (d) detail B (e) detail C

Stirrups required to resist shear shall be hoops. Throughout the length of flexural members where hoops are not required, stirrups shall be spaced at no more than $d/2$ and shall have seismic hooks [ACI Sec. 21.5.3.4].

Tension lap splices are prohibited in flexural members (1) within beam-column joints, (2) within a distance from the face of a joint equal to twice the member depth, and (3) where flexural yielding is anticipated.

Lap splices are not permitted where flexural yielding is expected (i.e., at plastic hinge points) because such splices are not reliable when the loading repeatedly exceeds the yield strength of the reinforcement. Where permitted, they are proportioned as tension lap splices.

P. Welded Splices and Mechanical Connections [ACI 318 Sec. 21.1.6 and Sec. 21.1.7]

Welded splices and approved mechanical connections can be used but must maintain the clearance and coverage requirements. For members resisting earthquake forces, welded splices are not permitted within (1) an anticipated plastic hinge region, (2) a distance of two beam depths on either side of the plastic hinge region, or (3) a joint.

Splices with mechanical connections are classified as type 1 and type 2 splices according to ACI 318 specified strength capacities. For type 1 splices, a full-welded splice and full mechanical connection should be capable of developing 125% of the specified yield strength, f_y, of the bar. For type 2 splices, mechanical connections should be capable of developing in tension both the ultimate tensile strength or 125% of specified yield strength, f_y, of the bar.

Welding of designed reinforcement for any purpose other than approved splicing is prohibited. Welding of stirrups, ties, inserts, and other elements to the longitudinal bars is prohibited [ACI 318 Sec. 21.1.7.2].

Q. Special Moment-Resisting Frame Members with Bending and Axial Loads [ACI 318 Sec. 21.6]

The minimum member dimension is 12 in (305 mm). The ratio of the shortest-to-longest dimension cannot be less than 0.4. Tension lap splices [ACI 318 Sec. 12.15.2] are allowed within the center half of the member's length [ACI 318 Sec. 21.6.3.3]. The total column moment-resisting strength at a joint must be greater than 6/5 of the total girder moment-resisting strength at a joint [ACI 318 Sec. 21.6.2.2]. The longitudinal reinforcement ratio must be between 0.01 and 0.06, inclusive. (The 6% limit on column reinforcement ratio is too high for most designs. Considering the difficulty in designing and fabricating a joint as well as the joint congestion caused by additional ductility requirements, a practical upper limit of 4% is more reasonable.) Special provisions for transverse reinforcement (confinement) in the member apply. (See Fig. 9.6 and Fig. 9.7.)

In Fig. 9.6, distance S_h is the maximum spacing of the transverse reinforcement. ACI 318 Sec. 21.6.4.3 specifies this as the smallest of one-quarter of the smallest member dimension (B_s in Fig. 9.6), six times the diameter of the longitudinal reinforcement, and $s_o = 4 + (14 - h_x)/3$, where 4 in $\leq s_o \leq$ 6 in. h_x is the horizontal spacing of crossties or legs of overlapping hoops, not to exceed 14 in.

Figure 9.6 *Special Beam-Column Reinforcement [ACI 318 Sec. 21.6]*

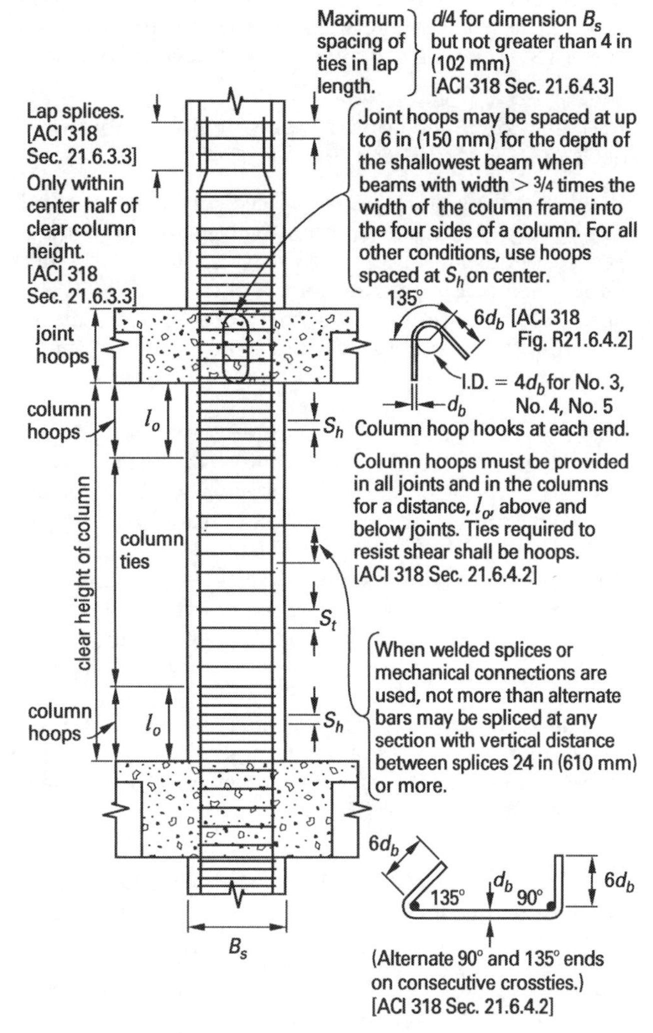

S_h = Hoop and supplementary crosstie spacing, not to exceed one-quarter of the smallest member dimension (B_s), six times the diameter of the longitudinal reinforcement, and $s_o = 4 + (14 - h_x)/3$, where 4 in $\leq s_o \leq$ 6 in. h_x is the horizontal spacing of crossties or legs of overlapping hoops, not to exceed 14 in [ACI 318 Sec. 21.6.4.3]

S_t = Column tie spacing, not to exceed $6d_b$ of longitudinal bars, or 6 in [ACI 318 Sec. 21.6.4.5]

B_s = Smallest dimension of compression member

l_o = Largest column dimension, but not less than the depth of the member at the joint face where flexural yielding is expected, one-sixth clear height, or 18 in (457 mm) [ACI 318 Sec. 21.6.4.1]

R. Columns Carrying Discontinuous Walls [ACI 318 Sec. 21.6.4.6]

Members supporting discontinuous elements (e.g., two columns supporting a short wall between them) must have special transverse reinforcement for their full height. (See Fig. 9.8.)

S. Diaphragms [ACI 318 Sec. 21.11]

A cast-in-place topping on a precast floor system can be considered as the diaphragm as if the cast-in-place

Figure 9.7 *Details of Transverse Confinement Reinforcement in Beam-Columns [ACI 318 Sec. 21.6.4]*

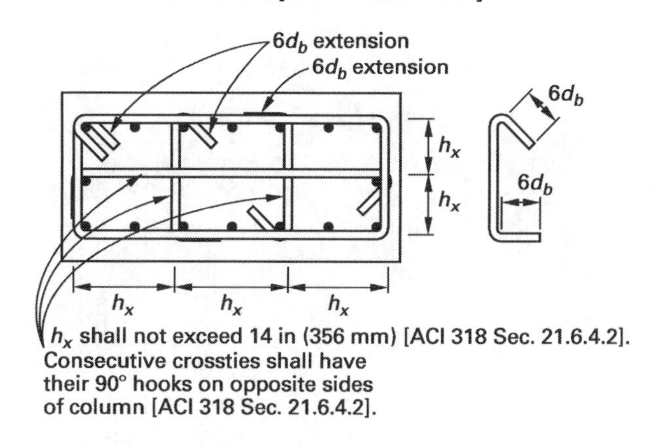

Figure 9.8 *Columns Carrying a Discontinuous Wall [ACI 318 Sec. 21.6.4.6]*

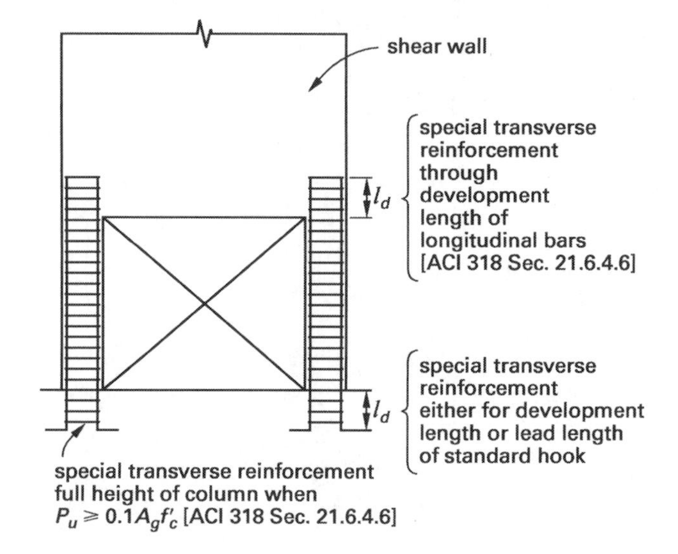

topping functioning alone is proportioned and detailed to resist the seismic forces. Monolithic concrete diaphragms and cast-in-place topping over precast elements used to resist seismic forces must be at least 2 in (51 mm) thick [ACI 318 Sec. 21.11.6].

Based on ACI 318 Sec. 21.11.7.4, when mechanical connectors are used to transfer forces between the diaphragm and the lateral system, the anchorage must be a Type 2 splice [ACI 318 Sec. 21.1.6]. Tendons that can be wire, cable, bar, rod, or strand, or a bundle of such elements used to impart prestress to concrete, should not be used as primary reinforcement in boundaries and collector elements.

T. Minimum Column Size [ACI 318 Sec. 21.7.2.3]

In order for a horizontal bar passing through a joint to develop its full bond strength, the minimum column size is 20 times the largest longitudinal bar extending through the joint in the direction of loading for normal weight concrete. For lightweight concrete, the minimum column size is 26 times the bar diameter.

U. Joints [ACI 318 Sec. 21.7]

Joints must be able to withstand forces of $1.25f_y$ in the longitudinal reinforcement. Flexural members framing into opposite sides of a column must have continuous reinforcement to pass through the column joint. Special provisions for confinement around the joint apply. (See Fig. 9.1.)

The shear strength of a joint increases significantly when all four sides of the joint are confined by members (beams) framing into the joint (column) [ACI 318 Sec. 21.7.4].

V. Concrete Cover [ACI 318 Sec. 7.7.1]

Concrete cover over reinforcement steel should generally satisfy the requirements of ACI 318 Sec. 7.7.1. For corrosive environments, including exterior members exposed to marine spray and deicing salts, ACI 318 Sec. R7.7.6 recommends up to $2^1/_2$ in (64 mm) of cover, depending on the type of member. Since some cover can be expected to spall during a seismic event, excessively thick cover concrete constitutes a falling debris hazard. For that reason, for special moment frame members subjected to bending and axial loading, cover thickness greater than 4 in (102 mm) is not permitted [ACI 318 Sec. 21.6.4.7]. Also because of spalling, column strength should not depend on the concrete cover.

10 Details of Seismic-Resistant Steel Structures

1. STEEL CONSTRUCTION DETAILS

The design, construction, and quality of steel components (members and connections) in buildings that resist seismic forces should conform to the requirements of IBC Chap. 22 provisions. The design forces resulting from earthquake motions are determined on the basis of energy dissipation in the nonlinear range of response.

A. Design and Construction Provisions

For design, fabrication, erection, and quality control of structural steel, the IBC permits the use of the load and resistance factor design (LRFD) method as well as the allowable stress design (ASD) method. Steel design based on the LRFD method should use the factored load combinations of IBC Sec. 1605.2 and the applicable requirements of IBC Sec. 2205. For steel design based on the ASD method, the factored load combinations of IBC Sec. 1605.3 and the applicable requirements of IBC Sec. 2205 apply.

B. Seismic Provisions for Structural Steel Buildings

In addition to the structural steel requirements in IBC Sec. 2205.2, steel structural elements resisting seismic forces should also comply with AISC 341, *Seismic Provisions for Structural Steel Buildings.*

Based on AISC 341 Sec. A1, buildings in seismic design category A, B, or C (see Sec. 6.18), do not need to be designed or detailed according to the structural steel seismic provisions.

C. Loads and Load Combinations

The appropriate critical combination of factored loads, as required by IBC Sec. 1605.2 at load and resistance factor limits or IBC Sec. 1605.3.2 at allowable stress limits with stress increases allowed, should be used in determining the required strength of a steel structure and its elements. Where the amplified horizontal earthquake load is required, use the overstrength factors from ASCE/SEI7 [AISC 341 Sec. B2].

D. Steel Frames

The most common structural system for high-rise buildings in seismically active areas is the ductile steel frame. Ductile frame behavior is much easier to achieve with steel construction than with concrete construction because steel is intrinsically a ductile material. Design emphasis is, therefore, shifted away from ensuring that the material itself will behave in a ductile manner to ensuring that the structural frame will behave in a ductile manner.

Ductile frame operation is accomplished by extensive use of moment-resisting connections known as *type I connections.*[1] (See Fig. 10.1.) These connections transmit column moments to beams and girders, forcing those members to carry the moments. It is easier for beams and girders to resist moments because their lengths (and, hence, their moment arms) are long. Without moment-resisting joints, the columns would most likely fail (by web crippling, for example) before the beams and girders became fully stressed. Therefore, a beam-column connection must be able to transmit without failure a moment equal to the plastic capacity of the beam. Frames with joints that do this are known as *special moment-resisting frames* (SMRF).[2]

E. Steel Properties [AISC 341 Sec. A3]

To ensure ductile behavior of beams in a strong column-weak beam design, AISC 341 Sec. A3.1 requires structural steel used in *special moment frames* in which inelastic behavior is expected (based on load combinations of the system overstrength factor, Ω_O, times the effect of horizontal seismic force, Q_E) to have a specified minimum yield strength less than or equal to 50 ksi (345 MPa), unless the material is tested in accordance with AISC 341 Sec. K2 or other rational criteria. Usable steels are limited to A36, A53, A500 (grades B and C), A501, A529, A572 (with some limitations, grades 42 and 50), A588, A913 (grade 50), A992, and A1043. Since grade 50 A572 steel is often manufactured with yield strengths greater than 50 ksi, A913 and A992 (which specify upper limits for yield strength) are preferred.

The 50 ksi limitation applies to structural components that are expected to yield during an earthquake. The limitation does not apply to columns of A588 and A913 grade 65 steel for which the only expected inelastic behavior is yielding at the column base (because columns are not considered to be yielding elements), to A283 (grade D) steel used for base plates, or to anchor

[1] *Type II joints* do not develop any appreciable moments. Steel structures in any seismic zone can use type II joints. This is typical of braced-frame designs.

[2] The term *ductile moment-resisting space frame* (DMRSF), once used extensively, is no longer commonly used.

Figure 10.1 *Moment-Resisting Joint for New Construction Incorporating SidePlate™ Connection Technology*

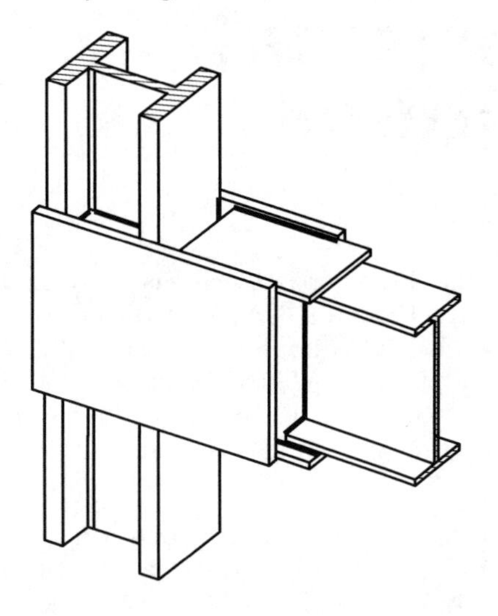

SidePlate™ is the trademark of SidePlate Systems, Inc., of Long Beach, CA.

bolts. AISC 360 Sec. A3.1c and AISC 341 Sec. A3.3 specify additional Charpy V-notch toughness requirements for hot-rolled shapes with flange thickness greater than or equal to 1.5 in (38 mm) and plates with thickness greater than or equal to 2 in (51 mm) that are used in steel special moment frames.

Ordinary moment frames (OMFs) and *ordinary concentrically braced frames* have a maximum specified minimum yield strength of 55 ksi (380 MPa), allowing the use of such steels as A572 grade 55, A913 (grades 60 and 65), and A1011 HSLAS grade 55.

Other steels and non-steel materials may be used in *buckling-restrained braced frames*, subject to other requirements.

F. Member Strength

The term *strength* (as in "full strength") means that the members must be capable of developing the following forces or moments.

- moment in flexural members $M_s = ZF_y$
  Z represents plastic modulus of the section, and F_y is specified minimum yield stress of the steel.
- shear in flexural members $V_s = 0.55F_y dt$
  d and t denote overall depth and thickness of the member, respectively.
- axial compression in flexural members $P_{sc} = 1.7F_a A$
  F_a signifies axial design strength, and A symbolizes cross-section area of the member.
- axial tension in flexural members flexural members $P_{st} = F_y A$
- bolts 1.7×allowable
- full-penetration welds $F_y A$
- partial-penetration welds $1.7F_s$
- fillet welds $1.7F_s$
  F_s represents the applicable allowable stress values. The one-third allowable stress increase is not permitted for the purpose of determining member or connection strength.

G. Column Requirements
[AISC 341 Sec. D1]

Load combinations required by IBC Sec. 1605.2 and Sec. 1605.3.2 must be supported at load and resistance factor limits or at allowable stress limits. For ASD, a one-third stress increase is permitted according to the IBC. Because the integrity of a structural steel system that has experienced ductile yielding is strongly dependent on the axial capacity of the columns, with limited exceptions, columns in seismic design categories D, E, or F must have the strength to support the amplified load combinations given in ASCE/SEI7 Sec. 12.4.3. Other requirements relating to column splices [AISC 341 Sec. D2] must also be met.

H. Ordinary Moment Frames
[AISC 341 Sec. E1]

Ordinary moment frames are permitted, but the *Blue Book* commentary discourages such use. It is likely, in any case, that OMFs will be heavier and thus costlier than SMRFs, since OMFs resist seismic forces elastically. OMFs may have applications where seismic forces are low and the design is controlled by wind. Code requirements for OMFs include verification of the moment connection strength and continuity plates [AISC 341 Sec. E1.6b].

I. Special Moment-Resisting Frames
[AISC 341 Sec. E3]

AISC 341 Sec. E3 is designed to ensure that joints and members behave in a ductile manner in SMRFs. The areas covered are beams and columns in which plastic hinges can form and the joint *panel zones* in which shear yielding can occur. (See Fig. 10.2.) The panel zone is normally the column web central to the beam-column connection. (SidePlate Connection Technology, Fig. 10.1, eliminates reliance on panel zone deformation by providing three panel zones: the two side plates, plus the column's own web.) With specific limitations, inelastic behavior in these areas is permitted.

For special moment-resisting frames, the adequacy of the beam-column connections is crucial to the integrity of the entire system. Previously, the beam-column connections were required to have minimum strength (i.e., strength equal to the beam in flexure). The code said this strength was achieved when beam flanges were full-penetration welded to columns and the beam web-to-column portion of the connection alone was able to support the gravity and seismic shear force. Bolted connections and other methods were also permitted if justified.

After the 1994 Northridge earthquake in California, new design guidelines were adopted regarding beam-column

Figure 10.2 *Panel Zone*

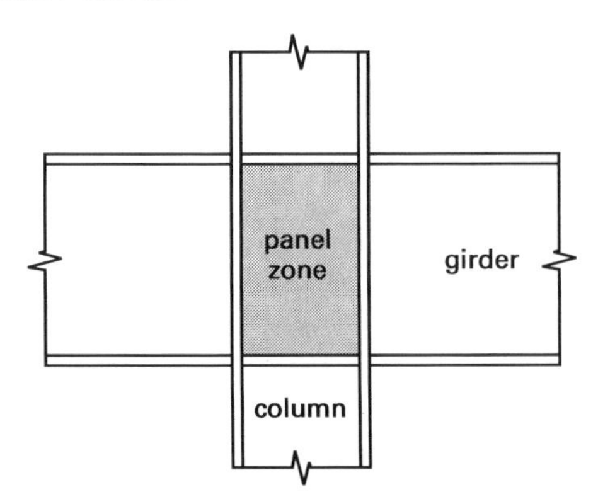

connection configurations, and the current AISC provisions reflect these changes.

Doubler plates can be used, with restrictions, to reduce panel zone shear stress or web depth/thickness ratio. Doubler plates should be welded across the plate width top and bottom. The minimum fillet weld is 3/16 in (4.7 mm). *Continuity plates* (column stiffeners at top and bottom of the panel zone) must match prequalified connection details or the thickness of the column flange must meet specified requirements [AISC 341 Sec. E3.6f].

A minimum *strength ratio* is given in order to prevent column failures at joints when the beams are "stronger" than the columns. This is the "strong column-weak beam" test. Some exceptions are permitted, however [AISC 341 Sec. E3.4a].

In order to withstand reverse bending moments, intersecting beams or bottom flange diagonal bracing is required.

To reduce the earthquake hazards of welded steel moment frame structures, it is implied that (1) plastic hinge location should be designed to develop within the beam span at a minimum distance of half the beam depth from the column face, (2) all elements of frame and their connections should be designed to develop plastic hinge moment strength, (3) weld filler metals should be specified with rated toughness values, (4) connections should be detailed with beam flange continuity plates, (5) column panel zones should be designed without doubler plate consideration where practical, (6) backing bars and weld tabs should be eliminated, (7) connection design should be proficient through prototype testing or by reference to tests of comparable connection configuration, and (8) the quality of material and construction is essential and should be considered crucial to total frame behavior.

J. Special Truss Moment Frames
[AISC 341 Sec. E4]

A *special truss moment frame* (STMF) is a combination of a strong column and a ductile girder. The STMF is a modern design concept based on extensive analytical and experimental research at the University of Michigan that provides ductile behavior of truss girders. STMF should be designed in accordance with AISC 341 Sec. E4. In Fig. 10.3, each horizontal truss that is part of the moment frame is shown with a *special segment* located within the middle one-half length of the truss.

Figure 10.3 *Special Truss Moment Frames (STMF-Steel)*

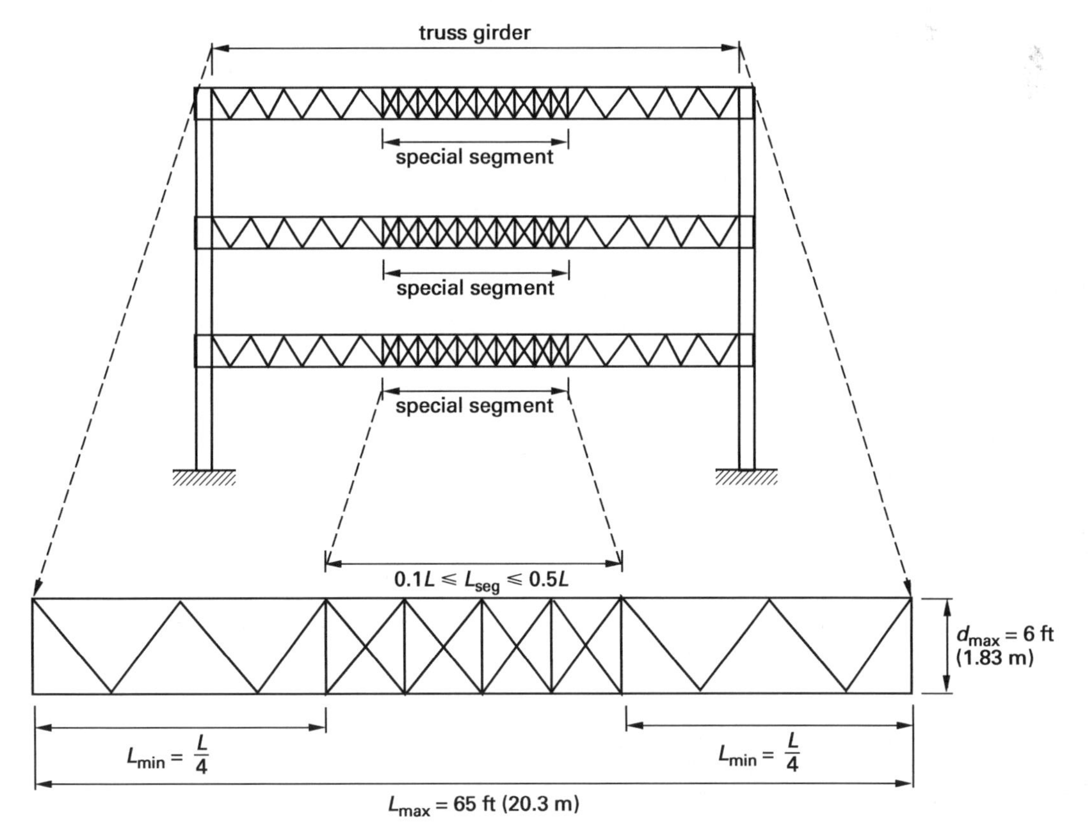

Special segments perform inelastically when induced by earthquake forces. In the fully yielded case, these special segments should develop vertical shear strength through flexure of the chord members and through axial tension and compression of diagonal web members.

The maximum span length between columns that include such trusses is 65 ft (20.3 m), and the overall depth should not exceed 6 ft (1.83 m). The length of the special segment ranges from 0.1 to 0.5 times the truss span length. Within the special segment, all panels should either be Vierendeel trusses or have X-bracing (see Fig. 10.4), and their length-to-depth ratio should be no less than 0.67 and no more than 1.5.

Figure 10.4 *X- and Vierendeel Trusses (elevations)*

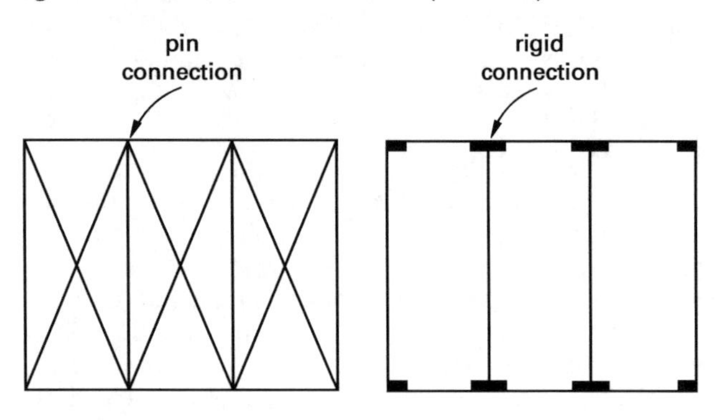

Within special segments, when diagonal members with identical sections are used, they should be arranged in an X-pattern and be separated by vertical members. AISC 341 requires that such vertical members be interconnected at points of crossing and that the connections have the strength to resist a minimum force equal to 0.25 times the diagonal member tension strength. Connections (bolted and welded) of all elements in the truss frames, including those within the truss, should be designed according to AISC 341 Sec. E4.6. Bolted connections and splicing of chord members within the special segment are not permitted [AISC 341 Sec. E4.4a].

The maximum axial stress in diagonal web members due to concentrated dead plus live loads acting within the special segment is $0.03F_y$ [AISC 341 Sec. E4.4a].

K. Welded and Bolted Connections

[AISC 341 Sec. E3]

In response to fractures in beam-column connections of steel moment frames during the 1994 Northridge earthquake in California, the AISC retains the "emergency code change" provisions as adopted on September 14, 1994.

The Northridge earthquake damaged a considerable number of welded beam-column connections in moment-resisting frames. The most common failure involved cracking of the welded connection from the beam bottom flange to the column flange. In some instances, the crack propagated into the column flange and/or beam web. In response, AISC 341 Sec. E3.6b specifies that the connection configurations (e.g., beam-column) should demonstrate, by approved cyclic load test results or calculation, the capability to uphold inelastic rotation up to 0.04 radians and develop the strength criterion for shear. Additionally, the effect of steel overstrength (expected value of yield strength) and strain hardening should be considered.

Reasons that contributed to the damage of welded steel moment frames can be summarized: (1) In resisting earthquake demands, the number of frame bays were limited. (2) The actual strengths versus code specified strengths were conflicting. (3) At beam-column connections, vast inelastic demands and stress concentrations occurred as a result of inappropriate detailing practice. (4) Column panel zones were weak. (5) Weld metal materials that were low in toughness were utilized. (6) In the welding procedures, levels of quality control were inadequate.

For welded steel moment frames that are expected to experience significant postelastic earthquake demands, connections should be designed so that the plastic hinge location develops within the *protected zone*, as shown in Fig. 10.5. (Special moment frame beams have protected zones just outside of the connections at each end.[3]) This can be ensured either by reinforcing the beam at the connection, as shown in Fig. 10.6, or by decreasing the cross section of the beam a distance of half the beam depth from the column face. The protected zone depends on the connection design, but generally extends from the end of the stiffening to one-half beam depth past the expected location of a plastic hinge.

Figure 10.5 *Desired Plastic Hinge Behavior*

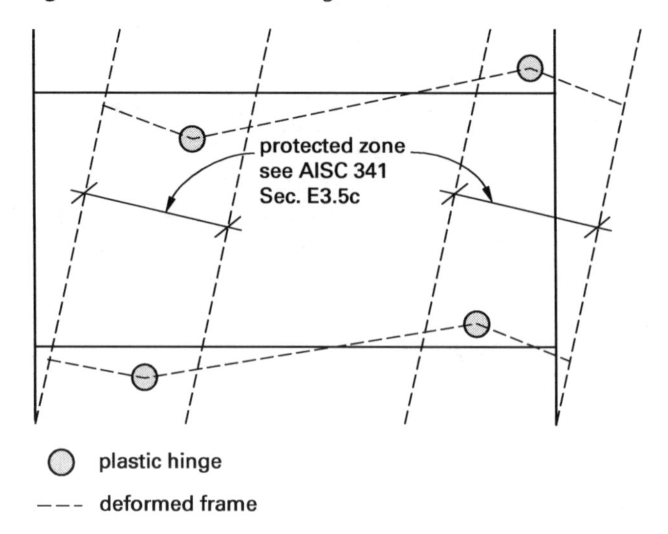

[3]Welded, bolted, screwed, and shot-in connections are not permitted within the protected zone. This includes shear studs or repair pins that penetrate the beam flange.

Figure 10.6 Connection Reinforcement Configurations

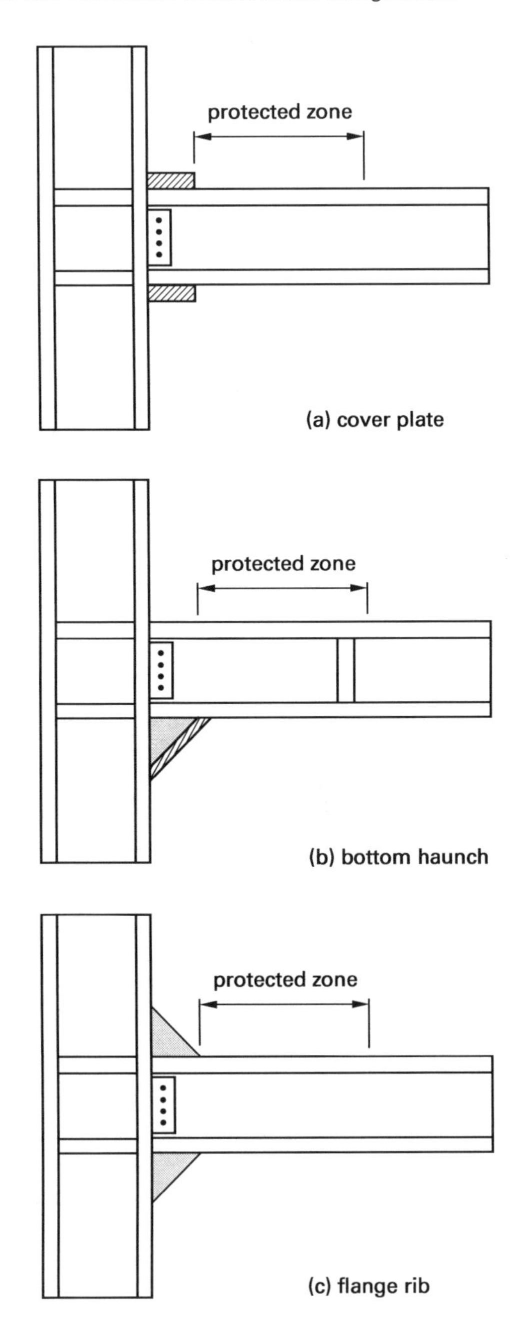

L. Girder-Column Joint Restraints
[AISC 341 Sec. E3.4]

Restrictions are given to ensure that moment frames are capable of reaching their seismic design capacity. Specifically, if the columns remain elastic, column flanges require lateral support only at the level of the top girder flange. AISC 341 Sec. E3.4c provides acceptable conditions for columns assumed to remain elastic. If the column does not remain elastic, column flange support is required at the tops and bottoms of the girder flanges.

M. Braced Frames
[AISC 341 Chap. F]

Various requirements, including slenderness ratios, are given for *ordinary concentrically braced frames* (OCBF) and *special concentrically braced frames* (SCBF). Figure 10.7 shows different systems of braced frames that are used: *diagonal bracing* (where diagonals connect the joints in adjacent levels); *chevron bracing* (where a pair of braces terminate at a single point within the clear beam span); *V-bracing* (a form of chevron bracing that intersects a beam from above); *inverted V-bracing* (a form of chevron bracing that intersects a beam from below); *K-bracing* (where a pair of braces located on one side of a column terminate at a single point within the clear column height); and *X-bracing* (where a pair of diagonal braces cross near midlength of the bracing members). Chevron bracing can be used for OCBF when certain conditions are met [AISC 341 Sec. F1.4a]. When chevron bracing is used in special concentrically braced frames, AISC 341 Sec. F1.4a requires that any member (beam) intersected by a brace be continuous between columns. K-bracing is not permitted for OCBF or SCBF, and tension-only frames cannot be used in SCBF [AISC 341 Sec. F1.4b, Sec. F2.4c, and Sec. F2.4d]. Tension-only frames are designed by ignoring the compression resistance of braces loaded in compression.

Figure 10.7 Types of Braced Frames

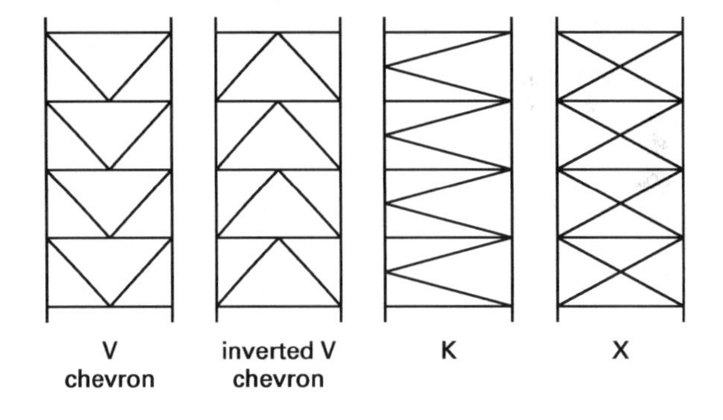

Based on App. H [ASCE/SEI7 Table 12.2-1], the response modification factors, R, for these frames are 3.25 and 6. The higher R value for special concentrically braced frames implies a more ductile system than ordinary concentrically braced frames. In special concentrically braced frames, all members and connections should be designed and detailed to resist shear and flexure. Also, any member that is intersected by a brace should be continuous through the connection.

Based on AISC 341 Sec. D1.1b, it is required that compression elements in special and ordinary concentrically braced frames be sized to minimize the potentiality of local buckling.

Figure 10.8 shows the width-thickness ratio (b/t) of angle sections used in SCBF, which should be limited to

$$\frac{b}{t} \leq 0.30\sqrt{\frac{E}{F_y}} \qquad 10.1$$

Figure 10.8 Angle Sections

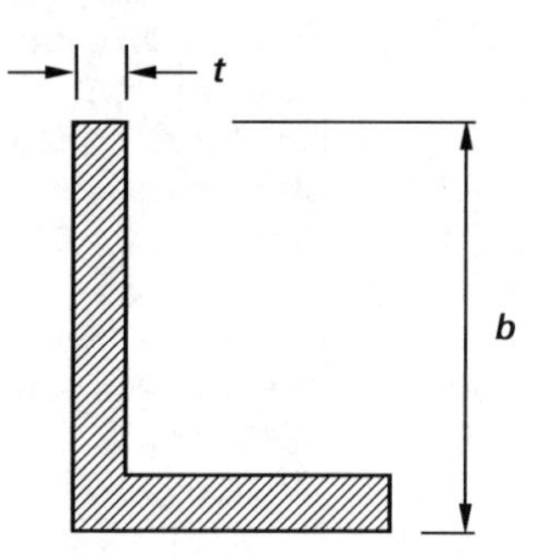

Figure 10.9 shows the outside diameter-wall thickness ratio (D/t) of circular sections used in SCBF, which should be limited to

$$\frac{D}{t} \leq 0.038\frac{E}{F_y} \qquad 10.2$$

Figure 10.9 Circular Sections

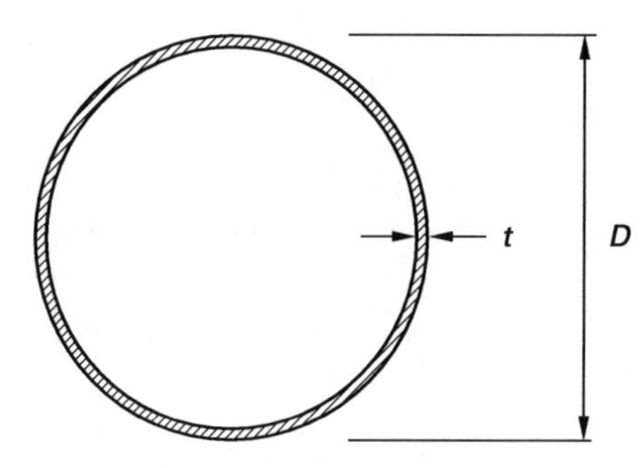

Figure 10.10 shows the outside width-wall thickness ratio (b/t) of rectangular tubes used in SCBF, which should be limited to

$$\frac{b}{t} \leq 0.55\sqrt{\frac{E}{F_y}} \qquad 10.3$$

Figure 10.10 Rectangular Tubes

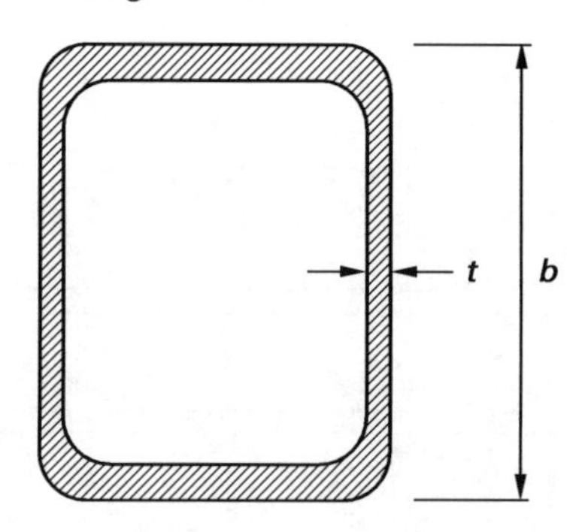

2. ECCENTRICALLY BRACED FRAMES

[AISC 341 Sec. F3]

Figure 10.11 illustrates the basic types of *eccentrically braced frames* (EBF) in which at least one end of each bracing member connects to a beam a short distance from a beam-to-column connection or from another beam-to-brace connection.[4] The short section of girder between the brace end and the column is known as the *link beam.* In Fig. 10.11, (a) and (b) show *end link EBFs*, and (c) shows a *center link EBF*. EBFs may be inverted, and different EBF systems can be mixed.

Figure 10.11 Types of Eccentrically Braced Frames

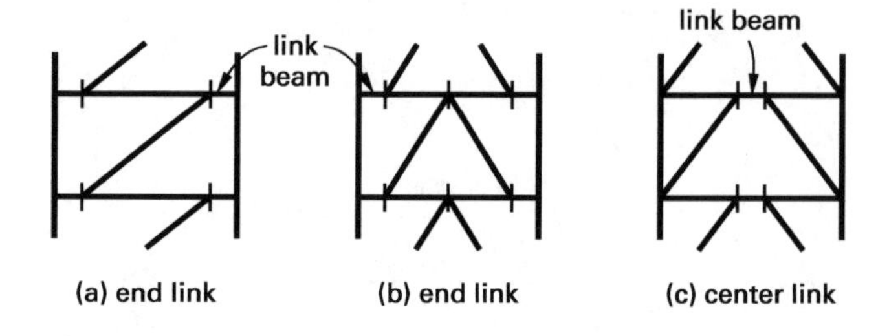

The concept behind EBFs is that in low-to-moderate ground shaking, a frame using them performs as a braced frame rather than as a moment frame. Therefore, the structure experiences small drifts, little if any architectural damage, and no structural damage. The link beam is specifically designed to yield in a major seismic event, thereby absorbing large quantities of seismic energy and preventing buckling of the other bracing members. To limit yielding to the link beam requires attention to detail at the connection. Figure 10.12 illustrates typical details. (Here, web stiffeners are used in the links to keep the web from buckling. These may not be needed in every instance.)

Figure 10.12 Detail of Original Eccentric Link Design

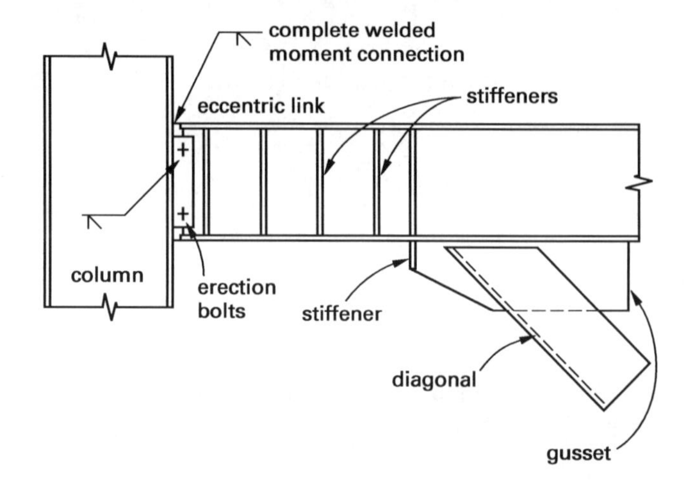

[4]This is exactly the type of design that engineers are taught to avoid in traditional classes on structures. However, intentional yielding of eccentric links is an area pioneered by the works of Egor P. Popov. One of the earliest California buildings to use this type of feature is the 16-floor Bank of America Regional Office and Branch Bank Building in San Diego, which Popov assisted in designing.

There is also some evidence that the beam-to-column connection should be at the lower end of the diagonal rather than at the top. Tests seem to show that the high ends are always stressed to yielding while the links at the low end remain elastic. This is illustrated in Fig. 10.13.

Figure 10.13 *Typical Deflections of Eccentric Links*

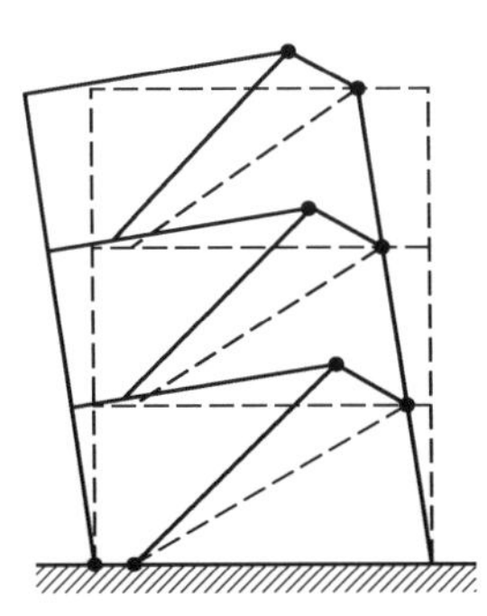

AISC 341 Sec. F3 covers the design of EBFs. Shear in the link beam and bay drift are the primary design variables. Shear stress in the link beam is usually limited to $(0.90)(0.60)F_y = 0.54F_y$ [AISC 341 Sec. F3.5b].

The shear force on a link beam—either end link or center link—is equal to the story shear times the story height divided by the distance between the column center lines (i.e., the *bay length*). For a given story height and distance between columns, frames with the same story shear will have the same link beam shear, regardless of their geometries (i.e., end link EBF or center link EBF). The load in each column is equal to the link beam shear. (There are two columns to carry the two link beam shears.)

$$V_{\text{link beam}} = \frac{F_x h}{L} \quad \textit{10.4}$$

$$P_{\text{column}} = V_{\text{link beam}} \quad \textit{10.5}$$

Figure 10.14 illustrates an *end link EBF*. The location of the link beam's inflection point initially must be assumed until all member sizes have been determined.[5] A location at mid-length of the link beam is a reasonable initial assumption. The fraction of the story shear carried by the inclined braces compared to that carried by direct moment frame action (approximately 0.75) is equal to the ratio of the distance between the points of inflection on the link beams (shown in Fig. 10.14 as distance L') to the bay length, b.

$$V_{\text{brace}} = \frac{V_{\text{link beam}} L'}{b} \quad \textit{10.6}$$

$$H_{\text{brace}} = \frac{F_x L'}{L} \quad \textit{10.7}$$

Figure 10.14 *End Link Beam*

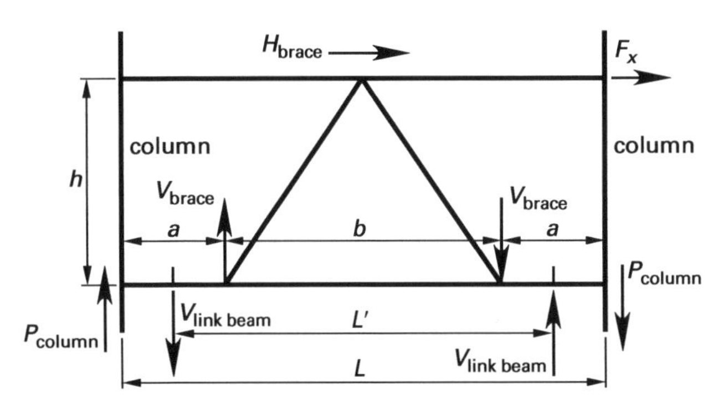

Figure 10.15 illustrates a *center link beam*. The beam-to-column connection may be "pinned" in the frame's plane, and all of the lateral shear is taken by the braces. The link beam point of inflection is at the (vertical) centerline of the bay.[6]

$$V_{\text{link beam}} = \frac{F_x h}{L}$$

$$V_{\text{brace}} = \frac{V_{\text{link beam}} L}{2a} \quad \textit{10.8}$$

$$H_{\text{brace}} = \frac{F_x}{2}$$

$$P_{\text{column}} = V_{\text{link beam}} \quad \textit{10.9}$$

Figure 10.15 *Center Link Beam*

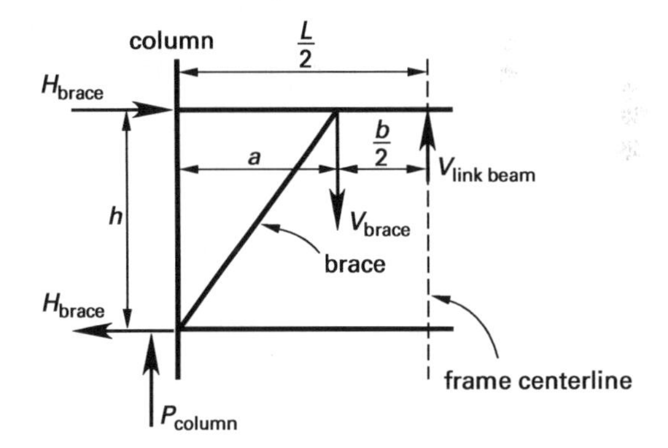

3. STEEL STUD WALL SYSTEMS

Steel stud wall systems used to resist lateral loads generated by wind or earthquake should comply with IBC Sec. 2211. Specific design and detailing requirements are provided by the American Iron and Steel Institute's *Standard for Cold-Formed Steel Framing Lateral Design.*

[5]This is a direct result of the need to weld the beam-column connection in order to obtain a moment-resisting condition.

[6]This is a direct result of the fact that the beam-column connection can be pinned.

11 Details of Seismic-Resistant Masonry Structures

1. MASONRY CONSTRUCTION DETAILS

The materials, design, construction, and quality assurance of masonry should be in accordance with IBC Chap. 21. Three different associations publish the *Building Code Requirements for Masonry Structures*, which provides the seismic guidelines for the IBC: ACI 530, ASCE5, and TMS 402.

A. Design and Construction Provisions

Masonry structures should be designed by one of the following methods.

1. *Allowable stress design.* For this method, masonry-designed structures should comply with the provisions of IBC Sec. 2106 and Sec. 2107.
2. *Strength design.* For this method, masonry-designed structures should comply with the provisions of IBC Sec. 2106 and Sec. 2108.
3. *Empirical design.* For this method, masonry-designed structures should comply with the provisions of IBC Sec. 2106 and Sec. 2109.

Masonry should also comply with the requirements of IBC Sec. 2101 through Sec. 2105, which relate to materials and construction. Masonry units should be dry at the time of placement. The IBC provides special requirements when masonry is constructed in cold weather. For various temperature ranges, there are special requirements outlined in IBC Sec. 2104. Metal reinforcement should be placed prior to grouting. Positioning bolts for reinforcement requires templates or approved equivalent means to prevent dislocation during grouting.

B. Seismic Provisions for Structural Masonry Buildings

IBC Sec. 2106 refers to ACI 530 for special provisions for seismic resistance. In particular, ACI 530 Sec. 1.18.4 is concerned with provisions for seismic design categories B, C, D, E, and F.

C. Loads and Load Combinations

Masonry buildings and structures should be designed to resist the load combinations specified in IBC Chap. 16. Where strength (load and resistance) factor design is used, structures and all portions should resist the most critical effects from the combinations of factored loads provided in IBC Sec. 1605.2. When allowable stress design is used, the most critical effects from the combinations of factored loads provided in IBC Sec. 1605.3 should be resisted by the masonry structures.

D. Masonry Construction Details

The strength of masonry structures is sensitive to materials, design, construction, and quality control. Quality assurance ensures that materials, construction, and workmanship comply with the plans and specifications and with IBC Sec. 2105. Masonry that is constructed under the watchful eye of an expert or other qualified person (as defined in IBC Sec. 1704) is expected to be of better quality. Therefore, the IBC specifically requires minimum levels of quality assurance tests and submittals [IBC Sec. 1704.5].

The following important points must be observed in masonry design for structures in seismic design categories D, E, and F.

1. For columns in seismic design category D or higher, all longitudinal bars must be encircled by lateral ties. The lateral ties give support to the longitudinal bars. Column ties should be spaced a maximum of 8 in (203 mm) for the full height of columns that are part of the lateral force-resisting system [ACI 530 Sec. 1.18.4.4.2.1].
2. Masonry shear walls that are part of the seismic force-resisting system must be reinforced with both vertical and horizontal reinforcement. Reinforcement should be continuous around wall corners and through intersections.
3. For special reinforced masonry shear walls, the minimum requirement for the sum of areas of vertical and horizontal reinforcement is 0.002 times the gross cross-sectional areas of wall. Also, the minimum area of reinforcement in either direction must not be less than 0.0007 times the gross cross-sectional area of the wall [ACI 530 Sec. 1.18.3.2.6].
4. For intermediate and special reinforced masonry shear walls, no. 4 bars (with a cross-sectional area of 0.2 in^2 (129 mm^2)) are the smallest that may be used for vertical reinforcement at each corner or wall intersection and at the edge of any openings. Maximum bar spacing is 48 in (1219 mm) [ACI 530 Sec. 1.18.3.2.3.1, Sec. 1.18.3.2.5, and Sec. 1.18.3.2.6].

5. For special reinforced masonry shear walls, two longitudinal wires of W1.7 (MW11) joint reinforcement are the smallest that may be used for horizontal reinforcement at the tops and bottoms of walls and at the tops and bottoms of any openings. Maximum spacing is 48 in (1219 mm) [ACI 530 Sec. 1.18.3.2.3.1 and Sec. 1.18.3.2.6].

6. In seismic design categories E and F, the minimum steel ratio (based on area bt) for horizontal reinforcement is 0.0015 for stack bond (open-end) masonry that is not part of the lateral force resisting system [ACI 530 Sec. 1.18.4.5.1].

7. Only Type S or Type M cement-lime mortar or mortar cement mortar are permitted for walls that are part of the seismic force-resisting system in seismic design categories D, E, and F [ACI 530 Sec. 1.18.4.4.2.2].

E. Masonry Wall Frames

ASCE/SEI7 Sec. 14.4 provides general requirements for the design of shear walls connected with coupling beams. In designing a masonry frame, the primary requirement is to ensure a ductile response to lateral forces.

Figure 11.1 illustrates piers (columns) and coupling beams in a CMU wall; Fig. 11.2 shows masonry wall frame details. Beams interconnect vertical elements of the lateral-load resisting system. For the beams, clear span should not be less than twice the depth. The minimum nominal depth of the beam should be 8 in (203 mm) [ACI 530 Sec. 3.3.4.2.5]. The ratio of nominal beam depth to nominal beam width should be less than six. For piers, the maximum nominal length is six times the nominal thickness, and the minimum nominal length is three times the nominal thickness [ACI 530 Sec. 3.3.4.3.3]. These dimensions ensure that flexural yielding will be limited

Figure 11.1 Masonry Wall Frame

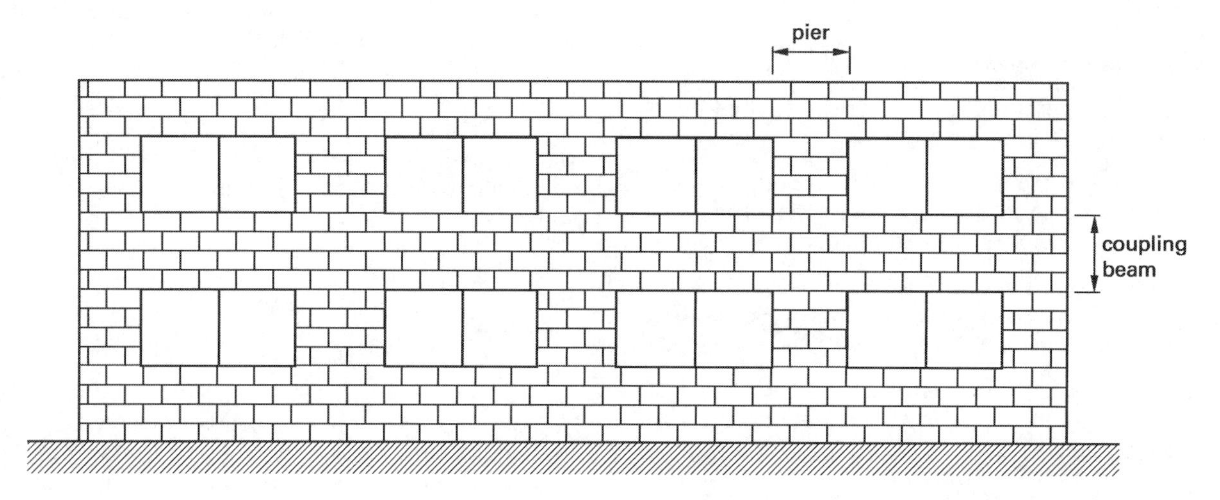

Figure 11.2 Masonry Wall Frame Details

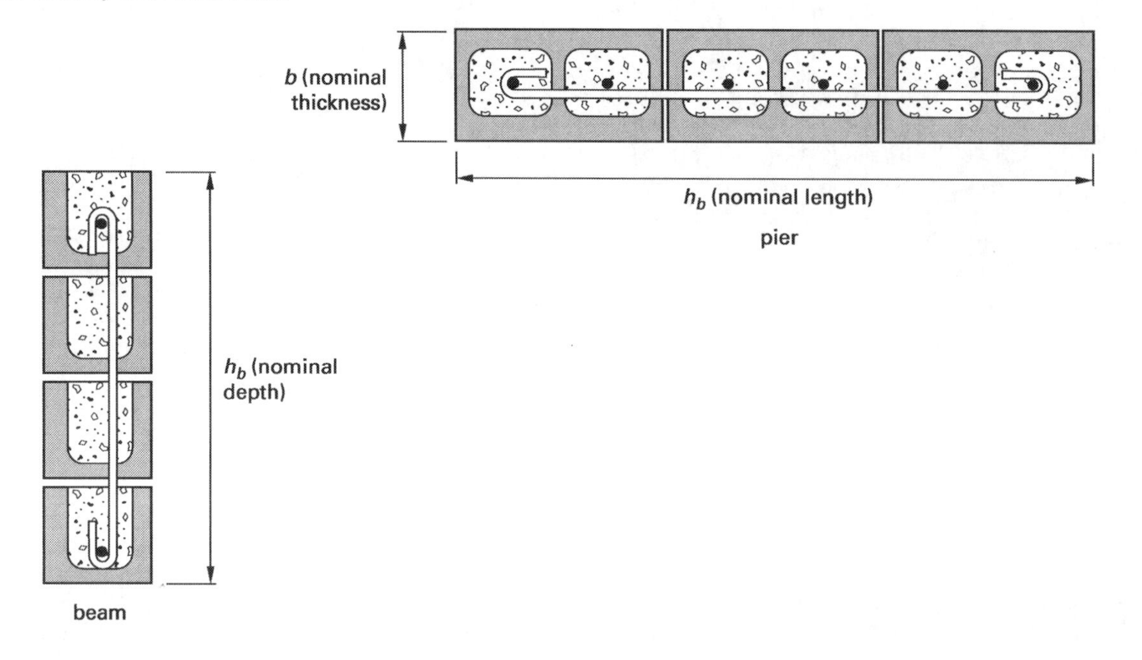

to the beams at the face of the piers and to the bottom of the columns at the base of the structure.

F. Allowable Shear Stress

The *allowable shear stress* in a masonry wall depends on the compressive strength of the masonry, f'_m. ACI 530 Sec. 2.3.6 provides the methodology for determining the allowable shear stress, F_v, considering both masonry and steel resistance.

It is normal practice to have the same reinforcement ratio horizontally and vertically. In seismic design categories B, C, D, E, or F, all masonry walls not part of the lateral force-resisting system must be self-supporting and isolated from the elements of the lateral force-resisting system [ACI 530 Sec. 1.18.3.1]. The empirical design method cannot be used in seismic design categories C or higher [ACI 530 Sec. 1.18.4.3.1].

12 Details of Seismic-Resistant Wood Structures

1. WOOD STRUCTURES

The quality and design of wood members and their fastenings should be in accordance with IBC Chap. 23.

A. Design and Construction Provisions

Wood structures should be designed by one of the following design methods.

1. *Allowable stress design.* For this method, wood designed structures should comply with the provisions of IBC Sec. 2304, Sec. 2305, and Sec. 2306.

2. *Load and resistance design.* For this method, wood designed structures should comply with the provisions of IBC Sec. 2304, Sec. 2305, and Sec. 2307.

3. *Conventional light-frame construction.* For this method, wood designed structures should comply with the provisions of IBC Sec. 2304 and Sec. 2308.

LRFD currently is an alternate design procedure to the traditional allowable stress design method as prescribed in the *National Design Specification for Wood Construction ASD/LRFD* (NDS). Provisions for the design of wood buildings in IBC Chap. 23 are based on the NDS. (NDS, published by the American Wood Council, is a nationally recognized guide for wood structural design.) The NDS method incorporates factored loads and design provisions and guidelines for structural lumber, glued-laminated timber, poles and piles, connections, I-joists, structural composite lumber, trusses, structural panels, shear walls and diaphragms, and structural framing connections.

B. Seismic Provisions for Structural Wood Buildings

Special provisions for seismic load-resisting systems for all engineered wood structures are provided in IBC Sec. 2305. In particular, the IBC includes significant requirements for seismic design categories D, E, or F.

C. Loads and Load Combinations

Wood buildings and structures should be designed to resist the load combinations specified in IBC Chap. 16. Where allowable stress design is used, the most critical effects from the combinations of loads provided in IBC Sec. 1605.3 should be resisted by the wood structures. When load and resistance factor design is applied, structures and all portions should resist the most critical effects from the combinations of factored loads provided in IBC Sec. 1605.2.

2. WOOD STRUCTURAL PANEL DIAPHRAGM DESIGN CRITERIA

Design of wood structural panel diaphragms requires consideration of diaphragm ratios, horizontal and vertical diaphragm shears, and connector/fastener values. Figure 12.1 illustrates the layouts of wood structural panel sheathing used to construct diaphragms.

In addition, there are special requirements for seismic design, introduced in the following subsections.

A. Framing

Collectors (i.e., drag struts) are required. Openings in diaphragms require perimeter framing. Such perimeter framing must be detailed to distribute shear along its length. Diaphragm plywood cannot be used to splice the perimeter members. Chords must be in the plane of the diaphragm unless it can be shown that chords in other locations of the walls will work. (See Fig. 12.2.)

B. Wood Structural Panels
[IBC Sec. 2306.3]

For layouts of *wood structural panel* diaphragm sheathing, see Fig. 12.1.

Wood structural panels must be of the exterior glue type. For wood structural panel diaphragms and shear

Figure 12.1 *Layouts of Wood Structural Panel Diaphragm Sheathing*

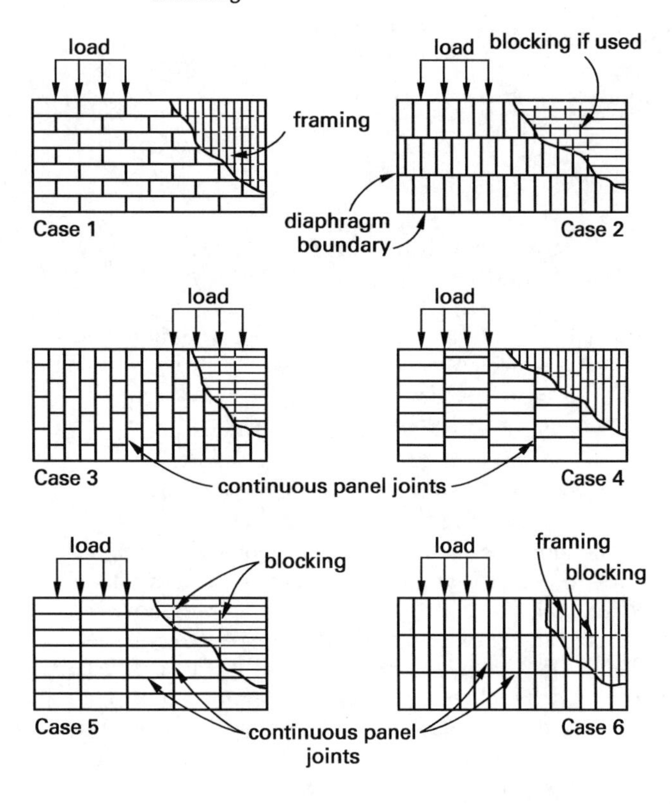

Figure 12.2 *Chord in the Plane of a Diaphragm*

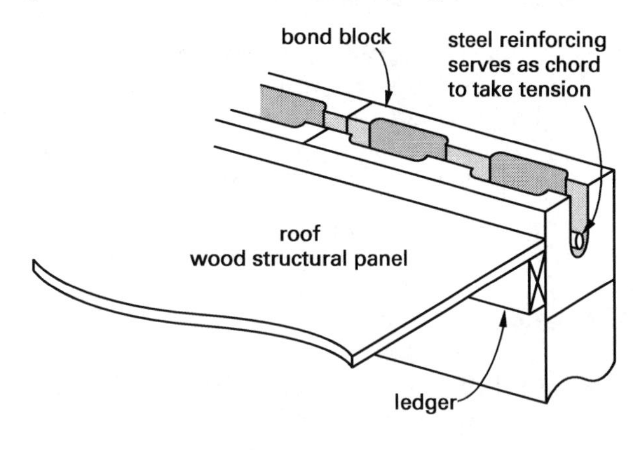

walls, sheets must be at least 4 ft by 8 ft (122 cm by 244 cm), except at edges of the diaphragm where pieces with a minimum dimension of 2 ft (61 cm) may be used.[1]

In resisting horizontal forces, horizontal diaphragms may be sheathed with wood structural panels; however, the horizontal forces should be equal to or less than those established in Table 12.1 (corresponding to IBC Table 2306.2(1)).

[1]Pieces even smaller than 2 ft (61 cm) may be used if all edges are properly blocked.

Specifications for wood structural panel material often refer to "DOC PS 1" or "DOC PS 2." These designations are U.S. Department of Commerce (DOC) voluntary product standards and are incorporated into the lower portion of the APA grademark on conforming panels.

Voluntary Product Standard PS 1 for Construction and Industrial Plywood specifies how plywood must be manufactured. PS 1 was developed by the U.S. Department of Commerce in conjunction with the construction plywood industry. *Voluntary Product Standard PS 2, Performance Standard for Wood-Based Structural Use Panels*, specifies how all-veneer (i.e., plywood) panel products must perform in a designated application, rather than how they must be manufactured. PS 2 allows manufacturers to create wood structural panel products from wood fiber resources and species not provided in PS 1. The APA has also developed a performance standard that covers composite and nonveneer panels, such as *oriented strand board* (OSB).

3. WOOD STRUCTURAL PANEL SHEAR WALLS

The design of wood structural panel-sheathed shear walls is covered in IBC Sec. 2306.3. Table 12.2 [IBC Table 2306.3(1)] gives the details for the design and analysis of such walls.

When both faces of shear walls are sheathed with the wood structural panels in accordance with Table 12.2, allowable shear for the wall may be taken as twice the tabulated shear for one side assuming the shear capacities are equal. Otherwise, the allowable shear should be taken as either the shear for the side with the higher capacity or twice the shear for the side with lower capacity, whichever is greater.

Unblocked shear walls are prohibited. Shear walls may suffer major damage or complete failure in the event of severe earthquakes due to inadequate staple edge distance into 2 in (51 mm) nominal framing members. For seismic design categories D, E, or F, where allowable shear for shear walls exceeds 350 lbf/ft (5.11 N/mm), foundation sill plates and all framing members receiving edge stapling from abutting panels should be 3 in (76 mm) nominal members (see Table 12.2 [IBC Table 2306.3(1), Ftn. g]).

4. SHEAR WALL ASPECT RATIOS

Maximum aspect ratios for shear walls are not explicitly defined by the *International Building Code*. Instead, IBC Sec. 2302, in the definition for shear walls, refers the engineer to the requirements that are defined by the American Wood Council's *Special Design Provisions for Wind and Seismic* (SDPWD). SDPWD Sec. 4.3.4 provides the maximum allowable aspect ratios, as listed in Table 12.3. h is the height of the perforated shear wall, and b_s is the minimum shear wall segment length.

Table 12.1 *Allowable Shear for Horizontal Wood Structural Panel Diaphragms with Douglas Fir-Larch or Southern Pine[a] Framing for Wind for Seismic Loading[f] (pounds per foot)*

(See Fig. 12.1 for cases.)
[Adapted from IBC Table 2306.2(1)]

PANEL GRADE	STAPLE LENGTH AND GAGE[d]	MINIMUM FASTENER PENETRATION IN FRAMING (in)	MINIMUM NOMINAL PANEL THICKNESS (in)	MINIMUM NOMINAL WIDTH OF FRAMING MEMBER[e] (in)	BLOCKED DIAPHRAGMS: Fastener spacing (in) at diaphragm boundaries (all cases), at continuous panel edges parallel to load (Cases 3 and 4) and at all panel edges (Cases 5 and 6)[b] × 25.4 for mm				UNBLOCKED DIAPHRAGMS: Fasteners spaced 6 in (152 mm) max. at supported edges[b]	
					6	4	2½[c]	2[c]	Case 1 (No unblocked edges or continous joints parallel to load)	All other configurations (Cases 2,3,4,5 and 6)
					Fastener spacing (in) at other panel edges[b] × 25.4 for mm					
					6	6	4	3		
		× 25.4 for mm			× 0.0146 for N/mm					
Structural I grades	1½ 16 gage	1	3/8	2	175	235	350	400	155	115
				3	200	265	395	450	175	130
			15/32	2	175	235	350	400	155	120
				3	200	265	395	450	175	130
Sheathing, and other grades covered in DOC PS 1 and PS 2	1½ 16 gage	1	3/8	2	160	210	315	360	140	105
				3	180	235	355	400	160	120
			7/16	2	165	225	335	380	150	110
				3	190	250	375	425	165	125
			15/32	2	160	210	315	360	140	105
				3	180	235	355	405	160	120
			19/32	2	175	235	350	400	155	115
				3	200	265	395	450	175	130

[a]For framing of other species: (1) Find specific gravity for species of lumber in AF&PA NDS. (2) For staples, find shear value from the table for Structural I panels (regardless of actual grade) and multiply value by 0.82 for species with specific gravity of 0.42 or greater, or 0.65 for all other species.
[b]Space fasteners maximum 12 in o.c. along intermediate framing members (6 in o.c. where supports are spaced 48 in o.c.).
[c]Framing at adjoining panel edges must be 3 in nominal or wider.
[d]Staples must have a minimum crown width of 7/16 in and must be installed with their crowns parallel to the long dimension of the framing member.
[e]The minimum nominal width of framing members not located at boundaries or adjoining panel edges must be 2 in.
[f]For shear loads of normal or permanent load duration as defined by the AF&PA NDS, the values in the table must be multiplied by 0.63 or 0.56, respectively.

5. WOOD STRUCTURAL PANEL DIAPHRAGM CONSTRUCTION DETAILS

A wood structural panel must be nailed continuously along its edges (*edge stapling*) and throughout its interior (*field stapling*) to achieve full development of the sheet. Table 12.1 [IBC Table 2306.2(1)] specifies the edge staple spacing required to carry a particular shear load. The plywood is stapled to joists, blocking, and ledgers, as shown in Fig. 7.5.

Each force element must have a full transmission path across the diaphragm. Collectors must frame into suitable walls or other collector elements. *Continuity ties* are required between adjacent edges of sheathing where edge stapling from the sheathing places the framing member below it in cross-grain tension (see Sec. 12.9).

Figure 12.3 illustrates how ties can be used to transmit tension and compression forces through a perpendicular girder.

6. BRIDGING/BLOCKING

Beams, rafters, and joists should be supported laterally to prevent rotation or lateral displacement. The shear transfer action is provided by blocking (see Fig. 12.1).

Bridging/blocking in accordance with IBC Sec. 2308.8.5 provides lateral support and prevents long joists from buckling—that is, "rolling out" or rotating out from under the members the joists support. When a joist buckles, its moment of inertia in the plane of support decreases significantly. Blocking is typically toe-nailed to the joists.

A blocked diaphragm is one in which all edges of the wood structural panels not falling on structural framing members are supported on and connected to blocking.

7. SUBDIAPHRAGMS

A *subdiaphragm*, as defined in IBC Sec. 2302, is "a portion of a larger wood diaphragm designed to anchor and transfer local forces to primary diaphragm struts

Table 12.2 *Allowable Shear for Wind or Seismic Loading in Pounds per Foot for Wood Structural Panel Shear Walls with Framing of Douglas Fir-Larch or Southern Pine*[a,b,f,g,i]

[Adapted from IBC Table 2306.3(1)]

PANEL GRADE	MINIMUM NOMINAL PANEL THICKNESS (in)	MINIMUM FASTENER PENETRATION IN FRAMING (in)	PANELS APPLIED DIRECTLY TO FRAMING					PANELS APPLIED OVER 1/2 IN (13 MM) OR 5/8 IN (16 MM) GYPSUM SHEATHING				
			Staple Size[h]	Fastener Spacing at Panel Edges (in) × 25.4 for mm				Staple Size[h]	Fastener Spacing at Panel Edges (in) × 25.4 for mm			
				6	4	3	2[d]		6	4	3	2[d]
	× 25.4 for mm			× 0.0146 for N/mm					× 0.0146 for N/mm			
Structural I sheathing	3/8	1	1 1/2 16 gage	155	235	315	400	2 16 gage	155	235	310	400
	7/16			170	260	345	440		155	235	310	400
	15/32			185	280	375	475		155	235	300	400
Sheathing, plywood siding[e], except Group 5 species, ANSI/APA PRP 210 siding	5/16[c] or 1/4[c]	1	1 1/2 16 gage	145	220	295	375	2 16 gage	110	165	220	285
	3/8			140	210	280	360		140	210	280	360
	7/16			155	230	310	395		140	210	280	360
	15/32			170	255	335	430		140	210	280	360
	19/32		1 3/4 16 gage	185	280	375	475	—	—	—	—	—

For SI: 1 inch = 25.4 mm, 1 pound per foot = 14.5939 N/m.

[a]For framing of other species: (1) Find specific gravity for species of lumber in AF&PA NDS. (2) For staples, find shear value from the table for Structural I panels (regardless of actual grade) and multiply value by 0.82 for species with specific gravity of 0.42 or greater, or 0.65 for all other species.

[b]Panel edges backed with 2 in nominal or wider framing. Install panels either horizontally or vertically. Space fasteners maximum 6 in o.c. along intermediate framing members for 3/8 in and 7/16 in panels installed on studs spaced 24 in o.c. For other conditions and panel thickness, space fasteners maximum 12 in o.c. on intermediate supports.

[c]3/8 in panel thickness or siding with a span rating of 16 in o.c. is the minimum recommended where applied direct to framing as exterior siding. For grooved panel siding, the nominal panel thickness is the thickness of the panel measured at the point of fastening.

[d]Framing at adjoining panel edges shall be 3 in nominal or wider.

[e]Values apply to all-veneer plywood. Thickness at point of fastening on panel edges governs shear values.

[f]Where panels applied on both faces of a wall and fastener spacing is less than 6 in o.c. on either side, panel joints shall be offset to fall on different framing members, or framing shall be 3 in nominal or thicker at adjoining panel edges.

[g]In seismic design categories D, E, or F, where shear design values exceed 350 pounds per linear foot, all framing members receiving edge fastening from abutting panels shall not be less than a single 3 in nominal member, or two 2 in nominal members fastened together in accordance with IBC Sec. 2306.1 to transfer the design shear value between framing members. Wood structural panel joint and sill plate fastening shall be staggered at all panel edges. See AF&PA SDPWS for sill plate size and anchorage requirements.

[h]Staples shall have a minimum crown width of 7/16 in and shall be installed with their crowns parallel to the long dimension of the framing members.

[i]For shear loads of normal or permanent load duration as defined by the AF&PA NDS, the values in the table shall be multiplied by 0.63 or 0.56, respectively.

and the main diaphragm." For example, in Fig. 12.4, the lateral forces from the masonry wall are transferred to the subdiaphragm through the anchor ties. The subdiaphragm span is the distance between the end shear wall and the center diaphragm strut. The anchor ties run the full depth of the subdiaphragm. The subdiaphragm should be designed as if it acts alone in transferring the tie forces developed in the anchor to the shear wall and strut.

The maximum length-to-width ratio for a wood structural subdiaphragm is usually limited to 2.5:1. Therefore, the subdiaphragm depth limits the length of anchor ties required. The lengths of diaphragm struts and cross ties at diaphragm discontinuities are similarly limited, as shown in Fig. 12.5.

8. CONNECTOR STRENGTHS

Connectors, or fasteners, for wood are typically nails, lag bolts (i.e., pointed bolts installed in pilot holes from one side), and machine bolts (i.e., nutted bolts). Connectors can fail in one of several ways. Connectors, particularly nails, can pull out; wood in shear connections can fail in bearing; connectors can fail in shear or bending. (Since they are much stronger than the wood pieces they connect, the connectors seldom fail in tension.)

It is generally unnecessary to deal with properties of the connectors such as yield strength in shear, longitudinal friction factor, and so on. The IBC does not provide tables of allowable loads for connectors. Section 2306.1 of the IBC requires designs to meet the requirements of other wood engineering standards (e.g., NDS).

A. Nails [NDS Chap. 11]

The IBC defers to the NDS for specific information regarding timber design, such as nail strengths. Appendix K.1 [NDS Table 11N] gives the allowable single-shear design values, Z, when a nail (box, common, or sinker) is driven the specified distance into the side of the lumber. Values depend on the wood species and specific gravity, G. The diameter is one critical dimension for a nail. Pennyweight, the common method of specifying nails, is not consistent with nail diameter. The allowable shear

Table 12.3 *Maximum Shear Wall Dimension Ratios*

shear wall sheathing material	maximum h/b_s ratio
1. wood structural panels, unblocked	2:1
2. wood structural panels, blocked	3:5:1[a]
3. particleboard, blocked	2:1
4. diagonal sheathing, conventional	2:1
5. gypsum wallboard	2:1[b]
6. portland cement plaster	2:1[b]
7. fiberboard	3:5:1[c]

[a]For design to resist seismic forces, the shear wall aspect ratio shall not exceed 2:1 unless the nominal unit shear capacity is multiplied by $2b_s/h$.
[b]Walls having aspect ratios exceeding 1.5:1 shall be blocked shear walls.
[c]For design to resist seismic forces, the shear wall aspect ratio shall not exceed 1:1 unless the nominal unit shear capacity is multiplied by the aspect ratio factor (seismic) $= 0.1 + 0.9b_s/h$. The value of the aspect ratio factor (seismic) shall not be greater than 1.0. For design to resist wind forces, the shear wall aspect ratio shall not exceed 1:1 unless the nominal unit shear capacity is multiplied by the aspect ratio factor (wind) $= 1.09 - 0.09h/b_s$. The value of the aspect ratio factor (wind) shall not be greater than 1.0.

Reproduced from the *Special Design Provisions for Wind and Seismic*, 2008 ed. Reprinted with courtesy, American Wood Council, Leesburg, VA.

Figure 12.3 *Typical Seismic Tie*

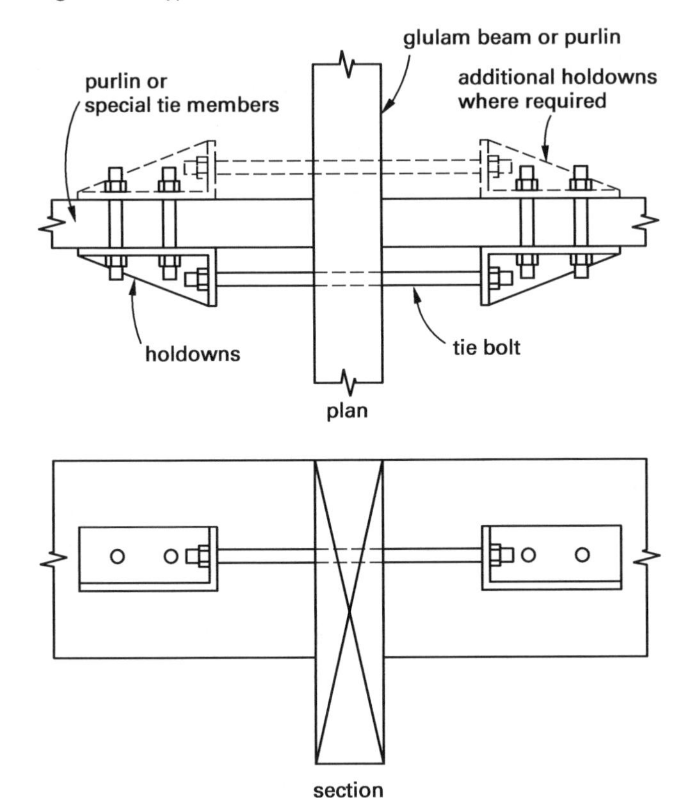

load is two-thirds of the table's values when nails are loaded in shear parallel to the grain [NDS Sec. 11.5.2]. Toe-nails also require a reduction of allowable stress by multiplying by a factor of 0.83 [NDS Sec. 11.5.4].

Appendix L [NDS Table 11.2C] gives the allowable withdrawal design strength for nails driven perpendicular to the grain (side-grain withdrawal). Nails driven parallel to the grain of the wood (end-grain withdrawal) are not permitted to resist withdrawal loads. Other restrictions for spacing and edge distances are also specified by NDS Chap. 11. When nails are used with structural sheathing, nails are to be driven so that their heads are flush with, but do not break, the surface of the sheathing.

B. Bolts [NDS Chap. 11]

Similar to nails, the IBC defers to the NDS for bolt strength calculations. Allowable shear design values, Z, for wood-to-wood connections are given in App. M and App. N [NDS Tables 11A and 11F] for single-shear or double-shear loading, respectively. Appendix O [NDS Table 11E] provides design strengths for bolts connecting sawn lumber to concrete.[2]

Allowable loads on proprietary structural straps and ties, as shown in Fig. 12.6, must be given by the manufacturer of those ties.

C. Adjustment Factors

The *nominal connector* (nail, bolt, etc.) *shear strengths*, Z, given in the NDS connector tables (see App. L through App. O) must be converted to *allowable connector design values*, Z', by multiplying by applicable *adjustment factors*.

$$Z' = C_D C_M C_t C_{eg} C_{di} C_{tn} Z \qquad 12.1$$

In Eq. 12.1, C_D is the *load duration factor*, C_M is the *wet service factor*, C_t is the *temperature factor*, C_{eg} is the *end-grain factor*, C_{di} is the *diaphragm factor*, and C_{tn} is the *toe-nail factor*. These values are 1.0 for "normal" usage, but NDS Table 10.3.1 should be consulted for values in particular situations. For example, the toe-nail factor is 0.67 for toe-nails experiencing withdrawal loads, but is 0.83 for toe-nails experiencing shear loads. There are also special rules that apply. For example, C_M is always 1.0 for toe-nails loaded in withdrawal, as is C_M for threaded fasteners regardless of moisture content. The end-grain factor, C_{eg}, is 0.67 and should be used when the connector's axis is parallel with the wood grain and the connector is loaded in shear. When lag screws are screwed in parallel to the grain and loaded in withdrawal, the end-grain factor is 0.75. When multiple connectors (e.g., nails) are used in diaphragm construction, the diaphragm factor, C_{di}, is 1.1.

[2]It is also important to recognize that connector forces used in wood-to-concrete and wood-to-masonry are limited by the strength of the concrete and masonry. Inasmuch as the wood is the weaker material, it seems logical that the wood provisions would determine the design, but there is no guarantee of this. Limitations are covered in ACI 318 for concrete and ACI 530 for masonry.

Figure 12.4 Subdiaphragm

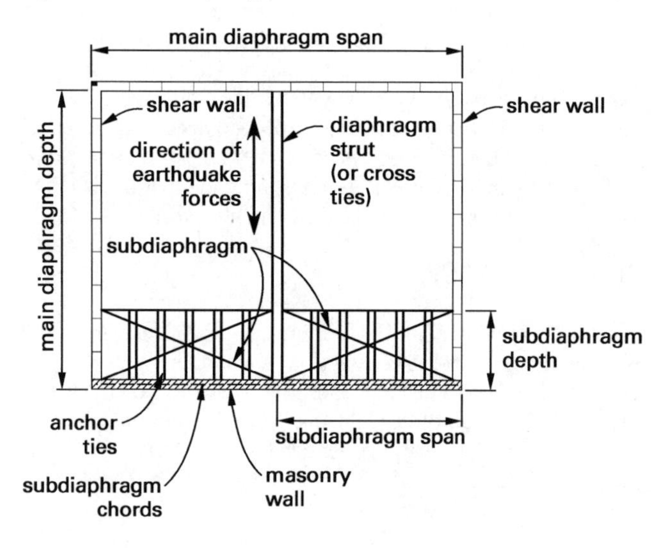

Figure 12.5 Subdiaphragm Limiting Strut Length

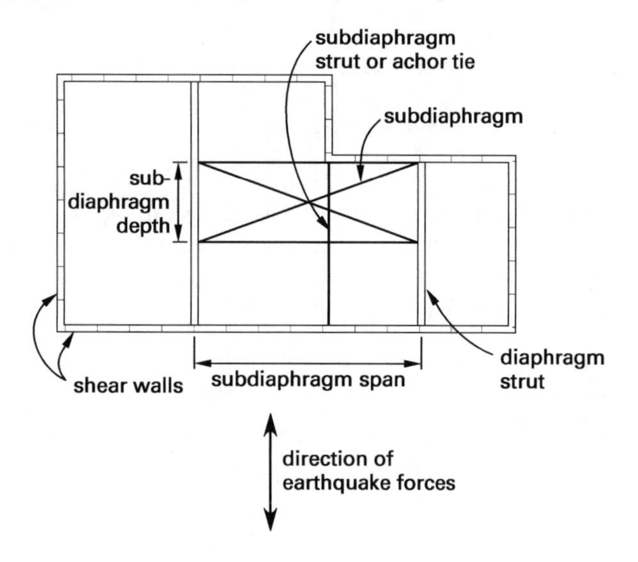

Figure 12.6 Typical Tie Installation

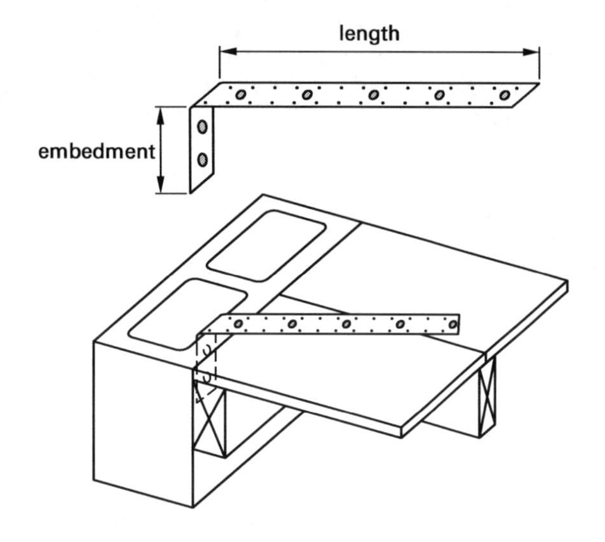

9. FLEXIBLE DIAPHRAGM TO WALL CONNECTION DETAILS

There are many acceptable ways that flexible diaphragms can be connected to masonry shear walls. (There are even more methods of making connections to wood-framed walls.) All acceptable methods provide an unbroken path for the force to follow from the diaphragm to the foundation. Any detail where the joists, ledgers, or diaphragm merely sit on supports without positive connection is unsatisfactory. Toe-nailing and nailing subject to withdrawal are not permitted in seismic design categories C, D, E, or F [ASCE/SEI7 Sec. 12.11.2.2.3].

Figure 12.7 shows a detail for a wall into which the joist (which may also be called a purlin or other framing member term) ends the floor framing. The lateral force from the diaphragm travels from the plywood through the edge nails into the *ledger*. (A ledger may also be called a *nailer* or *sill*.) A diaphragm shear force parallel to the masonry wall will continue through the ledger bolts into the masonry wall and through the parallel wall to the foundation and ground.

Figure 12.7 Inadequate Joist-Wall Framing

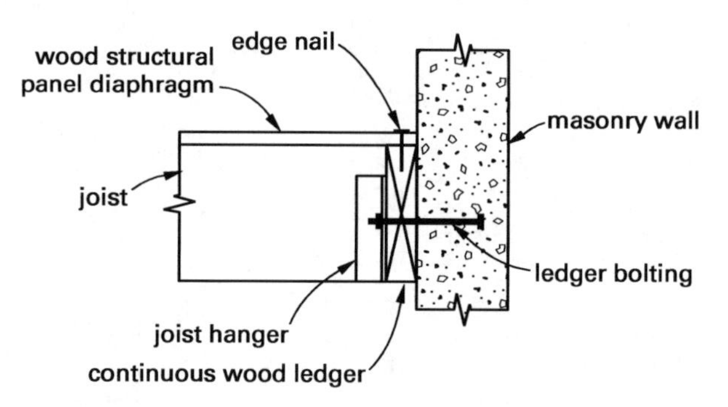

The framing method shown in Fig. 12.7 is not adequate (nor is it permitted [ASCE/SEI7 Sec. 12.11.2.2.3]) for forces perpendicular to the wall because of cross-grain bending. (See Sec. 12.10.) Additional tension connection straps, as shown in Fig. 12.8 and Fig. 12.9, must be added [ASCE/SEI7 Sec. 12.11.2.2.3]. The tension connection should be continuously nailed back (i.e., strapped) a considerable distance into the diaphragm to eliminate the high local tensile stress that would otherwise occur.

Figure 12.8 Adequate Joist-Wall Framing for Tension Forces

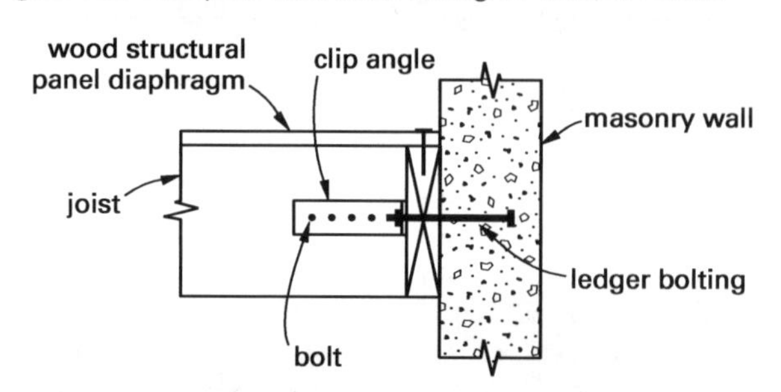

Figure 12.9 Ledger and Joist Tie

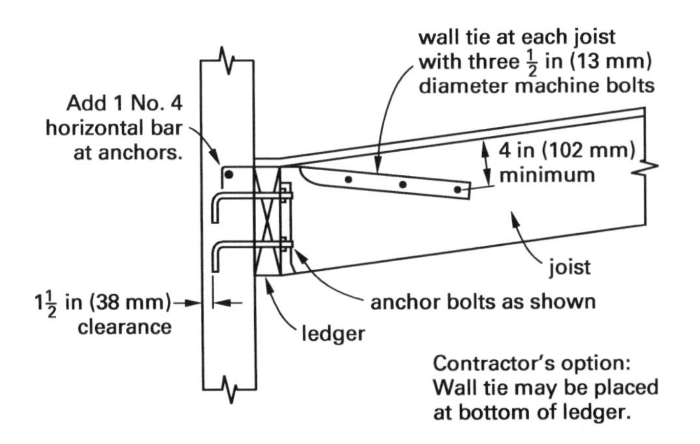

Figure 12.9 illustrates the typical details of a connection using both anchor bolts (for the ledger) and an embedded tie (for the joist[3]). Figure 12.6 illustrates yet another option.

The framing for a wall parallel to the joists is shown in Fig. 12.10. The design (i.e., using a ledger) is basically the same as for the connections at the ends of the joists. (The detail as shown places the ledger in cross-grain bending. (See Fig. 12.11.) Ties would also be required between the diaphragm and wall.)

Figure 12.10 Joist-Parallel Wall Framing

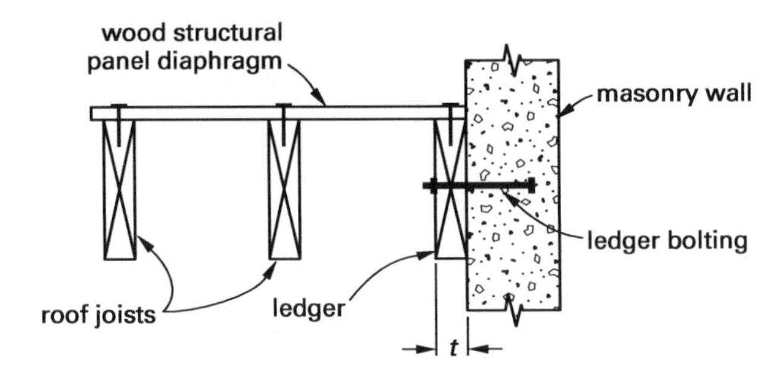

ASCE/SEI7 Sec. 12.11 requires that the connection between concrete and masonry walls and floor and roof diaphragms must be designed to resist the greatest of 10% of the weight of the wall, or $0.4S_{DS}I_eW_p$. In addition, the anchor spacing cannot exceed 4 ft (122 cm) unless the wall is designed to resist bending between anchors.[4] Toe-nailing may not be used to attach diaphragms to wood ledgers [ASCE/SEI7 Sec. 12.11.2.2.3]. Flexible diaphragms (usually wood-framed diaphragms) must create connections capable of resisting the force given by $F_p = 0.4S_{DS}k_aI_eW_p$ [ASCE/SEI7 Eq. 12.11-1]. (See Sec. 6.45 for special provisions regarding the connector design force required to maintain continuity between portions of the diaphragm.)

[3]As Sec. 7.7 implies, the term *joist* may be replaced by *purlin* or other framing member term.

[4]It is usually easier to specify a 4 ft (122 cm) anchor spacing than to design the wall connection for bending.

Figure 12.11 Cross-Grain Bending and Tension

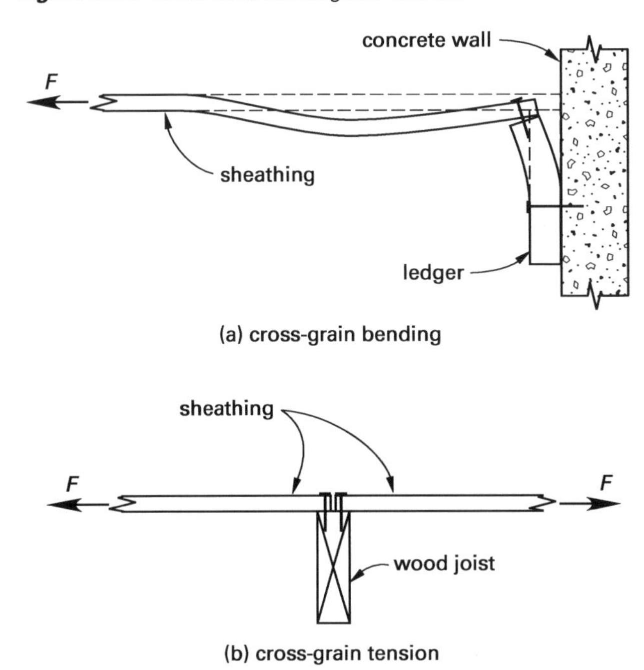

10. CROSS-GRAIN LOADING

Cross-grain bending and *cross-grain tension* in ledgers and joists, as illustrated in Fig. 12.11, are not permitted [ASCE/SEI7 Sec. 12.11.2.2.3].

11. FRAMING FOR DIAPHRAGM OPENINGS

It is not uncommon for a diaphragm to have openings for skylights, furnace flues, stairwells, and so on. IBC Sec. 2308.11.3.3 requires that such openings have blocking and metal ties installed along the edges that are perpendicular to the opening. Metal ties with a minimum tensile yield strength of 33,000 psi (227 MPa), a minimum size of 0.058 in by 1.5 in (1.47 mm by 38 mm), and attached with at least eight 16d common nails on each side of the header-joist intersection must be used. The purpose of the framing is to redistribute shears from areas adjacent to the openings around, or past, the openings to other collection elements.

An opening in the diaphragm imposes the following two requirements if the diaphragm is to operate correctly.

- The blocking around the perimeter of the opening must transfer the unequal loads on each side of the opening.
- The members around the perimeter of the opening must run from wall to wall, with tension straps used at corner connections to maintain continuity of the members. (See Fig. 12.12.) It is common to use double members for perimeter blocking. This eliminates the prohibited practice of using the diaphragm wood structural panel to splice perimeter wood.

Figure 12.12 *Framing an Opening*

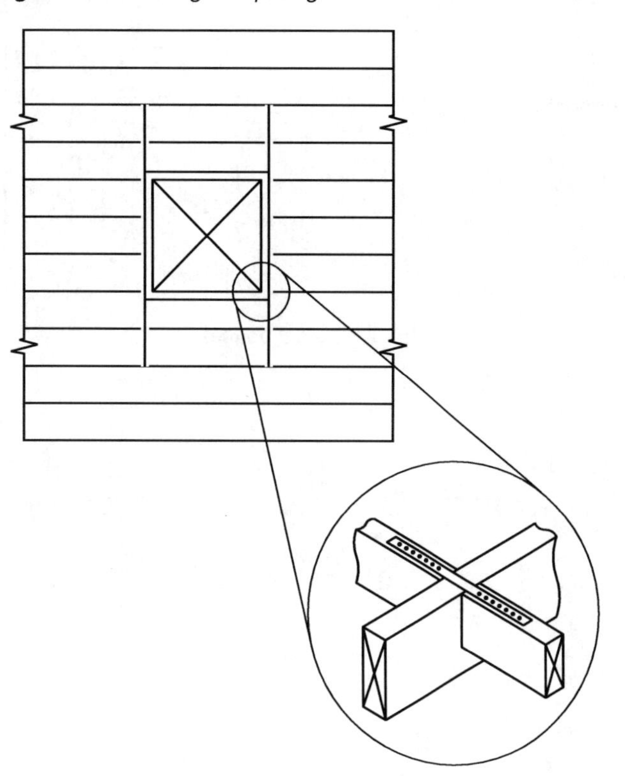

12. CRIPPLE WALLS

[IBC Sec. 2308.9.4]

When the first level of a structure is constructed above the original ground (i.e., raised foundation), the space between the first level and the ground is referred to as *crawl space*. The created crawl space varies in size depending on the length of wooden stud walls used on the top of an exterior foundation to support a building. These stud walls are termed *cripple walls*. Provisions for the cripple walls are given in IBC Sec. 2308.9.4. Bracing these short walls in accordance with IBC Sec. 2308.9.4.1 prevents their failures during earthquakes, which would cause damage to the entire structure.

Solid blocking is required to brace cripple walls having a cripple stud height of 14 in (356 mm) or less. For those cripple walls with stud heights exceeding 14 in (356 mm), the IBC requires that they be braced in accordance with IBC Table 2308.9.3(1) for seismic design categories A, B, or C. For structures in seismic design categories D or E, the provisions of IBC Sec. 2308.12.4 are required.

13 Tilt-Up Construction

1. TILT-UP DETAIL DESIGN

A *tilt-up building* typically uses precast structural panels, a wood structural panel diaphragm roof supported on wood joists or purlins, and either steel or glued-laminated (glulam) wood girders. Figure 13.1 gives details of tilt-up construction connections.

Analysis of tilt-up concrete shear walls is essentially the same as for cast-in-place concrete walls except that the panel-to-panel, panel-to-ceiling, and panel-to-floor details become critical. Shear between two panels must be developed by shear keys, dowels, or welded inserts. Contact joints between adjacent panels are assumed to develop no strength in shear or in tension due to friction, grouting, or architectural detailing.

Tilt-up wall construction used in one-story industrial and commercial buildings fared poorly in the 1971 San Fernando and 1987 Whittier earthquakes. The main weaknesses were found in the connections, or *anchorages*, between the roof and walls, particularly between main girders, purlins, and joists and the walls, the connection of the perimeter wood *ledger* to the wall, and the nailing of the wood structural panel diaphragm to the ledger. Basically, the walls moved outward, the girders and purlins detached from the ledgers, and the roofs fell to the ground.

Another problem with tilt-up construction occurs because each panel in a line (i.e., as part of a wall) is separate from the other panels and, therefore, resists a seismic load parallel to the panel in proportion to its relative rigidity. Since all of the panels are connected at their tops and bottoms, all will deflect the same amount. However, solid panels will resist the seismic load in shear (i.e., as a shear wall) while panels with large openings such as windows or doors will resist the seismic load in bending (i.e., as a beam).

Connections, such as those at weld plates, to shear walls should be numerous and regular, with a maximum spacing of approximately 4 ft (122 cm) along the top of the shear wall, and such connections should be capable of transferring three times the expected lateral load. It is also necessary for each panel to be attached to the floor to counteract the overturning moment. (See Sec. 7.19.) These details will prevent the poor performance that has been experienced in some previous earthquakes.

Figure 13.1 *Details of Tilt-Up Construction Connections*

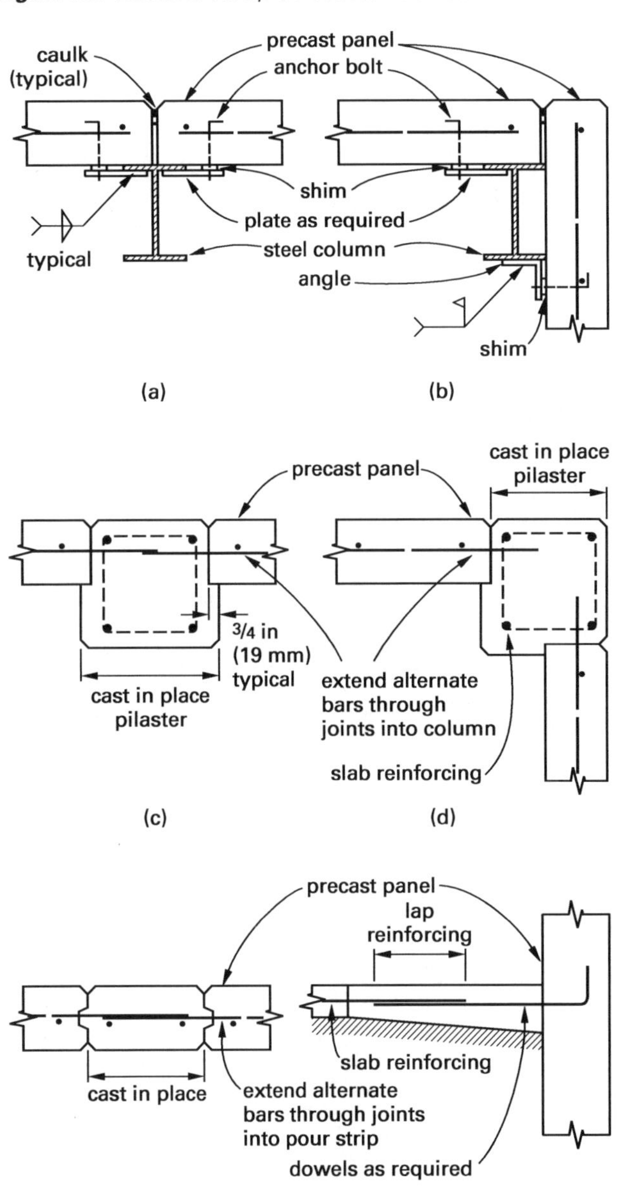

Openings in tilt-up walls have become so numerous that a wall, with its openings and spandrels, is sometimes more like a frame. A vertical wall segment (commonly known as a wall pier) is a short length of wall bounded by windows or the edge of the tilt-up unit [ACI 318 Sec. 21.7.4]. The shear strength and required reinforcement vary depending on the aspect ratio of the wall height to the wall length. If this aspect ratio is less than 2.5, the code requires the pier to be designed as a column [IBC Sec. 1908.1.8 modifying ACI 318 Sec. 21.7].

In the 1989 Loma Prieta earthquake, a number of "well-designed" tilt-up buildings experienced differential movement between the panel pilasters and the glulam roof beams that sat atop the pilasters. Greater attention should be given to detailing the tie spacing at the top of the pilaster to keep the anchor bolts from spalling the pilaster concrete.

14 Special Design Features

1. ENERGY DISSIPATION SYSTEMS

Various active and passive devices that reduce the magnitude or duration (or both) of the seismic force are in use or evaluation. These devices include active mass systems, passive visco-elastic dampers, tendon devices, and base isolation. Such devices may be incorporated into a design when approved by the building official.

2. BASE ISOLATION

The base shear experienced by a structure is, simplistically, the product of the structure mass and the acceleration (i.e., $F = ma$). Little can be done to reduce the mass of a structure in an earthquake, but the acceleration can be reduced if the structure is not attached rigidly to its foundation. Application of this concept is known as *base isolation* or *decoupling*, as shown in Fig. 14.1. The connections between the structure and the foundation are known as *isolation bearings*. Base isolation is applicable to bridges as well as to buildings.[1]

In effect, the ground is allowed to move back and forth under a building during an earthquake, leaving the building "stationary." Since the building theoretically does not accelerate, it does not experience a seismic force. In most cases of base isolation, the building is partially constrained, but the concept is the same. This can be done by "skewering" the base isolator with a vertical rod surrounded by a clearance hole. It may be

Figure 14.1 *Base Isolation (Foothill Center, San Bernardino County, California)*

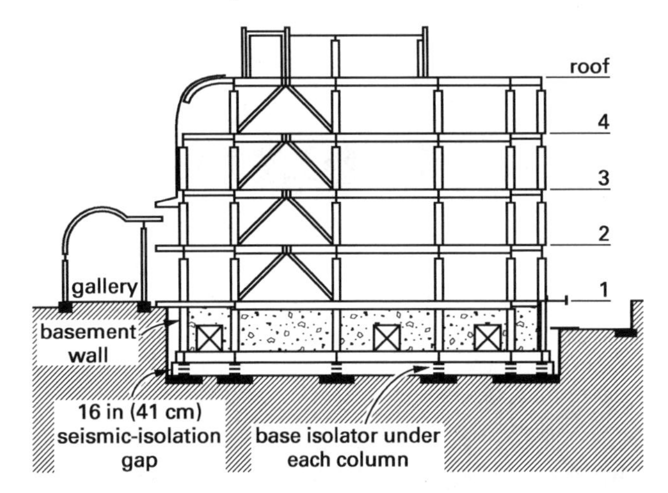

necessary to excavate a trench (or "moat") around the building to allow for differential movement.[2] Other nonstructural considerations include attaching utility service with flexible pipes and cables, as well as suspending elevator pits from the basement.

Bearings consisting of elastomeric (e.g., neoprene) alone may be suitable for absorbing horizontal (thrusting) loads, but metal is incorporated into designs that support vertical loads. There are three primary methods of base isolation; Fig. 14.2 illustrates one. These are supporting the building on (1) large ball bearings sandwiched between plates, (2) elastomeric bearings consisting of alternating layers of steel (or lead) and rubber, and (3) traditional structural expansion joints consisting of a layer of Teflon and a layer of rubber sandwiched between two steel plates. Combinations of these three methods (e.g., some bearings and some expansion joints) are desirable from a cost standpoint, since true bearings are costly. Some combinations work nearly as well as "pure" base isolation systems.

Most base isolators consist of alternate layers of some elastomer (i.e., rubber) and steel plate. Another type of isolator is a Teflon-coated slider. Teflon sliders, though capable of isolation, do not provide significant damping, so it must be provided elsewhere.

[1]It is generally accepted that bridges supported on box-girders less than 300 ft (91.4 m) long, and whose superstructures are supported at every pier (rather than being monolithic), are the best candidates for base isolation. Longer monolithic bridges are already so flexible that their natural period is long enough to reduce stresses significantly. It is also usually cheaper to design the bridge pier foundations as moment-resisting members than to specify base isolation in new construction (as opposed to retro-fitting old bridges). The first new bridge to use base isolation was the Sexton Creek Bridge near Cairo, Illinois, installed by the Illinois Department of Transportation. In 1986, the Metropolitan Water District (MWD) of Southern California used base isolation on its Santa Ana River crossing of the Upper Feeder pipeline. California's Department of Transportation (CALTRANS) has retrofitted several bridge structures in this manner.

[2]Every building is different, but a seismic isolation gap of approximately 12 in to 16 in (31 cm to 41 cm) should be used, although differential motion may need to be limited to less than the full gap size. At larger displacements, the deflection control element takes over and the building follows the earthquake motion. Refer to ASCE/SEI7 Chap. 17.

Figure 14.2 *One Type of Isolation Bearing*

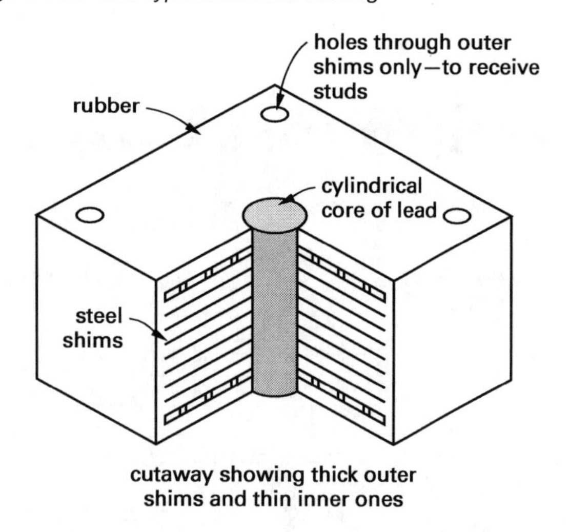

The number of isolators necessary in an existing building depends on the type of foundation. A structure with a continuous foundation around its periphery may require hundreds of small isolators to spread out the building weight. However, a structure with a column-like support system might require only one isolator per column.

While base isolation is not yet in widespread routine use, there is significant evidence that the technique is successful.[3] Natural building periods of structures using base isolation of the earliest designs have been more than doubled, moving the structures into areas on the response spectrum of lower acceleration. (See Sec. 5.1.) In several recent earthquakes, most buildings experienced maximum accelerations at the roof well in excess of the ground acceleration. However, buildings using base isolation experienced accelerations from 25% to 50% lower than the ground acceleration. Building and bridge periods were 2 to 10 times larger than their original values.

It has been suggested that a building fitted with base isolation will experience a Richter magnitude 8 earthquake as a magnitude 5 or 5.5 earthquake.[4] Instrumentation in the four-story Foothill Center illustrated in Fig. 14.1 was operational during the April 1990 magnitude 5.5 earthquake epicentered approximately 7 mi (11 km) away. (See App. C.) The base isolation system reduced the seismic forces by almost half. The foundation acceleration was 0.15 g but was only 0.08 g above the isolators. Acceleration was 0.16 g at the roof. The acceleration at two similar buildings approximately 6 mi (9.7 km) from the epicenter was 0.13 g at the ground and 0.39 g at the roof.

Because few base-isolated buildings have experienced a truly significant earthquake, there is concern that the elastomer will deteriorate, perhaps due to atmospheric ozone or other contaminants in the building, or that the bearing will "freeze up" after many years.

Another possible problem with base isolation derives from its very benefit—that of lengthening the natural period of the building. A natural soil period on the order of 3 or more seconds (as in cases where the soil is "soft") may coincide with the lengthened period of the building. This will produce resonance effects, such as those that occurred in the 1985 Mexico City and 1989 Loma Prieta earthquakes.

Base isolation is not suitable, economically or practically, for every building. One reason is that the base isolation bearings must all, within reason, be located at the same elevation. Buildings with footings that step down hillsides, for example, are poor candidates for base isolation.

In previous years, seismic isolators were "outside" the provisions of the common building codes and were considered exotic and experimental. Now they are installed routinely as retrofit devices to add seismic protection to buildings and bridges. Base isolation is now specifically permitted (subject to the approval of the building official).

3. SEISMIC-ISOLATED DESIGN DETAILS

ASCE/SEI7 Chap. 17 provides requirements for every seismic-isolated structure. This code defines the design lateral force for buildings and elements of structures based on a *design displacement*. Design displacement is described as the design-basis earthquake lateral displacement, excluding additional displacement due to actual and accidental torsion required for design of the isolation system. In addition, the seismic-isolated structures must be designed to withstand the *maximum earthquake*.

The codification of design of base isolated structures has resulted in a different approach than that for traditional building design. Similar to fixed-base building design, an equivalent lateral force method [ASCE/SEI7 Sec. 17.5], response spectrum procedure [ASCE/SEI7 Sec. 17.6.3.3], or response history procedure [ASCE/SEI7 Sec. 17.6.3.4] is allowed. Due to the complexity of engineering for isolated structures, the code is more stringent about the building configurations that can be designed using the equivalent lateral force method [ASCE/SEI7 Sec. 17.4.1].

The equivalent lateral force method is significantly different than the method defined for traditional buildings. The behavior of the building is dominated by the performance of the isolators, in particular their lateral stiffnesses, $k_{D,\mathrm{min}}$ and $k_{D,\mathrm{max}}$, which are determined by testing. The period of vibration is determined based on the weight of the structure and the stiffness of the isolation system [ASCE/SEI7 Eq. 17.5-2]. From this period of vibration, a design displacement, D_D, can be calculated [ASCE/SEI7 Eq. 17.5-1]. The total force expected

[3]The first new building to use base isolation was the 1983 Foothill Communities Law and Justice Center in San Bernardino, California. (See Fig. 14.1.) Base isolation is commonly used in the "seismic-retrofitting" of existing historical structures, such as the Salt Lake City and County Building (which opened in 1989), since it is one of the few methods that does not require extensive exterior work.

[4]Time will tell.

for the isolated structure to resist, V_b, can then be determined from Eq. 14.1 [ASCE/SEI7 Eq. 17.5-7].

$$V_b = k_{D,\max} D_D \quad 14.1$$

The total base shear is dependent on the stiffness of the isolation system, rather than the structural system. This base shear is then distributed to the different floors of the building similar to the methods used for traditional building design [ASCE/SEI7 Eq. 17.5-9].

Likewise, dynamic analysis procedures are defined by ASCE/SEI7 Sec. 17.6. These procedures are more similar to the methods used for traditional building design, but they require more sophisticated models of the isolation system. ASCE/SEI7 Sec. 17.6.2.1 specifies that models of the isolators must include

- the spatial distribution of the isolators
- the most critical location of any eccentric mass
- an assessment of the overturning and/or uplift of the isolator units
- an account of the effect of vertical load, bilateral load, and/or the rate of loading of the isolator

Based on ASCE/SEI7 Sec. 17.5.4.3, the total design base shear of the seismic-isolated structures, V_s, determined by the equivalent lateral force procedure should be equal to or greater than the base shear for a fixed-based structure having the same weight and period, the design wind load, or the lateral seismic force required to completely activate the isolated system factored by 1.5.

The maximum interstory drift ratio of the structure above the seismic-isolated structure should not exceed the limits given by ASCE/SEI7 Sec. 17.5.6.

4. DAMPING SYSTEMS

Passive and active damping systems, like base isolation, are in their infancy, at least in terms of large-scale use. These systems increase the damping ratio of the building and, in so doing, decrease the amplitude of swaying. Some involve moving blocks and counterweights, while others, such as *passive visco-elastic dampers* or *friction dampers*, are not much more than large shock absorbers.[5] Those that require power for motors and information from sensors for computers are known as *active systems*; those that do not are *passive systems*. Design procedures are defined in Chap. 18 of ASCE/SEI7.

Already used in some high-rise buildings to reduce wind drift, *active mass dampers* are nothing more than multi-ton blocks, usually of concrete or steel, suspended like a pendulum by a cable or mounted on tracks in one of the building's upper stories. When the wind or an earthquake makes the building sway, a computer sensing the motion signals a motor to move the weight in the opposite direction, thereby minimizing or neutralizing the motion.

Only a specific size of block will work in a building, because the weight of the block depends on the building's weight, the location of the block, the lag time, and the mode to be counteracted. Therefore, the mass is "tuned" to the structure, and the systems are also known as *tuned mass dampers* (TMD).

Active tendon and *active pulse systems* are similar, except that the building is moved by hydraulic pistons in the foundation or between stories instead of by a mass at the top. The energy pulse usually only needs to be applied once or twice each building motion cycle.

These devices typically reduce the lateral forces by one-third to one-half while increasing the building weight approximately 1%. They are also suitable for torsion control when placed off-center in a structure.

The most significant drawback to active systems is the fact that they require external power not only for the computer, but also for the motors driving the masses. Further, the large masses currently in use ride on oil bearings that take up to four minutes to pressurize. Thus, while active systems are useful in reducing drift during a predicted and slowly increasing windstorm, such devices are not yet substitutes for proper seismic design.

Active damping systems require detailed engineering design and analysis, and are generally restricted to structures with unique requirements. An example is located in Tokyo in the Kyobashi Seiwa Building, known for its extraordinary shape (11 stories high and only 13 ft (396 cm) wide).

Most damping devices installed in buildings are passive. (One of the best-known passive systems is located on the top floor of the 59-story Citicorp Building in Manhattan, where a 400 ton concrete-tuned block is located.)

An *added damping and stiffness* (ADAS) element is a passive damping system that generally consists of a combination of steel plates and spacers. The plates are bolted to structural bracing at the plate tops and bottoms. As the top and bottom structural bracing members displace relative to one another, the ADAS plates bend (i.e., yield) and dampen vibrations. The advantage of ADAS elements over conventional damping systems or shock absorbers is that ADAS elements contain no moving parts and require no maintenance.

5. ARCHITECTURAL CONSIDERATIONS

Due to planned yielding, the inter-story deflections in a major earthquake will be several times larger than the elastic deflections that are calculated from the base shear equation in the ASCE/SEI7. Therefore, damage to architectural (i.e., nonstructural) items is likely. Even for lesser-magnitude events, however, nonstructural items must be properly detailed.[6]

[5]Friction dampers have more in common with devices used to absorb coupling shocks in railway rolling stock than with automobile shock absorbers.

[6]This sentence probably should begin "Particularly for lesser-magnitude events" because nobody wants architectural damage, even in the more common small events.

Proper architectural detailing means providing proper clearances for exterior cladding, glazing (i.e., glass), wall finishes (e.g., marble veneers), interior partitions, and wall panels. Chimneys in residential buildings must be properly reinforced internally and securely strapped to the building at the roof line. Some elements can be free-floating—that is, they can move independently of the building. Proper attention must be given to the connection of these elements to the building.

To ensure that the occupants are able to get out of the building, doors should be designed to remain functional.

Floor coverings must be capable of three-dimensional movements.

All elements capable of falling and causing damage or injury must be rigidly attached to structural members. For example, suspended ceilings (e.g., tee-bar) and lights in drop-in ceilings must be tied to ceiling members above. Partitions, particularly those that do not run floor-to-ceiling, require special attention.

Columns in traditional moment-resisting frames are typically spaced 10 ft to 20 ft (3 m to 7 m) apart. This spacing presents a challenge to architects when they try to provide unobstructed occupant space in high-rise buildings. Designers who want exterior column spacings greater than this must use other techniques to increase the strength of their buildings. Use of high-strength concrete, braced cores, ductile frames, ductile outrigger framing, and bandages are some of the techniques used.

Most trusses contain numerous axial members that are arranged in triangular sections. These triangular sections, though effective, make it difficult to include corridors (when in the core of the building) and windows (when on the perimeter of the building) in the design.

With the ability to create efficient moment-resisting joints, an increasing number of engineers are designing trusses comprised of rectangular sections. Bridge-like trusses comprised of rectangular sections with moment-resisting joints are known as *quadrangular girders*, *open-web girders* or *trusses*, *ductile frames*, or *Vierendeel girders*. (See Fig. 14.3.) In addition to providing unobstructed access through the openings in the truss, these structures are economical to build.

In-fill panels can either be given sufficient clearance so that they are not crushed by the deflection of adjacent structural elements, as in a panel forming a wall between two columns, or short sections of the panel (usually at its vertical edges) can be designed to be flexible or much weaker and replaceable.[7] Buttresses—short panel tees or wall returns—should be used to prevent the panel from falling over.

Computer room floors that are raised on pedestals and placed on unbraced stringers have fared poorly in past earthquakes.

[7]Care must be taken when using flexible materials that they remain flexible indefinitely. Foamed polyethylene and polysulfide products appear *not* to meet this requirement.

Figure 14.3 *Vierendeel (Quadrangular) Girders*

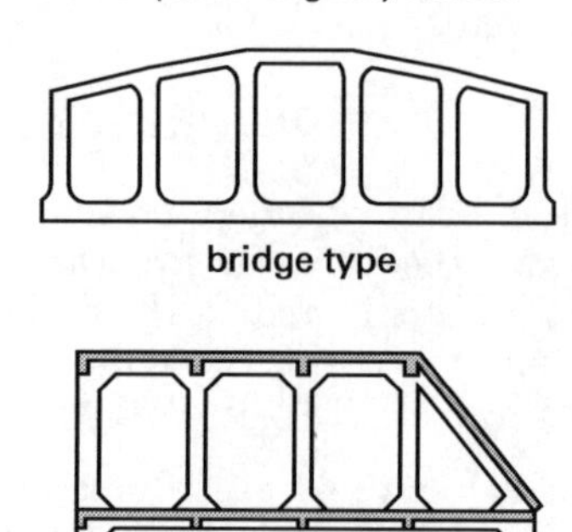

Equipment such as air-conditioning devices, motors, pumps, tanks, piping, and air ducts must be securely bolted to their foundations to prevent sliding and overturning. Mere use of clamps or clips to prevent overturning is inadequate because the equipment can slide out from under the clips during an earthquake.

6. UPGRADING EXISTING CONSTRUCTION

Structures built in prior years can be upgraded with new features to make them more resistant to seismic forces. Such an activity is known as *retrofitting*. Bridges (some of which were heavily damaged in the 1971 San Fernando and 1989 Loma Prieta earthquakes) can be fitted with cable restraints and increased-capacity shear keys to restrict longitudinal motion.

Nonductile concrete columns with rectangular cross sections, built before the ductility requirements were added to the UBC in 1973, have been upgraded by being wrapped in steel plates. Columns with round or nonrectangular cross sections can be wrapped with steel or composite fiber wire to give them greater ductility. Carbon fibers are probably too brittle to be used to wrap columns as a means of achieving spiral strand confinement. However, Kevlar fibers (the same material used in most bullet-proof vests) show promise to wrap columns.

Steel jackets are comprised of two semicircular portions welded up the seams. Grout is injected into the space between the jacket and column. Thickness of the steel jacket depends on the loads expected.

The proper application of flat steel plates ("Vierendeel bandages," "Vlasovian bandages," or just "bandages") at selected exterior locations on a structure can increase resistance to seismic forces. Bandages do this by increasing the torsional moment of inertia of the structure (when the bandages are connected to core framing members) and by providing alternate vertical load paths (when the bandages are connected to load-carrying members at different levels). Inasmuch as only portions of a structure are covered, bandages are not effective in containing concrete. Therefore, this technique should not be confused with the encasement of concrete columns in steel jackets.

Belting, as a means of upgrading an unreinforced masonry building, is a technique that extends long steel rods horizontally from corner to corner, where the rods are attached to plates on the building's corners. The resulting "girdled" nature of the building is supposed to keep the building's walls from pulling away from the building core during an earthquake. For this technique to work, the rods must placed relatively close together, and they must extend all the way up the building.

Although California has adopted seismic retrofit standards for hospitals, there are no national codes specifically governing the methods of retrofitting or upgrading existing structures. For that matter, no one really knows how successfully retrofitted structures will fare in an earthquake. Approximately 40% of the unreinforced masonry buildings that had been retrofitted sustained damage in the 1987 Whittier earthquake. During the 1989 Loma Prieta earthquake, a four-story unreinforced masonry building in San Francisco that had been seismically retrofitted by belting suffered major seismic damage.

15 Practice Problems

1. EARTHQUAKES IN GENERAL

1. What is the lowest acknowledged numerical Richter magnitude that would identify a major earthquake?

Solution

This is an ambiguous question since different people would interpret the word "major" differently. In general, a moderate earthquake would have a Richter magnitude of 5, a strong earthquake would have a magnitude of 7, and a great earthquake would have a magnitude of 8 or higher. (See Sec. 2.3.)

2. What is the theoretical upper numerical limit on the Richter magnitude scale?

Solution

There is no theoretical upper limit on the Richter magnitude scale. (See Sec. 2.3.)

3. What does the Richter magnitude scale measure?

Solution

It is not clear what is intended by the word *measure.* The Richter apparatus *detects* earth movement. The numerical magnitude *describes* earthquake strength. The numerical value *represents* a measure of energy release on a logarithmic scale. (See Sec. 2.3.)

4. Which of the following locates the center of an earthquake in three dimensions?

(A) tectonic boundary

(B) focal depth

(C) epicenter

(D) hypocenter

Solution

The hypocenter is the three dimensional location of the center of an earthquake in the earth. The epicenter is the projection of this point on the earth's surface, thus it is a two-dimensional location. The distance between the epicenter and hypocenter is the focal depth.

The correct answer is D.

5. Which of the following measures of earthquakes is most closely tied to the economic loss after an earthquake?

(A) Richter magnitude

(B) moment magnitude

(C) Modified Mercalli Intensity (MMI)

(D) peak ground acceleration

Solution

Although damage is correlated to all of the measurement scales, only intensity readings are related to whether any manufactured infrastructures exist in the region of the earthquake.

The correct answer is C.

6. A transmission tower is being designed with mechanical damping, using a response spectra from a design earthquake to complete the design. The transmission tower is expected to have a fundamental period of 2.0 sec and a damping ratio of 8%. Which point on the response spectra relates to the peak ground acceleration of the design earthquake?

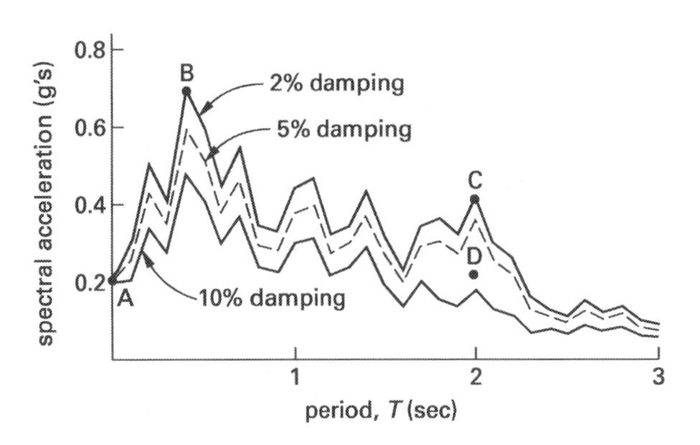

period	acceleration (g's) damping 2%	5%	10%
0.00	0.200	0.200	0.200
0.10	0.300	0.255	0.204
0.20	0.500	0.425	0.340
0.30	0.400	0.340	0.272
0.40	0.700	0.595	0.476
0.50	0.600	0.510	0.408
0.60	0.440	0.374	0.299
0.70	0.540	0.459	0.367
0.80	0.340	0.289	0.240
0.90	0.330	0.281	0.224
1.00	0.440	0.374	0.299
1.10	0.460	0.391	0.313
1.20	0.320	0.272	0.218
1.30	0.340	0.300	0.240
1.40	0.430	0.366	0.292
1.50	0.320	0.272	0.190
1.60	0.230	0.196	0.137
1.70	0.340	0.289	0.202
1.80	0.360	0.306	0.153
1.90	0.320	0.272	0.136
2.00	0.420	0.357	0.179
2.10	0.300	0.255	0.128
2.20	0.260	0.221	0.111
2.30	0.160	0.136	0.068
2.40	0.130	0.111	0.077
2.50	0.110	0.094	0.065
2.60	0.150	0.128	0.089
2.70	0.120	0.102	0.071
2.80	0.140	0.119	0.083
2.90	0.100	0.085	0.060
3.00	0.090	0.077	0.054

(A) point A

(B) point B

(C) point C

(D) point D

Solution

The response spectra are graphs of the peak acceleration of single-degree-of-freedom systems accelerated using the earthquake ground motion. However, the question asks for peak ground acceleration, which corresponds to the "infinitely rigid" soil and a period of zero seconds.

The correct answer is A.

2. BUILDINGS

7. What is the difference between *stiffness* and *rigidity* as used in seismic consideration?

Solution

Stiffness is the force that will deflect a structure elastically a unit amount in a given direction. *Rigidity*—strictly *relative rigidity*—is a normalized stiffness. Whereas the stiffness of a single member can be used in numerical calculations, rigidities can only be used when forces are being distributed among several members. (See Sec. 4.3 and Sec. 4.4.)

8. What is the difference between *ductility* and *flexibility* as used in seismic consideration?

Solution

Flexibility is the reciprocal of *stiffness*. It is the elastic deflection obtained when a unit force is applied. (See Sec. 4.3.) *Ductility* is the ability of a material to distort and yield without fracture or collapse. (See Sec. 5.5.) Since flexibility deals with elastic deformation and ductility deals with inelastic deformation, there is little connection between the two concepts.

9. What is the relationship between *rigidity* and the variables of pier height, depth, and thickness?

Solution

Roughly, *rigidity* is proportional to the first power of thickness and to the cube of pier depth and is inversely proportional to the cube of pier height. (See Sec. 4.5.)

10. What is *ductility*?

Solution

Ductility is the ability of a material to distort and yield without fracture or collapse. (See Sec. 5.5.)

11. What is the *ductility factor*?

Solution

The *ductility factor* of a material is the ratio of its strain energy at fracture to its strain energy at yield. There are other similar and related definitions. (See Sec. 5.6.)

12. What factors influence the ductility factor?

Solution

From a metallurgical perspective, temperature and previous stress-strain history influence the ductility of a ductile material such as steel. The higher the temperature, the greater the ductility. The more the material has been worked or stressed in previous cycles or events, the more brittle (the opposite of ductile) it becomes. From a structural perspective, ductility depends on the type of construction (i.e., steel or concrete), the structural system, the quality of construction, the detailing, and the redundancy. (See Sec. 5.5 and Sec. 5.6.)

The strain rate can embrittle certain steels and welds, as can the welding procedure and post-weld heat treatments. Member orientation during rolling processes can alter stress-strain behavior, particularly in rolled plates.

13. What is the minimum recommended ductility factor?

Solution

It is not possible to specify a minimum recommended ductility factor exactly because it depends on the type of structure, construction material used, intended use of the structure, and many other factors. However, the ductility factor should be well in excess of 1.0 and is generally no less than approximately 2.2–2.5 for modern structures. (See Sec. 5.5 and Sec. 5.6.)

14. What is *ductile framing*?

Solution

In its simplest interpretation, a structure with ductile framing will not collapse even though its structural frame has sustained significant distortion, misalignment, and other yielding damage. (See Sec. 9.1.)

15. What is the principle reason for specifying a minimum ductility factor?

Solution

The principle reason for specifying a minimum ductility factor is to obtain a *ductility margin* (i.e., the ductility between yield and collapse) sufficient to ensure survivability in a design earthquake.

16. Why will a theoretical analysis of elastic response of a structure usually overestimate the stresses resulting from an earthquake?

Solution

A structure will not behave totally elastically during an earthquake. Local yielding at high stress locations reduces the seismic energy (i.e., the energy of oscillation) initially present in the structure.

17. Describe the two components of *drift*.

Solution

Shear drift is the sideways deflection of a building due to lateral (sideways) loads. *Chord drift* is the sideways deflection due to axial (vertical) loads. (See Sec. 5.12.)

18. What is the P-delta effect?

Solution

The P-delta effect is an additional column bending stress caused by eccentric vertical loads. (See Sec. 5.13.)

19. How are drift and the P-delta effect related?

Solution

When a structure drifts, its vertical loads become eccentric. The eccentric loading increases the column stress, and the stress increase is called the P-delta effect. (See Sec. 5.12 and Sec. 5.13.)

20. What is the *natural* (*fundamental*) *period* of a building?

Solution

The *natural period*, or *fundamental period*, of a building is the time it takes the building to complete one full swing in its primary mode of oscillation. (See Sec. 3.8 and Sec. 4.6.)

21. What does the term *redundancy* mean as it is used in the context of modern high-rise buildings?

Solution

Redundancy, as used in a seismic context, is synonymous with *distributed excess capacity* and *multiple load paths*. A *redundant design* has a safety factor, but the converse statement is not necessarily true. For example, if a vertical 100 kip (445 kN) load is supported by a single column having a 120 kip (534 kN) capacity, the design will have excess capacity but no redundancy since the structure will collapse if the column fails. If the 100 kip (445 kN) load is supported by 12 columns, each with a 10 kip (45 kN) capacity, the design will have both redundancy and excess capacity.

22. Has the recent trend in high-rise buildings been toward increased or decreased redundancy? Why?

Solution

Redundancy is increasingly seen as a crucial characteristic of high-rise designs. Multiple redundant load paths greatly increase the reliability of a structure.

Since building members (e.g., columns, girders, and shear walls) and details (i.e., column-girder joints) do not always behave as intended (due to our meager knowledge of the behavior of so-called ductile designs, design or construction errors, and higher-than-expected loading), the design should allow for the unintended loss of all capacity in a small fraction of the members. The tolerable loss of this excess capacity is the principle of redundant design.

23. What causes *torsional shear*?

Solution

Torsional shear occurs when an earthquake acts on a structure whose centers of mass and rigidity do not coincide. (See Sec. 5.14.)

24. What is *negative torsional shear*?

Solution

Negative torsional shear is the torsional shear on one side of a structure that is opposite in sign to the shear force, or direct shear, induced by the base shear. (See Sec. 5.15.)

25. How should negative torsional shear be treated?

Solution

Inasmuch as the direction of an earthquake is not known in advance, negative torsional shear should be disregarded—it may not be used to decrease the size of a wall or other member. (See Sec. 5.15.)

26. What is the *Rayleigh method*, and where would it be used?

Solution

The Rayleigh method is a method determining the mode shape of a multiple-degree-of-freedom system through an iterative process. (See Sec. 4.20.)

27. Explain *critical damping.*

Solution

Critical damping is the amount of structural damping that causes oscillation to die out and return to the equilibrium position faster than any other amount of damping. (See Sec. 4.8.)

28. What is the *damping ratio*?

Solution

The *damping ratio* is the ratio of the actual damping coefficient to the critical damping coefficient. (See Sec. 4.8.)

29. What is the practical range of damping ratios?

Solution

Damping ratios of typical buildings range from approximately 0.02 for steel-frame construction to approximately 0.15 for wood-frame construction. (See Sec. 4.10.)

30. To what extent does damping affect the natural period of vibration of a structural frame?

Solution

Damping increases the actual period of vibration slightly, compared to the natural period of vibration. However, even with highly damped structures, the increase is usually 1% or less. Therefore, the natural period is used in the building code calculations and the effect of damping is disregarded. (See Sec. 4.11.)

31. What is a *response spectrum*?

Solution

A *response spectrum* is a graph of the maximum response (acceleration, velocity, or displacement) to a specified excitation of a single-degree-of-freedom system plotted as a function of the SDOF system's natural period. (See Sec. 5.1.)

32. How does the *portal method* deal with the effects of column lengthening and shortening?

Solution

The *portal method* disregards changes in column length. (See Sec. 8.2.)

33. A 20-story building has stories that are all equal height (12 ft or 3.75 m). The building is laterally supported only by reinforced concrete frames that are well detailed and considered to be special moment-resisting frame status. The structures have been engineered to ensure that no rigid components enclose or adjoin to the frames that would prevent the frames from deflecting. Which of the answers is most nearly the approximate fundamental period of the structure?

(A) 0.8 sec

(B) 1.2 sec

(C) 2.2 sec

(D) 21 sec

Customary U.S. Solution

Because the building is a concrete moment frame and is detailed to make sure that no nonstructural components impede the lateral deflection, the building meets the requirements of the second row of Table 6.11 [ASCE/SEI7 Table 12.8-2]. From this table, the value of C_t is 0.016 and the value of x is 0.9. The height of the roof is 240 ft. Using these values, the period of the structure can be calculated from Eq. 6.11 [ASCE/SEI7 Eq. 12.8-7].

$$\begin{aligned} T_a &= C_t h_n^x \\ &= (0.016)(240 \text{ ft})^{0.9} \\ &= 2.2 \text{ sec} \end{aligned}$$

SI Solution

Because the building is a concrete moment frame and is detailed to make sure that no nonstructural components impeded the lateral deflection, the building meets the requirements of the second row of Table 6.11 [ASCE/SEI7 Table 12.8-2]. From this table, the value of C_t is 0.0466 and the value of x is 0.9. The height of the roof is 75 m. Using these values, the period of the structure can be calculated from Eq. 6.11 [ASCE/SEI7 Eq. 12.8-7].

$$\begin{aligned} T_a &= C_t h_n^x \\ &= (0.0466)(75 \text{ m})^{0.9} \\ &= 2.27 \text{ s} \quad (2.2 \text{ s}) \end{aligned}$$

The correct answer is C.

3. STRUCTURAL SYSTEMS

34. Distinguish between a *moment-resisting frame* and a *special moment-resisting frame.*

Solution

A *moment-resisting frame* has rigid connections between members (e.g., between girders and columns) such that moments applied to columns are partially resisted by girder bending and vice versa. With sufficiently large moments, however, even the elastic capacity of such structural systems that share loads between girders and columns can be exceeded. The integrity and load-carrying ability of a *special moment-resisting frame* (previously known as a *ductile moment-resisting frame*) will remain intact even after yielding has been experienced. (See Sec. 6.21.)

35. What are *bearing wall systems* and *box systems*?

Solution

Box system is another name for *bearing wall system.* A bearing wall system relies on shear and load-bearing walls to carry dead, live, and seismic loads. (See Sec. 6.21.)

36. Generally speaking, which is more likely to have a smaller natural frequency of vibration, a steel moment-resisting frame or a concrete moment-resisting frame, given equal heights and moments of inertia?

Solution

While the performance of both steel and concrete frames are similar in this regard, some theoretical generalizations are possible. The steel frame may be slightly more flexible (smaller stiffness), producing a smaller frequency and longer period. Also, the concrete frame may have greater mass, producing a smaller frequency and longer period. (The effect of damping on the period is minimal.) The ASCE/SEI7 equation for the period (see Eq. 6.11) clearly indicates that steel buildings generally are expected to have longer periods (smaller frequencies). (See Sec. 6.26.)

37. Which is more likely to have a larger damping ratio, a steel or concrete moment-resisting frame?

Solution

Concrete construction generally has a greater damping ratio. (See Sec. 4.10.)

38. Are plastic hinges designed in columns, in girders, or in both?

Solution

A *plastic hinge* forms when a member yields. The yielding of a girder or of a girder-column joint may produce distortion, floor and roof sagging, and misalignment without collapse. The yielding of a column, however, may lead to structural collapse. Therefore, unlike girders, columns should not be designed to form plastic hinges.

39. What is the structural system called that does not have a complete vertical load-carrying space frame?

Solution

Bearing wall systems, or box systems, use walls, not frame members, to carry the vertical loads. (See Sec. 6.21.)

40. If the slab roof is supported by the wall shown, what is the correct value of R for the structure?

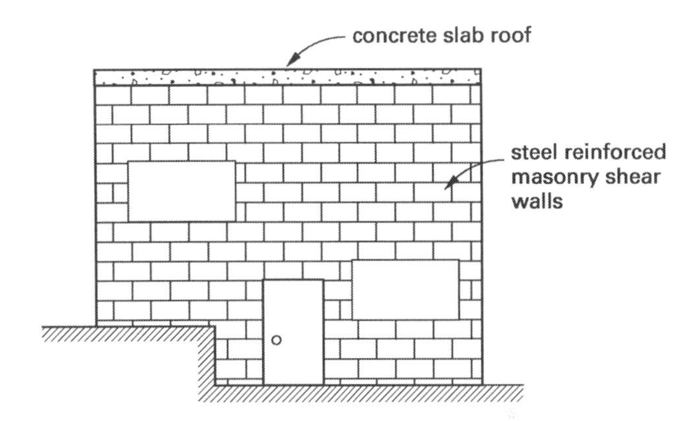

(A) 1.5

(B) 3.5

(C) 4.0

(D) 5.5

Solution

The shear walls will be the only support for the roof, making the walls a bearing system. Thus, according to App. H [ASCE/SEI7 Table 12.2-1], the value of R for an intermediate reinforced masonry shear wall is 3.5. Both special and ordinary reinforced masonry wall systems are also listed in App. H, but have different R values than the answer options provided.

The correct answer is B.

41. A structure with a 1.4 sec fundamental period of vibration is to be built with different structural systems at different levels. The base is a five-story concrete shear wall structure with concrete columns independent of the walls supporting the gravity load. Above this will be a 10-story, special moment-resisting steel frame. The steel frame is 15 times more flexible than the concrete base. If the upper 10 stories were a structure by itself, the period of vibration would be 1.5 sec. What is the correct value of R for the structure?

(A) The entire building must be designed for $R = 6.0$.

(B) The entire building may be designed for $R = 8.0$.

(C) The steel frame may be designed using $R = 8.0$, excluding the weight of the base when determining the loading of the steel frame.

(D) The steel frame may be designed using $R = 8.0$, but the weight of the base must be included when determining the loading of the steel frame.

Solution

ASCE/SEI7 allows different values of R to be used when structures are built from different structural systems at different levels as long as the structure meets certain criteria in the code. The building meets parts (a) and (b) of the two-stage procedure defined in ASCE/SEI7 Sec. 12.2.3.2. When buildings are designed this way, the base shear of the upper levels can be calculated without including the weight of the lower portion of the structure.

The correct answer is C.

4. IBC AND ASCE/SEI7

42. Describe how seismic base shear is calculated according to the IBC.

Solution

The IBC determines the *seismic base shear* by multiplying the weight of the structure by a seismic response coefficient. This coefficient is based on expected ground acceleration of the site, building occupancy, building period, soil-profile type, and structural system. (See Sec. 6.30.)

43. Which of the terms in the seismic coefficient equation can be equal to 1.0?

Solution

The importance factor, I_e, can have a value of 1.0. It is also possible for the natural structure period, T, to have a calculated value of 1.0 sec. The variables of S_{D1} or S_{DS} can have a value of 1.0 for certain sites. (See Sec. 6.14, Sec. 6.17, Sec. 6.18, and Sec. 6.26.)

44. In the application of the base shear formula, what factor of R should apply if a structure is a special steel concentrically braced frame in one direction and an ordinary reinforced concrete shear wall building in the orthogonal direction? The building is *not* considered a bearing wall system in either direction.

Solution

Since the structure's performance is analyzed independently for earthquakes in the two orthogonal directions, the structure is treated as a special steel concentrically braced frame building ($R = 6.0$) for an earthquake in one direction and as a shear wall building ($R = 5.0$) for the orthogonal direction. (See Sec. 6.21.)

45. Explain in general terms how the base shear is distributed in a horizontal plane to the various resisting elements.

Solution

Most structures have indeterminant structural systems. For this common situation, the base shear is distributed to the resisting elements in proportion to their rigidities. (See Sec. 4.5.) Structures with determinant structural systems have the base shear distributed to the resisting elements according to static equilibrium.

46. Draw the seismic force diagram acting on a multistory building.

Solution

Refer to Fig. 6.14(a), Sec. 6.35, or Fig. 6.19, Sec. 6.42.

47. Draw the cumulative shear diagram acting on a multistory building.

Solution

Refer to Fig. 6.14(b), Sec. 6.35.

48. What is the IBC building height limit for special reinforced concrete shear wall construction in a bearing-wall structural system that is located in seismic design category D?

Solution

The IBC refers to the requirements of ASCE/SEI7, which allows special reinforced concrete shear wall construction in buildings as high as 160 ft (49 m). (See App. H.)

49. Which has a smaller R value, a steel or concrete special moment-resisting frame?

Solution

Steel and concrete structures with special moment-resisting frames have the same R value, 8.0. (See App. H.)

50. What possible values can R take on for a steel or reinforced concrete moment-resisting frame?

Solution

R can have values of 8.0 (SMRF of steel or concrete), 7.0 (special steel truss), 5.0 (concrete IMRF), 4.5 (steel IMRF), 3.5 (steel OMRF), and 3.0 (concrete OMRF). (See App. H.)

51. What R value would be used for (a) a large football grandstand with bleachers and (b) a tall vertical tank supported on a raised platform supported by a tower similar to a special steel concentrically braced structure framework?

Solution

(a) A football grandstand with bleachers is a self-supporting nonbuilding structure falling under the jurisdiction of the local building official. Its mass can be considered to be lumped at the various spectator levels, and the supporting system continues between floors. Therefore, it is covered by the IBC and ASCE/SEI7. However, it is not specifically mentioned in ASCE/SEI7 Table 15.4-2. (See Sec. 6.46.) It should be considered to fail in the "other" category at the end of the table and have an R value of 1.25.

(b) The tank is on a raised platform supported by a braced framework, so this is a nonbuilding structure supported by another structure. ASCE/SEI7 Sec. 13.1.5 says that nonbuilding structures (including storage racks and tanks) that are supported by other structures shall be designed in accordance with Chap. 15.

Based on ASCE/SEI7 Table 15.4-2, for elevated tanks supported on structural towers similar to buildings, use ASCE/SEI7 Table 12.2-1. (See App. H.) From ASCE/SEI7 Table 12.2-1, for a building frame consisting of special steel, concentrically braced, $R=6$.

52. What is the absolute IBC limitation on drift?

Solution

The key word in this question is "absolute." There is no absolute limitation on drift in the IBC, as any drift that can be shown to be "tolerable" is permitted. Other limitations apply, however, when drift is not tolerable. (See Sec. 6.40.)

53. What percentage of buildings with moment-resisting frame systems in seismic design category F should be detailed to qualify as special moment frames?

Solution

All buildings with moment-resisting frame systems must be detailed to qualify as special moment-resisting frames. (See Sec. 6.21.) Intermediate moment frames and ordinary moment frames are not permitted in seismic design category F.

54. What percentage of the live load in a warehouse should be added to the dead load when calculating base shear?

Solution

A minimum of 25% of the warehouse live load should be added. (See Sec. 6.29.)

55. What is the range of mapped maximum considered earthquake ground motion for Arizona and New Mexico of 0.2 second spectral response acceleration?

Solution

From IBC Fig. 1613.3.1(1), the highest value is 50% of g (near latitude 35° and longitude 106°), and the lowest value is 7.8% of g (near latitude 35° and longitude 104°). (See Sec. 3.2.)

56. Considering IBC Fig. 1613.3.1(1) and Fig. 1613.3.1(2), what regions of the country have values of S_1 larger than S_S?

Solution

No site on the map (or any of the other maps) has a value of S_1 larger than S_S.

57. What is the correct value to use in Eq. 6.6 [IBC Eq. 16-37] for S_S for a site located at latitude 45° and longitude 105° (northeastern Wyoming)?

(A) 0.043

(B) 0.160

(C) 61.8 in/sec^2 (1570 mm/s^2)

(D) 16%

Solution

The location on IBC Fig. 1613.3.1(1) is partway between the contours for 15% g and 20% g. The value should be interpolated (or the higher value contour should be used). The value used in Eq. 6.6 should be the decimal reading, not the percentage, and it should not be converted to customary U.S. or SI units. (The base shear and structure weight in the ASCE/SEI7 formulas have the same units.)

The correct answer is B.

58. A special steel moment frame structure is located at a site where S_S is 110% from IBC Fig. 1613.3.1(1) and S_1 is 43% from IBC Fig. 1613.3.1(2). The site class is D. The importance factor is 1.0. What is the seismic response coefficient, C_s, for the structure if the structure's period of vibration is 0.5 sec?

(A) 0.027

(B) 0.034

(C) 0.070

(D) 0.097

Solution

The site coefficient values of F_a and F_v are obtained from interpolating the provided values from IBC Table 1613.3.3(1) and Table 1613.3.3(2) using the rows for site class D. For F_a, the values from Table 1613.3.3(1) are 1.1 for S_S of 1.00 and 1.0 for S_S of 1.25. Interpolating for the value when S_S is 1.10 results in the value of F_a being 1.06. Likewise, for F_v, the values from Table 1613.3.3(2) are 1.6 for S_1 of 0.40 and 1.5 for S_1 of 0.50. Interpolating for the value when S_1 is 0.43 results in the value of F_v being 1.57.

Using these results, the maximum considered earthquake spectral response acceleration parameters are determined using IBC Sec. 1613.3.3. From Eq. 6.6 [IBC Eq. 16-37],

$$\begin{aligned} S_{MS} &= F_a S_S = (1.06)(1.1) \\ &= 1.17 \end{aligned}$$

From Eq. 6.7 [IBC Eq. 16-38],

$$\begin{aligned} S_{M1} &= F_v S_1 = (1.57)(0.43) \\ &= 0.68 \end{aligned}$$

The design spectral response acceleration parameters are determined using IBC Sec. 1613.3.4. From Eq. 6.8 [IBC Eq. 16-39],

$$\begin{aligned} S_{DS} &= \tfrac{2}{3}S_{MS} = \left(\tfrac{2}{3}\right)(1.17) \\ &= 0.78 \end{aligned}$$

From Eq. 6.9 [IBC Eq. 16-40],

$$\begin{aligned} S_{D1} &= \tfrac{2}{3}S_{M1} = \left(\tfrac{2}{3}\right)(0.68) \\ &= 0.45 \end{aligned}$$

The value of the structure response modification factor, R, for the structure is obtained from App. H [ASCE/SEI7 Table 12.2-1]. Under category C.1, special steel moment-resisting frames have an R value of 8.

With these design parameters, the equations for the seismic coefficient are obtained from ASCE/SEI7 Sec. 12.8. The first equation provides an upper limit on the seismic coefficient. From Eq. 6.20 [ASCE/SEI7 Eq. 12.8-2],

$$\begin{aligned} C_s &= \frac{S_{DS}}{\frac{R}{I_e}} = \frac{0.78}{\frac{8}{1}} \\ &= 0.0975 \end{aligned}$$

This value will be correct for all structures where $T < T_s$. To determine T_s, set Eq. 6.20 [ASCE/SEI7 Eq. 12.8-2] equal to Eq. 6.21(a) [ASCE/SEI7 Eq. 12.8-3].

$$\frac{S_{DS}}{\frac{R}{I_e}} = \frac{S_{D1}}{T_s\left(\frac{R}{I_e}\right)}$$

$$\begin{aligned} T_s &= \frac{S_{D1}}{S_{DS}} = \frac{0.45}{0.78} \\ &= 0.58 \text{ sec} \end{aligned}$$

Since the period $T = 0.5$ sec is below T_s, the value calculated for C_s is correct.

The correct answer is D.

5. CONCRETE AND MASONRY STRUCTURES

59. What are the possible modes of failure due to seismic forces if the lateral force-resisting system of an older high-rise building is constructed of reinforced concrete?

Solution

(See Sec. 9.1.) This question does not specify whether the concrete is specially reinforced or whether the concrete is used in a frame or shear wall structure. In general, a reinforced concrete frame will have failed if the concrete spalls or crushes before plastic yielding of the steel reinforcing occurs, or if the steel reinforcing is stressed plastically. Failure can be expected to occur:

(a) at the ends of well-designed columns when there is insufficient shear resistance (i.e., such that the column breaks out of its supports)

(b) in poorly designed columns with insufficient confinement

(c) in shear walls due to inadequate vertical reinforcing

(d) at construction joints due to inadequate bonding between members

(e) in beams due to inadequate shear reinforcing

(f) in columns due to excessive drift and overturning moment

60. What are the most important considerations in achieving ductility in concrete frames?

Solution

The most important considerations are confinement and continuity. (The steel in confined, or specially reinforced, concrete should yield before the concrete crushes.) Adequate bonding between steel and concrete must be ensured. Steel must be capable of developing its full tensile strength. Members must be adequately tied together at joints. (See Sec. 9.1.)

61. What is meant by *confined concrete*?

Solution

Confined concrete is also called *ductile concrete* or *specially reinforced concrete.* The steel in ductile concrete will yield before the concrete crushes. This enables the concrete member to develop its full compressive strength without yielding. (See Sec. 9.1.)

62. Why is concrete confined at joints and in members?

Solution

The confining steel in ductile concrete enables the concrete to develop its full compressive strength while the longitudinal reinforcing steel yields. (See Sec. 9.1.)

63. What are some of the construction methods used to ensure ductile behavior of concrete?

Solution

To confine concrete, columns are spiral wrapped at closer intervals and additional hoops are used at joints and other locations. Continuity of reinforcement is achieved by special attention to splices. Special attention is given to reinforcement of shear walls. Hooks, ties, stirrups, and hoops are detailed to prevent pull-out. (See Sec. 9.1.)

64. With regard to resistance to seismic forces, which is better in steel-reinforced concrete columns: spiral ties or horizontal hoops? Why?

Solution

Spiral transverse reinforcement is the most efficient confinement, but it may not be possible to use it. From a construction standpoint, spiral reinforcement for smaller columns is easier to form in the field. From a seismic standpoint, spiral reinforcement provides slightly better confinement. Larger columns, however, cannot be wrapped in the field, and extending spirals into beam-column joints is difficult, so individual factory-fabricated hoops must be used.

65. What is the function of the spiral and individual ties used in a concrete column?

Solution

Ties confine the concrete and keep it from crushing. (See Sec. 9.1.)

66. Do special reinforcement hoops replace regular ties in beams and columns?

Solution

Special reinforcement (primarily in the form of additional hoops) is used in addition to regular beam stirrups and column ties. In beams, special reinforcement is required at points of expected yielding (i.e., at plastic hinges). In columns, hoops are required at column-girder connections. (See Sec. 9.1.)

67. What is the effectiveness of stirrups in deep concrete beams?

Solution

After inclined cracks form at the ends of deep beams, the load is carried in a "tied arch" configuration that has considerable remaining strength. Stirrups in the center of a deep beam are not particularly effective.

68. If a masonry wall is part of the lateral force resisting system in a structure with seismic design category D, what types of cement mortar may be used?

Solution

Only Type S or Type M may be used. (See Sec. 11.1(D).)

69. Which of the shear reinforcement patterns shown can be used as hoops for seismic resistance? Assume all bend radii and tails meet code requirements.

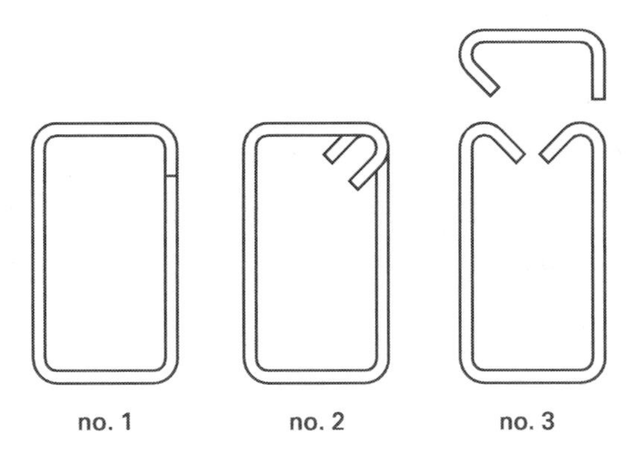

(A) no. 1 and no. 2

(B) no. 1 and no. 3

(C) no. 2 and no. 3

(D) All of the patterns are acceptable.

Solution

Closed ties must have tails and 135° bends, so no. 1 does not qualify. Stirrups with 135° bends with J-hooks across the top are acceptable replacements for closed ties.

The correct answer is C.

70. A 12 ft high, special concrete moment-resisting frame column is detailed similarly to Fig. 9.6. The column is a 14 in square and will have eight no. 7 bars (0.875 in diameter) longitudinal reinforcement. To resist the shear force, no. 3 hoops (0.375 in diameter) will be used. Crossties divide the column cross-section into quarters. To reduce cost, the minimum number of hoops will be used. Excluding the joint region, which of the following spacing layouts will be suitable for the column?

(A) hoops spaced 12 in on center the full height of the column

(B) hoops spaced 7 in on center the full height of the column

(C) hoops spaced 4 in on center the full height of the column

(D) hoops spaced 3 in on center the full height of the column

Solution

The horizontal spacing of crossties, h_x, is 7 in. In the hinge regions, the maximum tie spacing, S_h, is limited to the smallest of

- $6d_b = (6)(0.875 \text{ in}) = 5.25 \text{ in}$
- $\frac{1}{4}B_s = \left(\frac{1}{4}\right)(14 \text{ in}) = 3.5 \text{ in}$
- $s_o = 4 \text{ in} + \dfrac{14 \text{ in} - h_x}{3}$

$$= 4 \text{ in} + \frac{14 \text{ in} - 7 \text{ in}}{3}$$

$$= 6.33 \text{ in} > 6 \text{ in}$$

The maximum tie spacing is 3.5 in (3 in).

The correct answer is D.

6. STEEL STRUCTURES

71. What are the possible modes of failure due to seismic forces if the lateral force-resisting system of a high-rise building is constructed of steel?

Solution

In general, steel will be considered to have "failed" if it severely yields or fractures. (See Sec. 10.1.)

72. Where is a steel-framed building with a properly designed special moment-resisting frame most likely to yield in an earthquake?

Solution

Yielding and formation of plastic hinges in a steel structure can be expected at points where the moments are greatest, such as at girder ends and at column-girder connections. Columns can buckle due to bending and eccentric effects. Flanges and webs of members can buckle from local stresses and fail from fatigue loading. (Girder-column connections should not fail, however, through weld and bolt failure. All connections should be able to sustain the full plastic moment of connected members.)

73. Which of the following types of steel structures are most likely to have doubler plates?

(A) ordinary concentrically braced frames

(B) special concentrically braced frames

(C) ordinary moment-resisting frames

(D) special moment-resisting frames

Solution

Doubler plates are used to increase the shear strength of the shear panel zone of steel moment connections between beams and columns. They are especially crucial when plastic hinge formation is expected to occur outside the beam-to-column joint. This is the intended failure mode for special moment-resisting frames, not the other three options.

The correct answer is D.

74. An eccentric braced frame is designed as an end link system, as shown. The story height is 12 ft (3.75 m), the bay length is 14 ft (4.4 m), and the link length is 4 ft (1.25 m). If the drift is 2%, what will be the rotation of the link?

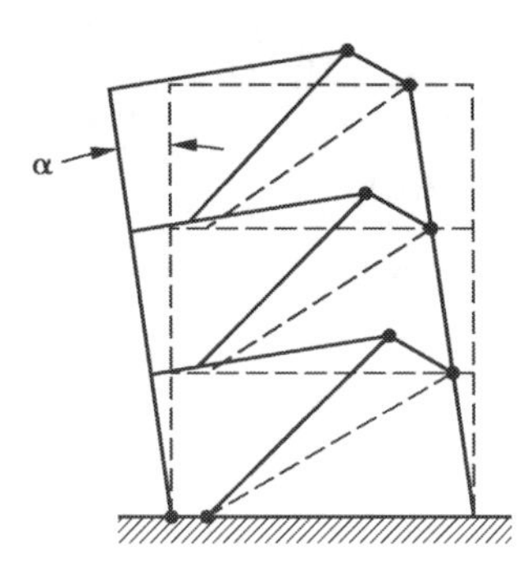

(A) 2.0%

(B) 5.0%

(C) 7.0%

(D) not enough information provided

Solution

Link rotation is calculated considering the vertical deflection resulting from the drift. In the illustration, the joint between the left column and the beams is assumed to remain 90° throughout the earthquake. A drift α of the left column will cause the rotation of the left end of the beam to be α, and the joint at the brace will deflect vertically a distance h. For compatibility, the link to the right of the brace must rotate enough to allow for the same vertical deflection. l is the length of the beam from the left column to the brace, and e is the length of the beam from the brace to the right column. θ is the rotation of the link.

Using small angle approximation, $h = \alpha l = \theta e$, so

$$\theta = \left(\frac{l}{e}\right)\alpha$$

Customary U.S. Solution

$$\theta = \left(\frac{l}{e}\right)\alpha = \left(\frac{10}{4}\right)(0.02) = 0.05 \quad (5\%)$$

SI Solution

$$\theta = \left(\frac{l}{e}\right)\alpha = \left(\frac{3.15}{1.25}\right)(0.02) = 0.05 \quad (5\%)$$

The correct answer is B.

7. SOILS AND FOUNDATIONS

75. Given a soil engineering report, what should be considered in order to improve the seismic response of the building?

Solution

Devastating resonance effects can be avoided if the natural building period does not coincide with the site period. (See Sec. 3.6, Sec. 3.8, and Sec. 4.14.) If the site has fine sand, soil liquefaction effects must be considered. (See Sec. 3.7.)

76. For the soil profile shown, what is the highest level that the water table can be for liquefaction not to be a potential problem?

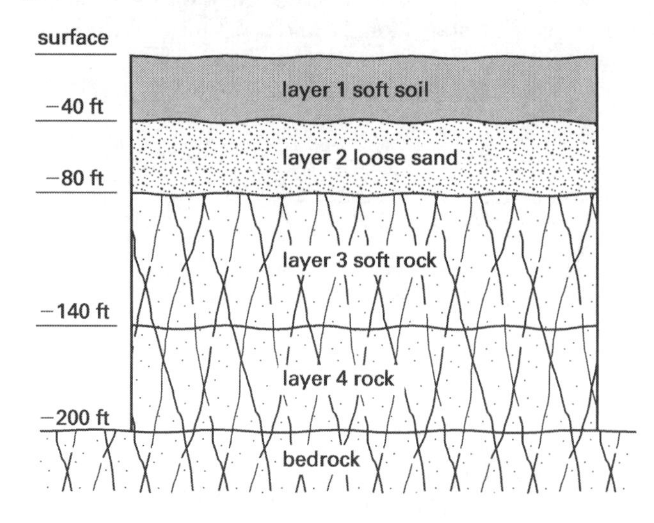

(A) 40 ft

(B) 80 ft

(C) 200 ft

(D) The site is not likely to have liquefaction no matter what the elevation of the water table.

Solution

If the loose sand layer becomes saturated, liquefaction may occur. To ensure that liquefaction does not occur, the water table must remain below this layer (i.e., below 80 ft).

The correct answer is B.

77. Use the soil profile shown in Prob. 76. The soil layers have the following characteristics.

layer	soil description	shear wave velocity (ft/sec)
1	soft soil	500
2	loose sand	1000
3	soft rock	2000
4	rock	4000

What is the correct site class for this location according to ASCE/SEI7?

(A) B

(B) C

(C) D

(D) E

Solution

ASCE/SEI7 provides Eq. 20.4-1 to determine site class when shear wave velocities of individual soil layers are known. The code defines that only the top 100 ft of the soil profile be used in the calculation.

Making a table for calculations,

layer	depth (ft)	shear wave velocity (ft/sec)	depth/velocity (sec)
1	40	500	0.08
2	40	1000	0.04
3	20	2000	0.01
	100		0.13

The average shear wave velocity of the top 100 ft of the soil profile is

$$v_s = \frac{100 \text{ ft}}{0.13 \text{ sec}} = 770 \text{ ft/sec}$$

Using this value with ASCE/SEI7 Table 20.3-1 establishes that the site class is D.

The correct answer is C.

8. ANALYSIS PROBLEMS

78. The roof of the structure shown is rigid. A 10 kip (44.48 kN) load is applied at the roof. What are the resisting shears in each column?

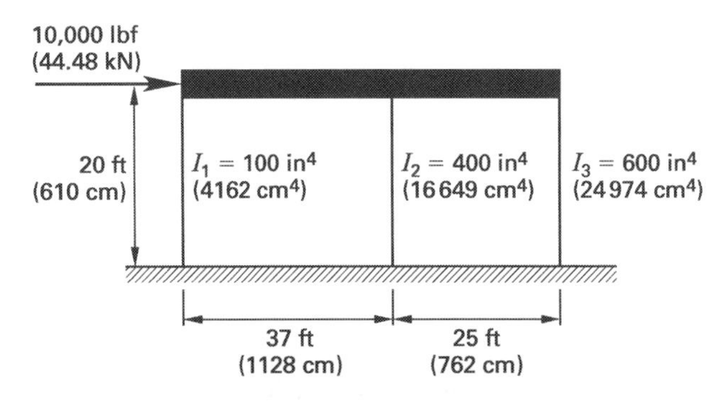

Customary U.S. Solution

Since column lengths and materials are the same, the total rigidity is proportional to the sum of the moments of inertia. The total is

$$\begin{aligned} I_{\text{total}} &= I_1 + I_2 + I_3 \\ &= 100 \text{ in}^4 + 400 \text{ in}^4 + 600 \text{ in}^4 \\ &= 1100 \text{ in}^4 \end{aligned}$$

The load carried by the first (left) column is

$$\begin{aligned} V_1 &= (10 \text{ kips})\left(\frac{100 \text{ in}^4}{1100 \text{ in}^4}\right) \\ &= 0.91 \text{ kips} \end{aligned}$$

The load carried by the second (middle) column is

$$\begin{aligned} V_2 &= (10 \text{ kips})\left(\frac{400 \text{ in}^4}{1100 \text{ in}^4}\right) \\ &= 3.64 \text{ kips} \end{aligned}$$

The load carried by the third (right) column is

$$\begin{aligned} V_3 &= (10 \text{ kips})\left(\frac{600 \text{ in}^4}{1100 \text{ in}^4}\right) \\ &= 5.45 \text{ kips} \end{aligned}$$

SI Solution

Since column lengths and materials are the same, the total rigidity is proportional to the sum of the moments of inertia. The total is

$$\begin{aligned} I_{\text{total}} &= I_1 + I_2 + I_3 \\ &= 4162 \text{ cm}^4 + 16\,649 \text{ cm}^4 + 24\,974 \text{ cm}^4 \\ &= 45\,785 \text{ cm}^4 \end{aligned}$$

The load carried by the first (left) column is

$$V_1 = (44.48\ \text{kN})\left(\frac{4162\ \text{cm}^4}{45\,785\ \text{cm}^4}\right) = 4.04\ \text{kN}$$

The load carried by the second (middle) column is

$$V_2 = (44.48\ \text{kN})\left(\frac{16\,649\ \text{cm}^4}{45\,785\ \text{cm}^4}\right) = 16.17\ \text{kN}$$

The load carried by the third (right) column is

$$V_3 = (44.48\ \text{kN})\left(\frac{24\,947\ \text{cm}^4}{45\,785\ \text{cm}^4}\right) = 24.24\ \text{kN}$$

79. A power transformer is mounted on a pedestal. The structural adequacy for the seismic environment is to be demonstrated. The horizontal component of the seismic force is to be defined by the El Centro north-south elastic response spectra. If the lateral frequency of the system exceeds 30 Hz, the response is defined as being equivalent to a 0.5 g static load. The vertical response is defined as 50% of the horizontal response, and both occur simultaneously. Free lateral vibration tests on the structure show that the ratio of successive swings of the structure is 0.882. Assume that the pedestal mass and shear flexibility, foundation stiffness, and torsion can be neglected. Specific properties are shown in the illustration.

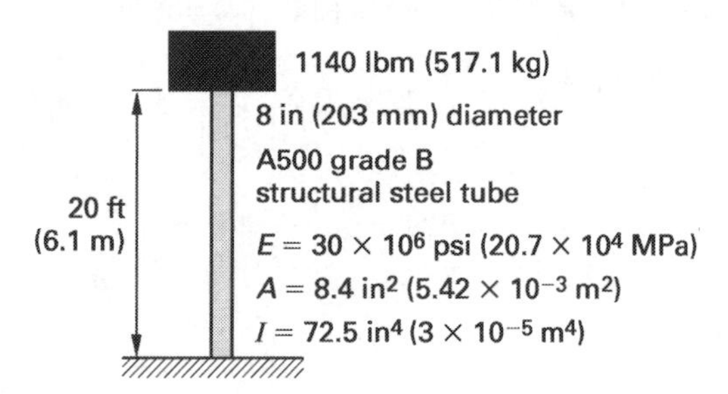

(a) Calculate the undamped lateral frequency and period.

(b) Calculate the undamped vertical frequency and period.

(c) Explain how the assumptions affect these frequencies and periods.

(d) What is the logarithmic decrement for damping of lateral vibrations?

(e) What is the damping ratio?

(f) What is the lateral acceleration?

(g) What is the vertical acceleration?

(h) What is the maximum shear force at the base using a quasistatic solution method?

(i) Are the assumptions used to calculate the frequencies conservative with respect to the lateral load?

(j) Assume that the pedestal fails by bending. Does the structure have high ductility?

Customary U.S. Solution

(a) From Table 4.1, the lateral stiffness of a vertical cantilever is

$$k = \frac{3EI}{h^3} = \frac{(3)\left(3\times 10^7\ \frac{\text{lbf}}{\text{in}^2}\right)(72.5\ \text{in}^4)}{(20\ \text{ft})^3\left(12\ \frac{\text{in}}{\text{ft}}\right)^2} = 5664\ \text{lbf/ft}$$

The mass in slugs (equivalent to lbf-sec^2/ft) can be calculated from the mass in pounds by dividing by the gravitational constant, g_c.

$$m_{\text{slugs}} = \frac{m_{\text{lbm}}}{g_c} = \frac{1140\ \text{lbm}}{32.2\ \frac{\text{ft-lbm}}{\text{lbf-sec}^2}} = 35.4\ \text{slugs}$$

From Eq. 4.14 and Eq. 4.15, the natural period for lateral vibrations is

$$\begin{aligned} T_{\text{lateral}} &= \frac{2\pi}{\omega} = 2\pi\sqrt{\frac{m}{k}} \\ &= 2\pi\sqrt{\frac{35.4\ \text{slugs}}{5664\ \frac{\text{lbf}}{\text{ft}}}} \\ &= 0.497\ \text{sec} \quad (0.5\ \text{sec}) \end{aligned}$$

From Eq. 4.13, the frequency is

$$f = \frac{1}{T} = \frac{1}{0.5\ \text{sec}} = 2\ \text{Hz}$$

(b) From Table 4.1, the deflection due to a compressive load is

$$x = \frac{Fh}{AE}$$

The stiffness is the force per unit deflection.

$$\begin{aligned} k &= \frac{F}{x} = \frac{AE}{h} \\ &= \frac{(8.4\ \text{in}^2)\left(30\times 10^6\ \frac{\text{lbf}}{\text{in}^2}\right)}{20\ \text{ft}} \\ &= 1.26\times 10^7\ \text{lbf/ft} \end{aligned}$$

Again from Eq. 4.14 and Eq. 4.15, the natural period for vertical vibrations is

$$\begin{aligned} T_{\text{vertical}} &= \frac{2\pi}{\omega} = 2\pi\sqrt{\frac{m}{k}} \\ &= 2\pi\sqrt{\frac{35.4\ \text{slugs}}{1.26\times 10^7\ \frac{\text{lbf}}{\text{ft}}}} \\ &= 1.05\times 10^{-2}\ \text{sec} \end{aligned}$$

From Eq. 4.13, the frequency is

$$f = \frac{1}{T} = \frac{1}{1.05 \times 10^{-2} \text{ sec}} = 95 \text{ Hz}$$

(c) It is assumed that all flexibility comes from the tube. The flexibility of other elements in the structural system decreases the stiffness, which increases the period. It is also assumed that the tube is massless. Increasing the vibrating mass also increases the period.

(d) From Eq. 4.20, the logarithmic decrement, δ, is

$$\delta = \ln\left(\frac{x_n}{x_{n+1}}\right) = \ln \frac{1}{0.882} = 0.126$$

(e) From Eq. 4.20, the damping ratio is solved directly.

$$\delta = 0.126 = \frac{2\pi\zeta}{\sqrt{1-\zeta^2}}$$

$$\zeta = 0.02 \quad (2\%)$$

(f) From Fig. 5.4 with 2% damping as determined in part (e),

$$S_d = 2.2 \text{ in}$$

$$S_v = 28 \text{ in/sec}$$

$$S_a = (0.9 \text{ g})\left(32.2 \ \frac{\text{ft}}{\text{sec}^2\text{-g}}\right) = 29 \text{ ft/sec}^2$$

(g) The vertical response is 50% of the horizontal response.

$$S_d = (0.5)(2.2 \text{ in}) = 1.1 \text{ in}$$

$$S_v = (0.5)\left(28 \ \frac{\text{in}}{\text{sec}}\right) = 14 \text{ in/sec}$$

$$S_a = (0.5)\left(29 \ \frac{\text{ft}}{\text{sec}^2}\right) = 14.5 \text{ ft/sec}^2$$

(h) The maximum shear by quasistatic approach is given by Eq. 3.1. (Since mass is already in slugs, it does not have to be divided by g_c.)

$$\begin{aligned} V_{\text{horizontal}} &= mS_a \\ &= (35.4 \text{ slugs})\left(29 \ \frac{\text{ft}}{\text{sec}^2}\right) \\ &= 1027 \text{ lbf} \end{aligned}$$

$$\begin{aligned} V_{\text{vertical}} &= mS_a \\ &= (35.4 \text{ slugs})\left(14.5 \ \frac{\text{ft}}{\text{sec}^2}\right) \\ &= 513 \text{ lbf} \end{aligned}$$

(i) As determined in part (c), the assumptions result in a smaller period. Since lower values of period give a higher acceleration (see Fig. 5.1), the assumptions result in the structure being designed for higher forces. Thus, the assumptions are conservative.

(j) Failing by inelastic bending, as opposed to fracture or collapse, is one of the indications of a ductile structure.

SI Solution

(a) From Table 4.1, the lateral stiffness of a vertical cantilever is

$$\begin{aligned} k &= \frac{3EI}{h^3} = \frac{(3)(20.7 \times 10^4 \text{ MPa})(3 \times 10^{-5} \text{ m}^4)}{(6.1 \text{ m})^3} \\ &= 8.2 \times 10^{-2} \text{ MN/m} \end{aligned}$$

From Eq. 4.14 and Eq. 4.15, the natural period for lateral vibrations is

$$\begin{aligned} T_{\text{lateral}} &= \frac{2\pi}{\omega} = 2\pi\sqrt{\frac{m}{k}} \\ &= 2\pi\sqrt{\frac{517.1 \text{ kg}}{\left(8.2 \times 10^{-2} \ \frac{\text{MN}}{\text{m}}\right)\left(1 \times 10^6 \ \frac{\text{N}}{\text{MN}}\right)}} \\ &= 0.499 \text{ s} \quad (0.5 \text{ s}) \end{aligned}$$

From Eq. 4.13, the frequency is

$$f = \frac{1}{T} = \frac{1}{0.5 \text{ s}} = 2 \text{ Hz}$$

(b) From Table 4.1, the deflection due to a compressive load is

$$x = \frac{Fh}{AE}$$

The stiffness is the force per unit deflection.

$$\begin{aligned} k &= \frac{F}{x} = \frac{AE}{h} \\ &= \frac{(5.42 \times 10^{-3} \text{ m}^2)(20.7 \times 10^4 \text{ MPa})}{6.1 \text{ m}} \\ &= 1.84 \times 10^2 \text{ MN/m} \end{aligned}$$

Again from Eq. 4.14 and Eq. 4.15, the natural period for vertical vibrations is

$$\begin{aligned} T_{\text{vertical}} &= \frac{2\pi}{\omega} = 2\pi\sqrt{\frac{m}{k}} \\ &= 2\pi\sqrt{\frac{517.1 \text{ kg}}{\left(1.84 \times 10^2 \ \frac{\text{MN}}{\text{m}}\right)\left(1 \times 10^6 \ \frac{\text{N}}{\text{MN}}\right)}} \\ &= 1.05 \times 10^{-2} \text{ s} \end{aligned}$$

From Eq. 4.13, the frequency is

$$f = \frac{1}{T} = \frac{1}{1.05 \times 10^{-2}\ \text{s}} = 95\ \text{Hz}$$

(c) It is assumed that all flexibility comes from the tube. The flexibility of other elements in the structural system decreases the stiffness, which increases the period. It is also assumed that the tube is massless. Increasing the vibrating mass also increases the period.

(d) From Eq. 4.20, the logarithmic decrement, δ, is

$$\delta = \ln\left(\frac{x_n}{x_{n+1}}\right) = \ln\frac{1}{0.882} = 0.126$$

(e) From Eq. 4.20, the damping ratio is solved directly.

$$\delta = 0.126 = \frac{2\pi\zeta}{\sqrt{1-\zeta^2}}$$

$$\zeta = 0.02 \quad (2\%)$$

(f) From Fig. 5.4 with 2% damping as determined in part (e),

$$S_d = (2.2\ \text{in})\left(0.0254\ \frac{\text{m}}{\text{in}}\right) = 5.6 \times 10^{-2}\ \text{m}$$

$$S_v = \left(28\ \frac{\text{in}}{\text{s}}\right)\left(0.0254\ \frac{\text{m}}{\text{s}}\right) = 0.71 \times 10^{-2}\ \text{m/s}$$

$$S_a = (0.9\ \text{g})\left(9.81\ \frac{\text{m}}{\text{s}^2\text{·g}}\right) = 8.83\ \text{m/s}^2$$

(g) The vertical response is 50% of the horizontal response.

$$S_d = (0.5)(5.6 \times 10^{-2}\ \text{m}) = 2.8 \times 10^{-2}\ \text{m}$$

$$S_v = (0.5)\left(0.71 \times 10^{-2}\ \frac{\text{m}}{\text{s}}\right) = 0.36 \times 10^{-2}\ \text{m/s}$$

$$S_a = (0.5)\left(8.83\ \frac{\text{m}}{\text{s}^2}\right) = 4.42\ \text{m/s}^2$$

(h) The maximum shear by quasistatic approach is given by Eq. 3.1.

$$\begin{aligned} V_{\text{horizontal}} &= mS_a \\ &= (517.1\ \text{kg})\left(8.83\ \frac{\text{m}}{\text{s}^2}\right) \\ &= 4566\ \text{N} \end{aligned}$$

$$\begin{aligned} V_{\text{vertical}} &= mS_a \\ &= (517.1\ \text{kg})\left(4.42\ \frac{\text{m}}{\text{s}^2}\right) \\ &= 2286\ \text{N} \end{aligned}$$

(i) As determined in part (c), the assumptions result in a smaller period. Since lower values of period give a higher acceleration (see Fig. 5.1), the assumptions result in the structure being designed for higher forces. Thus, the assumptions are conservative.

(j) Failing by inelastic bending, as opposed to fracture or collapse, is one of the indications of a ductile structure.

80. A two-story jail uses concrete shear walls as shown. None of the shear walls has openings, but only the east wall covers the entire length. The story height is 12 ft (366 cm). The thickness of the roof, wall, and floor slabs are 5 in, 10 in, and 5 in (127 mm, 254 mm, and 127 mm), respectively. Assume all walls shown are fixed piers.

(a) What are the relative rigidities of the second-story walls A, B, C, and D?

(b) Where is the center of rigidity?

(c) How would the shears in the second floor be determined? (Do not actually calculate the shears.)

(d) What is the effect on the first floor of offsetting wall A as shown?

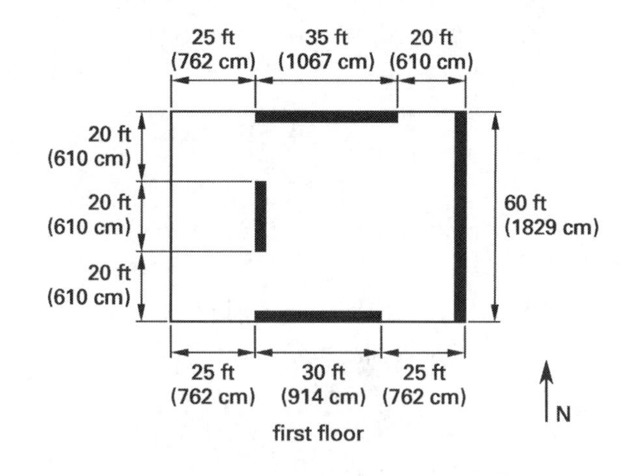

first floor

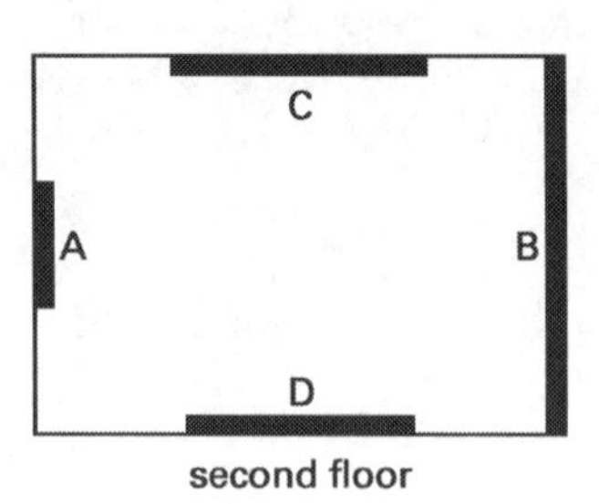

second floor

Customary U.S. Solution

(a) As stated in the problem, the walls are fixed piers. Use App. D to determine the tabulated rigidities of the fixed piers. The rigidities of perpendicular walls are taken as zero without regard to the h/d values.

wall	$\left(\frac{h}{d}\right)_{\text{E-W}}$	$\left(\frac{h}{d}\right)_{\text{N-S}}$	$R_{\text{E-W}}$	$R_{\text{N-S}}$
A	–	0.6	0	4.96
B	–	0.2	0	16.447
C	0.34	–	9.44	0
D	0.40	–	7.911	0

(b) The center of rigidity is located at (x_R, y_R). Distances are measured from the southwest corner.

$$x_R = \frac{(0\ \text{ft})(4.96) + (80\ \text{ft})(16.447)}{4.96 + 16.447} = 61.5\ \text{ft}$$

$$y_R = \frac{(0\ \text{ft})(7.911) + (60\ \text{ft})(9.44)}{7.911 + 9.44} = 32.6\ \text{ft}$$

(c) The wall shears are distributed in proportion to the second-floor wall rigidities.

(d) The second-story shear from the outside (west) wall will have to be transferred through the second-story floor (first-story ceiling) slab to wall A below. Failure may occur in the slab if it is not properly detailed.

SI Solution

(a) As stated in the problem, the walls are fixed piers. Use App. D to determine the tabulated rigidities of the fixed piers. The rigidities of perpendicular walls are taken as zero without regard to the h/d values.

wall	$\left(\frac{h}{d}\right)_{\text{E-W}}$	$\left(\frac{h}{d}\right)_{\text{N-S}}$	$R_{\text{E-W}}$	$R_{\text{N-S}}$
A	–	0.6	0	4.96
B	–	0.2	0	16.447
C	0.34	–	9.44	0
D	0.40	–	7.911	0

(b) The center of rigidity is located at (x_R, y_R). Distances are measured from the southwest corner.

$$x_R = \frac{(0\ \text{cm})(4.96) + (2439\ \text{cm})(16.447)}{4.96 + 16.447} = 1874\ \text{cm}$$

$$y_R = \frac{(0\ \text{cm})(7.911) + (1830\ \text{cm})(9.44)}{7.911 + 9.44} = 996\ \text{cm}$$

(c) The wall shears are distributed in proportion to the second-floor wall rigidities.

(d) The second-story shear from the outside (west) wall will have to be transferred through the second-story floor (first-story ceiling) slab to wall A below. Failure may occur in the slab if it is not properly detailed.

81. The structure shown is subjected to a lateral loading of 0.3 g. The columns are set in hard rock concrete footings, but the tops can be considered to be pinned to the slab. The reinforced concrete slab is 6 in (152 mm) thick and has a finished density of 150 lbm/ft^3 (2400 kg/m^3). The columns are rectangular A500 grade B steel tubing, HSS $5 \times 5 \times 5/16$ in. Disregard the dead load of the columns, all live load, slab rotation, axial column compression, combined stresses, and column deflection. For the purpose of this problem, assume the bending stress is limited to 37 kips/in^2 (255 MPa). Determine if the columns are adequate.

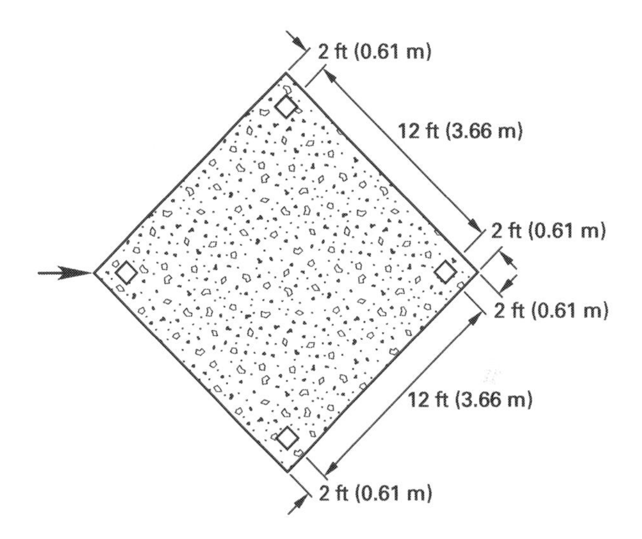

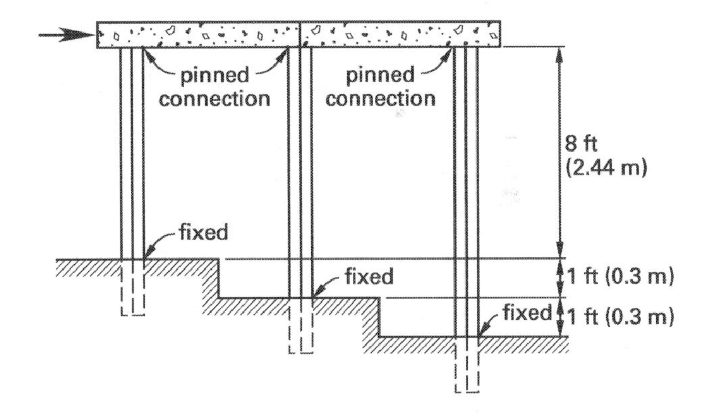

Customary U.S. Solution

The properties of A500 grade B HSS $5 \times 5 \times 5/16$ in square structural tubing are found in the AISC *Steel Construction Manual.* The modulus of elasticity is approximately 29×10^6 psi.

$$A = 5.26\ \text{in}^2$$

$$I_x = I_y = 19.0\ \text{in}^4$$

Although the tubing is oriented 45° from the plane of the principal axis, the moment of inertia per tube is still 19.0 in^4 in the direction of bending, because the tube is symmetrical and the product of inertia is zero. Although the lower ends of the columns are fixed, the tops are pinned. Therefore, simple cantilever curvature occurs. From Table 4.1, the stiffness is

$$k = \frac{3EI}{h^3}$$

There are three different column lengths, so there are three stiffnesses.

$$k_{8\text{ ft}} = \frac{(3)\left(29\times10^6\ \frac{\text{lbf}}{\text{in}^2}\right)(19.0\ \text{in}^4)}{(8\ \text{ft})^3\left(12\ \frac{\text{in}}{\text{ft}}\right)^3} = 1868\ \text{lbf/in}$$

$$k_{9\text{ ft}} = \frac{(3)\left(29\times10^6\ \frac{\text{lbf}}{\text{in}^2}\right)(19.0\ \text{in}^4)}{(9\ \text{ft})^3\left(12\ \frac{\text{in}}{\text{ft}}\right)^3} = 1312\ \text{lbf/in}$$

$$k_{10\text{ ft}} = \frac{(3)\left(29\times10^6\ \frac{\text{lbf}}{\text{in}^2}\right)(19.0\ \text{in}^4)}{(10\ \text{ft})^3\left(12\ \frac{\text{in}}{\text{ft}}\right)^3} = 957\ \text{lbf/in}$$

The total stiffness is the sum of the stiffnesses of the four columns.

$$\begin{aligned} k_{\text{total}} &= k_{8\text{ ft}} + 2k_{9\text{ ft}} + k_{10\text{ ft}} \\ &= 1868\ \frac{\text{lbf}}{\text{in}} + (2)\left(1312\ \frac{\text{lbf}}{\text{in}}\right) + 957\ \frac{\text{lbf}}{\text{in}} \\ &= 5449\ \text{lbf/in} \end{aligned}$$

The slab mass is

$$m = \frac{V\rho}{g_c} = \frac{(16\ \text{ft})(16\ \text{ft})(6\ \text{in})\left(150\ \frac{\text{lbm}}{\text{ft}^3}\right)}{\left(12\ \frac{\text{in}}{\text{ft}}\right)\left(32.2\ \frac{\text{ft-lbm}}{\text{lbf-sec}^2}\right)} = 596\ \text{slugs}$$

The seismic force is

$$F = ma = (596\ \text{slugs})(0.3\ \text{g})\left(32.2\ \frac{\text{ft}}{\text{sec}^2\text{-g}}\right) = 5760\ \text{lbf}$$

The resisting force in each tube is proportional to its relative rigidity (stiffness). The shortest tube has the highest stiffness, so the shortest tube experiences the highest stress. The prorated portion of the force carried by the shortest tube is

$$F_{8\text{ ft}} = \left(\frac{1868\ \frac{\text{lbf}}{\text{in}}}{5449\ \frac{\text{lbf}}{\text{in}}}\right)(5760\ \text{lbf}) = 1975\ \text{lbf}$$

The moment at the base of the shortest tube is

$$\begin{aligned} M = FL &= (1975\ \text{lbf})(8\ \text{ft})\left(12\ \frac{\text{in}}{\text{ft}}\right) \\ &= 1.896\times10^5\ \text{in-lbf} \end{aligned}$$

The bending stress is

$$\begin{aligned} f_b = \frac{Mc}{I} &= \frac{(1.896\times10^5\ \text{in-lbf})\left(\frac{5\ \text{in}}{\sqrt{2}}\right)}{19.0\ \text{in}^4} \\ &= 35{,}281\ \text{lbf/in}^2\ \text{(psi)} \\ F_b &= 37{,}000\ \text{lbf/in}^2\quad \text{[given]} \end{aligned}$$

Since $f_b < F_b$, the design is acceptable.

SI Solution

The properties of A500 grade B HSS $5\times5\times{}^5\!/_{16}$ in square structural tubing are found in the AISC *Steel Construction Manual*. The modulus of elasticity is approximately 2×10^5 MPa.

$$A = 3.39\times10^{-3}\ \text{m}^2$$

$$I_x = I_y = 7.91\times10^{-6}\ \text{m}^4$$

Although the tubing is oriented 45° from the plane of the principal axis, the moment of inertia per tube is still 7.91×10^{-6} m^4 in the direction of bending, because the tube is symmetrical and the product of inertia is zero. Although the lower ends of the columns are fixed, the tops are pinned. Therefore, simple cantilever curvature occurs. From Table 4.1, the stiffness is

$$k = \frac{3EI}{h^3}$$

There are three different column lengths, so there are three stiffnesses.

$$k_{2.44\text{ m}} = \frac{(3)(2\times10^5\ \text{MPa})(7.91\times10^{-6}\ \text{m}^4)}{(2.44\ \text{m})^3} = 0.327\ \text{MN/m}$$

$$k_{2.74\text{ m}} = \frac{(3)(2\times10^5\ \text{MPa})(7.91\times10^{-6}\ \text{m}^4)}{(2.74\ \text{m})^3} = 0.231\ \text{MN/m}$$

$$k_{3.04\text{ m}} = \frac{(3)(2\times10^5\ \text{MPa})(7.91\times10^{-6}\ \text{m}^4)}{(3.04\ \text{m})^3} = 0.169\ \text{MN/m}$$

The total stiffness is the sum of the stiffnesses of the four columns.

$$\begin{aligned} k_{\text{total}} &= k_{2.44\,\text{m}} + 2k_{2.74\,\text{m}} + k_{3.04\,\text{m}} \\ &= 0.327\ \frac{\text{MN}}{\text{m}} + (2)\left(0.231\ \frac{\text{MN}}{\text{m}}\right) + 0.169\ \frac{\text{MN}}{\text{m}} \\ &= 0.958\ \text{MN/m} \end{aligned}$$

The slab mass is

$$\begin{aligned} m &= V\rho \\ &= \frac{(4.88\ \text{m})(4.88\ \text{m})(152\ \text{mm})\left(2400\ \frac{\text{kg}}{\text{m}^3}\right)}{1000\ \frac{\text{mm}}{\text{m}}} \\ &= 8687\ \text{kg} \end{aligned}$$

The seismic force is

$$\begin{aligned} F = ma &= (8687\ \text{kg})(0.3\ \text{g})\left(9.81\ \frac{\text{m}}{\text{s}^2\cdot\text{g}}\right) \\ &= 25\,566\ \text{N} \quad (25.6\ \text{kN}) \end{aligned}$$

The resisting force in each tube is proportional to its relative rigidity (stiffness). The shortest tube has the highest stiffness, so the shortest tube experiences the highest stress. The prorated portion of the force carried by the shortest tube is

$$F_{2.44\,\text{m}} = \left(\frac{0.327\ \frac{\text{MN}}{\text{m}}}{0.958\ \frac{\text{MN}}{\text{m}}}\right)(25.6\ \text{kN}) = 8.74\ \text{kN}$$

The moment at the base of the shortest tube is

$$\begin{aligned} M = FL &= (8.74\ \text{kN})(2.44\ \text{m}) \\ &= 21.33\ \text{kN·m} \end{aligned}$$

The bending stress is

$$\begin{aligned} f_b &= \frac{Mc}{I} \\ &= \frac{(21.33\ \text{kN·m})\left(\frac{127\ \text{mm}}{\sqrt{2}}\right)}{(7.91\times10^{-6}\ \text{m}^4)\left(1000\ \frac{\text{mm}}{\text{m}}\right)} \\ &= 242\,160\ \text{kPa} \quad (242.2\ \text{MPa}) \end{aligned}$$

$$F_b = 255\ \text{MPa} \quad \text{[given]}$$

Since $f_b < F_b$, the design is acceptable.

82. An eight-story office building, including a penthouse, is supported by a special steel chevron-braced frame that is intended to carry no gravity load, but resist all seismic loads. The weights and heights of each floor are as given in the illustration. The building is for risk category II, and is located at a site where the maximum considered earthquake ground motion is 1.0 g for short periods and 0.4 g for a 1 sec period. The site's soil profile matches site class A. T_L = 12 sec. Perform a seismic analysis consistent with the IBC, and determine (a) the base shear, and (b) the seismic force at the third floor.

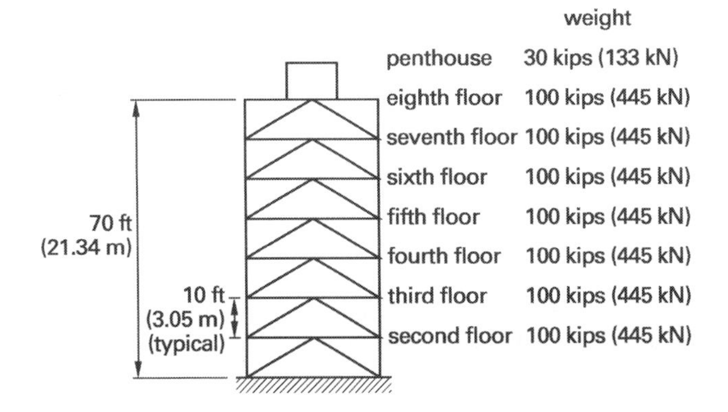

Customary U.S. Solution

(a) The natural period, T, can be determined from Eq. 6.11 [ASCE/SEI7 Eq. 12.8-7] (i.e., Method A).

$$T_a = C_t h_n^x$$

From Table 6.11, for an ordinary steel-braced frame system, C_t = 0.02 and x = 0.75. h_n = 70 ft.

$$\begin{aligned} T_a &= (0.02)(70\ \text{ft})^{0.75} \\ &= 0.48\ \text{sec} \end{aligned}$$

Assuming the penthouse is not part of the structural system and is more an "element" than a building, this building is really a seven-story office building. Therefore, the building weight is

$$\begin{aligned} W &= (7\ \text{stories})\left(100\ \frac{\text{kips}}{\text{story}}\right) + 30\ \text{kips} \\ &= 730\ \text{kips} \end{aligned}$$

According to the IBC, for this site S_S = 1.00 and S_1 = 0.40. For site class A, the value of F_a and F_v is 0.8.

$$\begin{aligned} S_{MS} &= F_a S_S = (0.8)(1.00) = 0.80 \\ S_{M1} &= F_v S_1 = (0.8)(0.40) = 0.32 \\ S_{DS} &= \tfrac{2}{3} S_{MS} = \left(\tfrac{2}{3}\right)(0.80) = 0.533 \\ S_{D1} &= \tfrac{2}{3} S_{M1} = \left(\tfrac{2}{3}\right)(0.32) = 0.213 \end{aligned}$$

The values of S_{DS} and S_{D1} result in the building being in seismic design category D.

Since the building performs no special function, I_e = 1.0. For a special steel concentrically braced building frame system, the value of R is 6.0.

From Eq. 6.21(a), the maximum seismic coefficient is

$$C_{s,\max} = \frac{S_{D1}}{T\left(\frac{R}{I_e}\right)} = \frac{0.213}{(0.48 \text{ sec})\left(\frac{6.0}{1.0}\right)}$$
$$= 0.0740$$

Since $S_1 < 0.6$, based on ASCE/SEI7 Sec. 12.8.1.1, the minimum base shear coefficient is the larger of 0.01 or $0.044S_{DS}I_e$.

$$0.044S_{DS}I_e = (0.44)(0.533)(1.0)$$
$$= 0.0235$$

From Eq. 6.20,

$$C_s = \frac{S_{DS}}{\frac{R}{I_e}} = \frac{0.533}{\frac{6}{1.0}} = 0.0888$$

Since $0.0235 < 0.0740 < 0.0888$, $C_s = 0.0740$.

From Eq. 6.23,

$$V = C_sW = (0.0740)(730 \text{ kips}) = 54.0 \text{ kips}$$

(b) For Eq. 6.26, the value of k is 1 because $T =$ 0.48 sec < 0.5 sec.

Forming a table is the easiest way to determine the floor forces.

level x	h_x (ft)	w_x (kips)	h_xw_x (ft-kips)
7	70	130	9100
6	60	100	6000
5	50	100	5000
4	40	100	4000
3	30	100	3000
2	20	100	2000
1	10	100	1000
		total	30,100

For the third floor (level 2),

$$\frac{h_xw_x}{\sum h_xw_x} = \frac{(20 \text{ ft})(100 \text{ kips})}{30{,}100 \text{ ft-kips}} = 0.0664$$

From Eq. 6.26 and Eq. 6.27,

$$F_2 = \frac{Vw_2h_2}{\sum w_ih_i} = (54.0 \text{ kips})(0.0664) = 3.59 \text{ kips}$$

SI Solution

(a) The natural period, T, can be determined from Eq. 6.11 [ASCE/SEI7 Eq. 12.8-7] (i.e., Method A).

$$T_a = C_th_n^x$$

From Table 6.11, for an ordinary steel-braced frame system, $C_t = 0.0488$ and $x = 0.75$. $h_n = 21.34$ m.

$$T_a = (0.0488)(21.34 \text{ m})^{0.75}$$
$$= 0.48 \text{ s}$$

Assuming the penthouse is not part of the structural system and is more an "element" than a building, the building is really only seven stories. Therefore, the building weight is

$$W = (7 \text{ stories})\left(445 \frac{\text{kN}}{\text{story}}\right) + 133 \text{ kN}$$
$$= 3248 \text{ kN}$$

According to the IBC, for this site $S_S = 1.00$ and $S_1 = 0.40$. For site class A, the value of F_a and F_v is 0.8.

$$S_{MS} = F_aS_S = (0.8)(1.00) = 0.80$$
$$S_{M1} = F_vS_1 = (0.8)(0.40) = 0.32$$
$$S_{DS} = \tfrac{2}{3}S_{MS} = \left(\tfrac{2}{3}\right)(0.80) = 0.533$$
$$S_{D1} = \tfrac{2}{3}S_{M1} = \left(\tfrac{2}{3}\right)(0.32) = 0.213$$

The values of S_{DS} and S_{D1} result in the building being in seismic design category D.

Since the building performs no special function, $I_e = 1.0$. For a special steel concentrically braced building frame system, the value of R is 6.0.

From Eq. 6.21(a), the maximum seismic coefficient is

$$C_s = \frac{S_{D1}}{T\left(\frac{R}{I_e}\right)} = \frac{0.213}{(0.48 \text{ sec})\left(\frac{6.0}{1.0}\right)}$$
$$= 0.0740$$

Since $S_1 < 0.6$, based on ASCE/SEI7 Sec. 12.8.1.1, the minimum base shear coefficient is the larger of 0.01 or $0.044S_{DS}I_e$.

$$0.044S_{DS}I_e = (0.044)(0.533)(1.0)$$
$$= 0.0235$$

From Eq. 6.20,

$$C_s = \frac{S_{DS}}{\frac{R}{I_e}} = \frac{0.533}{\frac{6}{1.0}}$$
$$= 0.0888$$

Since $0.0235 < 0.0740 < 0.0888$, $C_s = 0.0740$.

From Eq. 6.23,

$$V = C_sW = (0.0740)(3248 \text{ kN})$$
$$= 240 \text{ kN}$$

(b) For Eq. 6.26, the value of k is 1 since $T =$ 0.48 s < 0.5 s.

Forming a table is the easiest way to determine the floor forces.

level x	h_x (m)	w_x (kN)	$h_x w_x$ (kN·m)
7	21.34	578	12 335
6	18.29	445	8139
5	15.24	445	6782
4	12.19	445	5425
3	9.14	445	4067
2	6.10	445	2715
1	3.05	445	1357
		total	40 820

For the third floor (level 2),

$$\frac{h_x w_x}{\sum h_x w_x} = \frac{(6.10 \text{ m})(445 \text{ kN})}{40\,820 \text{ kN·m}} = 0.0665$$

From Eq. 6.26 and Eq. 6.27,

$$\begin{aligned} F_2 &= \frac{V w_2 h_2}{\sum w_i h_i} = (240 \text{ kN})(0.0665) \\ &= 16.0 \text{ kN} \end{aligned}$$

83. The owner of a building is considering adding a north wing to increase the size of the building. The existing floor plan and planned wing are shown in the following illustration, along with structural walls and tabulated rigidities. All walls have the same thickness and height. The roof mass is to be disregarded. The owner is concerned about the lack of symmetry that the remodeled building will have. The base shear on the existing building is 90,000 lbf (400.3 kN), parallel to the long dimension (150 ft (45.72 m)). An additional 15,000 lbf (66.7 kN) of base shear will be added by the new wing.

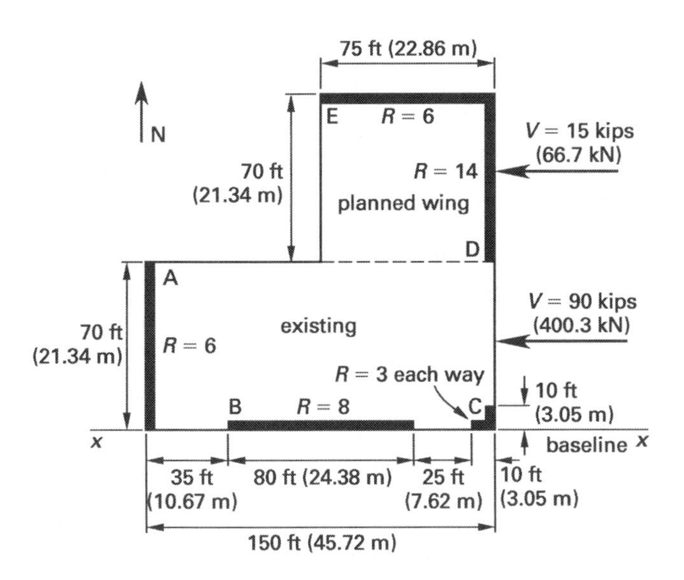

(a) Where will the center of mass be located when the wing is installed?

(b) Where will the center of rigidity be located when the wing is installed?

(c) What will be the torsional moment?

(d) Which active walls experience a negative torsional shear?

Customary U.S. Solution

(a) The wall masses are proportional to their lengths since their thicknesses and heights are all the same. Using the 150 ft east-west wall as the baseline, for earthquakes in the east-west direction, the center of mass, CM, is located at a distance of

$$\begin{aligned} y_{\text{CM}} &= \frac{\sum m_i y_i}{\sum m_i} \\ &= \frac{\begin{array}{l}(70 \text{ ft})(35 \text{ ft}) + (80 \text{ ft})(0) + (10 \text{ ft})(0) \\ \quad + (10 \text{ ft})(5 \text{ ft}) + (70 \text{ ft})(70 \text{ ft} + 35 \text{ ft}) \\ \quad + (75 \text{ ft})(70 \text{ ft} + 70 \text{ ft})\end{array}}{70 \text{ ft} + 80 \text{ ft} + 10 \text{ ft} + 10 \text{ ft} + 70 \text{ ft} + 75 \text{ ft}} \\ &= \frac{20{,}350 \text{ ft}^2}{315 \text{ ft}} \\ &= 64.6 \text{ ft (up from the baseline)} \end{aligned}$$

(b) Walls running north-south do not contribute to rigidity for east-west earthquakes. Only walls B, C (part), and E are active. The center of rigidity, CR, is located at a distance of

$$\begin{aligned} y_{\text{CR}} &= \frac{\sum R_i y_i}{\sum R_i} \\ &= \frac{(8)(0) + (3)(0) + (6)(70 \text{ ft} + 70 \text{ ft})}{8 + 3 + 6} \\ &= \frac{840 \text{ ft}}{17} \\ &= 49.4 \text{ ft (up from the baseline)} \end{aligned}$$

(c) The actual eccentricity is

$$e_{\text{actual}} = y_{\text{CM}} - y_{\text{CR}} = 64.6 \text{ ft} - 49.4 \text{ ft} = 15.2 \text{ ft}$$

The accidental eccentricity required by ASCE/SEI7 Sec. 12.8.4.2 is 5% of the transverse building direction.

$$e_{\text{accidental}} = (0.05)(70 \text{ ft} + 70 \text{ ft}) = 7 \text{ ft}$$

The total eccentricity is

$$\begin{aligned} e &= e_{\text{actual}} \pm e_{\text{accidental}} = 15.2 \text{ ft} \pm 7 \text{ ft} \\ e_{\max} &= 22.2 \text{ ft} \\ e_{\min} &= 8.2 \text{ ft} \end{aligned}$$

The torsional moment is

$$M_{max} = Ve = (90 \text{ kips} + 15 \text{ kips})(22.2 \text{ ft}) = 2331 \text{ ft-kips}$$

$$M_{min} = Ve = (90 \text{ kips} + 15 \text{ kips})(8.2 \text{ ft}) = 861 \text{ ft-kips}$$

(d) Walls B, C, and E resist the direct shear and are active. The base shear acts through the center of mass, tending to cause counterclockwise rotation about the center of rigidity. This is resisted by all walls (A, B, C, D, and E with clockwise direction forces). For walls B and C, the clockwise direction opposes forces due to direct shear. Walls B and C have negative torsional shear components.

SI Solution

(a) The wall masses are proportional to their lengths since their thicknesses and heights are all the same. Using the 45.72 m east-west wall as the baseline, for earthquakes in the east-west direction, the center of mass, CM, is located at a distance of

$$y_{CM} = \frac{\sum m_i y_i}{\sum m_i}$$

$$= \frac{\begin{array}{l}(21.34 \text{ m})(10.67 \text{ m}) + (24.38 \text{ m})(0) \\ \quad + (3.05 \text{ m})(0) + (3.05 \text{ m})(1.52 \text{ m}) \\ \quad + (21.34 \text{ m})(21.34 \text{ m} + 10.67 \text{ m}) \\ \quad + (22.86 \text{ m})(21.34 \text{ m} + 21.34 \text{ m})\end{array}}{\begin{array}{l}21.34 \text{ m} + 24.38 \text{ m} + 3.05 \text{ m} + 3.05 \text{ m} \\ \quad + 21.34 \text{ m} + 22.86 \text{ m}\end{array}}$$

$$= \frac{1891.09 \text{ m}^2}{96.02 \text{ m}}$$

$$= 19.69 \text{ m (up from the baseline)}$$

(b) Walls running north-south do not contribute to rigidity for east-west earthquakes. Only walls B, C (part), and E are active. The center of rigidity, CR, is located at a distance of

$$y_{CR} = \frac{\sum R_i y_i}{\sum R_i}$$

$$= \frac{(8)(0) + (3)(0) + (6)(21.34 \text{ m} + 21.34 \text{ m})}{8 + 3 + 6}$$

$$= \frac{256.08 \text{ m}}{17}$$

$$= 15.06 \text{ m (up from the baseline)}$$

(c) The actual eccentricity is

$$e_{actual} = y_{CM} - y_{CR} = 19.69 \text{ m} - 15.06 \text{ m} = 4.63 \text{ m}$$

The accidental eccentricity required by ASCE/SEI7 Sec. 12.8.4.2 is 5% of the transverse building direction.

$$e_{accidental} = (0.05)(21.34 \text{ m} + 21.34 \text{ m}) = 2.13 \text{ m}$$

The total eccentricity is

$$e = e_{actual} \pm e_{accidental} = 4.63 \text{ m} \pm 2.13 \text{ m}$$

$$e_{max} = 6.76 \text{ m}$$

$$e_{min} = 2.5 \text{ m}$$

The torsional moment is

$$M_{max} = Ve = (400.3 \text{ kN} + 66.7 \text{ kN})(6.76 \text{ m}) = 3157 \text{ kN·m}$$

$$M_{min} = Ve = (400.3 \text{ kN} + 66.7 \text{ kN})(2.5 \text{ m}) = 1168 \text{ kN·m}$$

(d) Walls B, C, and E resist the direct shear and are active. The base shear acts through the center of mass, tending to cause counterclockwise rotation about the center of rigidity. This is resisted by all walls (A, B, C, D, and E with "clockwise" direction forces). For walls B and C, the clockwise direction opposes forces due to direct shear. Walls B and C have negative shear due to torsion.

84. A one-story, 55 ft by 70 ft (16.76 m by 21.34 m) masonry-walled building with a rigid diaphragm roof is constructed in the shape of a rectangle. One side of the building has two doors. There are no other openings. The tabulated rigidities of each wall are shown. Each wall is 10 in (254 mm) thick. Disregard the mass of the roof diaphragm.

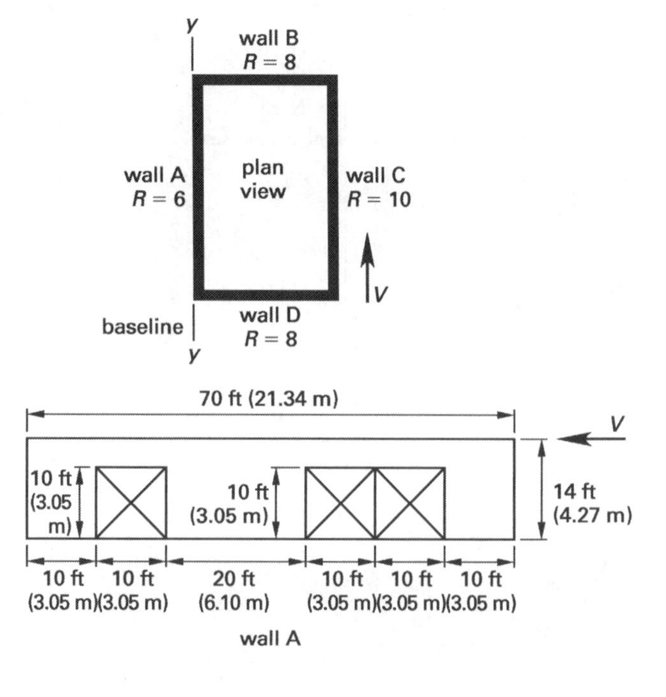

(a) What is the location of the center of rigidity?

(b) What is the torsional moment due to a total lateral force of 50,000 lbf (222.4 kN)?

(c) If 19,000 lbf (84.5 kN) of the lateral load are distributed to wall A, what loads are carried by each of the piers in wall A?

Customary U.S. Solution

(a) For earthquakes in the direction shown, only walls A and C contribute to rigidity. The center of rigidity, CR, is located along a line parallel to baseline y-y and located a distance away of

$$x_{CR} = \frac{\sum R_i x_i}{\sum R_i} = \frac{(6)(0) + (10)(55 \text{ ft})}{6 + 10} = \frac{550 \text{ ft}}{16} = 34.4 \text{ ft}$$

(b) Although the openings could be disregarded, enough information is given to determine the location of the center of mass considering the openings. The mass of each wall is proportional to its area.

wall	area (mass)
A	(70 ft)(14 ft) − (3)(10 ft)(10 ft) = 680 ft^2
B	(55 ft)(14 ft) = 770 ft^2
C	(70 ft)(14 ft) = 980 ft^2
D	(55 ft)(14 ft) = 770 ft^2

The center of mass, CM, is located at

$$x_{CM} = \frac{\sum m_i x_i}{\sum m_i}$$

$$= \frac{(680 \text{ ft}^2)(0) + (770 \text{ ft}^2)\left(\frac{55 \text{ ft}}{2}\right) + (980 \text{ ft}^2)(55 \text{ ft}) + (770 \text{ ft}^2)\left(\frac{55 \text{ ft}}{2}\right)}{680 \text{ ft}^2 + 770 \text{ ft}^2 + 980 \text{ ft}^2 + 770 \text{ ft}^2}$$

$$= \frac{96{,}250 \text{ ft}^3}{3200 \text{ ft}^2} = 30.1 \text{ ft}$$

The actual eccentricity is

$$e_{\text{actual}} = x_{CR} - x_{CM} = 34.4 \text{ ft} - 30.1 \text{ ft} = 4.3 \text{ ft}$$

The accidental eccentricity is

$$e_{\text{accidental}} = (0.05)(55 \text{ ft}) = 2.8 \text{ ft}$$

The total eccentricity is

$$e = e_{\text{actual}} \pm e_{\text{accidental}}$$

$$e_{\max} = 4.3 \text{ ft} + 2.8 \text{ ft} = 7.1 \text{ ft}$$

$$e_{\min} = 4.3 \text{ ft} - 2.8 \text{ ft} = 1.5 \text{ ft}$$

The torsional moment is

$$M_{\max} = Ve = (50{,}000 \text{ lbf})(7.1 \text{ ft}) = 355{,}000 \text{ ft-lbf}$$

$$M_{\min} = Ve = (50{,}000 \text{ lbf})(1.5 \text{ ft}) = 75{,}000 \text{ ft-lbf}$$

(c) There are three piers in wall A. Due to the effect of the beam running along the top, assume the piers are fixed. The tabulated rigidities are found from App. D.

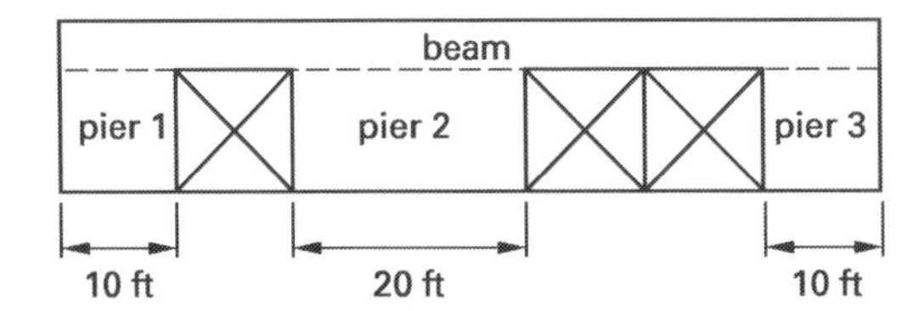

pier	h (ft)	d (ft)	$\frac{h}{d}$	R_{tab}
1	10	10	1	2.5
2	10	20	0.5	6.154
3	10	10	1	2.5

The fraction of the total wall shear taken by piers 1 and 3 each is

$$\frac{2.5}{2.5 + 6.154 + 2.5} = 0.22$$

The fraction taken by pier 2 is

$$1.00 - (2)(0.22) = 0.56$$

The pier shears are

$$V_1 = V_3 = (0.22)(19{,}000 \text{ lbf}) = 4180 \text{ lbf}$$

$$V_2 = (0.56)(19{,}000 \text{ lbf}) = 10{,}640 \text{ lbf}$$

SI Solution

(a) For earthquakes in the direction shown, only walls A and C contribute to rigidity. The center of rigidity, CR, is located along a line parallel to baseline y-y and located a distance away of

$$x_{CR} = \frac{\sum R_i x_i}{\sum R_i} = \frac{(6)(0) + (10)(16.76 \text{ m})}{6 + 10} = \frac{167.60 \text{ m}}{16} = 10.48 \text{ m}$$

(b) Although the openings could be disregarded, enough information is given to determine the location of the center of mass considering the openings. The mass of each wall is proportional to its area.

wall	area (mass)
A	$(21.34\ \text{m})(4.27\ \text{m}) - (3)(3.05\ \text{m})(3.05\ \text{m}) = 63.21\ \text{m}^2$
B	$(16.76\ \text{m})(4.27\ \text{m}) = 71.57\ \text{m}^2$
C	$(21.34\ \text{m})(4.27\ \text{m}) = 91.12\ \text{m}^2$
D	$(16.76\ \text{m})(4.27\ \text{m}) = 71.57\ \text{m}^2$

The center of mass, CM, is located at

$$x_{\text{CM}} = \frac{\sum m_i x_i}{\sum m_i}$$

$$= \frac{(63.21\ \text{m}^2)(0) + (71.57\ \text{m}^2)\left(\frac{16.76\ \text{m}}{2}\right) + (91.12\ \text{m}^2)(16.76\ \text{m}) + (71.57\ \text{m}^2)\left(\frac{16.76\ \text{m}}{2}\right)}{63.21\ \text{m}^2 + 71.57\ \text{m}^2 + 91.12\ \text{m}^2 + 71.57\ \text{m}^2}$$

$$= \frac{2726.68\ \text{m}^3}{297.47\ \text{m}^2}$$

$$= 9.17\ \text{m}$$

The actual eccentricity is

$$e_{\text{actual}} = x_{\text{CR}} - x_{\text{CM}} = 10.48\ \text{m} - 9.17\ \text{m} = 1.31\ \text{m}$$

The accidental eccentricity is

$$e_{\text{accidental}} = (0.05)(16.76\ \text{m}) = 0.84\ \text{m}$$

The total eccentricity is

$$e = e_{\text{actual}} \pm e_{\text{accidental}}$$

$$e_{\text{max}} = 1.31\ \text{m} + 0.84\ \text{m} = 2.15\ \text{m}$$

$$e_{\text{min}} = 1.31 - 0.84\ \text{m} = 0.47\ \text{m}$$

The torsional moment is

$$M_{\text{max}} = Ve = (222.4\ \text{kN})(2.15\ \text{m}) = 478.2\ \text{kN·m}$$

$$M_{\text{min}} = Ve = (222.4\ \text{kN})(0.47\ \text{m}) = 104.5\ \text{kN·m}$$

(c) There are three piers in wall A. Due to the effect of the beam running along the top, assume the piers are fixed. The tabulated rigidities are found from App. D.

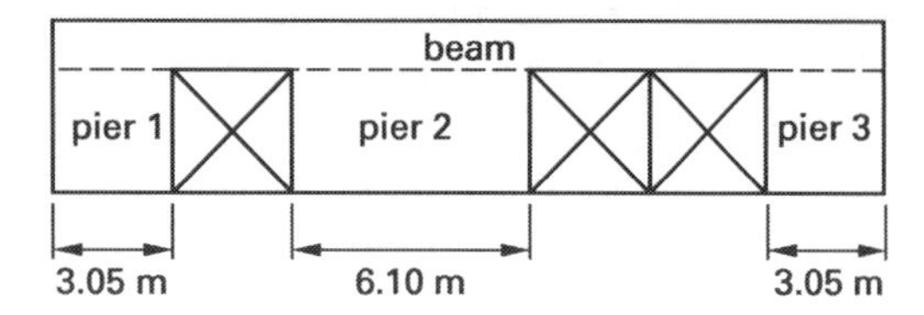

pier	h (m)	d (m)	$\frac{h}{d}$	R_{tab}
1	3.05	3.05	1	2.5
2	3.05	6.10	0.5	6.154
3	3.05	3.05	1	2.5

The fraction of the total wall shear taken by piers 1 and 3 each is

$$\frac{2.5}{2.5 + 6.154 + 2.5} = 0.22$$

The fraction taken by pier 2 is

$$1.00 - (2)(0.22) = 0.56$$

The pier shears are

$$V_1 = V_3 = (0.22)(84.5\ \text{kN}) = 18.6\ \text{kN}$$

$$V_2 = (0.56)(84.5\ \text{k}) = 47.3\ \text{kN}$$

85. A one-story building with masonry walls is constructed in a box shape. All walls are 24 ft (7.32 m) high. The walls have a weight of 50 lbf/ft^2 (2.4 kN/m^2). The wood structural panel roof and roof skylights have a weight of 25 lbf/ft^2 (1.2 kN/m^2). The roof carries a 20 lbf/ft^2 (0.96 kN/m^2) live load of permanent air-conditioning equipment. The seismic load in the north-south direction is 15% of the participating building weight. Do not consider east-west earthquake performance.

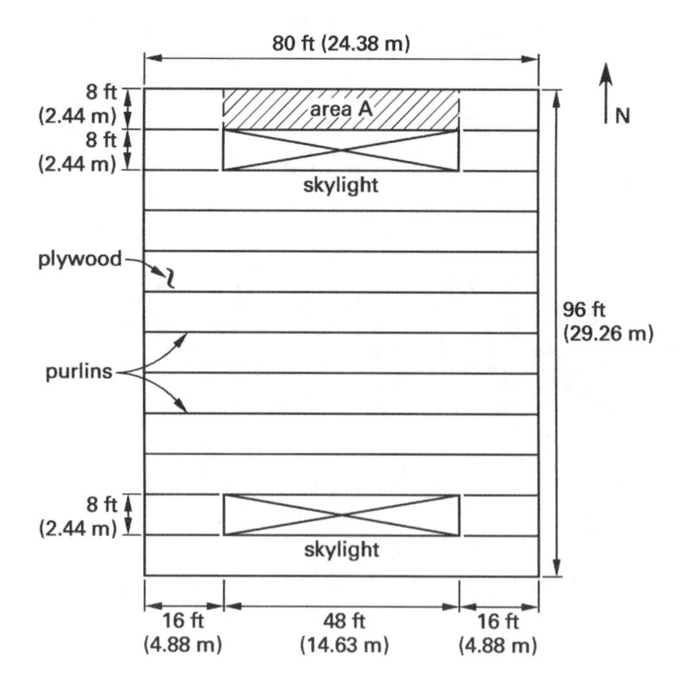

(a) What is the diaphragm force?

(b) What is the maximum wall shear force?

(c) What is the maximum chord force in the diaphragm for a north-south earthquake?

Customary U.S. Solution

(a) For a north-south earthquake, the diaphragm force results from the acceleration of the roof mass and half of the short (80 ft) wall masses. The roof weight (including the air-conditioning equipment weight) is

$$\begin{aligned} W_{\text{roof}} &= \text{area} \times \text{loading} \\ &= (80\ \text{ft})(96\ \text{ft})\left(20\ \frac{\text{lbf}}{\text{ft}^2} + 25\ \frac{\text{lbf}}{\text{ft}^2}\right) \\ &= 345{,}600\ \text{lbf} \end{aligned}$$

The weight of the perpendicular walls is

$$\begin{aligned} W_{\perp\text{walls}} &= (80\ \text{ft})(24\ \text{ft})(2\ \text{walls})\left(50\ \frac{\text{lbf}}{\text{ft}^2}\right) \\ &= 192{,}000\ \text{lbf} \end{aligned}$$

Only the top half of the perpendicular walls contributes to the diaphragm force. Since the seismic load is given as 15% of the building weight,

$$\begin{aligned} F_{\text{diaphragm}} &= (0.15)\left(W_{\text{roof}} + \tfrac{1}{2}W_{\perp\text{walls}}\right) \\ &= (0.15)\left(345{,}600\ \text{lbf} + \frac{192{,}000\ \text{lbf}}{2}\right) \\ &= 66{,}240\ \text{lbf} \end{aligned}$$

(b) The weight of the parallel walls is

$$\begin{aligned} W_{\|\text{walls}} &= (96\ \text{ft})(24\ \text{ft})(2\ \text{walls})\left(50\ \frac{\text{lbf}}{\text{ft}^2}\right) \\ &= 230{,}400\ \text{lbf} \end{aligned}$$

Only the top half of all the walls contributes to shear in the parallel walls.

$$\begin{aligned} F_{\text{total}} &= (0.15)\left(W_{\text{roof}} + \tfrac{1}{2}W_{\perp\text{walls}} + \tfrac{1}{2}W_{\|\text{walls}}\right) \\ &= (0.15)\left(345{,}600\ \text{lbf} + \frac{192{,}000\ \text{lbf}}{2} + \frac{230{,}400\ \text{lbf}}{2}\right) \\ &= 83{,}520\ \text{lbf} \end{aligned}$$

Since there are two parallel walls, the maximum wall shear force is

$$V_{\|\text{wall}} = \frac{83{,}520\ \text{lbf}}{2\ \text{walls}} = 41{,}760\ \text{lbf/wall}$$

(c) From Eq. 7.10, the chord force in the perpendicular walls is

$$\begin{aligned} C &= \frac{F_{\text{diaphragm}}L}{8b} = \frac{(66{,}240\ \text{lbf})(80\ \text{ft})}{(8)(96\ \text{ft})} \\ &= 6900\ \text{lbf} \end{aligned}$$

SI Solution

(a) For a north-south earthquake, the diaphragm force results from the acceleration of the roof mass and half of the short (24.38 m) wall masses. The roof weight (including the air-conditioning equipment weight) is

$$\begin{aligned} W_{\text{roof}} &= \text{area} \times \text{loading} \\ &= (24.38\ \text{m})(29.26\ \text{m})\left(0.96\ \frac{\text{kN}}{\text{m}^2} + 1.2\ \frac{\text{kN}}{\text{m}^2}\right) \\ &= 1540.9\ \text{kN} \end{aligned}$$

The weight of the perpendicular walls is

$$\begin{aligned} W_{\perp\text{walls}} &= (24.38\ \text{m})(7.32\ \text{m})(2\ \text{walls})\left(2.4\ \frac{\text{kN}}{\text{m}^2}\right) \\ &= 856.6\ \text{kN} \end{aligned}$$

Only the top half of the perpendicular walls contributes to the diaphragm force. Since the seismic load is given as 15% of the building weight,

$$\begin{aligned} F_{\text{diaphragm}} &= (0.15)\left(W_{\text{roof}} + \tfrac{1}{2}W_{\perp\text{walls}}\right) \\ &= (0.15)\left(1540.9\ \text{kN} + \frac{856.6\ \text{kN}}{2}\right) \\ &= 295.4\ \text{kN} \end{aligned}$$

(b) The weight of the parallel walls is

$$\begin{aligned} W_{\|\text{walls}} &= (29.26\ \text{m})(7.32\ \text{m})(2\ \text{walls})\left(2.4\ \frac{\text{kN}}{\text{m}^2}\right) \\ &= 1028.1\ \text{kN} \end{aligned}$$

Only the top half of all the walls contributes to shear in the parallel walls.

$$\begin{aligned} F_{\text{total}} &= (0.15)\left(W_{\text{roof}} + \tfrac{1}{2}W_{\perp\text{walls}} + \tfrac{1}{2}W_{\|\text{walls}}\right) \\ &= (0.15)\left(1540.9\ \text{kN} + \frac{856.6\ \text{kN}}{2} + \frac{1028.1\ \text{kN}}{2}\right) \\ &= 372.5\ \text{kN} \end{aligned}$$

Since there are two parallel walls, the maximum wall shear force is

$$V_{\|\text{wall}} = \frac{372.5\ \text{kN}}{2\ \text{walls}} = 186.3\ \text{kN/wall}$$

(c) From Eq. 7.10, the chord force in the perpendicular walls is

$$\begin{aligned} C &= \frac{F_{\text{diaphragm}}L}{8b} = \frac{(295.4\ \text{kN})(24.38\ \text{m})}{(8)(29.26\ \text{m})} \\ &= 30.8\ \text{kN} \end{aligned}$$

86. The plan view of a one-story, 12 ft (3.66 m) high masonry-walled retail shop is shown. Windows cover most of the front and half of one side. The building has a flexible wood structural panel roof diaphragm with continuous roof struts (shown as lighter lines) criss-crossing and dividing the roof into small diaphragms. Each strut is designed to serve as a chord, if necessary. The masonry shear walls have a dead weight of 55 lbf/ft^2 (2.6 kN/m^2). The wood structural panel roof has a dead weight of 15 lbf/ft^2 (0.7 kN/m^2). For the purpose of the ASCE/SEI7 base shear equation, $V = 0.183W$.

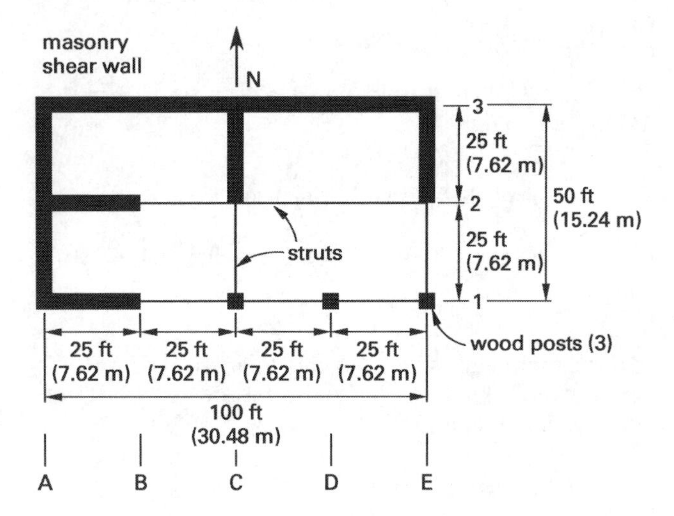

(a) What is the maximum diaphragm shear along lines A and E?

(b) What is the maximum chord or drag force at point C-1?

(c) What is the maximum drag force at point B-1 due to an east-west earthquake?

Customary U.S. Solution

The earthquake direction is not given for parts (a) and (b) in this problem and must be considered variable.

(a) For north-south earthquakes, the roof is effectively divided into two subdiaphragms. One diaphragm is bounded by points (moving clockwise) A-1, A-3, C-3, and C-1. The other is bounded by points C-1, C-3, E-3, and E-1. Although each diaphragm is the same size, the accelerating masses are different.

For line A,

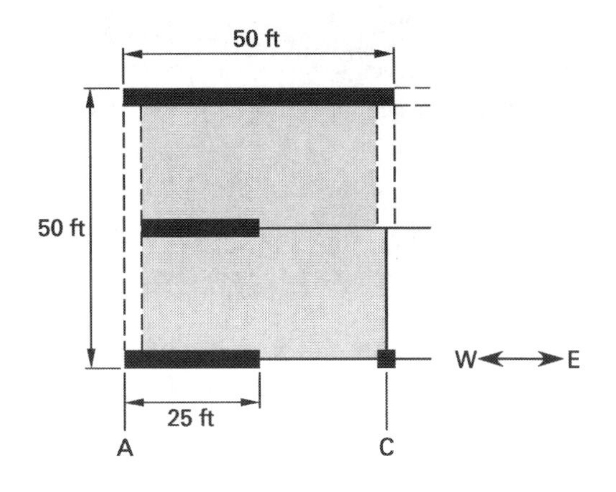

The weight of the roof is

$$W_{\text{roof}} = (50\ \text{ft})(50\ \text{ft})\left(15\ \frac{\text{lbf}}{\text{ft}^2}\right) = 37{,}500\ \text{lbf}$$

The weight of the 50 ft wall (full-height) is

$$W_{50\ \text{ft}} = (12\ \text{ft})(50\ \text{ft})\left(55\ \frac{\text{lbf}}{\text{ft}^2}\right) = 33{,}000\ \text{lbf}$$

The weight of the two 25 ft walls totals 33,000 lbf also.

$$W_{25\ \text{ft}} = 33{,}000\ \text{lbf}$$

The seismic effect from the roof and the 50 ft wall is shared equally between walls A and C. Since the two remaining walls extend only halfway between A and C, a rational method of allocating their seismic effects to walls A and C must be used. Consider a simply supported beam loaded uniformly along the first half of its length. The reaction closest to the uniform load will carry $^3/_4$ of the total load. Therefore, assume that wall A carries $^3/_4$ of the seismic effect from the two short walls. (Other assumptions may be valid, depending on construction details.) Finally, assume that only the top half of the walls contribute to diaphragm shear.

The diaphragm reaction at line A is

$$\begin{aligned} F_{\text{A}} &= 0.183W \\ &= (0.183) \\ &\quad \times \left(\frac{37{,}500\ \text{lbf}}{2} + \frac{\left(\frac{1}{2}\right)(33{,}000\ \text{lbf})}{2} + \frac{\left(\frac{3}{4}\right)(33{,}000\ \text{lbf})}{2}\right) \\ &= 7206\ \text{lbf} \end{aligned}$$

For line E,

50 ft
50 ft
C
E

$$W_{\text{roof}} = 37{,}500\ \text{lbf}$$

$$W_{\text{E-W wall}} = (12\ \text{ft})(50\ \text{ft})\left(55\ \frac{\text{lbf}}{\text{ft}^2}\right) = 33{,}000\ \text{lbf}$$

The east-west wall weight tributary to the roof level is

$$\frac{33{,}000\ \text{lbf}}{2} = 16{,}500\ \text{lbf}$$

$$\begin{aligned} F_E &= 0.183W \\ &= (0.183)\left(\frac{37{,}500 \text{ lbf}}{2} + \frac{16{,}500 \text{ lbf}}{2}\right) \\ &= 4941 \text{ lbf} \end{aligned}$$

(b) The earthquake direction is not given.

North-south earthquake

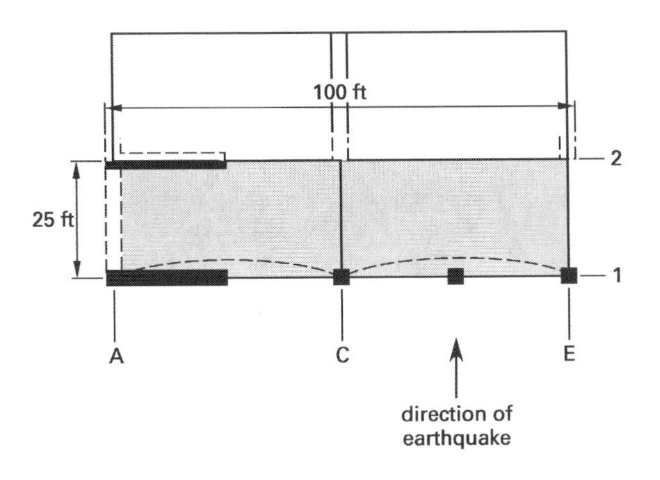

Drag force at point C-1 (line C)

All diaphragm shear would frame into point C-2 since the wooden post at point C-1 can be assumed to have no lateral stiffness. Therefore, the maximum drag force at point C-1 due to a north-south earthquake is zero.

Chord force at point C-1 (line 1)

Half of the roof mass (corresponding to the tributary area bounded by points A-1, A-2, E-2, and E-1) and the upper half of the east-west walls in the tributary area contribute to the chord forces along lines 1 and 2. (Other interpretations might also be justified. For example, the entire roof might be considered.)

$$W_{\text{roof}} = (25 \text{ ft})(100 \text{ ft})\left(15 \ \frac{\text{lbf}}{\text{ft}^2}\right) = 37{,}500 \text{ lbf}$$

$$\begin{aligned} W_{\text{E-W walls}} &= (12 \text{ ft})\left(25 \text{ ft} + \left(\tfrac{1}{2}\right)(25 \text{ ft})\right)\left(55 \ \frac{\text{lbf}}{\text{ft}^2}\right) \\ &= 24{,}750 \text{ lbf} \end{aligned}$$

Only half of the east-west wall along line 2 is used. This is because the other half is tributary to the adjacent area bounded by points A-2, A-3, C-3, and C-2.

Only the upper half of the east-west walls is effective in loading the diaphragm.

$$\frac{24{,}750 \text{ lbf}}{2} = 12{,}375 \text{ lbf}$$

The diaphragm force is

$$\begin{aligned} F_{\text{diaphragm}} &= 0.183W \\ &= (0.183)(37{,}500 \text{ lbf} + 12{,}375 \text{ lbf}) \\ &= 9127 \text{ lbf} \end{aligned}$$

From Eq. 7.10, the chord force is

$$C = \frac{F_{\text{diaphragm}} L}{8b} = \frac{(9127 \text{ lbf})(100 \text{ ft})}{(8)(25 \text{ ft})} = 4564 \text{ lbf}$$

East-west earthquake

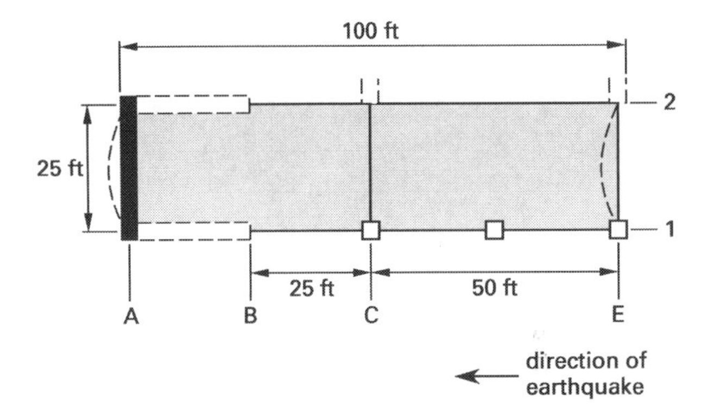

Drag force at point C-1 (line 1)

The collector along line 1 transfers half the diaphragm loading from the area bounded by points C-1, C-2, E-2, and E-1 into point B-1. (The other half is transferred to point B-2.) Since there are no north-south walls in this area, the only contribution to collector force is the tributary roof weight. Between lines C and E, the roof weight is

$$W_{\text{roof}} = (25 \text{ ft})(50 \text{ ft})\left(15 \ \frac{\text{lbf}}{\text{ft}}\right) = 18{,}750 \text{ lbf}$$

The diaphragm force is

$$\begin{aligned} F &= 0.183W = (0.183)(18{,}750 \text{ lbf}) \\ &= 3431 \text{ lbf} \end{aligned}$$

Half of this is transferred along the strut on line 1. At point C-1, the drag force is

$$D = \frac{3431 \text{ lbf}}{2} = 1716 \text{ lbf}$$

Chord force at point C-1 (line C)

While there is a chord force along line C, the chord force at C-1 is zero because point C-1 is at the end of the chord.

(c) This is similar to part (b), except that the tributary roof area is 75 ft long instead of 50 ft long. Scaling up from the answer derived in part (b),

$$D = \left(\frac{75 \text{ ft}}{50 \text{ ft}}\right)(1716 \text{ lbf}) = 2574 \text{ lbf}$$

SI Solution

The earthquake direction is not given for parts (a) and (b) in this problem and must be considered variable.

(a) For north-south earthquakes, the roof is effectively divided into two diaphragms. One diaphragm is bounded by points (moving clockwise) A-1, A-3, C-3, and C-1. The other is bounded by points C-1, C-3, E-3, and E-1. Although each diaphragm is the same size, the accelerating masses are different.

For line A,

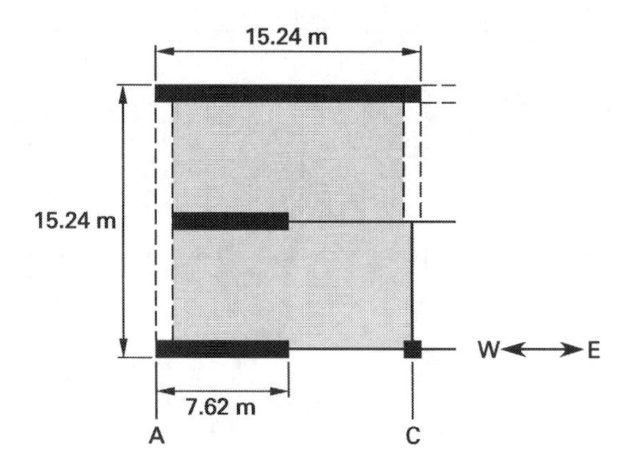

The weight of the roof is

$$W_{\text{roof}} = (15.24\text{ m})(15.24\text{ m})\left(0.7\ \frac{\text{kN}}{\text{m}^2}\right) = 162.6\text{ kN}$$

The weight of the 15.24 m wall (full height) is

$$W_{15.24\text{ m}} = (3.66\text{ m})(15.24\text{ m})\left(2.6\ \frac{\text{kN}}{\text{m}^2}\right) = 145.0\text{ kN}$$

The weight of the two 7.62 m walls totals 145.0 kN also.

$$W_{7.62\text{ m}} = 145.0\text{ kN}$$

The seismic effect from the roof and the 15.24 m wall is shared equally between walls A and C. Since the two remaining walls extend only halfway between A and C, a rational method of allocating their seismic effects to walls A and C must be used. Consider a simply supported beam loaded uniformly along the first half of its length. The reaction closest to the uniform load will carry $^3/_4$ of the total load. Therefore, assume that wall A carries $^3/_4$ of the seismic effect from the two short walls. (Other assumptions may be valid, depending on construction details.) Finally, assume that only the top half of the walls contribute to diaphragm shear. The diaphragm reaction at line A is

$$\begin{aligned} F_{\text{A}} &= 0.183W \\ &= (0.183) \\ &\quad \times \left(\frac{162.6\text{ kN}}{2} + \frac{\left(\frac{1}{2}\right)(145.0\text{ kN})}{2} + \frac{\left(\frac{3}{4}\right)(145.0\text{ kN})}{2}\right) \\ &= 31.5\text{ kN} \end{aligned}$$

For line E,

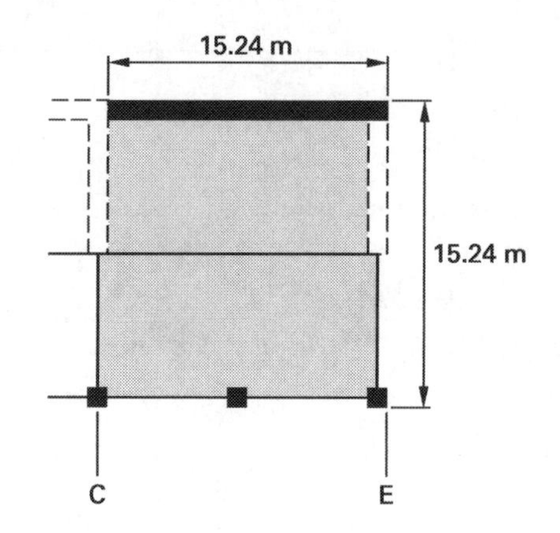

$$W_{\text{roof}} = 162.6\text{ kN}$$

$$W_{\text{E-W wall}} = (3.66\text{ m})(15.24\text{ m})\left(2.6\ \frac{\text{kN}}{\text{m}^2}\right) = 145.0\text{ kN}$$

The east-west wall weight tributary to the roof level is

$$\frac{145.0\text{ kN}}{2} = 72.5\text{ kN}$$

$$\begin{aligned} F_{\text{E}} &= 0.183W \\ &= (0.183)\left(\frac{162.6\text{ kN}}{2} + \frac{72.5\text{ kN}}{2}\right) \\ &= 21.5\text{ kN} \end{aligned}$$

(b) The earthquake direction is not given.

North-south earthquake

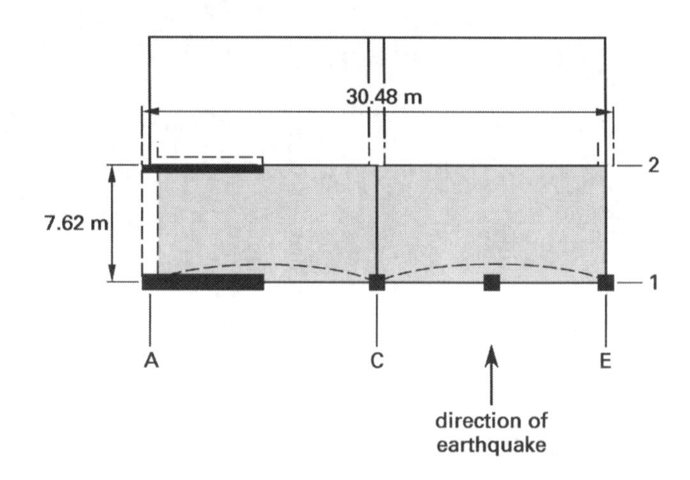

Drag force at point C-1 (line C)

All diaphragm shear would frame into point C-2 since the wooden post at point C-1 can be assumed to have no lateral stiffness. Therefore, the maximum drag force at point C-1 due to a north-south earthquake is zero.

Chord force at point C-1 (line 1)

Half of the roof mass (corresponding to the tributary area bounded by points A-1, A-2, E-2, and E-1) and the upper half of the east-west walls in the tributary area contribute to the chord forces along lines 1 and 2. (Other interpretations might also be justified. For example, the entire roof might be considered.)

$$W_{\text{roof}} = (7.62 \text{ m})(30.48 \text{ m})\left(0.7 \ \frac{\text{kN}}{\text{m}^2}\right) = 162.6 \text{ kN}$$

$$\begin{aligned} W_{\text{E-W walls}} &= (3.66 \text{ m})\left(7.62 \text{ m} + \left(\tfrac{1}{2}\right)(7.62 \text{ m})\right)\left(2.6 \ \frac{\text{kN}}{\text{m}^2}\right) \\ &= 108.8 \text{ kN} \end{aligned}$$

Only half of the east-west wall along line 2 is used. This is because the other half is tributary to the adjacent area bounded by points A-2, A-3, C-3, and C-2.

Only the upper half of the east-west walls is effective in loading the diaphragm.

$$\frac{108.8 \text{ kN}}{2} = 54.4 \text{ kN}$$

The diaphragm force is

$$\begin{aligned} F_{\text{diaphragm}} &= 0.183W \\ &= (0.183)(162.6 \text{ kN} + 54.4 \text{ kN}) \\ &= 39.7 \text{ kN} \end{aligned}$$

From Eq. 7.10, the chord force is

$$\begin{aligned} C &= \frac{F_{\text{diaphragm}}L}{8b} = \frac{(39.7 \text{ kN})(30.48 \text{ m})}{(8)(7.62 \text{ m})} \\ &= 19.9 \text{ kN} \end{aligned}$$

East-west earthquake

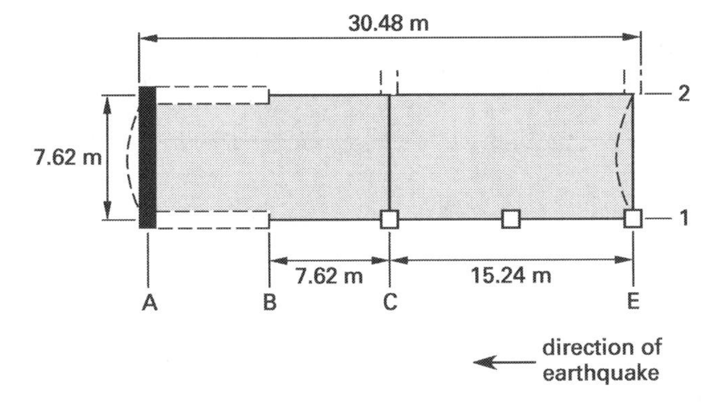

Drag force at point C-1 (line 1)

The collector along line 1 transfers half the diaphragm loading from the area bounded by points C-1, C-2, E-2, and E-1 into point B-1. (The other half is transferred to point B-2.) Since there are no north-south walls in this area, the only contribution to collector force is the tributary roof weight. Between lines C and E, the roof weight is

$$\begin{aligned} W_{\text{roof}} &= (7.62 \text{ m})(15.24 \text{ m})\left(0.7 \ \frac{\text{N}}{\text{m}^2}\right) \\ &= 81.3 \text{ kN} \end{aligned}$$

The diaphragm force is

$$\begin{aligned} F &= 0.183W = (0.183)(81.3 \text{ kN}) \\ &= 14.9 \text{ kN} \end{aligned}$$

Half of this is transferred along the strut on line 1. At point C-1, the drag force is

$$\begin{aligned} D &= \frac{14.9 \text{ kN}}{2} \\ &= 7.45 \text{ kN} \end{aligned}$$

Chord force at point C-1 (line C)

While there is a chord force along line C, the chord force at C-1 is zero because point C-1 is at the end of the chord.

(c) This is similar to part (b), except that the tributary roof area is 22.86 m long instead of 15.24 m long. Scaling up from the answer derived in part (b),

$$\begin{aligned} D &= \left(\frac{22.86 \text{ m}}{15.24 \text{ m}}\right)(7.45 \text{ kN}) \\ &= 11.2 \text{ kN} \end{aligned}$$

87. A 120 ft (36.58 m) wide by 240 ft (73.15 m) long warehouse in seismic design category E is oriented with its long dimension in the north-south direction. The risk category is II. The natural period is 0.38 sec. This structure site has a value of $S_{DS} = 0.70$, $S_{D1} = 0.32$, $S_S = 1.50$, and $S_1 = 0.60$. The walls are solid cast-in-place concrete, 10 ft (3.05 m) high and 12 in (305 mm) thick, with no significant openings. The warehouse floor has a live load of 200 lbf/ft^2 (9.6 kN/m^2). The roof uses $^{19}/_{32}$ in (15 mm) plywood sheathing with a grade consistent with DOC PS 1. The diaphragm is blocked and has an average weight of 15 lbf/ft^2 (0.7 kN/m^2) fastened to a 3 in (76 mm) (nominal) ledger and framing (Douglas fir-larch) with $1^1/_2$ in 16 gage staples. Do not use the simplified lateral force procedure, and consider only north-south earthquake motions. Use allowable stress design with basic load combinations.

(a) What diaphragm edge staple spacing is required along the 240 ft (73.15 m) wall?

(b) Sketch the method of interconnecting the diaphragm, ledger, and wall.

(c) Determine the force on each ledger bolt if they are spaced every 3 ft (91.4 cm).

(d) What size ledger bolt should be used to connect the ledger to the concrete walls, assuming the bolt strength is controlled by the timber?

(e) What size grade 60 steel rebar should be used as the diaphragm chord if the allowable stress is 24,000 psi (165.5 MPa)?

(f) Where should the chord be located?

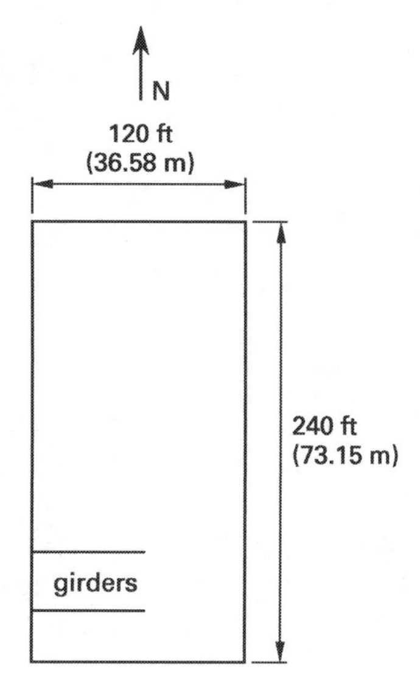

Solution

From Table 6.2, $I_e = 1.00$. In seismic design category E, walls must be specially reinforced. From App. H for a special reinforced concrete shear wall (bearing wall) system, the response modification factor $R = 5.0$.

From Eq. 6.21(a) and Eq. 6.23, the total design base shear is

$$V = \left(\frac{S_{D1}}{T\left(\frac{R}{I_e}\right)} \right) W = \left(\frac{0.32}{(0.38 \text{ sec})\left(\frac{5.0}{1.0}\right)} \right) W = 0.168W$$

V_{design} should be greater than the minimum total design base shear obtained from Eq. 6.22(a) and Eq. 6.23 [ASCE/SEI7 Eq. 12.8-6] or $0.010W$ [ASCE/SEI7 Eq. 12.8-5].

$$V_{\min} = \left(\frac{0.5S_1}{\frac{R}{I_e}} \right) W = \left(\frac{(0.5)(0.60)}{\frac{5.0}{1.0}} \right) W = 0.06W$$

The calculated $V_{\text{design}} = 0.168W$ can be used since it exceeds the controlling value of $V_{\min} = 0.06W$. Based on ASCE/SEI7 Sec. 12.8.1.1, however, this design base shear value should not exceed the maximum total design base shear calculated from Eq. 6.20 and Eq. 6.23 [ASCE/SEI7 Eq. 12.8-2]. $V_{\max}$ is

$$V_{\max} = \left(\frac{S_{DS}}{\frac{R}{I_e}} \right) W = \left(\frac{0.70}{\frac{5.0}{1.0}} \right) W = 0.14W$$

The calculated $V_{\text{design}} = 0.168W$ is greater than $V_{\max} = 0.14W$. Thus, the maximum required design base value for this warehouse is

$$V_{\text{design}} = 0.14W$$

The building is a combination of concrete shear walls with a timber diaphragm. The concrete shear walls would traditionally be designed using ACI design methods and would use the basic combination load factors of ASCE/SEI7 Sec. 2.3.2. The timber diaphragm would traditionally be designed using NDS design methods, which are based on allowable stress design. Therefore, to verify the materials needed for the roof and its attachment to the walls, the load combinations of ASCE/SEI7 Sec. 2.4.1 are used. The load combinations of ASCE/SEI7 Sec. 2.4.1 reduce the value of earthquake loads by 30% for use with allowable stresses.

Customary U.S. Solution

(a) The roof weight is

$$W_{\text{roof}} = \frac{(120 \text{ ft})(240 \text{ ft})\left(15 \ \frac{\text{lbf}}{\text{ft}^2}\right)}{1000 \ \frac{\text{lbf}}{\text{kip}}} = 432 \text{ kips}$$

The east-west walls contribute to diaphragm loading. Concrete has a weight density of approximately 150 lbf/ft^3.

$$W_{\text{E-W walls}} = \frac{(10\text{ ft})(2\text{ walls})(120\text{ ft})(12\text{ in})\left(150\ \frac{\text{lbf}}{\text{ft}^3}\right)}{\left(12\ \frac{\text{in}}{\text{ft}}\right)\left(1000\ \frac{\text{lbf}}{\text{kip}}\right)} = 360\text{ kips}$$

Only the top half of the east-west walls are effective in loading the diaphragm.

$$\frac{360\text{ kips}}{2} = 180\text{ kips}$$

Since this is a warehouse, a minimum of 25% of the live load must be added to the building weight. (See Sec. 6.29.) It could also be argued that the storage sits on grade and does not add to the inertial mass. Consider instead the case where racks are braced at the top and a portion of the live load is tributary to the roof.

$$W_{\text{live}} = (0.25)(120\text{ ft})(240\text{ ft})\left(200\ \frac{\text{lbf}}{\text{ft}^2}\right) = 1.44\times 10^6\text{ lbf}\quad (1440\text{ kips})$$

The base shear is

$$V = 0.14W = (0.14)(432\text{ kips} + 180\text{ kips} + 1440\text{ kips}) = 287\text{ kips}$$

Using the load combinations of ASCE/SEI7 Sec. 2.4.1, the base shear can be reduced for the roof diaphragm for use with allowable stress.

$$F_{\text{diaphragm}} = 0.7V = (0.7)(287\text{ kips}) = 201\text{ kips}$$

The shear force is resisted along two sides of the diaphragm, each 240 ft in length. The shear force per unit length along each of the north-south walls is

$$V = \frac{(201\text{ kips})\left(1000\ \frac{\text{lbf}}{\text{kip}}\right)}{(2\text{ walls})\left(240\ \frac{\text{ft}}{\text{wall}}\right)} = 419\text{ lbf/ft}\quad (420\text{ lbf/ft})$$

From Table 12.1, 2 in staple spacing will provide at least 450 lbf/ft of shear resistance, regardless of the framing or the orientation of the plywood.

(b)

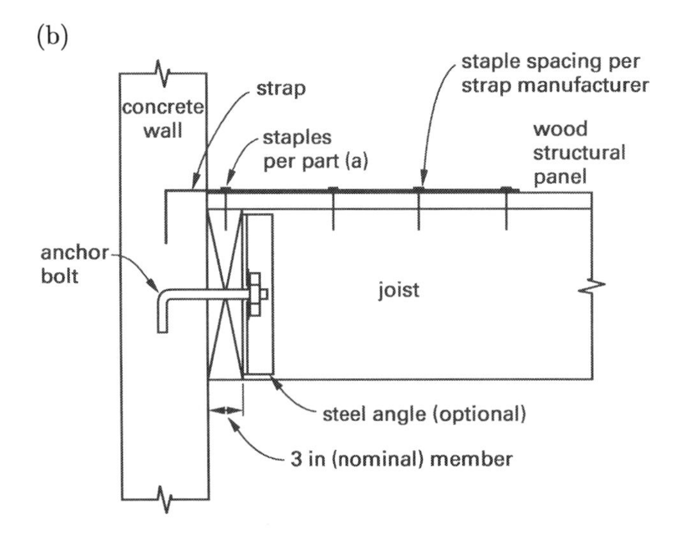

(c) The ledger is nominally 3 in thick. The shear loading is parallel to the grain. Since the bolts are spaced every 3 ft, the shear load is

$$V_{\text{bolt}} = \left(3\ \frac{\text{ft}}{\text{bolt}}\right)\left(419\ \frac{\text{lbf}}{\text{ft}}\right) = 1257\text{ lbf/bolt}$$

(d) From App. O, use the value of $Z_{\parallel}$ since the bolt is being pushed in the long direction of the ledger. For lumber with $G=0.50$ and side member thickness of $2^1/_2$ in, select the $^3/_4$ in diameter bolt with a strength of 1540 lbf. (The LRFD time effect factor for seismic events is 1.0. If ASD is used, a load duration factor of 1.6 could be used to reduce the bolt size to $^5/_8$ in.)

(e) From Eq. 7.10, the maximum chord force along the short walls will be

$$C = \frac{F_{\text{diaphragm}}L}{8b} = \frac{(201\text{ kips})\left(1000\ \frac{\text{lbf}}{\text{kip}}\right)(120\text{ ft})}{(8)(240\text{ ft})} = 12{,}563\text{ lbf}$$

Since basic load combinations are used, the one-third increase in allowable stress cannot be used. Since the allowable stress for grade 60 rebar is 32,000 psi (ACI 530 Sec. 2.3.3.1), the required bar size is given by Eq. 7.13.

$$A = \frac{C}{\text{allowable tensile stress}} = \frac{12{,}563\text{ lbf}}{32{,}000\ \frac{\text{lbf}}{\text{in}^2}} = 0.39\text{ in}^2$$

Use a no. 6 bar (diameter, $^3/_4$ in; area, 0.44 in^2).

(f) The chord should be located in the plane of the diaphragm along the north and south walls. (See Fig. 12.2.)

SI Solution

(a) The roof weight is

$$W_{\text{roof}} = (36.58\ \text{m})(73.15\ \text{m})\left(0.7\ \frac{\text{kN}}{\text{m}^2}\right) = 1873\ \text{kN}$$

The east-west walls contribute to diaphragm loading. Concrete has a density of approximately 2400 kg/m^3.

$$\begin{aligned} W_{\text{E-W walls}} &= \frac{\begin{array}{c}(3.05\ \text{m})(2\ \text{walls})(36.58\ \text{m}) \\ \times\ (305\ \text{mm})\left(2400\ \frac{\text{kg}}{\text{m}^3}\right)\left(9.81\ \frac{\text{m}}{\text{s}^2}\right)\end{array}}{\left(1000\ \frac{\text{mm}}{\text{m}}\right)\left(1000\ \frac{\text{N}}{\text{kN}}\right)} \\ &= 1602\ \text{kN} \end{aligned}$$

Only the top half of the east-west walls are effective in loading the diaphragm.

$$\frac{1602\ \text{kN}}{2} = 801\ \text{kN}$$

Since this is a warehouse, a minimum of 25% of the live load must be added to the building weight. (See Sec. 6.29.) It could also be argued that the storage sits on grade and does not add to the inertial mass. Consider instead the case where racks are braced at the top and a portion of the live load is tributary to the roof.

$$\begin{aligned} W_{\text{live}} &= (0.25)(36.58\ \text{m})(73.15\ \text{m})\left(9.6\ \frac{\text{kN}}{\text{m}^2}\right) \\ &= 6422\ \text{kN} \end{aligned}$$

The base shear is

$$\begin{aligned} V &= 0.14W \\ &= (0.14)(1873\ \text{kN} + 801\ \text{kN} + 6422\ \text{kN}) \\ &= 1273\ \text{kN} \end{aligned}$$

Using the load combinations of ASCE/SEI7 Sec. 2.4.1, the base shear can be reduced for the roof diaphragm for use with allowable stress.

$$F_{\text{diaphragm}} = 0.7V = (0.7)(1273\ \text{kN}) = 891\ \text{kN}$$

The shear force is resisted along two sides of the diaphragm, each 73.15 m in length. The shear force per unit length along each of the north-south walls is

$$\begin{aligned} V &= \frac{(891\ \text{kN})\left(1000\ \frac{\text{N}}{\text{kN}}\right)}{(2\ \text{walls})\left(73.15\ \frac{\text{m}}{\text{wall}}\right)} \\ &= 6090\ \text{N/m} \quad (6.0\ \text{N/mm}) \end{aligned}$$

From Table 12.1, 50 mm staple spacing will provide at least 6.6 N/mm of shear resistance, regardless of the framing or the orientation of the plywood.

(b)

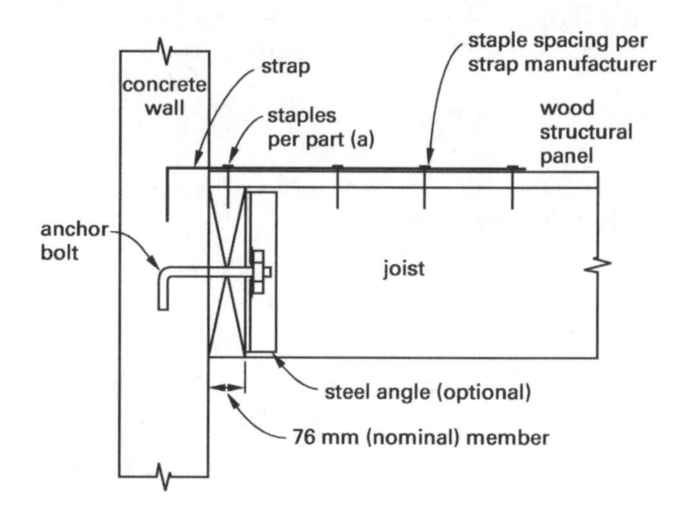

(c) The ledger is nominally 76 mm thick. The shear loading is parallel to the grain. Since the bolts are spaced every 91.4 cm, the shear load is

$$\begin{aligned} V_{\text{bolt}} &= \frac{\left(91.4\ \frac{\text{cm}}{\text{bolt}}\right)\left(6090\ \frac{\text{N}}{\text{m}}\right)}{100\ \frac{\text{cm}}{\text{m}}} \\ &= 5566\ \text{N/bolt} \end{aligned}$$

(d) From App. O, use the value of $Z_{||}$ since the bolt is being pushed in the long direction of the ledger. For lumber with $G = 0.50$ and side member thickness of 64 mm, select a 19 mm diameter bolt with a strength of 6364 N. (The LRFD time effect factor for seismic events is 1.0. If ASD is used, a load duration factor of 1.6 could be used to reduce the bolt size to 16 mm.)

(e) From Eq. 7.10, the maximum chord force along the short walls will be

$$\begin{aligned} C &= \frac{F_{\text{diaphragm}}L}{8b} \\ &= \frac{(891\ \text{kN})\left(1000\ \frac{\text{N}}{\text{kN}}\right)(36.58\ \text{m})}{(8)(73.15\ \text{m})} \\ &= 55\,695\ \text{N} \end{aligned}$$

Since basic load combinations are used, the one-third increase in allowable stress cannot be used. Since the allowable stress for grade 60 rebar is 220.7 MPa

(ACI 530 Sec. 2.3.3.1), the required bar size is given by Eq. 7.13.

$$\begin{aligned} A &= \frac{C}{\text{allowable tensile stress}} \\ &= \frac{55\,695 \text{ N}}{(220.7 \text{ MPa})\left(1\,000\,000 \; \frac{\text{Pa}}{\text{MPa}}\right)} \\ &= 2.5 \times 10^{-4} \text{ m}^2 \quad (250 \text{ mm}^2) \end{aligned}$$

Use a no. 6 bar (diameter, 19.05 mm; area, 284 mm^2).

(f) The chord should be located in the plane of the diaphragm along the north and south walls. (See Fig. 12.2.)

9. QUALITATIVE PROBLEMS

88. List the numbers corresponding to the types of individuals who have the legal authority in California to perform the functions listed.

1. any licensed civil engineer
2. civil engineer licensed before January 1, 1982
3. civil engineer licensed after January 1, 1982
4. licensed civil engineer specializing in structures
5. licensed (California) soils engineer
6. licensed structural engineer
7. engineer licensed in any field
8. unlicensed civil engineer specializing in structures, under the responsible charge of a licensed civil engineer
9. any unlicensed civil engineer
10. licensed architect
11. licensed land surveyor
12. licensed photogrammetric surveyor
13. licensed contractor
14. licensed building designer
15. any member of the general public
16. no one

(a) use the title "consulting engineer" publicly

(b) use the title "civil engineer" publicly

(c) use the title "structural engineer" publicly

(d) use the title "soils engineer" publicly

(e) use the title "land surveyor" publicly

(f) personally perform civil engineering work

(g) solicit civil engineering work for others

(h) perform architectural work

(i) sign, stamp, and seal civil engineering design plans developed by an unlicensed engineering subordinate under the individual's direct engineering control

(j) sign, stamp, and seal civil engineering design plans developed by an unlicensed engineer in the individual's company whose paycheck the individual signs

(k) sign, stamp, and seal civil engineering design plans developed by a qualified, unlicensed, moonlighting engineer who pays the individual

(l) sign, stamp, and seal civil engineering design plans developed by a qualified, unlicensed engineer who does not pay the individual

(m) allow a qualified, unlicensed person to use a registered civil engineer's stamp, seal, or registration number

(n) design a single-story, wood-framed residence

(o) design a two-story, wood-framed residence

(p) design a five-story, concrete-framed building

(q) design a five-story, steel-framed building

(r) design an above-ground water tower structure

(s) design a hospital building

(t) design a new public school building

(u) design a new private school building

(v) design a steel bridge

(w) inspect an earthquake-damaged building within 30 days of the event, without payment, when requested by the local building official

(x) supervise the construction of designed structures

(y) perform land surveying work for hire

(z) solicit land surveying work for others

(aa) perform a survey of public lands

(ab) perform a survey of private lands to be subdivided

(ac) lay out a construction site using surveying knowledge, methods, and equipment

(ad) gather in the field information to be placed on a deed or record-of-survey map

(ae) file a record-of-survey map with the county

(af) administer oaths, certify oaths, and take testimony under oath to identify lost corners

Solution

Unless noted otherwise, references in parentheses are to the *California Business and Professions Code*, Div. 3,

Chap. 7 (Professional Engineers) and Chap. 15 (Land Surveyors).

(a) 1, 5, 6, 7, 12 (6704, 6732)
(b) 1, 5, 6 (6704, 6732, 6734)
(c) 6 (6703, 6704, 6732, 6736)
(d) 5 (6704, 6732, 6736.1, 6763)
(e) 11 (6731, 8708, 8725, 8731)
(f) 1, 5, 6 (6731.2, 8726.1)
(g) 15
(h) 10 (6737)
(i) 1, 5, 6 (6730.2, 6735, 6740)
(j) 16 (6703, 6735)
(k) 16 (6703, 6735)
(l) 16 (6703, 6735)
(m) 16 (6732, 6735)
(n) 15 (6737, 6737.1)
(o) 15 (6737, 6737.1)
(p) 1, 5, 6, 8, 10 (6737)
(q) 1, 5, 6, 8, 10 (6737)
(r) 1, 5, 6, 8, 10 (6737)
(s) 6 (Refer to *Health and Safety Code*, Div. 107, Part 7, Chap. 1 (Health Facilities), Sec. 129805), 10
(t) 6, 10
(u) 6, 10 (6737)
(v) 1, 5, 6, 8
(w) 1, 5, 6, 7 (6706)
(x) 1, 5, 6, 10, 13 (6731, 6731.3, 6735.1)
(y) 2, 11, 12 (6731.2, 8726.1, 8731, 8775)
(z) 15
(aa) 2, 11, 12 (8708, 8731, 8775)
(ab) 2, 11, 12 (8708, 8726, 8731, 8775)
(ac) 12, 15
(ad) 2, 11, 12 (8726, 8731, 8775)
(ae) 2, 11, 12 (8731, 8762, 8775)
(af) 1, 2, 5, 6, 11, 12 (8760, 8775)

89. How does the IBC cover the design of bridges?

Solution

The IBC covers only the design of buildings and some other building-like structures. Design of bridges is not covered in the IBC. This subject is covered in CALTRANS and American Association of State and Highway Transportation Officials (AASHTO) publications.

90. What provisions does the IBC make for buildings subject to landslides, liquefaction, subsidence, gross differential settlement, or for those built close to a major ground-breaking fault?

Solution

None. The IBC provides rules for the design of buildings that will resist typical ground shaking. The IBC assumes the engineer will use good judgment in avoiding inherently dangerous locations.

91. What type of information will generally be supplied by the geotechnical engineer working on a building design team?

Solution

The geotechnical engineer will determine the (a) type of soil (i.e., sand, clay, or rock); (b) depth of water table; (c) depth to bedrock; (d) proximity to a fault; (e) maximum credible earthquake; and (f) likelihood of liquefaction, slides, subsidence, and differential settlement.

92. Which structural system resists lateral loads by flexure in members and joints?

Solution

Only moment-resisting frames resist lateral loads in this manner.

93. What is the meaning of the term "secondary stress" as it relates to a moment-resisting frame?

Solution

Primary stresses are the compressive and tensile forces that act uniformly on the cross section of the member. Secondary stresses are bending stresses that result from distortion of the frame when resisting lateral loads by flexure.

94. (a) What structural elements transfer lateral loads to vertical elements? (b) What structural elements transfer lateral loads to lower levels and the foundation?

Solution

(a) Horizontal elements such as diaphragms, horizontal bracing, and beams in moment-resisting frames transfer horizontal loads to vertical elements. (b) Vertical elements such as shear walls, braced frames, and columns in moment-resisting frames transfer lateral loads to lower levels.

95. A building site with soil site class B is located at 45° latitude and 120° longitude. In calculating the seismic source of the building, (a) What is the maximum considered earthquake ground motion for short period (0.2 sec) and 1.0 sec buildings? (b) What site coefficient should be used?

Solution

(a) From IBC Fig. 1613.3.1(1), the value of S_S is 0.40, and from IBC Fig. 1613.3.1(2), the value of S_1 is 0.15.

(b) From Table 6.4 [IBC Table 1613.3.3(1)], the value of F_a is 1.0, and from Table 6.5 [IBC Table 1613.3.3(2)], the value of F_v is also 1.0. (The maps of IBC Fig. 1613 are drawn for site class B, so F is unity for all levels of ground acceleration.)

96. A site contains soil that is vulnerable to potential failure or collapse under seismic loading.

(a) What types of soils are vulnerable to potential failure or collapse under seismic loading?

(b) What is the site class for this site?

(c) Should a site-specific evaluation be conducted for this site?

(d) Under what circumstance can site class D be matched to a site?

Solution

(a) Under seismic loading, liquefiable soil, quick and highly sensitive clays, collapsible weakly cemented soils, peats and highly organic clays of 10 ft (305 cm) or more in thickness, very high plasticity clays of 25 ft (762 cm) or more in thickness and having plasticity index greater than 75, and very thick soft/medium stiff clays of 120 ft (36.6 m) or more in thickness are types of soils vulnerable to potential failure or collapse.

(b) According to Table 6.3 [ASCE/SEI7 Table 20.3-1], sites vulnerable to potential failure or collapse correspond to site class F.

(c) Based on the criteria given, site class F requires a site-specific evaluation [ASCE/SEI7 Sec. 20.3.1].

(d) Each site should be assigned a site class according to appropriately documented geotechnical data. The site categorization procedure of ASCE/SEI7 Chap. 20 should be used for that purpose. In determining the site class when the soil properties are not identified in adequate detail, site class D can be presumed. The building official or others may "determine" that the soil profile matches site class E or F.

97. What is the meaning of the term *soft* (*weak*) *story*? Give an example.

Solution

A soft story does not have as much lateral force resistance as the stories above. An example is a moment-resisting frame supported by long columns over an open plaza below.

98. What restriction does the IBC and ASCE/SEI7 place on situations where the type of structural system is different for different levels of a multistory building? What is the exception?

Solution

The value of R used in the design of one level must be less than or equal to the value of R used to design the level above. An exception is where the story above constitutes less than 10% of the total structure weight (i.e., is very light).

99. In determining the total design lateral seismic force, F_p, on elements of structures, nonstructural components, and equipment supported by structures, the ASCE/SEI7 uses the a_p coefficient.

(a) What does this coefficient represent?

(b) How are the values of this coefficient obtained?

(c) When determining the anchorage force for the connection of a concrete nonstructural wall to a diaphragm, what value for the a_p coefficient should be applied?

Solution

(a) a_p is a numerical coefficient representing the in-structure component amplification factor that varies from 1.0 to 2.5. The minimum value of this factor is equal to 1.0. (See Sec. 6.44.)

(b) Table 6.15 [ASCE/SEI7 Table 13.5-1] provides values for this coefficient. Dynamic properties or empirical data of the component and the structure that supports it can determine this factor as well.

(c) Based on Table 6.15 [ASCE/SEI7 Table 13.5-1], the design force for the fasteners of the connecting system is to use $a_p = 1.25$. The materials of the wall element itself (i.e., the concrete and reinforcing steel) may use $a_p = 1.0$.

100. What is the meaning of the term *irregular building*? Give two examples.

Solution

For the purpose of the IBC, an irregular building meets one or more of the characteristics in ASCE/SEI7 Table 12.3-1 or Table 12.3-2. Examples are (1) a three-story, L-shaped building and (2) a five-story, square building with an open plaza comprising 60% of the floor area on level 3. (See Sec. 6.23.)

101. What is the meaning of the word *pounding*?

Solution

Pounding refers to adjacent buildings coming into contact with each other. One building can sway into another and pound it. The danger is greater when floor slabs of one building pound the columns of another; the danger is less when the floor slabs are at the same elevation. Up to 20% of the building failures in the 1985 Mexico City earthquake are thought to have been caused by pounding. Of the buildings damaged in the 1989 Loma Prieta earthquake in the Watsonville-Santa Cruz area, some were only 6 in (152 mm) apart and were damaged because they pounded each other. (See Sec. 6.41.)

102. Consider determining the seismic force on a building using the ASCE/SEI7's simplified static lateral force procedure.

(a) Which structures qualify for use of this design method?

(b) When using this design procedure for structures in regions of the country with severe earthquakes, what soil-profile type should be used, assuming the soil properties are not known in sufficient detail?

Solution

(a) The simplified static lateral force procedure is given in ASCE/SEI7 Sec. 12.14. This design method can be used for structures of risk category I or II with the maximum height of three stories excluding basements. Single-family dwellings can be included when they conform to this criterion. In addition, the structure must have at least two lines of lateral resistance in each of two major axis directions. (See Sec. 6.34.)

(b) In determining the seismic force on a structure using the ASCE/SEI7's simplified static lateral force procedure, soil site class D should be used when the soil properties are not known in sufficient detail to classify the soil-profile type.

103. When can the ASCE/SEI7's dynamic analysis method be used to determine the seismic force on a building? When can it not?

Solution

The dynamic method, as described by ASCE/SEI7 Sec. 12.9, can always be used. It is the static method that is limited and that must satisfy certain conditions. (See Sec. 6.33.)

104. What type of tie will be required for adjacent sheathing whose edge nailing places the framing member below it in cross-grain tension?

Solution

Cross-grain tension is not allowed in wood framing members. Therefore, *continuity ties* must be used to transfer tension across the joint in wood sheathing. (See Sec. 12.5.)

105. When using ASD, what factor should be applied to the dead load when designing for overturning effects caused by earthquake forces?

Solution

Every structure should be designed to resist the overturning effects caused by earthquake forces. Based on IBC Sec. 1605.3.1, when designing for overturning effects, a factor of 0.6 should be applied to the dead load when using ASD.

106. It is generally stated and understood that flexible diaphragms cannot transmit torsional shear stress to vertical resisting elements. Is this true for a flexible diaphragm that is cantilevered off of a vertical wall?

Solution

This is a tricky question. Any eccentric mass can cause torsion. A cantilevered flexible diaphragm, when acted upon by a seismic force perpendicular to its cantilevered dimension, will cause the wall to twist. However, this is different than transmitting torsion caused by one component to another. A cantilevered flexible diaphragm can cause torsion; it cannot transmit torsion.

107. It is generally stated and understood that the lateral loads resisted by vertical elements attached to rigid diaphragms are proportional to the element rigidities, and the lateral loads resisted by vertical elements attached to flexible diaphragms are proportional to tributary areas. How are lateral loads resisted by closely placed vertical elements that are arranged in-line, are parallel to an earthquake's motion, and are attached to a single flexible diaphragm?

Solution

Since all of the elements have the same tributary area, they will resist the lateral load in proportion to their relative rigidities.

108. What is the maximum allowable height-to-width ratio for a blocked vertical wood structural shear wall panel?

Solution

According to ANSI/AF&PA *Special Design Provisions for Wind and Seismic* (SDPWS) Table 4.3.4, the maximum allowable height-to-width ratio for a blocked wood structural shear wall panel is 3.5:1.

109. What are the minimum and maximum limits on force F_{px} that floors and diaphragms should be designed for?

Solution

Based on ASCE/SEI7 Sec. 12.10.1.1, floors and diaphragms should be designed to resist forces, F_{px}, calculated from Eq. 7.5 [ASCE/SEI7 Eq. 12.10-1]. F_{px} should not be less than $0.2S_{DS}I_e w_{px}$, but F_{px} need not exceed $0.4S_{DS}I_e w_{px}$.

110. Which is more life-threatening: shear cracking in a seismically detailed concrete column or flexural cracking of a seismically detailed concrete shear wall?

Solution

Cracking in a shear wall is probably more serious than cracking in a column. A properly-detailed column should not lose its loadbearing capacity merely because of cracking. The strict seismic detailing is intended to keep concrete in a column intact and confined even if it cracks. However, such confinement is not as complete in shear walls.

111. A tank on the roof of a building contains hazardous chemicals. What importance value, I_e, should be used in calculating the seismic force on the building?

Solution

From Table 6.2 [ASCE/SEI7 Table 1.5-2], $I_e = 1.25$.

112. What is the basic distinction between ordinary and special moment-resisting frames?

Solution

A special moment-resisting frame has been carefully detailed to remain ductile. An ordinary moment-resisting frame does not have this detailing. (See Sec. 6.21.)

113. Two buildings have the same mass, but one building has a shorter natural period than the other building. All other factors being equal, which building will experience the larger seismic force?

Solution

Most response spectra show that the lower the natural period, the higher the acceleration experienced by the building. Therefore, the building with the shorter period will probably experience the larger seismic force.

114. In IBC seismic design category D, which material is most likely to be less expensive when building a 30-story moment-resisting frame: steel or concrete?

Solution

This is a controversial question whose answer depends on location and may also depend on loyalties and specialties. In seismically active zones (e.g., California), most high-rise office buildings are built from steel, rather than concrete. Long-span open spaces are needed in offices, and steel beams can provide such space. However, smaller spans fit the requirements of hotels quite nicely, so low- to medium-height hotels may be constructed of concrete. Certainly, in developing countries where concrete is less expensive than imported rolled steel, there may be more concrete structures than steel. Nevertheless, for the parameters of this question (seismic design category D, 30 stories, and moment-resisting frame), steel is the obvious choice.

115. What consideration should be given to the design of a building that resists lateral force by a combination of braced frame and shear wall action?

Solution

Braced frames and shear walls have different stiffnesses and may deflect different amounts. This will cause a separation where the two resisting systems meet. The resisting elements must be proportioned so that the deflections are equal for both resisting systems.

116. In an extreme earthquake, what type of fascia would sustain the most damage: glass or concrete?

Solution

This is a fairly vague question since only the type of fascia material (and not the mounting method) is indicated. Glass has no ductility, so glass fascia probably would not fare well in an extreme earthquake. Concrete fascia would probably have been cast with continuous bar or mesh reinforcing. This reinforcing would help the concrete fascia remain intact when flexed.

117. For small buildings with only one or two floors, which of the different structural systems are more cost-effective? (Limit discussion to wood structural panel shear wall construction, masonry wall systems, steel braced frames, stiff-redundant steel systems, concrete moment-resisting frames, steel moment-resisting frames, and dual systems.)

Solution

Small buildings with only one or two floors can be built using any of the structural systems listed, although dual, redundant, and moment-resisting frame systems probably would not be used. The systems in order of increasing cost are

1. wood structural panel shear wall construction
2. masonry wall systems
3. ordinary steel braced frames
4. dual systems
5. stiff redundant steel systems
6. concrete moment-resisting frames
7. steel moment-resisting frames

118. For tall buildings with more than 10 floors, which of the different structural systems are more cost-effective? (Limit discussion to wood structural panel shear wall construction, masonry shear wall (box) systems, ordinary steel braced frames, stiff-redundant steel systems, concrete moment-resisting frames, steel moment-resisting frames, and dual systems.)

Solution

Wood structural panel and masonry systems would not be used for a building with 10 floors. Exceptionally tall buildings must be built either exceptionally stiff (e.g., the Empire State Building) or must use moment-resisting frames. Most modern tall buildings in California are constructed of steel. The logical conclusion is that these are less expensive than concrete buildings. Stiffness achieved through multiple redundancy is the most expensive. Though necessary in the early history of tall building construction, designing stiffness through redundancy is no longer practiced.

119. The floors in the top half of a tall multistory building are much smaller (in plan view) than the floors in the bottom half of the building. (a) What are the problems associated with this design? (b) How could the problems be counteracted?

Solution

(a) This question is essentially about setbacks. The main problem is that the upper half would have a different period and different mode shape than the lower half. The upper floors could oscillate out-of-phase with the upper floors. This is referred to as *whipping action.* Large stresses would be generated when the two sections were 180° out-of-phase. The stress would be most severe at the setback points.

(b) The upper half of the building must be designed so that, though smaller, it is as stiff or flexible as the lower half. There are many ways of increasing stiffness, including adding bracing, changing the spacing or number of interior members, and increasing member sizes. (Since the mass of the upper stories is reduced, just keeping the column and beam sizes the same as in the lower stories would help.) In some cases, a different construction material could be used. It is not generally practical to add stiffness by starting new columns at an upper floor.

120. An air conditioning unit is placed on a wood structural panel roof diaphragm. What effect does the new unit have on the damping ratio of the roof?

Solution

None. The damping ratio of the roof is a function of the roof material, design, and quality of construction.

121. Draw simple diagrams that show how a three-story frame would fail in (a) beam-hinge mode, (b) column-hinge mode, and (c) soft-story (also known as *weak-story*) mode. Show all plastic hinge points.

Solution

Plastic hinges are shown as solid bullets.

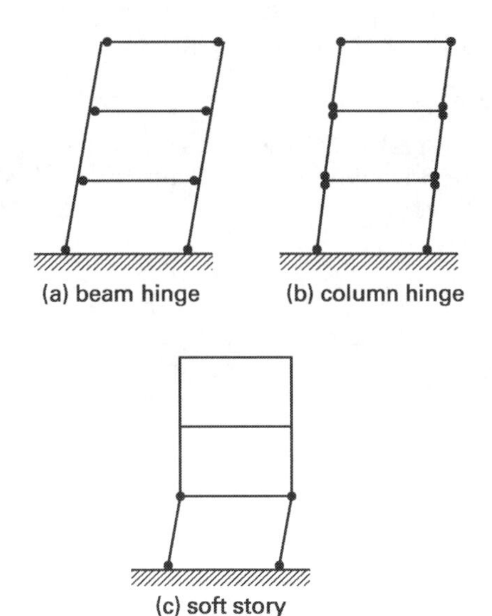

122. Draw simple diagrams that show how a four-story frame would fail in (a) weak-column mode (i.e., when the beams were stronger than the columns) and (b) weak-beam mode (i.e., when the columns were stronger than the beams). Show all plastic hinge points.

Solution

Plastic hinges are shown as solid bullets.

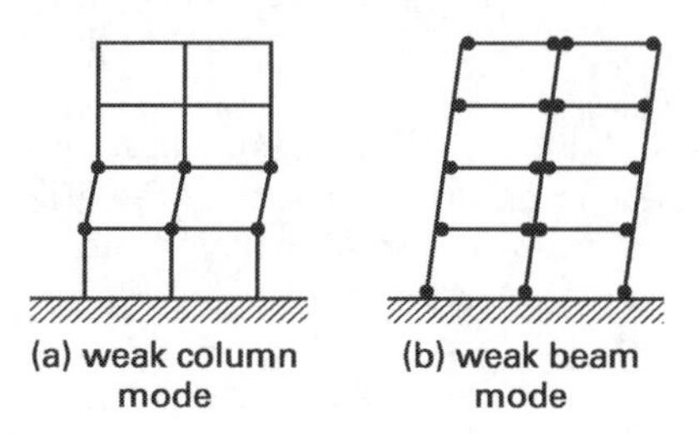

123. Draw the connections necessary to anchor the floor diaphragms shown to the side of a concrete masonry unit (CMU) wall. Show and label all connectors and other elements. No calculations are necessary and no specific spacings need to be specified. Assume positive attachment to the wall is spaced approximately every 4 ft (102 mm).

(a) wood structural panel floor on 2 in (nominal size) joists attached to 4 × 10 ledger

(b) wood structural panel supported directly by 4 × 10 ledger

(c) corrugated steel decking supported by steel ledger angle

(d) steel plate supported by steel ledger angle

(e) poured gypsum deck on metal form deck supported on steel ledger angle

Solution

(a)

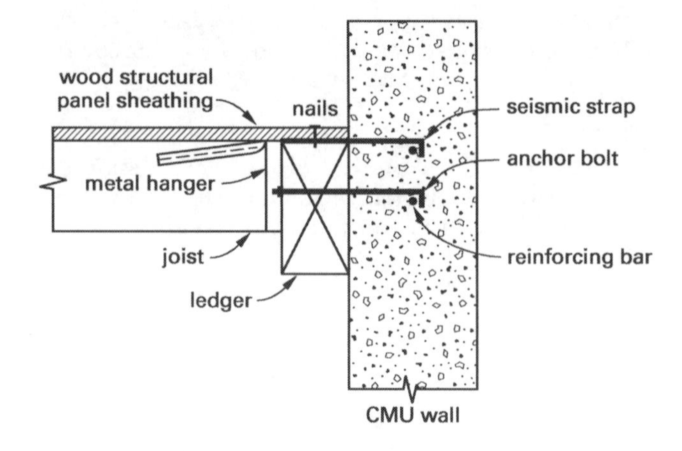

(b)

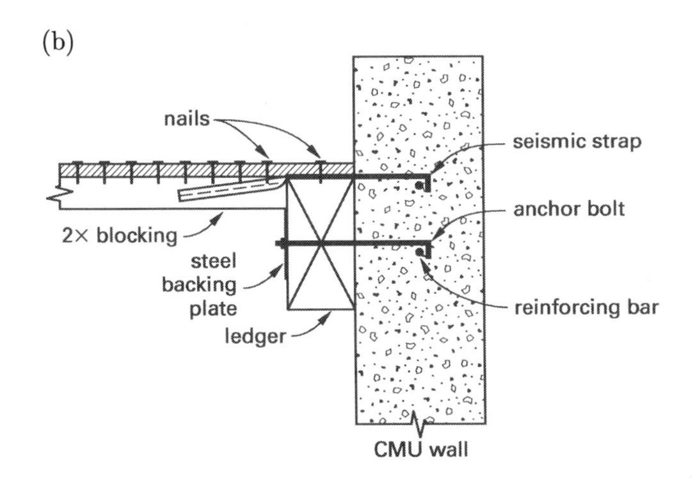

(c)

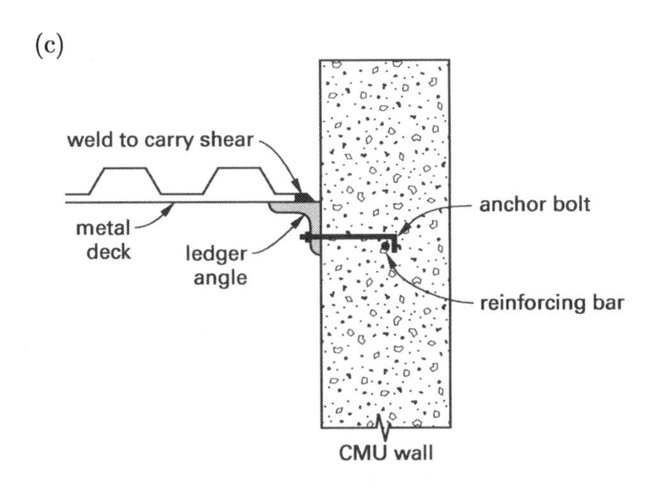

(d)

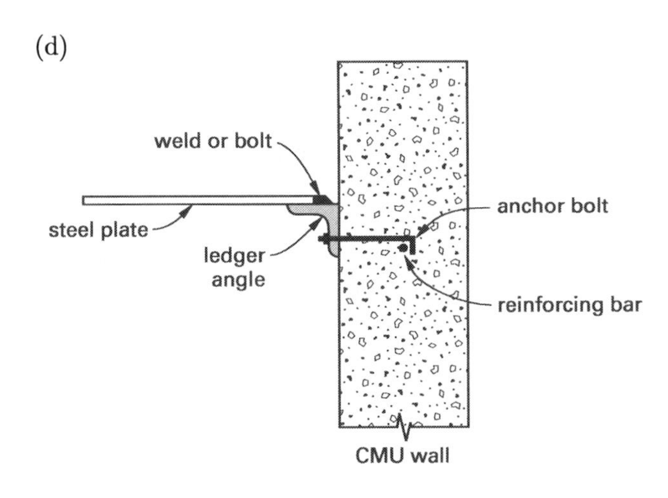

(e)

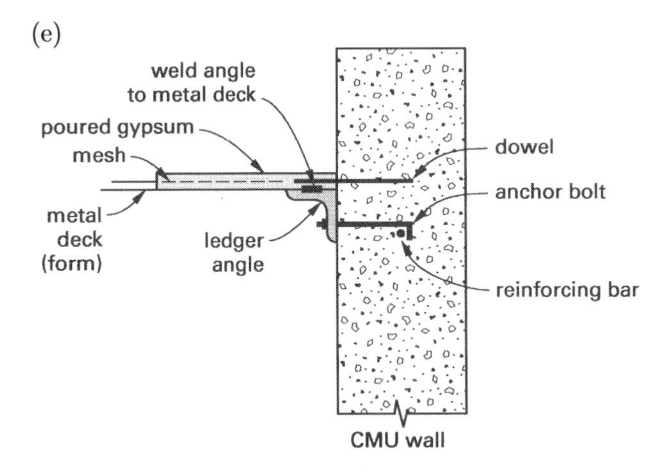

124. A wood structural panel diaphragm is supported by 2 in (nominal size) joists. The joists are supported by a wood ledger attached to a concrete masonry unit (CMU) wall. Detail the connection between a joist and the ledger if the joist does not coincide with a wall anchor bolt or seismic strap. Assume the ledger strength is adequate.

Solution

This problem is somewhat contrived because if the application is really critical, the joint should be connected directly to the wall. The intent of this problem is to design a positive connection between the joist and ledger. In doing so, it must be recognized that (a) provisions must be made to avoid tension splits in the joist, and (b) connector pull-out strengths must be considered in attaching the joist to the ledger. Toe-nailing is obviously inadequate and not permitted. If the joist is attached to the ledger with a commercial hanger having a row of vertical nails, the transmitted force will be limited by edge distance. The detail shown uses (a) nailing or bolting along the joist to avoid tension tear-out and (b) lag bolting to avoid connector pull-out.

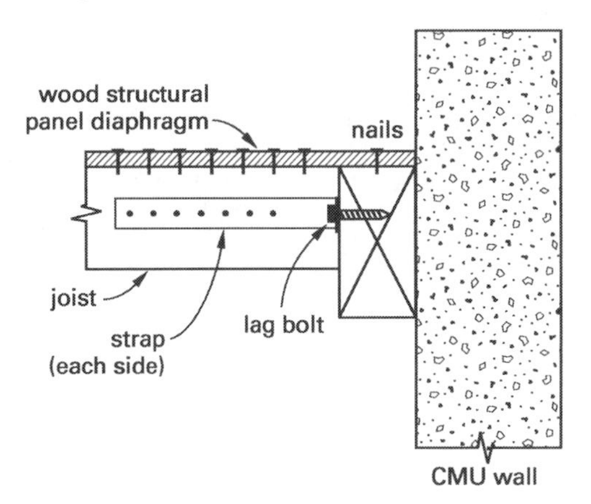

125. Detail a connection for a 2 in (nominal size) joist supporting a wood structural panel floor diaphragm sitting directly on a wall plate on top of a concrete masonry unit (CMU) wall. How could the connection avoid cross-grain tension in the wall plate?

Solution

The connection shown avoids cross-grain tension in the wall plate by avoiding any connection to the wall plate. Lateral joist loads are transmitted in shear through the anchor bolt. The wall plate remains in vertical compression at all times.

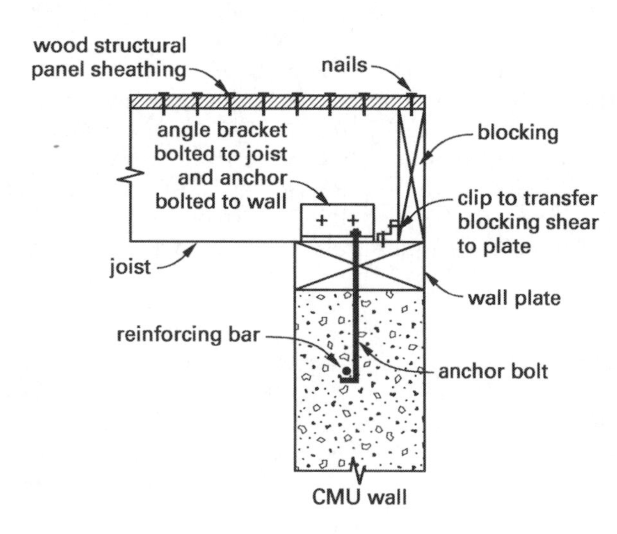

126. Describe how the connection between the joist and masonry unit wall shown may fail in cross-grain tension. How could the basic design be retained while eliminating the cross-grain tension?

Solution

If the lateral load is from left to right, the plate will be placed in compression by the anchor bolt reaction acting to the left. However, if the lateral load is from right to left, the plate will be placed in cross-grain tension by the anchor bolt reaction acting to the right. This design can be "fixed" by adding a framing clip to the right of the bolt (or by using one framing clip that extends to either side of the bolt).

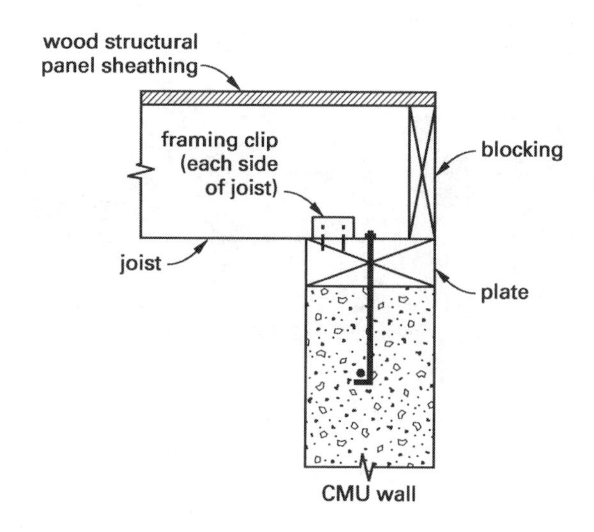

127. Illustrate how wood structural panel shear walls directly above each other on two different levels could be interconnected. (Exterior sheathing cannot be used to interconnect the walls.)

Solution

The most common method is to use a connecting tierod (tie down), as shown.

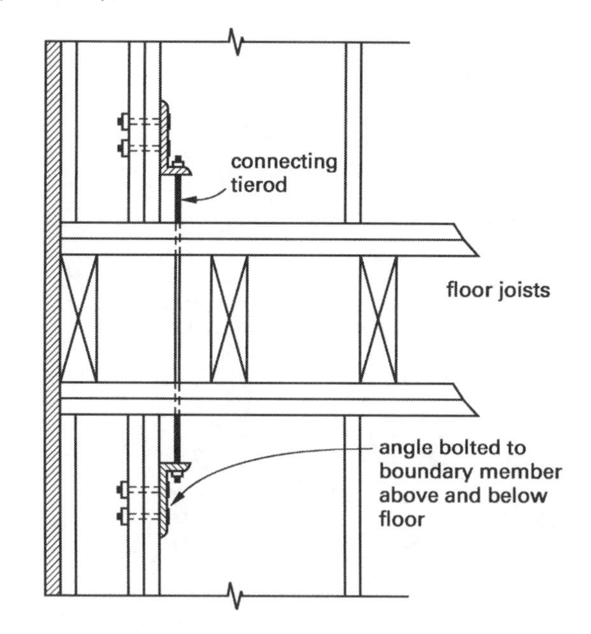

128. Draw and detail two common types of column-girder joints for special moment-resisting frames constructed from steel.

Solution

The two common types shown are (a) a fillet-welded joint, and (b) a bolted joint.

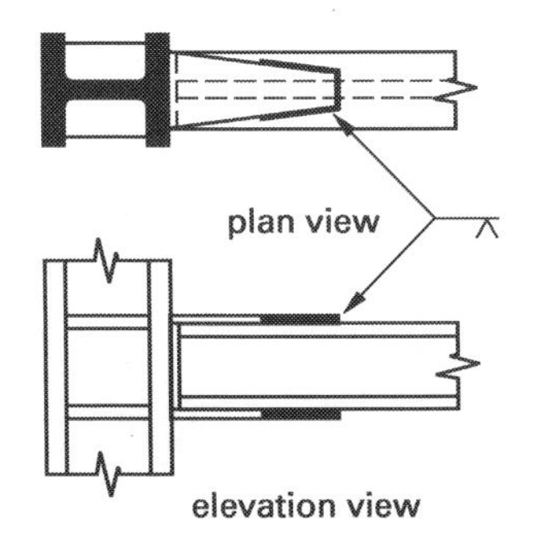

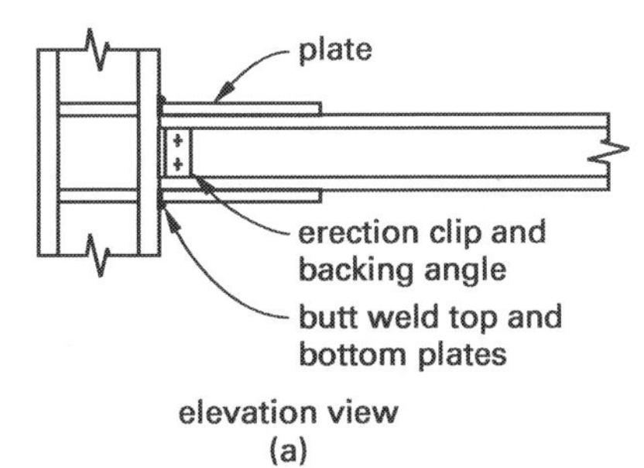

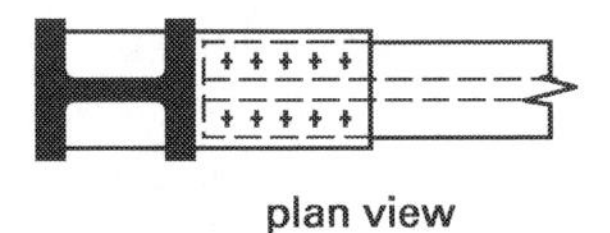

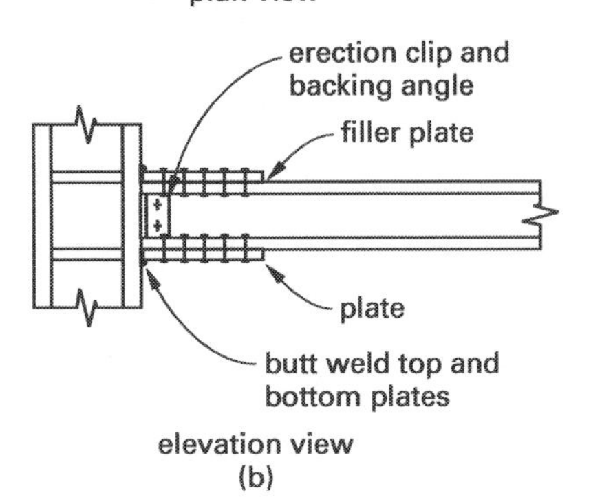

129. Draw a typical section showing how the end of a bridge section would be supported by an abutment.

Solution

The most important element in a bridge-to-abutment connection is a positive connection that prevents the two pieces from separating. Elements of secondary importance are the bearing, expansion joint, and shock-absorbing element (i.e., the rubber rings). Secondary cabling (not shown) may be provided as a back-up to the primary positive connection.

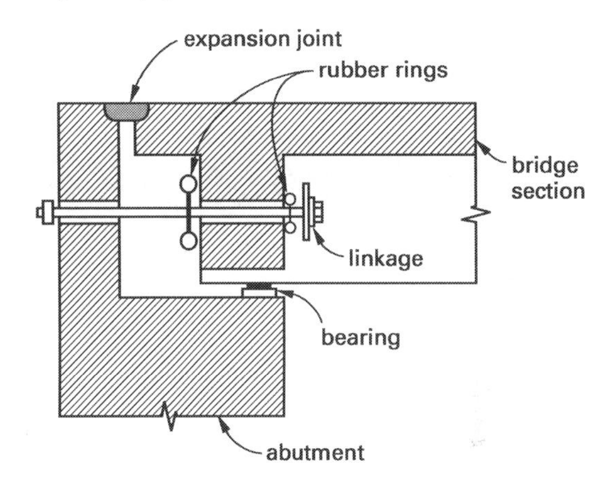

Appendices Table of Contents

APPENDIX A
Conversion Factors

multiply	by	to obtain	multiply	by	to obtain
acre	43,560	ft^2	kg	2.20462	lbm
Btu	778.17	ft-lbf	kg	0.06852	slug
Btu	1.055	kJ	kip	1000	lbf
Btu/hr	0.293	W	kJ	0.9478	Btu
Btu/lbm	2.326	kJ/kg	kJ	737.56	ft-lbf
Btu/lbm-°R	4.1868	kJ/kg·K	kJ/kg	0.42992	Btu/lbm
cm	0.3937	in	kJ/kg·K	0.23885	Btu/lbm-°R
cm^3	0.061024	in^3	km	3280.8	ft
erg	7.376×10^{-8}	ft-lbf	km	0.6214	mi
ft	0.3048	m	km/h	0.62137	mi/hr
ft^3	7.481	gal	kPa	0.14504	lbf/in^2
ft^3	0.028317	m^3	kW	737.6	ft-lbf/sec
ft-lbf	1.356×10^7	erg	kW	1.341	hp
ft-lbf	1.35582	J	L	0.03531	ft^3
ft/sec^2	0.0316	gravities	L	0.001	m^3
in/sec^2	0.002591	gravities	lbf	4.4482	N
gal	0.13368	ft^3	lbf/ft	14.5938	N/m
gal	3.7854×10^{-3}	m^3	lbf/ft^2	144	lbf/in^2
gal/min	0.002228	ft^3/sec	lbf/in^2	6894.8	Pa
g/cm^3	1000	kg/m^3	lbm	0.4536	kg
g/cm^3	62.428	lbm/ft^3	lbm/ft^3	0.016018	g/cm^3
gravities	32.2	ft/sec^2	lbm/ft^3	16.018	kg/m^3
gravities	386	in/sec^2	m	3.28083	ft
gravities	9.81	m/s^2	m^3	35.3147	ft^3
hp	2545	Btu/hr	mm	0.03937	in
hp	33,000	ft-lbf/min	m/s^2	0.1019	gravities
hp	550	ft-lbf/sec	mi	1.609	km
hp	0.7457	kW	mi/hr	1.6093	km/h
in	2.54	cm	N	0.22481	lbf
in	25.4	mm	Pa	1.4504×10^{-4}	lbf/in^2
in^3	16.387	cm^3	slug	32.2	lbm
J	0.73756	ft-lbf	W	3.413	Btu/hr

APPENDIX B
Definitions

Base: The level at which the earthquake motions are imparted to the structure.

Bearing wall system: A structural system without a complete vertical load-carrying frame. In this system, the lateral forces are resisted by shear walls or braced frames.

Braced frame: A vertical truss system that is provided to resist lateral forces and in which the members are subjected primarily to axial stresses.

Chord: A horizontal member that resists tension due to bending of a diaphragm.

Collector: A horizontal member that carries force from a diaphragm to a frame or wall providing lateral resistance.

Confined concrete: Concrete confined by closely-spaced ties restraining it in a direction perpendicular to the applied stress.

Critical damping: The amount of damping that results in the system recovering from an initial deflection in the minimum amount of time, without an amplitude reversal.

Damping: The characteristic that reduces the vibrational energy, primarily by friction.

Dip: The angle that a stratum or fault makes with the horizontal.

Dip slip: The component of the slip parallel with the dip of the fault.

Doubler plate: A steel plate added to the side of a column to provide additional strength to resist joint shear. It is often required for special steel moment-resisting frames.

Ductile moment-resisting frame: A frame with rigid connections between columns and girders that is ductile at potential yielding points. See also *Moment-resisting frame.*

Edge nailing: Closely spaced nails along the edges of plywood or structural sheathing to develop shear strength for diaphragms or shear walls.

Elastic rebound theory: A seismic theory, based on the tectonic plate concept, that proposes that stresses are created in fault lines by shifting of the tectonic plates, and that faults resist motion until the accumulated stress overcomes the internal friction.

Equivalent static load: A single horizontal load for which an earthquake-resistant building should be designed.

Essential facility: A facility that must remain functional after a major earthquake.

Fault: A fracture or fracture zone along which the sides can move relative to one another and parallel to the fracture.

Fault creep: Continuous displacement along a fault at a slow and varying rate that is usually not accompanied by noticeable earthquakes.

Fault displacement: Relative movement of the two sides of a fault, measured in any specified direction (usually parallel to the fault).

Fault gouge: Filler material that forms between two plates sliding against each other.

Fault sag: A narrow tectonic (generally earth-filled) depression common in strike-slip fault zones, less than a few hundred feet wide and approximately parallel to the fault zone. See also *Sag pond.*

Fault scarp: A cliff or steep slope formed by displacement of the ground surface.

Field nailing: Nails that are not along the edges of plywood or structural walls. These nails do not require the short spacing.

Fracture: A general term for a break, joint, or fault in the earth.

Frame: A two-dimensional structural system without bearing walls that is composed of interconnected laterally-supported members and that functions as a self-contained unit.

Gouge: See *Fault gouge.*

Graben (plural, graben): A fault block, generally long and narrow, that has dropped down relative to the adjacent blocks.

Holdown: A metal fastener connecting the bottom of a wood shear wall to the concrete foundation. The holdown resists uplift due to lateral loads on a structure. Usually a proprietary product is used.

Hoop: A one-piece closed tie or continuously wound tie that encloses longitudinal reinforcement.

Hypocenter: The actual location of the earthquake beneath the earth's surface.

Igneous rock: Rocks formed by the solidification of molten magma.

Intensity: A measurement of an earthquake related to the level of damage and effect on humans. The most common intensity rating system is the Modified Mercalli Intensity (MMI).

Isoseismal: Contours showing the location of equivalent levels of seismic intensity after an earthquake.

Joint shear: In a moment resisting frame, this term refers to the reverse shear force developed in the joint defined by the intersection of the beam and column. Verification of the adequacy of joint shear is often required in both concrete and steel moment frames.

Lateral force-resisting system: The part of the building that resists earthquake and wind forces.

Left-normal slip: Fault displacement consisting of nearly equal components of left and normal slips.

Left slip: Strike-slip displacement in which the block across the fault moves to the left.

Liquefaction: The loss of load-carrying ability in loose, usually saturated, soil or sand.

Moment-resisting frame: A vertical load-carrying frame in which the members and joints are capable of resisting forces primarily by flexure.

Normal fault: Any fault (including those with vertical slip) in which the block above an inclined surface moves downward relative to the block below the fault surface.

Normal slip: Vertical displacement of a fault.

Oblique slip: A combination of strike slip and reverse slip.

(continued)

APPENDIX B *(continued)*
Definitions

Panel zone: In a steel moment resisting frame, the panel zone is the intersection of the frame beams and columns.

Parapet: A low wall or railing (including decorative panels) extending, usually vertically, above the roof line.

Plastic hinge: A region where the yield moment strength of a flexural member is exceeded and that experiences significant rotation.

Reserve energy: Energy that a ductile system is capable of absorbing in the inelastic region.

Resonance: A condition existing when the frequency of excitation is the same as the natural frequency of the building or soil.

Reverse fault: A fault in which the block above an inclined fault surface moves upward relative to the block below the fault surface.

Right-normal slip: Fault displacement consisting of nearly equal components of right and normal slips.

Right slip: Strike-slip displacement in which the block across the fault from an observer moves to the right.

Rigid frame: A vertical load-carrying frame in which the members and joints resist forces by rotation and flexure. See also *Moment-resisting space frame.*

Sag pond: A fault sag that has filled with water.

Seismic hook: Shear reinforcing steel for concrete beams of columns with a 135° bend and a tail length at least six times the bar diameter.

Shear wall: A wall designed to resist lateral forces parallel to the plane of the wall.

Sill plate: The horizontal framing member at the base of a wood shear wall.

Slip: The relative displacement, measured on the surface, of two points on opposite sides of a fault.

Space frame: A three-dimensional structural system without bearing walls that is composed of interconnected laterally supported members and that functions as a self-contained unit.

Special ductile frame: A structural frame designed to remain vertically functional after the formation of plastic hinges from reversed lateral displacements.

Special inspection: A requirement for various inspection work to be conducted during construction and installation. Often required when higher stresses are allowed on members, particularly with masonry. (See IBC Sec. 1704.)

Special shear wall: A reinforced concrete shear wall designed and detailed in accordance with the special IBC provisions.

Stirrup tie: A closed stirrup completely encircling the longitudinal members of a beam or column and conforming to the definition of a hoop.

Strike: The horizontal direction or bearing of the fault on the surface.

Strike-slip: The horizontal component of slip, parallel to the strike of the fault.

Strike-slip fault: A fault in which the slip is approximately in the direction of the strike.

Structural observation: A level of inspection requiring an engineer to observe various aspects of the construction work. (See IBC Sec. 1704.)

Supplementary crosstie: A tie with a standard 180° hook at each end.

Tectonic: Pertaining to or designating the internal and external rock structures and features caused from crustal and subcrustal activity deep in the earth.

Tectonic creep: Fault creep of tectonic origin.

Transcurrent fault: See *Strike-slip fault.*

Wrench fault: See *Strike-slip fault.*

APPENDIX C
Chronology of Important California Earthquakes

date	fault	Richter magnitude	surface effects and significance
1836	Hayward	7.0 (est.)	Ground breakage
1838	San Andreas	7.0 (est.)	Ground breakage
1852	Big Pine		Possible ground breakage
1857	San Andreas	8.0 (est.)	Right-lateral slip, possibly as much as 30 ft (914 cm)
1861	Calaveras		Ground breakage
1868	San Andreas		Long fissure in earth at Dos Palmas
1868	Hayward	7.0 (est.)	Strike-slip
1872	Owens Valley zone	8.3 (est.)	Right-lateral slip of 16–20 ft (488–610 cm). Left-lateral slip may also have occurred. Vertical slip, down to east, of 23 ft (701 cm).
1890	San Andreas		Fissures in fault zone; railroad tracks and bridge displaced
1899	San Jacinto	6.6 (est.)	Possible surface evidence
1906	San Andreas	7.7–8.3 (est.)	Known as the "San Francisco earthquake." Right-lateral slip up to 21 ft (640 cm). Resulted in the formation of the California State Earthquake Investigation Commission.
1922	San Andreas	6.5	Ground breakage
1925	Mesa/Santa Ynez	6.3	Known as the "Santa Barbara earthquake of 1925." U.S. Coast and Geodetic Society was directed to study the field of seismology.
1927	Santa Ynez	7.5	Occurred offshore in a submarine trench and was felt on land
1933	Newport-Inglewood	6.3	Known as the "Long Beach earthquake of 1933." Extensive property damage and loss of life. Many school buildings were destroyed. Resulted in the passage of the Field Act. The Division of Architecture of the State Department of Public Works was assigned responsibility to approve new buildings used for schools. The Riley Act was also passed, which set minimum requirements for lateral force design.
1934	San Andreas	6.0	Ground breakage
1934	San Jacinto	7.1	Ground breakage
1940	Imperial	7.1	Known as the "El Centro earthquake." 40 mi (64.4 km) of surface faulting. 80% of Imperial buildings were damaged. However, no Field Act school buildings were damaged. This was the first major earthquake to yield accelerograph data on building periods. A maximum acceleration (ground) of 0.33 g was experienced.
1947	Manix	6.4	Left-lateral slip of 3 in (76 mm)
1950	(unnamed)	5.6	Vertical slip, down to west, of 5–8 in (127–203 mm) along the west edge of Fort Sage Mountains
1951	Superstition Hills	5.6	Slight right-lateral slip
1952	White Wolf	7.7	Known as the "Kern County earthquake" and the "Arvin-Tehachapi earthquake." Extensive building damage to old buildings. Little or none to properly designed and Field Act buildings. Confirmed the requirement for proper design.
1956	San Miguel	6.8	Right-lateral slip, 3 ft (91 cm); vertical slip, down to southwest, 3 ft (91 cm)
1966	Imperial	3.6	Right-lateral slip, 1/2 in (13 mm)
1966	San Andreas	5.5	Known as the "Parkfield earthquake." Right-lateral slip of several inches. The maximum ground acceleration of 0.50 g was the highest recorded to date.
1968	Coyote Creek	6.4	Right-lateral slip up to 15 in (381 mm)
1971	San Fernando	6.4–6.6	Known as the "San Fernando earthquake" or the "Sylmar earthquake." Left-lateral slip up to 5 ft (152 cm); north-side thrusting up 3 ft (91 cm). Massive instrumentation due to 1965 Los Angeles building code resulted in more than 300 accelerograph plots. 1.24 g experienced at Pacoima dam.
1979	Imperial	6.6	Known as the "Imperial Valley earthquake." Right-lateral slip up to 21.6 in (55 cm) with more than 18.6 mi (30 km) of surface rupture. Extensive accelerograph data collected. Resulted in the first accelerograph from an extensively damaged building (Imperial County Services Building).

(continued)

APPENDIX C *(continued)*
Chronology of Important California Earthquakes

date	fault	Richter magnitude	surface effects and significance
1987	Whittier	6.1	Known as the "Whittier earthquake." Epicenter 10 mi (16 km) east of downtown Los Angeles. 0.45 g maximum lateral acceleration; 0.20 g vertical acceleration typical. Strong shaking duration of 4 sec. Six fatalities; unreinforced masonry structures damaged significantly.
1989	Loma Prieta	7.1	Primarily noted as causing the collapse of the Oakland Interstate 880 Cypress structure and homes in the San Francisco Marina district, both due to soil amplification effects despite the large distance from the epicenter. Ground breakage at epicenter located in Santa Cruz Mountains; 62 fatalities.
1990	Upland	5.5	Acceleration of 0.23 g horizontally and 0.13 g vertically. Dozens of aftershocks. Occurred on the San Antonio Canyon fault, east of downtown Los Angeles. Most damage in Pomona; some damage to reinforced masonry structures.
1992	Yucca Valley	7.4	One fatality, hundreds of injuries. Buckled and displaced roads, damaged 1400 structures. 43 mi (69.2 km) ground rupture. Followed by magnitude 6.5 quake at Big Bear resort area.
1992	Cape Mendocino	7.0	Offshore quake notable for largest-yet recorded ground acceleration of 1.85 g.
1994	Northridge	6.6	Previously unknown thrust fault. 40 sec of shaking. Repeated ground acceleration of over 1 g for 7 sec to 8 sec. Extensive damage to sections of the Santa Monica freeway and freeway overpasses not yet retrofitted. Unreinforced masonry buildings damaged, as expected. 60 fatalities, thousands of injuries. Numerous failures in steel moment-resisting connections.
1999	Lavic Lake (central part of Bullion fault)	7.1	Known as the "Bullion Mountains earthquake" and the "Hector Mine earthquake." Minimal injuries and damage due to sparse population in affected area.
2000	West Napa	5.2	Known as the "Yountville earthquake" and the "Napa earthquake." Source was 3 mi west of the West Napa fault on an unknown northwest-oriented, right-lateral stroke-slip fault. Ground acceleration approaching 0.5 g horizontally recorded in town was higher than expected from this magnitude and is attributed to amplification by young sediments along the Napa river.
2003	Oceanic fault zone (west of San Andreas fault)	6.5	Known as the "San Simeon earthquake." Two people killed and 40 injured in Paso Robles-Templeton area. At least 40 buildings collapsed or were severely damaged in Paso Robles. The San Simeon earthquake was documented by a large group of scientists and engineers. There was a surprising amount of liquefaction and related ground damage at relatively low levels of ground shaking away from the epicenter.
2004	San Andreas	6.0	Known as the "Parkfield-San Bernardino earthquake."
2005	Gorda Plate	7.2 (June 15); 6.6 (June 17)	Known as the "Gorda Plate earthquakes." Two large earthquakes off the coast of northern California near the border with Oregon. Only minor shaking noticed. Small wave motions (26 cm peak-to-trough) noticed at Crescent City.
2007	Calaveras (east of San Andreas fault)	5.6	Known as the "Alum Rock earthquake." Centered east of San Jose. No surface rupture. Strong shaking noted by over 60,000 reports.
2008	Yorba Linda Trend	5.5	Known as the "2008 Chino Hills earthquake." Epicenter 3 miles from the center of Chino Hills, but little damage was caused. No increase in seismic activity had preceded the earthquake.
2010	San Andreas (north of Mendocino Triple Junction)	6.5	Known as the "2010 Eureka earthquake." Offshore fault. Several buildings in Eureka were destroyed beyond repair. Largest earthquake to hit Eureka since 1999 Hector Mine earthquake. No tsunami.
2010	Laguna Salada	7.2	Known as the "El Mayor-Cucapah earthquake" or the "2010 Easter earthquake." Centered on Mexicali and Calexico in Northern Baja California. 4 fatalities; widespread damage, power outages, and broken water and gas mains in Southern California.
2012	Near Catalina Island	6.3	Known as the "2012 Avalon earthquake." Caused by Pacific and North American intraplate faulting, not associated with the San Andreas fault. No injuries or damage.

(continued)

APPENDIX C *(continued)*
Chronology of Important California Earthquakes

date	fault	Richter magnitude	surface effects and significance
2014	San Andreas	6.8	Offshore, 50 miles west of Eureka (Arcata, Ferndale). San Andreas Fault near Mendocino Triple Junction, where the Pacific, North American, and Gordo plates meet. No injuries or damage.
2014	West Napa (previously unmapped)	6.0	Known as the "South Napa earthquake." Damage localized to the Napa Valley area. Unanchored structures shifted off foundations; significant damage to city structures with masonry facades, including recently retrofitted buildings. No fatalities.

APPENDIX D
Rigidity of Fixed Piers

Calculated with $F = 100{,}000$ lbf (445 000 N), $t = 1.0$ in (25 mm), and $E = 100{,}000$ psi (6.9×10^6 kPa).

(*h*/*d*)	0.00	0.01	0.02	0.03	0.04	0.05	0.06	0.07	0.08	0.09
0.10	33.223	30.181	27.645	25.497	23.655	22.057	20.657	19.421	18.321	17.335
0.20	16.447	15.643	14.911	14.242	13.267	13.061	12.538	12.053	11.602	11.181
0.30	10.788	10.419	10.073	9.747	9.440	9.150	8.876	8.616	8.369	8.135
0.40	7.911	7.699	7.496	7.302	7.117	6.939	6.769	6.606	6.449	6.299
0.50	6.154	6.015	5.880	5.751	5.626	5.506	5.389	5.277	5.168	5.062
0.60	4.960	4.862	4.766	4.673	4.583	4.495	4.410	4.328	4.247	4.169
0.70	4.093	4.019	3.948	3.877	3.809	3.743	3.678	3.615	3.553	3.493
0.80	3.434	3.377	3.321	3.266	3.213	3.160	3.109	3.060	3.011	2.963
0.90	2.916	2.871	2.826	2.782	2.739	2.697	2.656	2.616	2.577	2.538
1.00	2.500	2.463	2.427	2.391	2.356	2.322	2.288	2.255	2.222	2.191
1.10	2.159	2.129	2.099	2.069	2.040	2.012	1.984	1.956	1.929	1.903
1.20	1.877	1.851	1.826	1.802	1.777	1.753	1.730	1.707	1.684	1.662
1.30	1.640	1.619	1.598	1.577	1.556	1.536	1.516	1.497	1.478	1.459
1.40	1.440	1.422	1.404	1.386	1.369	1.352	1.335	1.318	1.302	1.286
1.50	1.270	1.254	1.239	1.224	1.209	1.194	1.180	1.166	1.152	1.138
1.60	1.124	1.111	1.098	1.085	1.072	1.059	1.047	1.034	1.022	1.010
1.70	0.999	0.987	0.976	0.965	0.954	0.943	0.932	0.921	0.911	0.901
1.80	0.890	0.880	0.870	0.861	0.851	0.842	0.832	0.823	0.814	0.805
1.90	0.796	0.788	0.779	0.771	0.762	0.754	0.746	0.738	0.730	0.722
2.00	0.714	0.707	0.699	0.692	0.685	0.677	0.670	0.663	0.656	0.649
2.10	0.643	0.636	0.629	0.623	0.617	0.610	0.604	0.598	0.592	0.586
2.20	0.580	0.574	0.568	0.562	0.557	0.551	0.546	0.540	0.535	0.530
2.30	0.525	0.519	0.514	0.509	0.504	0.499	0.495	0.490	0.485	0.480
2.40	0.476	0.471	0.467	0.462	0.458	0.453	0.449	0.445	0.441	0.437
2.50	0.432	0.428	0.424	0.420	0.417	0.413	0.409	0.405	0.401	0.398
2.60	0.394	0.391	0.387	0.383	0.380	0.377	0.373	0.370	0.367	0.363
2.70	0.360	0.357	0.354	0.350	0.347	0.344	0.341	0.338	0.335	0.332
2.80	0.330	0.327	0.324	0.321	0.318	0.316	0.313	0.310	0.307	0.305
2.90	0.302	0.300	0.297	0.295	0.292	0.290	0.287	0.285	0.283	0.280
3.00	0.278	0.276	0.273	0.271	0.269	0.267	0.264	0.262	0.260	0.258
3.10	0.256	0.254	0.252	0.250	0.248	0.246	0.244	0.242	0.240	0.238
3.20	0.236	0.234	0.232	0.231	0.229	0.227	0.225	0.223	0.222	0.220
3.30	0.218	0.217	0.215	0.213	0.212	0.210	0.208	0.207	0.205	0.204
3.40	0.202	0.201	0.199	0.198	0.196	0.195	0.193	0.192	0.190	0.189
3.50	0.187	0.186	0.185	0.183	0.182	0.181	0.179	0.178	0.177	0.175
3.60	0.174	0.173	0.172	0.170	0.169	0.168	0.167	0.166	0.164	0.163
3.70	0.162	0.161	0.160	0.159	0.157	0.156	0.155	0.154	0.153	0.152
3.80	0.151	0.150	0.149	0.148	0.147	0.146	0.145	0.144	0.143	0.142
3.90	0.141	0.140	0.139	0.138	0.137	0.136	0.135	0.134	0.133	0.132
4.00	0.132	0.131	0.130	0.129	0.128	0.127	0.126	0.126	0.125	0.124
4.10	0.123	0.122	0.122	0.121	0.120	0.119	0.118	0.118	0.117	0.116
4.20	0.115	0.115	0.114	0.113	0.112	0.112	0.111	0.110	0.110	0.109
4.30	0.108	0.108	0.107	0.106	0.106	0.105	0.104	0.104	0.103	0.102
4.40	0.102	0.101	0.100	0.100	0.099	0.099	0.098	0.097	0.097	0.096
4.50	0.096	0.095	0.094	0.094	0.093	0.093	0.092	0.092	0.091	0.091
4.60	0.090	0.089	0.089	0.088	0.088	0.087	0.087	0.086	0.086	0.085
4.70	0.085	0.084	0.084	0.083	0.083	0.082	0.082	0.081	0.081	0.080
4.80	0.080	0.080	0.079	0.079	0.078	0.078	0.077	0.077	0.076	0.076
4.90	0.076	0.075	0.075	0.074	0.074	0.073	0.073	0.073	0.072	0.072
5.00	0.071	0.071	0.071	0.070	0.070	0.069	0.069	0.069	0.068	0.068
5.10	0.068	0.067	0.067	0.066	0.066	0.066	0.065	0.065	0.065	0.064
5.20	0.064	0.064	0.063	0.063	0.063	0.062	0.062	0.062	0.061	0.061
5.30	0.061	0.060	0.060	0.060	0.059	0.059	0.059	0.058	0.058	0.058
5.40	0.058	0.057	0.057	0.057	0.056	0.056	0.056	0.056	0.055	0.055
5.50	0.055	0.054	0.054	0.054	0.054	0.053	0.053	0.053	0.052	0.052
5.60	0.052	0.052	0.051	0.051	0.051	0.051	0.050	0.050	0.050	0.050
5.70	0.049	0.049	0.049	0.049	0.048	0.048	0.048	0.048	0.048	0.047
5.80	0.047	0.047	0.047	0.046	0.046	0.046	0.046	0.045	0.045	0.045
5.90	0.045	0.045	0.044	0.044	0.044	0.044	0.044	0.043	0.043	0.043
6.00	0.043	0.043	0.042	0.042	0.042	0.042	0.042	0.041	0.041	0.041
6.10	0.041	0.041	0.040	0.040	0.040	0.040	0.040	0.039	0.039	0.039
6.20	0.039	0.039	0.039	0.038	0.038	0.038	0.038	0.038	0.038	0.037
6.30	0.037	0.037	0.037	0.037	0.037	0.036	0.036	0.036	0.036	0.036
6.40	0.036	0.035	0.035	0.035	0.035	0.035	0.035	0.034	0.034	0.034
6.50	0.034	0.034	0.034	0.034	0.033	0.033	0.033	0.033	0.033	0.033
6.60	0.033	0.032	0.032	0.032	0.032	0.032	0.032	0.032	0.031	0.031
6.70	0.031	0.031	0.031	0.031	0.031	0.031	0.030	0.030	0.030	0.030
6.80	0.030	0.030	0.030	0.029	0.029	0.029	0.029	0.029	0.029	0.029
6.90	0.029	0.029	0.028	0.028	0.028	0.028	0.028	0.028	0.028	0.028
7.00	0.027	0.027	0.027	0.027	0.027	0.027	0.027	0.027	0.027	0.026
7.10	0.026	0.026	0.026	0.026	0.026	0.026	0.026	0.026	0.026	0.025
7.20	0.025	0.025	0.025	0.025	0.025	0.025	0.025	0.025	0.025	0.024
7.30	0.024	0.024	0.024	0.024	0.024	0.024	0.024	0.024	0.024	0.023
7.40	0.023	0.023	0.023	0.023	0.023	0.023	0.023	0.023	0.023	0.023
7.50	0.023	0.022	0.022	0.022	0.022	0.022	0.022	0.022	0.022	0.022
7.60	0.022	0.022	0.021	0.021	0.021	0.021	0.021	0.021	0.021	0.021
7.70	0.021	0.021	0.021	0.021	0.021	0.020	0.020	0.020	0.020	0.020
7.80	0.020	0.020	0.020	0.020	0.020	0.020	0.020	0.020	0.019	0.019
7.90	0.019	0.019	0.019	0.019	0.019	0.019	0.019	0.019	0.019	0.019
8.00	0.019	0.019	0.019	0.018	0.018	0.018	0.018	0.018	0.018	0.018
8.10	0.018	0.018	0.018	0.018	0.018	0.018	0.018	0.018	0.017	0.017
8.20	0.017	0.017	0.017	0.017	0.017	0.017	0.017	0.017	0.017	0.017

APPENDIX E
Rigidity of Cantilever Piers

Calculated with $F = 100{,}000$ lbf (445 000 N), $t = 1.0$ in (25 mm), and $E = 100{,}000$ psi (6.9×10^6 kPa).

(h/d)	0.00	0.01	0.02	0.03	0.04	0.05	0.06	0.07	0.08	0.09
0.10	32.895	29.822	27.255	25.076	23.203	21.575	20.146	18.880	17.752	16.738
0.20	15.823	14.992	14.233	13.538	12.898	12.308	11.761	11.252	10.778	10.335
0.30	9.921	9.531	9.165	8.820	8.495	8.187	7.895	7.618	7.356	7.106
0.40	6.868	6.642	6.425	6.219	6.021	5.833	5.652	5.479	5.313	5.153
0.50	5.000	4.853	4.712	4.576	4.445	4.319	4.197	4.080	3.968	3.859
0.60	3.754	3.652	3.555	3.460	3.369	3.280	3.195	3.112	3.032	2.955
0.70	2.880	2.808	2.738	2.670	2.604	2.540	2.478	2.418	2.360	2.303
0.80	2.248	2.195	2.143	2.093	2.045	1.997	1.952	1.907	1.864	1.822
0.90	1.781	1.741	1.702	1.665	1.628	1.593	1.558	1.524	1.492	1.460
1.00	1.429	1.398	1.369	1.340	1.312	1.285	1.259	1.233	1.208	1.183
1.10	1.160	1.136	1.114	1.092	1.070	1.049	1.028	1.008	0.989	0.970
1.20	0.951	0.933	0.916	0.898	0.881	0.865	0.849	0.833	0.818	0.803
1.30	0.788	0.774	0.760	0.746	0.733	0.720	0.707	0.695	0.683	0.671
1.40	0.659	0.648	0.636	0.626	0.615	0.604	0.594	0.584	0.575	0.565
1.50	0.556	0.546	0.537	0.529	0.520	0.512	0.503	0.495	0.487	0.480
1.60	0.472	0.465	0.457	0.450	0.443	0.436	0.430	0.423	0.417	0.410
1.70	0.404	0.398	0.392	0.386	0.380	0.375	0.369	0.364	0.358	0.353
1.80	0.348	0.343	0.338	0.333	0.329	0.324	0.319	0.315	0.310	0.306
1.90	0.302	0.298	0.294	0.290	0.286	0.282	0.278	0.274	0.270	0.267
2.00	0.263	0.260	0.256	0.253	0.250	0.246	0.243	0.240	0.237	0.234
2.10	0.231	0.228	0.225	0.222	0.219	0.216	0.214	0.211	0.208	0.206
2.20	0.203	0.201	0.198	0.196	0.194	0.191	0.189	0.187	0.184	0.182
2.30	0.180	0.178	0.176	0.174	0.172	0.170	0.168	0.166	0.164	0.162
2.40	0.160	0.158	0.156	0.155	0.153	0.151	0.149	0.148	0.146	0.145
2.50	0.143	0.141	0.140	0.138	0.137	0.135	0.134	0.132	0.131	0.129
2.60	0.128	0.127	0.125	0.124	0.123	0.121	0.120	0.119	0.118	0.116
2.70	0.115	0.114	0.113	0.112	0.111	0.109	0.108	0.107	0.106	0.105
2.80	0.104	0.103	0.102	0.101	0.100	0.099	0.098	0.097	0.096	0.095
2.90	0.094	0.093	0.092	0.091	0.091	0.090	0.089	0.088	0.087	0.086
3.00	0.086	0.085	0.084	0.083	0.082	0.082	0.081	0.080	0.079	0.079
3.10	0.078	0.077	0.076	0.076	0.075	0.074	0.074	0.073	0.072	0.072
3.20	0.071	0.071	0.070	0.069	0.069	0.068	0.067	0.067	0.066	0.066
3.30	0.065	0.065	0.064	0.063	0.063	0.062	0.062	0.061	0.061	0.060
3.40	0.060	0.059	0.059	0.058	0.058	0.057	0.057	0.056	0.056	0.055
3.50	0.055	0.055	0.054	0.054	0.053	0.053	0.052	0.052	0.052	0.051
3.60	0.051	0.050	0.050	0.050	0.049	0.049	0.048	0.048	0.048	0.047
3.70	0.047	0.046	0.046	0.046	0.045	0.045	0.045	0.044	0.044	0.044
3.80	0.043	0.043	0.043	0.042	0.042	0.042	0.041	0.041	0.041	0.040
3.90	0.040	0.040	0.040	0.039	0.039	0.039	0.038	0.038	0.038	0.038
4.00	0.037	0.037	0.037	0.037	0.036	0.036	0.036	0.035	0.035	0.035
4.10	0.035	0.034	0.034	0.034	0.034	0.034	0.033	0.033	0.033	0.033
4.20	0.032	0.032	0.032	0.032	0.031	0.031	0.031	0.031	0.031	0.030
4.30	0.030	0.030	0.030	0.030	0.029	0.029	0.029	0.029	0.029	0.028
4.40	0.028	0.028	0.028	0.028	0.028	0.027	0.027	0.027	0.027	0.027
4.50	0.026	0.026	0.026	0.026	0.026	0.026	0.025	0.025	0.025	0.025
4.60	0.025	0.025	0.024	0.024	0.024	0.024	0.024	0.024	0.024	0.023
4.70	0.023	0.023	0.023	0.023	0.023	0.023	0.022	0.022	0.022	0.022
4.80	0.022	0.022	0.022	0.021	0.021	0.021	0.021	0.021	0.021	0.021
4.90	0.021	0.020	0.020	0.020	0.020	0.020	0.020	0.020	0.020	0.020
5.00	0.019	0.019	0.019	0.019	0.019	0.019	0.019	0.019	0.019	0.018
5.10	0.018	0.018	0.018	0.018	0.018	0.018	0.018	0.018	0.017	0.017
5.20	0.017	0.017	0.017	0.017	0.017	0.017	0.017	0.017	0.017	0.016
5.30	0.016	0.016	0.016	0.016	0.016	0.016	0.016	0.016	0.016	0.016
5.40	0.015	0.015	0.015	0.015	0.015	0.015	0.015	0.015	0.015	0.015
5.50	0.015	0.015	0.015	0.014	0.014	0.014	0.014	0.014	0.014	0.014
5.60	0.014	0.014	0.014	0.014	0.014	0.014	0.013	0.013	0.013	0.013
5.70	0.013	0.013	0.013	0.013	0.013	0.013	0.013	0.013	0.013	0.013
5.80	0.013	0.012	0.012	0.012	0.012	0.012	0.012	0.012	0.012	0.012
5.90	0.012	0.012	0.012	0.012	0.012	0.012	0.012	0.012	0.011	0.011
6.00	0.011	0.011	0.011	0.011	0.011	0.011	0.011	0.011	0.011	0.011
6.10	0.011	0.011	0.011	0.011	0.011	0.011	0.010	0.010	0.010	0.010
6.20	0.010	0.010	0.010	0.010	0.010	0.010	0.010	0.010	0.010	0.010
6.30	0.010	0.010	0.010	0.010	0.010	0.010	0.010	0.009	0.009	0.009
6.40	0.009	0.009	0.009	0.009	0.009	0.009	0.009	0.009	0.009	0.009
6.50	0.009	0.009	0.009	0.009	0.009	0.009	0.009	0.009	0.009	0.009
6.60	0.009	0.009	0.008	0.008	0.008	0.008	0.008	0.008	0.008	0.008
6.70	0.008	0.008	0.008	0.008	0.008	0.008	0.008	0.008	0.008	0.008
6.80	0.008	0.008	0.008	0.008	0.008	0.008	0.008	0.008	0.008	0.008
6.90	0.007	0.007	0.007	0.007	0.007	0.007	0.007	0.007	0.007	0.007
7.00	0.007	0.007	0.007	0.007	0.007	0.007	0.007	0.007	0.007	0.007
7.10	0.007	0.007	0.007	0.007	0.007	0.007	0.007	0.007	0.007	0.007
7.20	0.007	0.007	0.007	0.007	0.006	0.006	0.006	0.006	0.006	0.006
7.30	0.006	0.006	0.006	0.006	0.006	0.006	0.006	0.006	0.006	0.006
7.40	0.006	0.006	0.006	0.006	0.006	0.006	0.006	0.006	0.006	0.006
7.50	0.006	0.006	0.006	0.006	0.006	0.006	0.006	0.006	0.006	0.006
7.60	0.006	0.006	0.006	0.006	0.006	0.006	0.005	0.005	0.005	0.005
7.70	0.005	0.005	0.005	0.005	0.005	0.005	0.005	0.005	0.005	0.005
7.80	0.005	0.005	0.005	0.005	0.005	0.005	0.005	0.005	0.005	0.005
7.90	0.005	0.005	0.005	0.005	0.005	0.005	0.005	0.005	0.005	0.005
8.00	0.005	0.005	0.005	0.005	0.005	0.005	0.005	0.005	0.005	0.005
8.10	0.005	0.005	0.005	0.005	0.005	0.005	0.005	0.005	0.005	0.005
8.20	0.004	0.004	0.004	0.004	0.004	0.004	0.004	0.004	0.004	0.004

APPENDIX F
Accelerogram of 1940 El Centro Earthquake
(North-South Component)

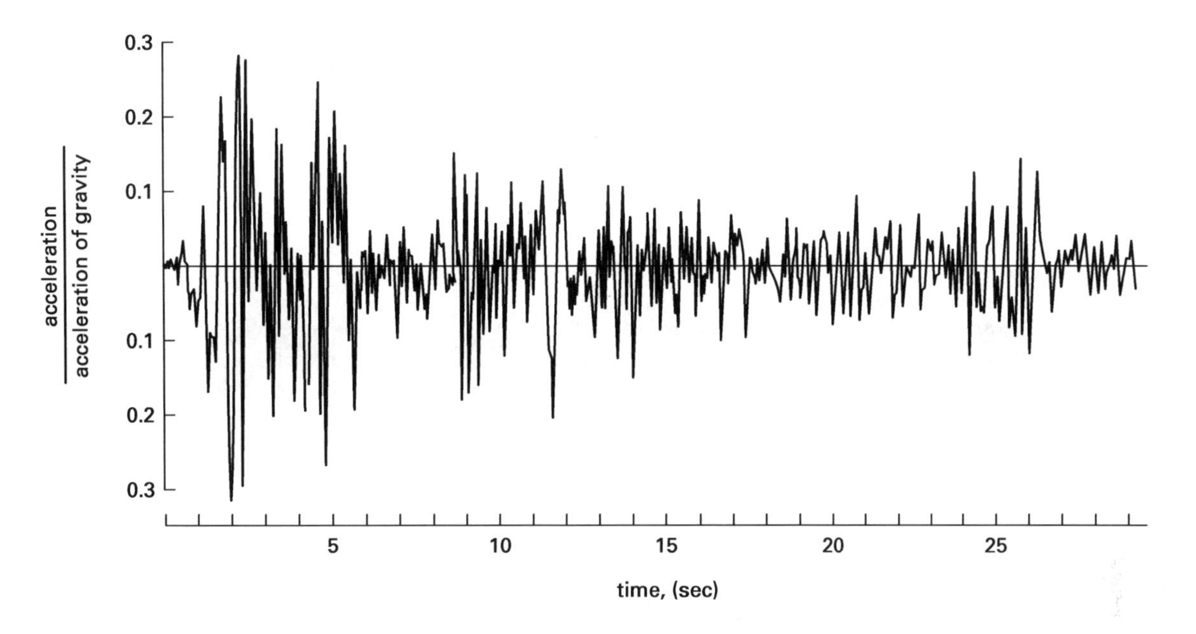

Source: Donald E. Hudson, "Ground Motion Measurements," in *Earthquake Engineering*, Robert L. Wiegel, ed., © 1970, p. 113.

APPENDIX G
Seismic Acronyms and Abbreviations

AASHTO American Association of State Highway Transportation Officials
ACI American Concrete Institute
ADAS added damping and stiffness (element)
AFM Air Force Manual (military)
AISC American Institute of Steel Construction
AISI American Iron and Steel Institute
ANSI American National Standards Institute
API American Petroleum Institute
ARS Acceleration Response Spectrum
ASCE American Society of Civil Engineers
ASD allowable stress design
ASTM American Society for Testing and Materials
AITC American Institute of Timber Construction
ATC Applied Technology Council
AWWA American Water Works Association
AZG acceleration zone graph
BOCA Building Officials and Code Administrators International
BRBF buckling resistant braced frame
CALTRANS California Department of Transportation
CIP cast-in-place
CMU concrete masonry unit
CQC complete quadratic combination
DAF dynamic amplification factor
DM design manual (military)
EBF eccentrically braced frame
EERI Earthquake Engineering Research Institute
EPA effective peak ground acceleration
EPV effective peak ground velocity
EQ earthquake
EUS eastern United States
FEA finite element analysis
FEMA Federal Emergency Management Agency
FP full-penetration (welding)
HVAC heating ventilating and air conditioning
IBC International Building Code
IC important classification
ICBO International Conference of Building Officials
ICC International Code Council
IDRS inelastic design response spectra
IEEE Institute of Electrical and Electronic Engineers
IMRF intermediate moment-resisting frame
IMRSF intermediate moment-resisting space frame (obsolete)
LRFD load resistance factor design
MDOF multiple degree of freedom
MM modified Mercalli (intensity)
MMRWF masonry moment-resisting wall frame
NAVFAC Naval Facilities Engineering Command
NBC National Building Code
NCMA National Concrete Masonry Association
NDS National Design Specification
NEHRP National Earthquake Hazards Reduction Program
NRC Nuclear Regulatory Commission
NSF National Science Foundation
OMF ordinary moment frame
OMRF ordinary moment-resisting frame
OMRSF ordinary moment-resisting space frame (obsolete)
OTM overturning moment
PCA Portland Cement Association
PGA peak ground acceleration
PGD peak ground displacement
PGV peak ground velocity
SBC Standard Building Code
SBCCI Southern Building Code Congress International
SCBF special concentrically braced frames
SD strength design
SDOF single degree of freedom
SEAOC Structural Engineers Association of California
SEAOCC Structural Engineers Association of Central California
SEAONC Structural Engineers Association of Northern California
SEAOSC Structural Engineers Association of Southern California
SH shear horizontal
SMRF special moment-resisting frame
SMRSF special moment-resisting space frame (obsolete)
SPC seismic performance category
SRSS square root of the sum of the squares
SSRC Structural Stability Research Council
STMF special truss moment frames
SV shear vertical
T&B top and bottom (welding)
TM training manual (military)
TMD tuned mass damper
TMS The Masonry Society
UBC Uniform Building Code
USGS United States Geological Society
VLLR vertical and lateral load-resisting elements
WUS western United States
YDS yield displacement spectrum

APPENDIX H
Design Coefficients and Factors
for Seismic Force-Resisting Systems
[ASCE/SEI7 Table 12.2-1]

seismic force-resisting system	ASCE 7 section where detailing requirements are specified	response modification coefficient, R[a]	overstrength factor, Ω_O[g]	deflection amplification factor, C_d[b]	structural system limitations including structural height, h_n (ft) limits[c]				
					seismic design category				
					B	C	D[d]	E[d]	F[e]
A. bearing wall systems									
1. special reinforced concrete shear walls[l,m]	14.2	5	2½	5	NL	NL	160	160	100
2. ordinary reinforced concrete shear walls[l]	14.2	4	2½	4	NL	NL	NP	NP	NP
3. detailed plain concrete shear walls[l]	14.2	2	2½	2	NL	NP	NP	NP	NP
4. ordinary plain concrete shear walls[l]	14.2	1½	2½	1½	NL	NP	NP	NP	NP
5. intermediate precast shear walls[l]	14.2	4	2½	4	NL	NL	40[k]	40[k]	40[k]
6. ordinary precast shear walls[l]	14.2	3	2½	3	NL	NP	NP	NP	NP
7. special reinforced masonry shear walls	14.4	5	2½	3½	NL	NL	160	160	100
8. intermediate reinforced masonry shear walls	14.4	3½	2½	2¼	NL	NL	NP	NP	NP
9. ordinary reinforced masonry shear walls	14.4	2	2½	1¾	NL	160	NP	NP	NP
10. detailed plain masonry shear walls	14.4	2	2½	1¾	NL	NP	NP	NP	NP
11. ordinary plain masonry shear walls	14.4	1½	2½	1¼	NL	NP	NP	NP	NP
12. prestressed masonry shear walls	14.4	1½	2½	1¾	NL	NP	NP	NP	NP
13. ordinary reinforced AAC masonry shear walls	14.4	2	2½	2	NL	35	NP	NP	NP
14. ordinary plain AAC masonry shear walls	14.4	1½	2½	1½	NL	NP	NP	NP	NP
15. light-frame (wood) walls sheathed with wood structural panels rated for shear resistance	14.5	6½	3	4	NL	NL	65	65	65
16. light-frame (cold-formed steel) walls sheathed with wood structural panels rated for shear resistance or steel sheets	14.1	6½	3	4	NL	NL	65	65	65
17. light-frame walls with shear panels of all other materials	14.1 and 14.5	2	2½	2	NL	NL	35	NP	NP
18. light-frame (cold-formed steel) wall systems using flat strap bracing	14.1	4	2	3½	NL	NL	65	65	65
B. building frame systems									
1. steel eccentrically braced frames	14.1	8	2	4	NL	NL	160	160	100
2. steel special concentrically braced frames	14.1	6	2	5	NL	NL	160	160	100
3. steel ordinary concentrically braced frames	14.1	3¼	2	3¼	NL	NL	35[j]	35[j]	NP[j]
4. special reinforced concrete shear walls[l,m]	14.2	6	2½	5	NL	NL	160	160	100
5. ordinary reinforced concrete shear walls[l]	14.2	5	2½	4½	NL	NL	NP	NP	NP
6. detailed plain concrete shear walls[l]	14.2 and 14.2.2.8	2	2½	2	NL	NP	NP	NP	NP
7. ordinary plain concrete shear walls[l]	14.2	1½	2½	1½	NL	NP	NP	NP	NP
8. intermediate precast shear walls[l]	14.2	5	2½	4½	NL	NL	40[k]	40[k]	40[k]
9. ordinary precast shear walls[l]	14.2	4	2½	4	NL	NP	NP	NP	NP
10. steel and concrete composite eccentrically braced frames	14.3	8	2½	4	NL	NL	160	160	100

(continued)

APPENDIX H *(continued)*
Design Coefficients and Factors
for Seismic Force-Resisting Systems
[ASCE/SEI7 Table 12.2-1]

seismic force-resisting system	ASCE 7 section where detailing requirements are specified	response modification coefficient, R^a	overstrength factor, $\Omega_O{}^g$	deflection amplification factor, $C_d{}^b$	structural system limitations including structural height, h_n (ft) limits[c]				
					seismic design category				
					B	C	D[d]	E[d]	F[e]
11. steel and concrete composite special concentrically braced frames	14.3	5	2	$4\frac{1}{2}$	NL	NL	160	160	100
12. steel and concrete composite ordinary braced frames	14.3	3	2	3	NL	NL	NP	NP	NP
13. steel and concrete composite plate shear walls	14.3	$6\frac{1}{2}$	$2\frac{1}{2}$	$5\frac{1}{2}$	NL	NL	160	160	100
14. steel and concrete composite special shear walls	14.3	6	$2\frac{1}{2}$	5	NL	NL	160	160	100
15. steel and concrete composite ordinary shear walls	14.3	5	$2\frac{1}{2}$	$4\frac{1}{2}$	NL	NL	NP	NP	NP
16. special reinforced masonry shear walls	14.4	$5\frac{1}{2}$	$2\frac{1}{2}$	4	NL	NL	160	160	100
17. intermediate reinforced masonry shear walls	14.4	4	$2\frac{1}{2}$	4	NL	NL	NP	NP	NP
18. ordinary reinforced masonry shear walls	14.4	2	$2\frac{1}{2}$	2	NL	160	NP	NP	NP
19. detailed plain masonry shear walls	14.4	2	$2\frac{1}{2}$	2	NL	NP	NP	NP	NP
20. ordinary plain masonry shear walls	14.4	$1\frac{1}{2}$	$2\frac{1}{2}$	$1\frac{1}{4}$	NL	NP	NP	NP	NP
21. prestressed masonry shear walls	14.4	$1\frac{1}{2}$	$2\frac{1}{2}$	$1\frac{3}{4}$	NL	NP	NP	NP	NP
22. light-frame (wood) walls sheathed with wood structural panels rated for shear resistance	14.5	7	$2\frac{1}{2}$	$4\frac{1}{2}$	NL	NL	65	65	65
23. light-frame (cold-formed steel) walls sheathed with wood structural panels rated for shear resistance or steel sheets	14.1	7	$2\frac{1}{2}$	$4\frac{1}{2}$	NL	NL	65	65	65
24. light-frame walls with shear panels of all other materials	14.1 and 14.5	$2\frac{1}{2}$	$2\frac{1}{2}$	$2\frac{1}{2}$	NL	NL	35	NP	NP
25. steel buckling-restrained braced frames	14.1	8	$2\frac{1}{2}$	5	NL	NL	160	160	100
26. steel special plate shear walls	14.1	7	2	6	NL	NL	160	160	100
C. moment-resisting frame systems									
1. steel special moment frames	14.1 and 12.2.5.5	8	3	$5\frac{1}{2}$	NL	NL	NL	NL	NL
2. steel special truss moment frames	14.1	7	3	$5\frac{1}{2}$	NL	NL	160	100	NP
3. steel intermediate moment frames	12.2.5.7 and 14.1	$4\frac{1}{2}$	3	4	NL	NL	35[h]	NP[h]	NP[h]
4. steel ordinary moment frames	12.2.5.6 and 14.1	$3\frac{1}{2}$	3	3	NL	NL	NP[i]	NP[i]	NP[i]
5. special reinforced concrete moment frames[n]	12.2.5.5 and 14.2	8	3	$5\frac{1}{2}$	NL	NL	NL	NL	NL
6. intermediate reinforced concrete moment frames	14.2	5	3	$4\frac{1}{2}$	NL	NL	NP	NP	NP
7. ordinary reinforced concrete moment frames	14.2	3	3	$2\frac{1}{2}$	NL	NP	NP	NP	NP
8. steel and concrete composite special moment frames	12.2.5.5 and 14.3	8	3	$5\frac{1}{2}$	NL	NL	NL	NL	NL
9. steel and concrete composite intermediate moment frames	14.3	5	3	$4\frac{1}{2}$	NL	NL	NP	NP	NP

(continued)

APPENDIX H *(continued)*
Design Coefficients and Factors
for Seismic Force-Resisting Systems
[ASCE/SEI7 Table 12.2-1]

seismic force-resisting system	ASCE 7 section where detailing requirements are specified	response modification coefficient, R^a	overstrength factor, $\Omega_O{}^g$	deflection amplification factor, $C_d{}^b$	structural system limitations including structural height, h_n (ft) limits[c] — seismic design category B	C	D^d	E^d	F^e
10. steel and concrete composite partially restrained moment frames	14.3	6	3	$5^1/_2$	160	160	100	NP	NP
11. steel and concrete composite ordinary moment frames	14.3	3	3	$2^1/_2$	NL	NP	NP	NP	NP
12. cold-formed steel—special bolted moment frame[p]	14.1	$3^1/_2$	3[o]	$3^1/_2$	35	35	35	35	35
D. dual systems with special moment frames capable of resisting at least 25% of prescribed seismic forces	12.2.5.1								
1. steel eccentrically braced frames	14.1	8	$2^1/_2$	4	NL	NL	NL	NL	NL
2. steel special concentrically braced frames	14.1	7	$2^1/_2$	$5^1/_2$	NL	NL	NL	NL	NL
3. special reinforced concrete shear walls[l,m]	14.2	7	$2^1/_2$	$5^1/_2$	NL	NL	NL	NL	NL
4. ordinary reinforced concrete shear walls[l]	14.2	6	$2^1/_2$	5	NL	NL	NP	NP	NP
5. steel and concrete composite eccentrically braced frames	14.3	8	$2^1/_2$	4	NL	NL	NL	NL	NL
6. steel and concrete composite special concentrically braced frames	14.3	6	$2^1/_2$	5	NL	NL	NL	NL	NL
7. steel and concrete composite plate shear walls	14.3	$7^1/_2$	$2^1/_2$	6	NL	NL	NL	NL	NL
8. steel and concrete composite special shear walls	14.3	7	$2^1/_2$	6	NL	NL	NL	NL	NL
9. steel and concrete composite ordinary shear walls	14.3	6	$2^1/_2$	5	NL	NL	NP	NP	NP
10. special reinforced masonry shear walls	14.4	$5^1/_2$	3	5	NL	NL	NL	NL	NL
11. intermediate reinforced masonry shear walls	14.4	4	3	$3^1/_2$	NL	NL	NP	NP	NP
12. steel buckling-restrained braced frames	14.1	8	$2^1/_2$	5	NL	NL	NL	NL	NL
13. steel special plate shear walls	14.1	8	$2^1/_2$	$6^1/_2$	NL	NL	NL	NL	NL
E. dual systems with intermediate moment frames capable of resisting at least 25% of prescribed seismic forces	12.2.5.1								
1. steel special concentrically braced frames[f]	14.1	6	$2^1/_2$	5	NL	NL	35	NP	NP
2. special reinforced concrete shear walls[l,m]	14.2	$6^1/_2$	$2^1/_2$	5	NL	NL	160	100	100
3. ordinary reinforced masonry shear walls	14.4	3	3	$2^1/_2$	NL	160	NP	NP	NP
4. intermediate reinforced masonry shear walls	14.4	$3^1/_2$	3	3	NL	NL	NP	NP	NP
5. steel and concrete composite special concentrically braced frames	14.3	$5^1/_2$	$2^1/_2$	$4^1/_2$	NL	NL	160	100	NP
6. steel and concrete composite ordinary braced frames	14.3	$3^1/_2$	$2^1/_2$	3	NL	NL	NP	NP	NP

(continued)

APPENDIX H *(continued)*
Design Coefficients and Factors
for Seismic Force-Resisting Systems
[ASCE/SEI7 Table 12.2-1]

seismic force-resisting system	ASCE 7 section where detailing requirements are specified	response modification coefficient, R^a	overstrength factor, $\Omega_O{}^g$	deflection amplification factor, $C_d{}^b$	structural system limitations including structural height, h_n (ft) limits[c] — seismic design category: B	C	D^d	E^d	F^e
7. steel and concrete composite ordinary shear walls	14.3	5	3	$4^1/_2$	NL	NL	NP	NP	NP
8. ordinary reinforced concrete shear walls[l]	14.2	$5^1/_2$	$2^1/_2$	$4^1/_2$	NL	NL	NP	NP	NP
F. shear wall-frame interactive system with ordinary reinforced concrete moment frames and ordinary reinforced concrete shear walls[l]	12.2.5.8 and 14.2	$4^1/_2$	$2^1/_2$	4	NL	NP	NP	NP	NP
G. cantilevered column systems detailed to conform to the requirements for:	12.2.5.2								
1. steel special cantilever column systems	14.1	$2^1/_2$	$1^1/_4$	$2^1/_2$	35	35	35	35	35
2. steel ordinary cantilever column systems	14.1	$1^1/_4$	$1^1/_4$	$1^1/_4$	35	35	NP[i]	NP[i]	NP[i]
3. special reinforced concrete moment frames[n]	12.2.5.5 and 14.2	$2^1/_2$	$1^1/_4$	$2^1/_2$	35	35	35	35	35
4. intermediate reinforced concrete moment frames	14.2	$1^1/_2$	$1^1/_4$	$1^1/_2$	35	35	NP	NP	NP
5. ordinary reinforced concrete moment frames	14.2	1	$1^1/_4$	1	35	NP	NP	NP	NP
6. timber frames	14.5	$1^1/_2$	$1^1/_2$	$1^1/_2$	35	35	35	NP	NP
H. steel systems not specifically detailed for seismic resistance, excluding cantilever column systems	14.1	3	3	3	NL	NL	NP	NP	NP

[a]Response modification coefficient, R, for use throughout the standard. Note R reduces forces to a strength level, not an allowable stress level.
[b]Deflection amplification factor, C_d, for use in Sections 12.8.6, 12.8.7, and 12.9.2.
[c]NL = Not Limited and NP = Not Permitted. For metric units use 30.5 m for 100 ft and use 48.8 m for 160 ft.
[d]See Section 12.2.5.4 for a description of seismic force-resisting systems limited to buildings with a structural height, h_n, of 240 ft (73.2 m) or less.
[e]See Section 12.2.5.4 for seismic force-resisting systems limited to buildings with a structural height, h_n, of 160 ft (48.8 m) or less.
[f]Ordinary moment frame is permitted to be used in lieu of intermediate moment frame for seismic design categories B or C.
[g]Where the tabulated value of the overstrength factor, Ω_O, is greater than or equal to $2^1/_2$, Ω_O is permitted to be reduced by subtracting the value of $^1/_2$ for structures with flexible diaphragms.
[h]See Section 12.2.5.7 for limitations in structures assigned to seismic design categories D, E, or F.
[i] See Section 12.2.5.6 for limitations in structures assigned to seismic design categories D, E, or F.
[j]Steel ordinary concentrically braced frames are permitted in single-story buildings up to a structural height, h_n, of 60 ft (18.3 m) where the dead load of the roof does not exceed 20 psf (0.96 kN/m^2) and in penthouse structures.
[k]An increase in structural height, h_n, to 45 ft (13.7 m) is permitted for single story storage warehouse facilities.
[l]In Section 2.2 of ACI 318. A shear wall is defined as a structural wall.
[m]In Section 2.2 of ACI 318. The definition of "special structural wall" includes precast and cast-in-place construction.
[n]In Section 2.2 of ACI 318. The definition of "special moment frame" includes precast and cast-in-place construction.
[o]Alternately, the seismic load effect with overstrength, E_{mh}, is permitted to be based on the expected strength determined in accordance with AISI S110.
[p]Cold-formed steel—special bolted moment frames shall be limited to one-story in height in accordance with AISI S110.

APPENDIX I
Standard Welding Symbols of the AISC/AWS

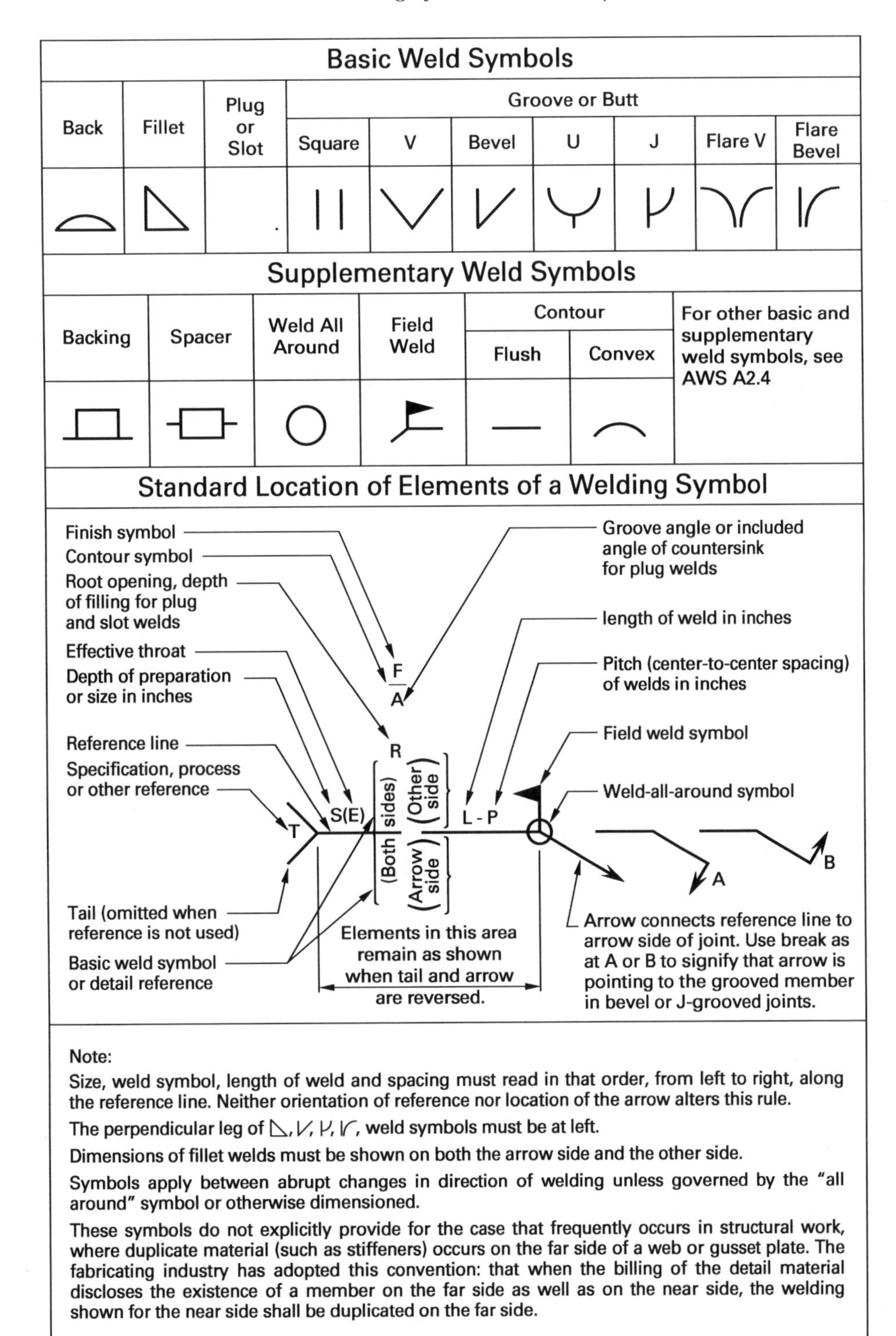

Note:

Size, weld symbol, length of weld and spacing must read in that order, from left to right, along the reference line. Neither orientation of reference nor location of the arrow alters this rule.

The perpendicular leg of ◺, ⩗, Ꮧ, ⟋, weld symbols must be at left.

Dimensions of fillet welds must be shown on both the arrow side and the other side.

Symbols apply between abrupt changes in direction of welding unless governed by the "all around" symbol or otherwise dimensioned.

These symbols do not explicitly provide for the case that frequently occurs in structural work, where duplicate material (such as stiffeners) occurs on the far side of a web or gusset plate. The fabricating industry has adopted this convention: that when the billing of the detail material discloses the existence of a member on the far side as well as on the near side, the welding shown for the near side shall be duplicated on the far side.

APPENDIX J

Simpson Strong-Tie Structural Connectors: Holdowns and Tension Ties

HDB/HD Holdowns*

Simpson Strong-Tie offers a wide variety of bolted holdowns offering low-deflection performance for a range of load requirements. All of these holdowns have been tested in accordance with ICC-ES's AC 155 acceptance criteria and are approved for use in vertical and horizontal applications.

The HD3B is a light-duty holdown designed for use in shearwalls and braced-wall panels, as well as other lateral applications.

The HD5B, HD7B and HD9B bolted holdowns incorporate the proven design of our HDQ8 SDS-style holdown and feature a unique seat design which greatly minimizes deflection under load. HDB holdowns are self jigging, ensuring that the code-required minimum of seven bolt diameters from the end of the post is met. They can be installed directly on the sill plate or raised above it and are suitable for back-to-back applications where eccentricity is a concern. HDBs are designed to provide loads for intermediate-load-range shearwalls, braced-wall panels and lateral applications.

HD holdowns offer the highest allowable loads, providing high capacity for both vertical and horizontal applications. The HD12 and HD19 are self jigging, ensuring that the code-required minimum of seven bolt diameters from the end of the post is met. They can be installed back-to-back when eccentricity is an issue.

MATERIAL: See table

FINISH: HD3B/HD5B/HD7B/HD9B – Galvanized;
HD – Simpson Strong-Tie® gray paint

INSTALLATION: • Use all specified fasteners. See General Notes.

- Bolt holes shall be a minimum of 1/32" to a maximum of 1/16" larger than the bolt diameter *(per NDS, section 11.1.2)*.
- Stud bolts should be snugly tightened with standard cut washers between the wood and nut *(BP's are required in the City and County of Los Angeles)*.
- The Designer must specify anchor bolt type, length, and embedment. See SB and SSTB Anchor bolts *(pages 33-37)*.
- To tie multiple 2x members together, the Designer must determine the fasteners required to join members without splitting the wood.

CODES: See page 13 for Code Reference Key Chart.

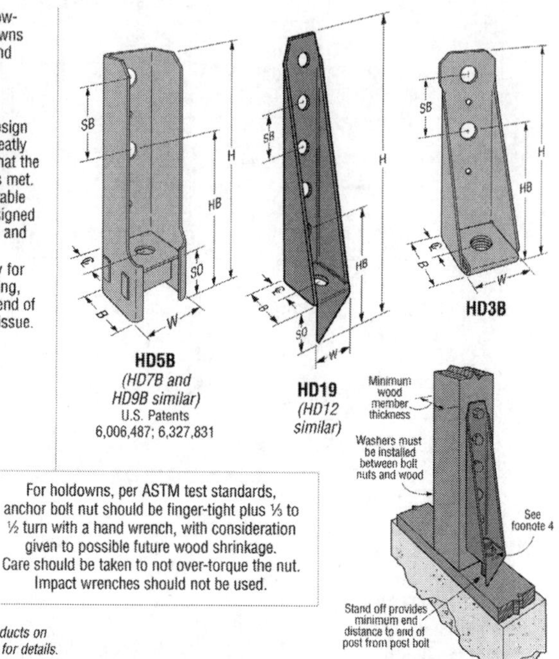

For holdowns, per ASTM test standards, anchor bolt nut should be finger-tight plus ⅓ to ½ turn with a hand wrench, with consideration given to possible future wood shrinkage. Care should be taken to not over-torque the nut. Impact wrenches should not be used.

These products are available with additional corrosion protection. Additional products on this page may also be available with this option, check with Simpson Strong-Tie for details.

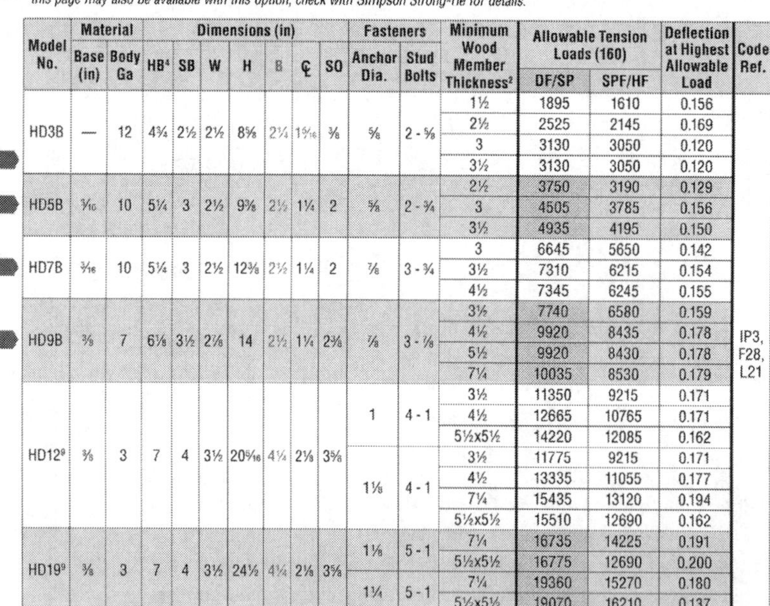

Model No.	Material: Base (in)	Material: Body Ga	Dimensions (in): HB[4]	SB	W	H	B	℄	SO	Fasteners: Anchor Dia.	Fasteners: Stud Bolts	Minimum Wood Member Thickness[2]	Allowable Tension Loads (160): DF/SP	SPF/HF	Deflection at Highest Allowable Load	Code Ref.
HD3B	—	12	4¾	2½	2½	8⅝	2¼	1 5/16	⅜	⅝	2 - ⅝	1½	1895	1610	0.156	IP3, F28, L21
												2½	2525	2145	0.169	
												3	3130	3050	0.120	
												3½	3130	3050	0.120	
HD5B	3/16	10	5¼	3	2½	9⅜	2½	1¼	2	⅝	2 - ¾	2½	3750	3190	0.129	
												3	4505	3785	0.156	
												3½	4935	4195	0.150	
HD7B	3/16	10	5¼	3	2½	12⅜	2½	1¼	2	⅞	3 - ¾	3	6645	5650	0.142	
												3½	7310	6215	0.154	
												4½	7345	6245	0.155	
HD9B	⅜	7	6⅛	3½	2⅞	14	2½	1¼	2⅜	⅞	3 - ⅞	3½	7740	6580	0.159	
												4½	9920	8435	0.178	
												5½	9920	8430	0.178	
												7¼	10035	8530	0.179	
HD12[9]	⅜	3	7	4	3½	20 5/16	4¼	2⅛	3⅝	1	4 - 1	3½	11350	9215	0.171	
												4½	12665	10765	0.171	
												5½x5½	14220	12085	0.162	
										1⅛	4 - 1	3½	11775	9215	0.171	
												4½	13335	11055	0.177	
												7¼	15435	13120	0.194	
												5½x5½	15510	12690	0.162	
HD19[9]	⅜	3	7	4	3½	24½	4¼	2⅛	3⅝	1⅛	5 - 1	7¼	16735	14225	0.191	
												5½x5½	16775	12690	0.200	
										1¼	5 - 1	7¼	19360	15270	0.180	
												5½x5½	19070	16210	0.137	

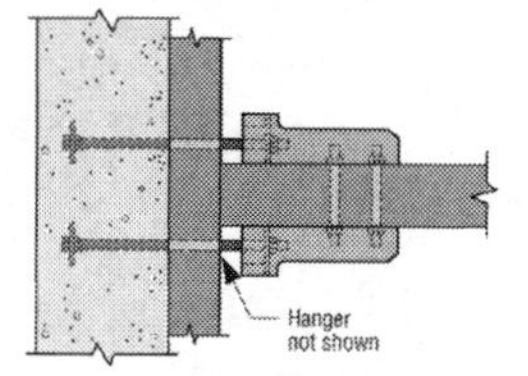

Horizontal HDB Installation
(Plan View)

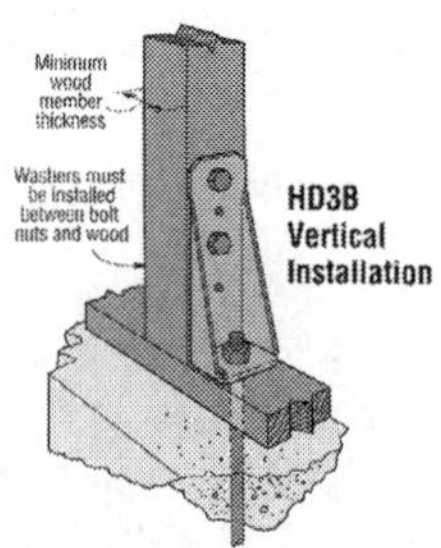

HD3B Vertical Installation

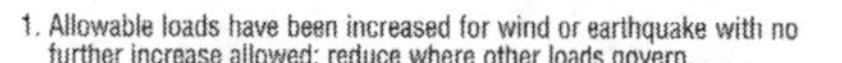

1. Allowable loads have been increased for wind or earthquake with no further increase allowed: reduce where other loads govern.
2. Post design by Specifier. Tabulated loads are based on 3½" wide member minimum, unless noted otherwise. Post may consist of multiple members provided they are connected independently of the holdown fasteners. See pages 226-227 for common post allowable loads.
3. Structural composite lumber columns have sides that show either the wide face or the edges of the lumber strands/veneers. Values in the tables reflect installation into the wide face. See technical bulletin T-SCLCOLUMN for values on the narrow face (edge) *(see page 232 for details)*.
4. HD and HDB holdowns are self-jigging and will ensure minumum bolt end distance, HB, when installed flush with the sill plate.
5. Deflection at Highest Allowable Tension Load includes fastener slip holdown deformation, and anchor bolt elongation for holdowns installed up to 6" above top of concrete. Holdowns may be installed raised up to 18" above top of concrete with no load reduction provided that additional elongation of the anchor rod is accounted for.
6. To achieve published loads, machine bolts shall be installed with the nut on the opposite side of the holdown. If reversed, the Designer shall reduce the allowable loads shown per NDS requirements when bolt threads are in the shear plane.
7. Lag bolts will not develop the listed loads.
8. Tabulated values may be doubled when the HD holdown is installed on opposite sides of the wood member. The Designer must evaluate the capacity of the wood member and the anchorage.
9. Standard cut washer is required under anchor nut for HD12 with 1" anchor and HD19 with 1⅛" anchors.

*Information presented is for illustrative purposes only and should not be used for design.

(continued)

APPENDIX J *(continued)*

Simpson Strong-Tie Structural Connectors: Holdowns and Tension Ties

HRS/ST/PS/HST/LSTA/LSTI/MST/MSTA/MSTC/MSTI Strap Ties

Straps are designed to transfer tension loads in a wide variety of applications.

HRS—A 12 gauge strap with a nailing pattern designed for installation on the edge of 2x members. The HRS416Z installs with Simpson Strong-Tie® Strong-Drive® SDS screws.

LSTA and MSTA—Designed for use on the edge of 2x members, with a nailing pattern that reduces the potential for splitting.

LSTI—Light straps that are suitable where pneumatic-nailing is necessary through diaphragm decking and wood chord open web trusses.

MST—Splitting may be a problem with installations on lumber smaller than 3½"; either fill every nail hole with 10dx1½" nails or fill every-other hole with 16d common nails. Reduce the allowable load based upon the size and quantity of fasteners used.

MSTC—High Capacity strap which utilizes a staggered nail pattern to help minimize wood splitting. Nail slots have been countersunk to provide a lower nail head profile.

FINISH: PS–HDG; HST3 and HST6–Simpson Strong-Tie® gray paint; all others–galvanized. Some products are available in stainless steel or ZMAX® coating; see Corrosion Information, page 14-15.

INSTALLATION: Use all specified fasteners. See General Notes.

OPTIONS: Special sizes can be made to order. Contact Simpson Strong-Tie.

CODES: See page 13 for Code Reference Key Chart.

MSTC and RPS meet code requirements for reinforcing cut members (16 gauge) at top plate and RPS at sill plate. International Residential Code®– 2000/2006 R602.6.1 International Building Code®– 2000/2006 2308.9.8 *(For RPS, refer to page 205.)*

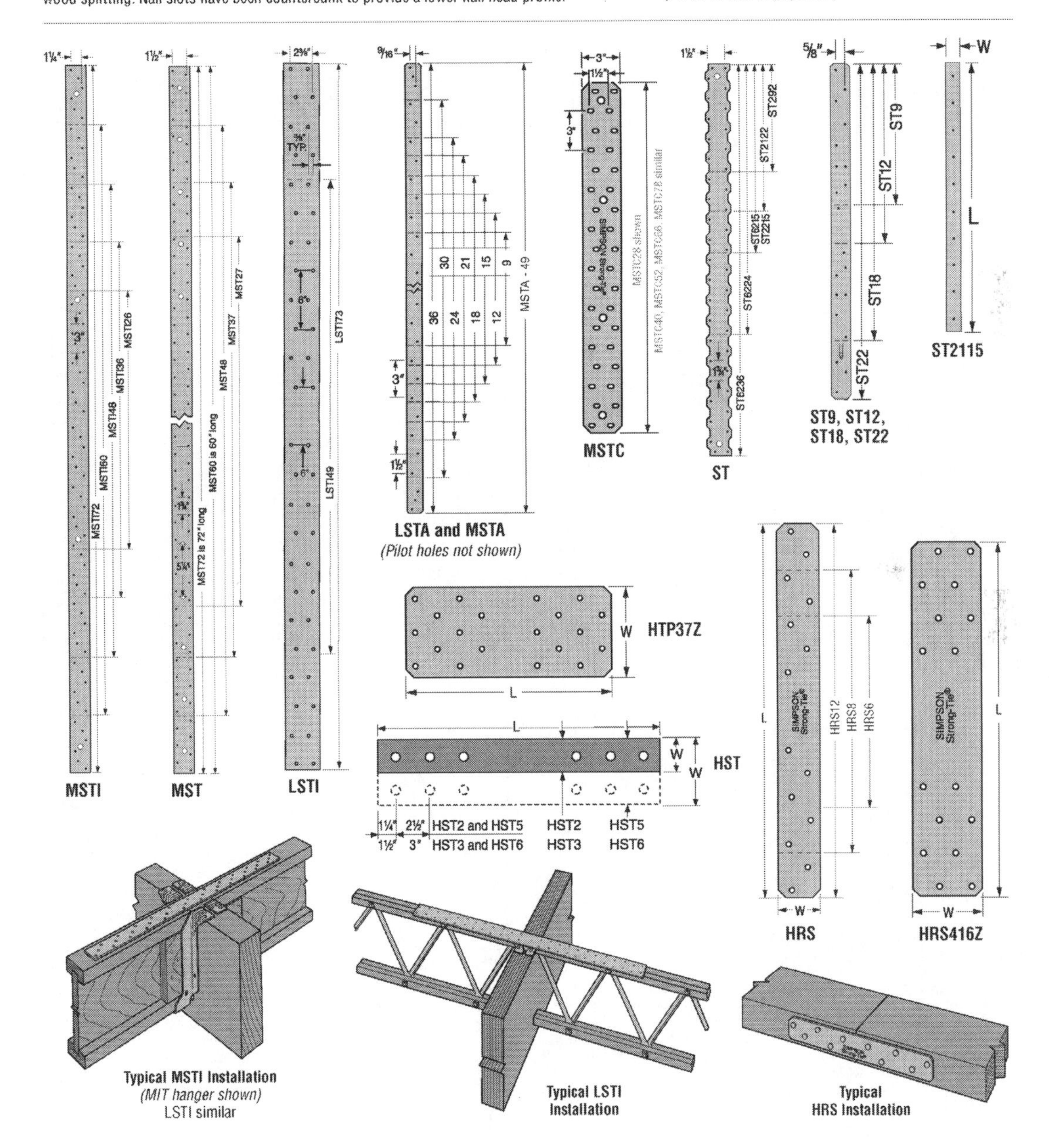

Reprinted with permission from *Connectors for Wood Construction* catalog, © 2013, Simpson Strong-Tie Company, Inc. Visit www.strongtie.com for the most up to date information.

(continued)

APPENDIX J *(continued)*

Simpson Strong-Tie Structural Connectors: Holdowns and Tension Ties

HRS/ST/PS/HST/LSTA/LSTI/MST/MSTA/MSTC/MSTI Strap Ties

These products are available with additional corrosion protection. Additional products on this page may also be available with this option, check with Simpson Strong-Tie for details.

These products are approved for installation with the Strong-Drive SD Structural-Connector screw. See page 27 for more information.

Model No.	Ga	Dimensions W	Dimensions L	Fasteners (Total)	Allowable Tension Loads (DF/SP) (160)	Allowable Tension Loads (SPF/HF) (160)	Code Ref.
LSTA9	20	1¼	9	8-10d	740	635	I4, L3, F2
LSTA12		1¼	12	10-10d	925	795	
LSTA15		1¼	15	12-10d	1110	950	
LSTA18		1¼	18	14-10d	1235	1110	
LSTA21		1¼	21	16-10d	1235	1235	
LSTA24		1¼	24	18-10d	1235	1235	
ST292		2 1/16	9 5/16	12-16d	1265	1120	
ST2122		2 1/16	12 13/16	16-16d	1530	1505	
ST2115		¾	16 5/16	10-16d	660	660	
ST2215		2 1/16	16 5/16	20-16d	1875	1875	
LSTA30	18	1¼	30	22-10d	1640	1640	
LSTA36		1¼	36	24-10d	1640	1640	
LSTI49		3¾	49	32-10dx1½	2975	2555	
LSTI73		3¾	73	48-10dx1½	4205	3830	
MSTA9		1¼	9	8-10d	750	645	
MSTA12		1¼	12	10-10d	940	810	
MSTA15		1¼	15	12-10d	1130	970	
MSTA18		1¼	18	14-10d	1315	1130	
MSTA21		1¼	21	16-10d	1505	1290	
MSTA24		1¼	24	18-10d	1640	1455	
MSTA30	16	1¼	30	22-10d	2050	1820	
MSTA36		1¼	36	26-10d	2050	2050	
MSTA49		1¼	49	26-10d	2020	2020	F2, L3
ST6215		2 1/16	16 5/16	20-16d	2095	1900	I4, IL14, L3, F2
ST6224		2 1/16	23 5/16	28-16d	2540	2540	I4, L3, F2
ST9		1¼	9	8-16d	885	760	I4, L3, F2
ST12		1¼	11⅝	10-16d	1105	950	
ST18		1¼	17¾	14-16d	1420	1330	
ST22		1¼	21⅝	18-16d	1420	1420	
MSTC28		3	28¼	36-16d sinkers	3455	2980	
MSTC40		3	40¼	52-16d sinkers	4745	4305	
MSTC52		3	52¼	62-16d sinkers	4745	4745	
HTP37Z		3	7	20-10dx1½	1850	1600	170
MSTC66	14	3	65¾	76-16d sinkers	5860	5860	I4, L3, F2
MSTC78		3	77¾	76-16d sinkers	5860	5860	
ST6236		2 1/16	33 13/16	40-16d	3845	3845	
HRS6	12	1⅜	6	6-10d	605	525	F26
HRS8		1⅜	8	10-10d	1010	880	
HRS12		1⅜	12	14-10d	1415	1230	
MSTI26		2 1/16	26	26-10dx1½	2745	2325	I4, L3, F2
MSTI36		2 1/16	36	36-10dx1½	3800	3220	
MSTI48		2 1/16	48	48-10dx1½	5065	4290	
MSTI60		2 1/16	60	60-10dx1½	5080	5080	
MSTI72		2 1/16	72	72-10dx1½	5080	5080	
HRS416Z		3¼	16	16-SDS ¼"x1½"	2835	2305	170

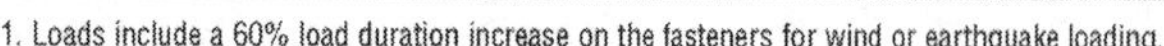

1. Loads include a 60% load duration increase on the fasteners for wind or earthquake loading.
2. 10dx1½" nails may be substituted where 16d sinkers or 10d are specified at 100% of the table loads except where straps are installed over sheathing.
3. 10d commons may be substituted where 16d sinkers are specified at 100% of table loads.
4. 16d sinkers (0.148" dia. x 3¼" long) or 10d commons may be substituted where 16d commons are specified at 0.84 of the table loads.
5. Use half of the nails in each member being connected to achieve the listed loads.
6. Tension loads apply for uplift when installed vertically.
7. **NAILS:** 16d = 0.162" dia. x 3½" long, 16d Sinker = 0.148" dia. x 3¼" long, 10d = 0.148" dia. x 3" long. 10dx1½ = 0.148" dia. x 1½" long. See page 22-23 for other nail sizes and information.

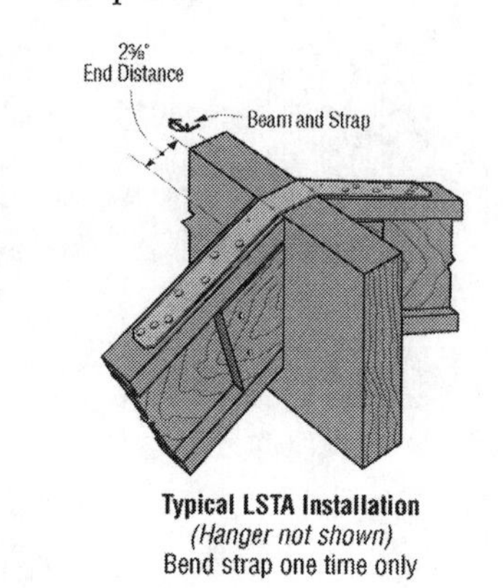

Typical LSTA Installation
(Hanger not shown)
Bend strap one time only

Model No.	Material Thickness Gauge	Dim. W	Dim. L	Bolts Qty	Bolts Dia	Code Ref.
PS218	7 ga	2	18	4	¾	180
PS418		4	18	4	¾	
PS720		6¾	20	8	½	

1. PS strap design loads must be determined by the Designer for each installation. Bolts are installed both perpendicular and parallel-to-grain. Hole diameter in the part may be oversized to accommodate the HDG. Designer must determine if the oversize creates an unacceptable installation.

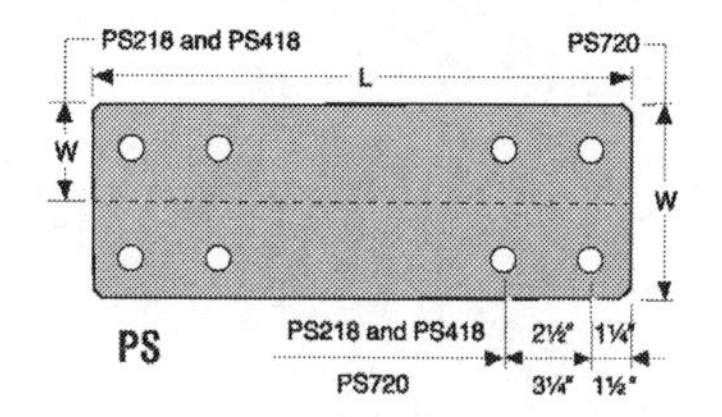

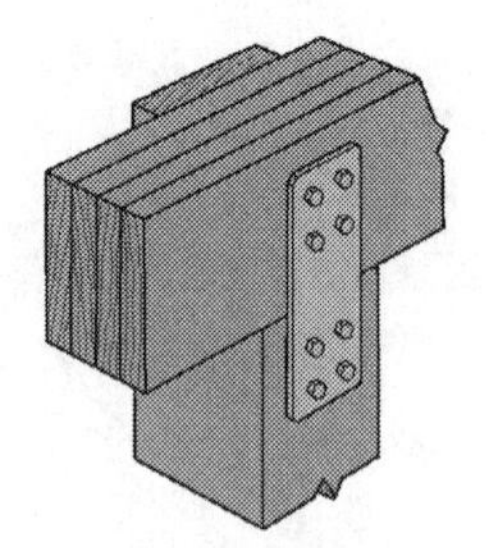

Typical PS720 Installation

(continued)

APPENDIX J *(continued)*

Simpson Strong-Tie Structural Connectors: Holdowns and Tension Ties

HST/MST/MSTC/MSTA Strap Ties and TS Twist Straps

These products are approved for installation with the Strong-Drive SD Structural-Connector screw. See page 27 for more information.

Floor-to-Floor Clear Span Table

Model No.	Clear Span	Fasteners (Total)	Allowable Tension Loads (DF/SP) (160)	Allowable Tension Loads (SPF/HF) (160)
MSTA49	18	26-10d	2020	2020
	16	26-10d	2020	2020
MSTC28	18	12-16d sinkers	1155	995
	16	16-16d sinkers	1540	1325
MSTC40	24	20-16d sinkers	2310	1985
	18	28-16d sinkers	2695	2320
	16	32-16d sinkers	3080	2650
MSTC52	24	36-16d sinkers	3465	2980
	18	44-16d sinkers	4235	3645
	16	48-16d sinkers	4620	3975
MSTC66	30	48-16d sinkers	4780	4120
	24	54-16d sinkers	5380	4640
	18	64-16d sinkers	5860	5495
	16	68-16d sinkers	5860	5840
MSTC78	30	64-16d sinkers	5860	5495
	24	72-16d sinkers	5860	5860
	18	76-16d sinkers	5860	5860
MST37	24	14-16d	1725	1495
	18	20-16d	2465	2135
	16	22-16d	2710	2345
MST48	24	26-16d	3215	2780
	18	32-16d	3960	3425
	16	34-16d	4205	3640
MST60	30	34-16d	4605	3995
	24	40-16d	5240	4700
	18	46-16d	6235	5405
MST72	30	48-16d	6505	5640
	24	54-16d	6730	6345
	18	62-16d	6730	6475

Nails are not required in the rim board area.

When nailing the strap over OSB/plywood, use a 2½" long nail minimum.

Typical Detail with Strap Installed Over Sheathing

STHD Shown

Stitch nailing of double studs by others

Clear Span

Floor-to-Floor Tie Installation showing a Clear Span

CODES: See page 13 for Code Reference Key Chart.

These products are available with additional corrosion protection. Additional products on this page may also be available with this option, check with Simpson Strong-Tie for details.

Model No.	Ga	Dimensions W	Dimensions L	Fasteners (Total) Nails	Fasteners Bolts Qty	Fasteners Bolts Dia	Allowable Tension Loads (DF/SP) Nails (160)	Allowable Tension Loads (DF/SP) Bolts (160)	Allowable Tension Loads (SPF/HF) Nails (160)	Allowable Tension Loads (SPF/HF) Bolts (160)	Code Ref.
MST27	12	2 1/16	27	30-16d	4	½	3700	2165	3200	2000	I4, L3, F2
MST37		2 1/16	37½	42-16d	6	½	5080	3025	4480	2805	
MST48		2 1/16	48	50-16d	8	½	5310	3675	5190	3410	
MST60	10	2 1/16	60	68-16d	10	½	6730	4485	6475	4175	
MST72		2 1/16	72	68-16d	10	½	6730	4485	6475	4175	
HST2	7	2½	21¼	—	6	⅝	—	5220	—	4835	
HST5		5	21¼	—	12	⅝	—	10650	—	9870	
HST3	3	3	25½	—	6	¾	—	7680	—	6660	
HST6		6	25½	—	12	¾	—	15470	—	13320	

1. Loads include a 60% load duration increase on the fasteners for wind or earthquake loading.
2. Install bolts or nails as specified by Designer. Bolt and nail values may not be combined.
3. Allowable bolt loads are based on parallel-to-grain loading and these minimum member thicknesses: MST–2½"; HST2 and HST5–4"; HST3 and HST6–4½".
4. Use half of the required nails in each member being connected to achieve the listed loads.
5. When installing strap over wood structural panel sheathing, use 2½" long nail minimum.
6. Tension loads apply for uplift as well when installed vertically.
7. **NAILS:** 16d = 0.162" dia. x 3½" long, 16d Sinker = 0.148" dia. x 3¼" long, 10dx1½ = 0.148" dia. x 1½" long. See page 22-23 for other nail sizes and information.

TS Twist Straps

Twist straps provide a tension connection between two wood members. An equal number of right and left hand units are supplied in each carton.

MATERIAL: 16 gauge. **FINISH:** Galvanized. See Corrosion Information, page 14-15.

INSTALLATION: • Use all specified fasteners. See General Notes.

• TS should be installed in pairs to reduce eccentricity.

CODES: See page 13 for Code Reference Key Chart.

Model No.	L	Fasteners (Total)	Allowable Loads (160)	Code Ref.
TS9	9	8-16d	530	170
TS12	11⅝	10-16d	665	
TS18	17¾	14-16d	930	
TS22	21⅝	18-16d	1215	

1. Install half of the fasteners on each end of the strap to achieve full loads.
2. Loads have been increased 60% for wind or earthquake loading with no further increase allowed; reduce where other loads govern.
3. 16d sinkers (0.148" dia. x 3¼") may be substituted for the specified 16d commons at 0.84 of the table loads.
4. Loads are for a single TS.
5. **NAILS:** 16d = 0.162" dia. x 3½" long, See page 22-23 for other nail sizes and information.

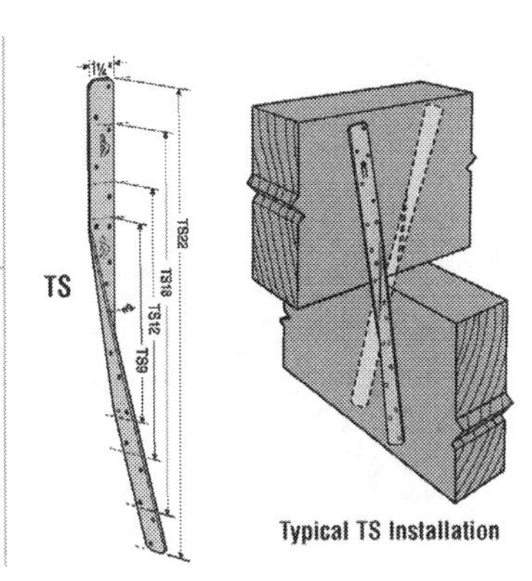

Typical TS Installation

(continued)

APPENDIX J *(continued)*
Simpson Strong-Tie Structural Connectors: Holdowns and Tension Ties

FTA/LFTA Floor Tie Anchors
T and L Strap Ties

Designed for use as a floor-to-floor tension tie, one FTA replaces two comparably sized holdowns and the threaded rod.

The LFTA Light Floor Tie Anchor is for nailed installations.

MATERIAL: See table

FINISH: LFTA–galvanized;
FTA–Simpson Strong-Tie® gray paint

INSTALLATION: • Use all specified fasteners. See General Notes.

- Washers required on side opposite FTA for full loads.
- Nail holes between floors allow preattachment to the joist during installation; these nails are not required.

OPTIONS:

- The standard model's clear span of 17" will accommodate up to a 12" joist. The clear span of the FTA may be increased with a corresponding increase in overall length.

CODES: See page 13 for Code Reference Key Chart.

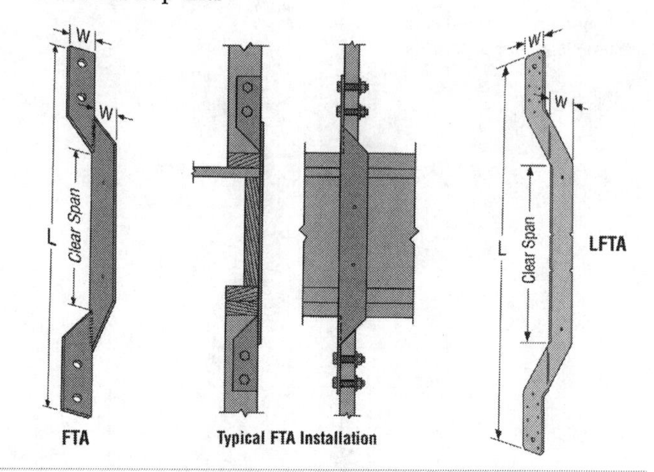

Model No.	Ga	Dimensions			Fasteners (Total)		Allowable Uplift Loads[1] (160)						Code Ref.
							Vertical Member Thickness					LFTA[2]	
		W	Clear Span	L	Qty	Dia	1½	2	2½	3	3½		
LFTA	16	2¼	17	38⅝	16-10d	—	—	—	—	—	—	1205	
FTA2	10	3	17	37½	4	⅝	1890	2515	3120	3385	3385	—	I17, L6, F16
FTA5	10	3½	17	45½	4	¾	2240	3000	3750	4400	4400	—	
FTA7	3	3½	17	56	6	⅞	3715	5020	6210	7600	7600	—	

1. Allowable loads have been increased for wind or earthquake loading with no further increase allowed. Reduce where other loads govern.
2. Reduce the allowable load for the LFTA according to the code when nails penetrate wood less than 1¾".
3. **NAILS:** 10d = 0.148" dia. x 3" long. See page 22-23 for other nail sizes and information.

T and L Strap Ties

T and L Strap Ties are versatile utility straps. See Architectural Products Group for aesthetically pleasing options with black powder-coated paint.

FINISH: Galvanized. See Corrosion Information, page 14-15.

CODES: See page 13 for Code Listing Key Chart.

Model No.	Ga	Dimensions			Fasteners			Code Ref.
						Bolts		
		L	H	W	Nails	Qty	Dia	
55L	16	4¾	4¾	1¼	5-10d	—	—	180
66L	14	6	6	1½	10-16d	3	⅜	
88L	14	8	8	2	12-16d	3	½	
1212L	14	12	12	2	14-16d	3	½	
66T	14	6	5	1½	8-16d	3	⅜	
128T	14	12	8	2	12-16d	3	½	
1212T	14	12	12	2	12-16d	3	½	

1. These connectors are not load-rated.
2. **NAILS:** 16d = 0.162" dia. x 3½" long, 10d = 0.148" dia. x 3" long. See page 22-23 for other nail sizes and information.

These products are available with additional corrosion protection. Additional products on this page may also be available with this option, check with Simpson Strong-Tie for details.

Model No.	Ga	Dimensions			Minimum Bolt End & Edge Distances		Bolts		Allowable Loads[1,2]		Code Ref.
									Tension/Uplift	F_1	
		W	H	L	d_1	d_2	Qty	Dia	(100/160)	(100/160)	
1212HL	7	2½	12	12	2½	4⅜	5	⅝	1535	565	170
1616HL	7	2½	16	16	2½	4⅜	5	⅝	1535	565	
1212HT	7	2½	12	12	2½	4⅜	6	⅝	2585	815	
1616HT	7	2½	16	16	2½	4⅜	6	⅝	2585	815	

1. 1212HL, 1616HL, 1212HT and 1616HT are to be installed in pairs with machine bolts in double shear. A single part with machine bolts in single shear is not load-rated.
2. Allowable loads are based on a minimum member thickness of 3½".
3. 1212HT, 1616HT loads assume a continuous beam.

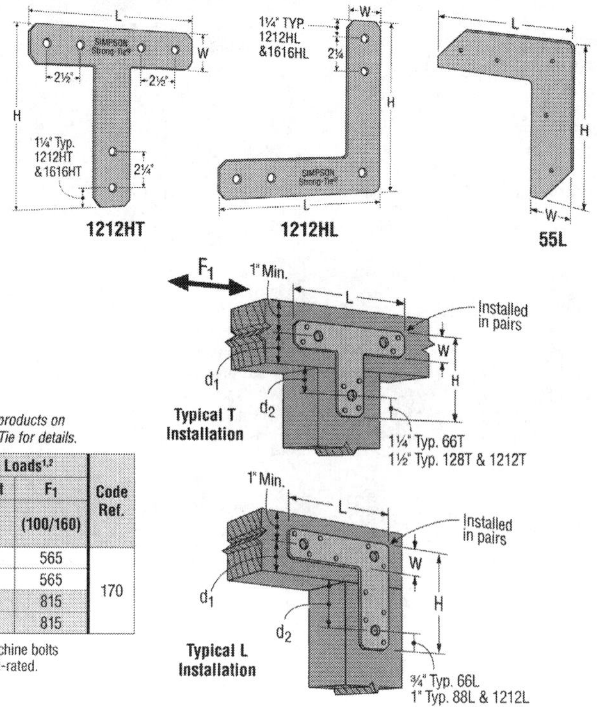

APPENDIX K.1

Common Wire, Box, or Sinker Steel Wire Nails: Reference Lateral Design Values, Z, for Single Shear (Two Member) Connections[1,2,3,4] for Sawn Lumber or SCL with Both Members of Identical Specific Gravity

[NDS Table 11N]

Side Member Thickness t_s (in.)	Nail Diameter D (in.)	Common Wire Nail (Pennyweight)	Box Nail (Pennyweight)	Sinker Nail (Pennyweight)	G=0.67 Red Oak (lbs.)	G=0.55 Mixed Maple Southern Pine (lbs.)	G=0.5 Douglas Fir-Larch (lbs.)	G=0.49 Douglas Fir-Larch (N) (lbs.)	G=0.46 Douglas Fir(S) Hem-Fir(N) (lbs.)	G=0.43 Hem-Fir (lbs.)	G=0.42 Spruce-Pine-Fir (lbs.)	G=0.37 Redwood (open grain) (lbs.)	G=0.36 Eastern Softwoods Spruce-Pine-Fir(S) Western Cedars Western Woods (lbs.)	G=0.35 Northern Species (lbs.)
3/4	0.099		6d	7d	73	61	55	54	51	48	47	39	38	36
	0.113	6d	8d	8d	94	79	72	71	65	58	57	47	46	44
	0.120			10d	107	89	80	77	71	64	62	52	50	48
	0.128		10d		121	101	87	84	78	70	68	57	56	54
	0.131	8d			127	104	90	87	80	73	70	60	58	56
	0.135		16d	12d	135	108	94	91	84	76	74	63	61	58
	0.148	10d	20d	16d	154	121	105	102	94	85	83	70	69	66
	0.162	16d	40d		183	138	121	117	108	99	96	82	80	77
	0.177			20d	200	153	134	130	121	111	107	92	90	87
	0.192	20d		30d	206	157	138	134	125	114	111	96	93	90
	0.207	30d		40d	216	166	147	143	133	122	119	103	101	97
	0.225	40d			229	176	158	154	144	132	129	112	110	106
	0.244	50d		60d	234	182	162	158	147	136	132	115	113	109
1	0.099		6d	7d	73	61	55	54	51	48	47	42	41	40
	0.113	6d[4]	8d	8d	94	79	72	71	67	63	61	55	54	51
	0.120			10d	107	89	81	80	76	71	69	60	59	56
	0.128		10d		121	101	93	91	86	80	79	66	64	61
	0.131	8d			127	106	97	95	90	84	82	68	66	63
	0.135		16d	12d	135	113	103	101	96	89	86	71	69	66
	0.148	10d	20d	16d	154	128	118	115	109	99	96	80	77	74
	0.162	16d	40d		184	154	141	137	125	113	109	91	89	85
	0.177			20d	213	178	155	150	138	125	121	102	99	95
	0.192	20d		30d	222	183	159	154	142	128	124	105	102	98
	0.207	30d		40d	243	192	167	162	149	135	131	111	109	104
	0.225	40d			268	202	177	171	159	144	140	120	117	112
	0.244	50d		60d	274	207	181	175	162	148	143	123	120	115
1-1/4	0.099		6d[4]	7d[4]	73	61	55	54	51	48	47	42	41	40
	0.113	6d[4]	8d	8d[4]	94	79	72	71	67	63	61	55	54	52
	0.120			10d	107	89	81	80	76	71	69	62	60	59
	0.128		10d		121	101	93	91	86	80	79	70	69	67
	0.131	8d[4]			127	106	97	95	90	84	82	73	72	70
	0.135		16d	12d	135	113	103	101	96	89	88	78	76	74
	0.148	10d	20d	16d	154	128	118	115	109	102	100	89	87	84
	0.162	16d	40d		184	154	141	138	131	122	120	103	100	95
	0.177			20d	213	178	163	159	151	141	136	113	110	105
	0.192	20d		30d	222	185	170	166	157	145	140	116	113	108
	0.207	30d		40d	243	203	186	182	169	152	147	123	119	114
	0.225	40d			268	224	200	193	177	160	155	130	127	121
	0.244	50d		60d	276	230	204	197	181	163	158	133	129	124
1-1/2	0.099			7d[4]	73	61	55	54	51	48	47	42	41	40
	0.113		8d[4]	8d[4]	94	79	72	71	67	63	61	55	54	52
	0.120			10d	107	89	81	80	76	71	69	62	60	59
	0.128		10d		121	101	93	91	86	80	79	70	69	67
	0.131	8d[4]			127	106	97	95	90	84	82	73	72	70
	0.135		16d	12d	135	113	103	101	96	89	88	78	76	74
	0.148	10d	20d	16d	154	128	118	115	109	102	100	89	87	84
	0.162	16d	40d		184	154	141	138	131	122	120	106	104	101
	0.177			20d	213	178	163	159	151	141	138	123	121	117
	0.192	20d		30d	222	185	170	166	157	147	144	128	126	120
	0.207	30d		40d	243	203	186	182	172	161	158	135	131	125
	0.225	40d			268	224	205	201	190	178	172	143	138	132
	0.244	50d		60d	276	230	211	206	196	181	175	146	141	135
1-3/4	0.113		8d[4]		94	79	72	71	67	63	61	55	54	52
	0.120			10d[4]	107	89	81	80	76	71	69	62	60	59
	0.128		10d[4]		121	101	93	91	86	80	79	70	69	67
	0.135		16d	12d	135	113	103	101	96	89	88	78	76	74
	0.148	10d[4]	20d	16d	154	128	118	115	109	102	100	89	87	84
	0.162	16d	40d		184	154	141	138	131	122	120	106	104	101
	0.177			20d	213	178	163	159	151	141	138	123	121	117
	0.192	20d		30d	222	185	170	166	157	147	144	128	126	122
	0.207	30d		40d	243	203	186	182	172	161	158	140	137	133
	0.225	40d			268	224	205	201	190	178	174	155	151	144
	0.244	50d		60d	276	230	211	206	196	183	179	159	154	147

1. Tabulated lateral design values, Z, shall be multiplied by all applicable adjustment factors (see Table 10.3.1).
2. Tabulated lateral design values, Z, are for common, box, or sinker steel wire nails (see Appendix Table L4) inserted in side grain with nail axis perpendicular to wood fibers; nail penetration, p, into the main member equal to 10D; and nail bending yield strengths, F_{yb}, of 100,000 psi for 0.099" ≤ D ≤ 0.142", 90,000 psi for 0.142" < D ≤ 0.177", 80,000 psi for 0.177" < D ≤ 0.236", and 70,000 psi for 0.236" < D ≤ 0.273".
3. Where the nail or spike penetration, p, is less than 10D but not less than 6D, tabulated lateral design values, Z, shall be multiplied by p/10D or lateral design values shall be calculated using the provisions of 11.3 for the reduced penetration.
4. Nail length is insufficient to provide 10D penetration. Tabulated lateral design values, Z, shall be adjusted per footnote 3.

APPENDIX K.2

Applicability of Adjustment Factors for Connections

[NDS Table 10.3.1]

		ASD Only	ASD and LRFD									LRFD Only		
		Load Duration Factor[1]	Wet Service Factor	Temperature Factor	Group Action Factor	Geometry Factor[3]	Penetration Depth Factor[3]	End Grain Factor[3]	Metal Side Plate Factor[3]	Diaphragm Factor[3]	Toe-Nail Factor[3]	Format Conversion Factor K_F	Resistance Factor ϕ	Time Effect Factor
Lateral Loads														
Dowel-type Fasteners (e.g. bolts, lag screws, wood screws, nails, spikes, drift bolts, & drift pins)	$Z' = Z$ x	C_D	C_M	C_t	C_g	C_Δ	-	C_{eg}	-	C_{di}	C_{tn}	3.32	0.65	λ
Split Ring and Shear Plate Connectors	$P' = P$ x	C_D	C_M	C_t	C_g	C_Δ	C_d	-	C_{st}	-	-	3.32	0.65	λ
	$Q' = Q$ x	C_D	C_M	C_t	C_g	C_Δ	C_d	-	-	-	-	3.32	0.65	λ
Timber Rivets	$P' = P$ x	C_D	C_M	C_t	-	-	-	-	C_{st}[4]	-	-	3.32	0.65	λ
	$Q' = Q$ x	C_D	C_M	C_t	-	C_Δ[5]	-	-	C_{st}[4]	-	-	3.32	0.65	λ
Spike Grids	$Z' = Z$ x	C_D	C_M	C_t	-	C_Δ	-	-	-	-	-	3.32	0.65	λ
Withdrawal Loads														
Nails, spikes, lag screws, wood screws, & drift pins	$W' = W$ x	C_D	C_M[2]	C_t	-	-	-	C_{eg}	-	-	C_{tn}	3.32	0.65	λ

1. The load duration factor, C_D, shall not exceed 1.6 for connections (see 10.3.2).
2. The wet service factor, C_M, shall not apply to toe-nails loaded in withdrawal (see 11.5.4.1).
3. Specific information concerning geometry factors C_Δ, penetration depth factors C_d, end grain factors, C_{eg}, metal side plate factors, C_{st}, diaphragm factors, C_{di}, and toe-nail factors, C_{tn}, is provided in Chapters 11, 12, and 13.
4. The metal side plate factor, C_{st}, is only applied when rivet capacity (P_r, Q_r) controls (see Chapter 13).
5. The geometry factor, C_Δ, is only applied when wood capacity, Q_w, controls (see Chapter 13).

APPENDIX K.3

Typical Dimensions for Dowel-Type Fasteners[1]
[NDS Table L1]

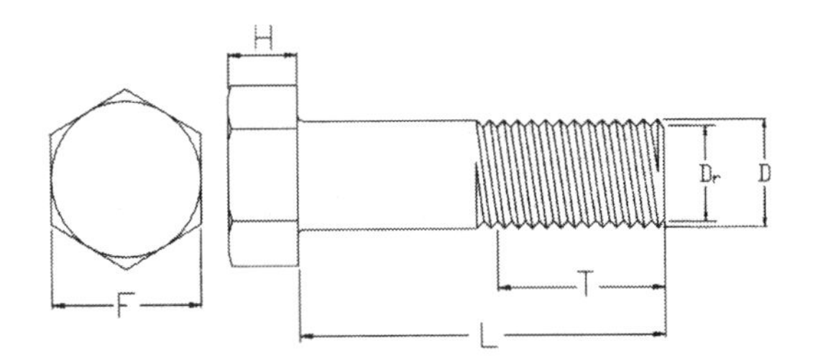

D = diameter
D_r = root diameter
T = thread length
L = bolt length
F = width of head across flats
H = height of head

		Diameter, D							
		1/4"	**5/16"**	**3/8"**	**1/2"**	**5/8"**	**3/4"**	**7/8"**	**1"**
D_r		0.189"	0.245"	0.298"	0.406"	0.514"	0.627"	0.739"	0.847"
F		7/16"	1/2"	9/16"	3/4"	15/16"	1-1/8"	1-5/16"	1-1/2"
H		11/64"	7/32"	1/4"	11/32"	27/64"	1/2"	37/64"	43/64"
T	L ≤ 6 in.	3/4"	7/8"	1"	1-1/4"	1-1/2"	1-3/4"	2"	2-1/4"
	L > 6 in.	1"	1-1/8"	1-1/4"	1-1/2"	1-3/4"	2"	2-1/4"	2-1/2"

1. Tolerances specified in ANSI B 18.2.1. Full body diameter bolt is shown. Root diameter based on UNC (coarse) thread series (see ANSI B1.1).

Reprinted with courtesy, American Wood Council, Leesburg, VA.

APPENDIX K.4
Standard Hex Lag Screws[1]
[NDS Table L2]

D = diameter
D_r = root diameter
S = unthreaded body length
T = minimum thread length[2]

E = length of tapered tip
L = lag screw length
N = number of threads/inch
F = width of head across flats
H = height of head

Reduced Body Diameter

Full-Body Diameter

Length, L		Diameter, D										
		1/4"	5/16"	3/8"	7/16"	1/2"	5/8"	3/4"	7/8"	1"	1-1/8"	1-1/4"
	D_r	0.173"	0.227"	0.265"	0.328"	0.371"	0.471"	0.579"	0.683"	0.780"	0.887"	1.012"
	E	5/32"	3/16"	7/32"	9/32"	5/16"	13/32"	1/2"	19/32"	11/16"	25/32"	7/8"
	H	11/64"	7/32"	1/4"	19/64"	11/32"	27/64"	1/2"	37/64"	43/64"	3/4"	27/32"
	F	7/16"	1/2"	9/16"	5/8"	3/4"	15/16"	1-1/8"	1-5/16"	1-1/2"	1-11/16"	1-7/8"
	N	10	9	7	7	6	5	4-1/2	4	3-1/2	3-1/4	3-1/4
	S	1/4"	1/4"	1/4"	1/4"	1/4"						
1"	T	3/4"	3/4"	3/4"	3/4"	3/4"						
	T-E	19/32"	9/16"	17/32"	15/32"	7/16"						
	S	1/4"	1/4"	1/4"	1/4"	1/4"						
1-1/2"	T	1-1/4"	1-1/4"	1-1/4"	1-1/4"	1-1/4"						
	T-E	1-3/32"	1-1/16"	1-1/32"	31/32"	15/16"						
	S	1/2"	1/2"	1/2"	1/2"	1/2"	1/2"					
2"	T	1-1/2"	1-1/2"	1-1/2"	1-1/2"	1-1/2"	1-1/2"					
	T-E	1-11/32"	1-5/16"	1-9/32"	1-7/32"	1-3/16"	1-3/32"					
	S	3/4"	3/4"	3/4"	3/4"	3/4"	3/4"					
2-1/2"	T	1-3/4"	1-3/4"	1-3/4"	1-3/4"	1-3/4"	1-3/4"					
	T-E	1-19/32"	1-9/16"	1-17/32"	1-15/32"	1-7/16"	1-11/32"					
	S	1"	1"	1"	1"	1"	1"	1"	1"	1"		
3	T	2"	2"	2"	2"	2"	2"	2"	2"	2"		
	T-E	1-27/32"	1-13/16"	1-25/32"	1-23/32"	1-11/16"	1-19/32"	1-1/2"	1-13/32"	1-5/16"		
	S	1-1/2"	1-1/2"	1-1/2"	1-1/2"	1-1/2"	1-1/2"	1-1/2"	1-1/2"	1-1/2"	1-1/2"	1-1/2"
4"	T	2-1/2"	2-1/2"	2-1/2"	2-1/2"	2-1/2"	2-1/2"	2-1/2"	2-1/2"	2-1/2"	2-1/2"	2-1/2"
	T-E	2-11/32"	2-5/16"	2-9/32"	2-7/32"	2-3/16"	2-3/32"	2"	1-29/32"	1-13/16"	1-23/32"	1-5/8"
	S	2"	2"	2"	2"	2"	2"	2"	2"	2"	2"	2"
5"	T	3"	3"	3"	3"	3"	3"	3"	3"	3"	3"	3"
	T-E	2-27/32"	2-13/16"	2-25/32"	2-23/32"	2-11/16"	2-19/32"	2-1/2"	2-13/32"	2-5/16"	2-7/32"	2-1/8"
	S	2-1/2"	2-1/2"	2-1/2"	2-1/2"	2-1/2"	2-1/2"	2-1/2"	2-1/2"	2-1/2"	2-1/2"	2-1/2"
6"	T	3-1/2"	3-1/2"	3-1/2"	3-1/2"	3-1/2"	3-1/2"	3-1/2"	3-1/2"	3-1/2"	3-1/2"	3-1/2"
	T-E	3-11/32"	3-5/16"	3-9/32"	3-7/32"	3-3/16"	3-3/32"	3"	2-29/32"	2-13/16"	2-23/32"	2-5/8"
	S	3"	3"	3"	3"	3"	3"	3"	3"	3"	3"	3"
7"	T	4"	4"	4"	4"	4"	4"	4"	4"	4"	4"	4"
	T-E	3-27/32"	3-13/16"	3-25/32"	3-23/32"	3-11/16"	3-19/32"	3-1/2"	3-13/32"	3-5/16"	3-7/32"	3-1/8"
	S	3-1/2"	3-1/2"	3-1/2"	3-1/2"	3-1/2"	3-1/2"	3-1/2"	3-1/2"	3-1/2"	3-1/2"	3-1/2"
8"	T	4-1/2"	4-1/2"	4-1/2"	4-1/2"	4-1/2"	4-1/2"	4-1/2"	4-1/2"	4-1/2"	4-1/2"	4-1/2"
	T-E	4-11/32"	4-5/16"	4-9/32"	4-7/32"	4-3/16"	4-3/32"	4"	3-29/32"	3-13/16"	3-23/32"	3-5/8"
	S	4"	4"	4"	4"	4"	4"	4"	4"	4"	4"	4"
9"	T	5"	5"	5"	5"	5"	5"	5"	5"	5"	5"	5"
	T-E	4-27/32"	4-13/16"	4-25/32"	4-23/32"	4-11/16"	4-19/32"	4-1/2"	4-13/32"	4-5/16"	4-7/32"	4-1/8"
	S	4-1/2"	4-1/2"	4-1/2"	4-1/2"	4-1/2"	4-1/2"	4-1/2"	4-1/2"	4-1/2"	4-1/2"	4-1/2"
10"	T	5-1/2"	5-1/2"	5-1/2"	5-1/2"	5-1/2"	5-1/2"	5-1/2"	5-1/2"	5-1/2"	5-1/2"	5-1/2"
	T-E	5-11/32"	5-5/16"	5-9/32"	5-7/32"	5-3/16"	5-3/32"	5"	4-29/32"	4-13/16"	4-23/32"	4-5/8"
	S	5"	5"	5"	5"	5"	5"	5"	5"	5"	5"	5"
11"	T	6"	6"	6"	6"	6"	6"	6"	6"	6"	6"	6"
	T-E	5-27/32"	5-13/16"	5-25/32"	5-23/32"	5-11/16"	5-19/32"	5-1/2"	5-13/32"	5-5/16"	5-7/32"	5-1/8"
	S	6"	6"	6"	6"	6"	6"	6"	6"	6"	6"	6"
12"	T	6"	6"	6"	6"	6"	6"	6"	6"	6"	6"	6"
	T-E	5-27/32"	5-13/16"	5-25/32"	5-23/32"	5-11/16"	5-19/32"	5-1/2"	5-13/32"	5-5/16"	5-7/32"	5-1/8"

1. Tolerances are specified in ANSI/ASME B18.2.1. Full-body diameter and reduced body diameter lag screws are shown. For reduced body diameter lag screws, the unthreaded body diameter may be reduced to approximately the root diameter, D_r.
2. Minimum thread length, T, for lag screw lengths, L, is 6" or 1/2 the lag screw length plus 0.5", whichever is less. Thread lengths may exceed these minimums up to the full lag screw length, L.

APPENDIX K.5
Standard Wood Screws
[NDS Table L3]

Table L3 Standard Wood Screws[1,5]

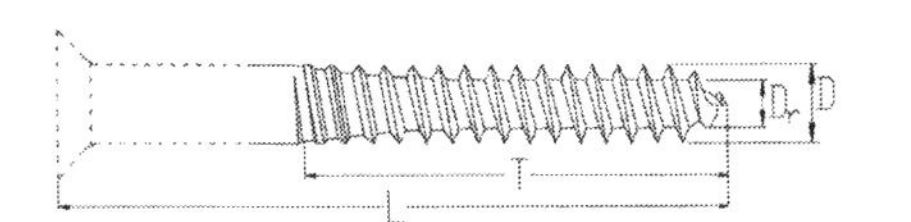

Cut Thread[2]

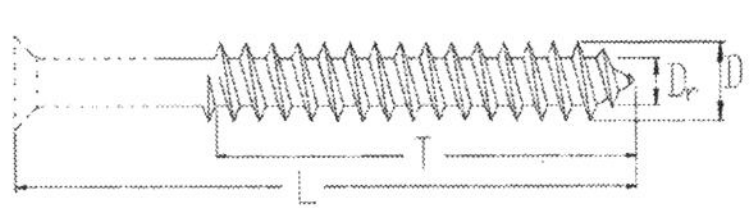

Rolled Thread[3]

D = diameter
D_r = root diameter
L = wood screw length
T = thread length

	Wood Screw Number										
	6	7	8	9	10	12	14	16	18	20	24
D	0.138"	0.151"	0.164"	0.177"	0.19"	0.216"	0.242"	0.268"	0.294"	0.32"	0.372"
D_r[4]	0.113"	0.122"	0.131"	0.142"	0.152"	0.171"	0.196"	0.209"	0.232"	0.255"	0.298"

1. Tolerances are specified in ANSI/ASME B18.6.1
2. Thread length on cut thread wood screws is approximately 2/3 of the wood screw length, L.
3. Single lead thread shown. Thread length is at least four times the screw diameter or 2/3 of the wood screw length, L, whichever is greater. Wood screws which are too short to accommodate the minimum thread length, have threads extending as close to the underside of the head as practicable.
4. Taken as the average of the specified maximum and minimum limits for body diameter of rolled thread wood screws.
5. It is permitted to assume the length of the tapered tip is 2D.

Reprinted with courtesy, American Wood Council, Leesburg, VA.

APPENDIX K.6
Standard Common, Box, and Sinker Steel Wire Nails
[NDS Table L4]

Table L4 Standard Common, Box, and Sinker Nails[1]

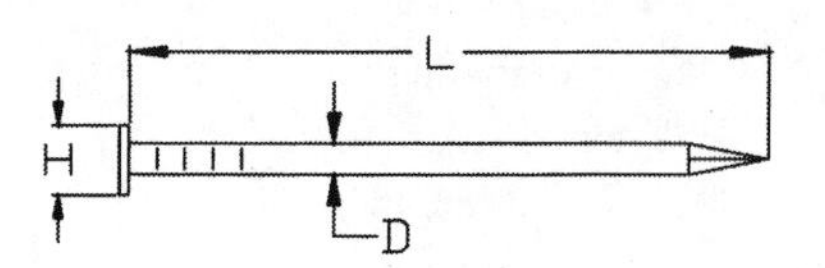

Common or Box

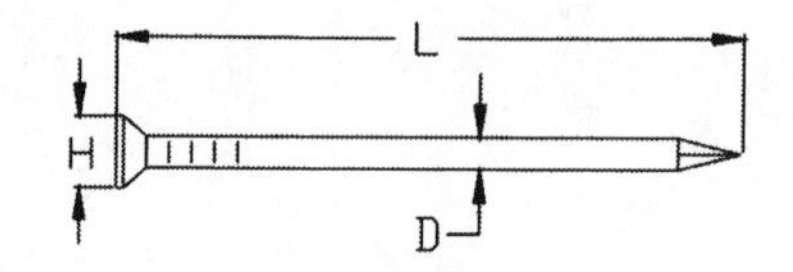

Sinker

D = diameter
L = length
H = head diameter

Type		Pennyweight										
		6d	7d	8d	10d	12d	16d	20d	30d	40d	50d	60d
Common	L	2"	2-1/4"	2-1/2"	3"	3-1/4"	3-1/2"	4"	4-1/2"	5"	5-1/2"	6"
	D	0.113"	0.113"	0.131"	0.148"	0.148"	0.162"	0.192"	0.207"	0.225"	0.244"	0.263"
	H	0.266"	0.266"	0.281"	0.312"	0.312"	0.344"	0.406"	0.438"	0.469"	0.5"	0.531"
Box	L	2"	2-1/4"	2-1/2"	3"	3-1/4"	3-1/2"	4"	4-1/2"	5"		
	D	0.099"	0.099"	0.113"	0.128"	0.128"	0.135"	0.148"	0.148"	0.162"		
	H	0.266"	0.266"	0.297"	0.312"	0.312"	0.344"	0.375"	0.375"	0.406"		
Sinker	L	1-7/8"	2-1/8"	2-3/8"	2-7/8"	3-1/8"	3-1/4"	3-3/4"	4-1/4"	4-3/4"		5-3/4"
	D	0.092"	0.099"	0.113"	0.12"	0.135"	0.148"	0.177"	0.192"	0.207"		0.244"
	H	0.234"	0.250"	0.266"	0.281"	0.312"	0.344"	0.375"	0.406"	0.438"		0.5"

1. Tolerances specified in ASTM F 1667. Typical shape of common, box, and sinker nails shown. See ASTM F 1667 for other nail types.

APPENDIX L

Nail and Spike Reference Withdrawal Design Values, W^1
[NDS Table 11.2C]

Table 11.2C Nail and Spike Reference Withdrawal Design Values, W[1]

Tabulated withdrawal design values, W, are in pounds per inch of fastener penetration into side grain of wood member (see 11.2.3.1).

Specific Gravity, G[2]	Plain Shank Nail and Spike Diameter, D															Threaded Nail Diameter, D				
	0.099"	0.113"	0.128"	0.131"	0.135"	0.148"	0.162"	0.192"	0.207"	0.225"	0.244"	0.263"	0.283"	0.312"	0.375"	0.120"	0.135"	0.148"	0.177"	0.207"
0.73	62	71	80	82	85	93	102	121	130	141	153	165	178	196	236	82	93	102	121	141
0.71	58	66	75	77	79	87	95	113	121	132	143	154	166	183	220	77	87	95	113	132
0.68	52	59	67	69	71	78	85	101	109	118	128	138	149	164	197	69	78	85	101	118
0.67	50	57	65	66	68	75	82	97	105	114	124	133	144	158	190	66	75	82	97	114
0.58	35	40	45	46	48	52	57	68	73	80	86	93	100	110	133	46	52	57	68	80
0.55	31	35	40	41	42	46	50	59	64	70	76	81	88	97	116	41	46	50	59	70
0.51	25	29	33	34	35	38	42	49	53	58	63	67	73	80	96	34	38	42	49	58
0.50	24	28	31	32	33	36	40	47	50	55	60	64	69	76	91	32	36	40	47	55
0.49	23	26	30	30	31	34	38	45	48	52	57	61	66	72	87	30	34	38	45	52
0.47	21	24	27	27	28	31	34	40	43	47	51	55	59	65	78	27	31	34	40	47
0.46	20	22	25	26	27	29	32	38	41	45	48	52	56	62	74	26	29	32	38	45
0.44	18	20	23	23	24	26	29	34	37	40	43	47	50	55	66	23	26	29	34	40
0.43	17	19	21	22	23	25	27	32	35	38	41	44	47	52	63	22	25	27	32	38
0.42	16	18	20	21	21	23	26	30	33	35	38	41	45	49	59	21	23	26	30	35
0.41	15	17	19	19	20	22	24	29	31	33	36	39	42	46	56	19	22	24	29	33
0.40	14	16	18	18	19	21	23	27	29	31	34	37	40	44	52	18	21	23	27	31
0.39	13	15	17	17	18	19	21	25	27	29	32	34	37	41	49	17	19	21	25	29
0.38	12	14	16	16	17	18	20	24	25	28	30	32	35	38	46	16	18	20	24	28
0.37	11	13	15	15	16	17	19	22	24	26	28	30	33	36	43	15	17	19	22	26
0.36	11	12	14	14	14	16	17	21	22	24	26	28	30	33	40	14	16	17	21	24
0.35	10	11	13	13	14	15	16	19	21	23	24	26	28	31	38	13	15	16	19	23
0.31	7	8	9	10	10	11	12	14	15	17	18	19	21	23	28	10	11	12	14	17

1. Tabulated withdrawal design values, W, for nail or spike connections shall be multiplied by all applicable adjustment factors (see Table 10.3.1).
2. Specific gravity, G, shall be determined in accordance with Table 11.3.3A.

APPENDIX M

Bolts: Reference Lateral Design Values, Z, for Single Shear (Two Member) Connections[1,2]
for Sawn Lumber or SCL with Both Members of Identical Specific Gravity
[NDS Table 11A]

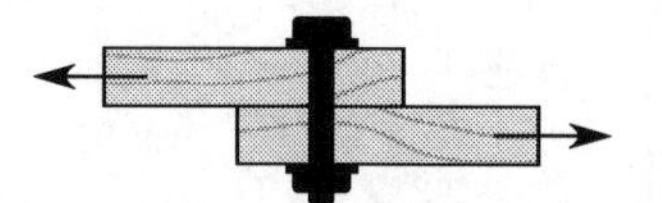

Thickness: Main Member t_m (in.)	Thickness: Side Member t_s (in.)	Bolt Diameter D (in.)	G=0.67 Red Oak $Z_{\parallel}$ (lbs.)	$Z_{s\perp}$ (lbs.)	$Z_{m\perp}$ (lbs.)	$Z_{\perp}$ (lbs.)	G=0.55 Mixed Maple Southern Pine $Z_{\parallel}$ (lbs.)	$Z_{s\perp}$ (lbs.)	$Z_{m\perp}$ (lbs.)	$Z_{\perp}$ (lbs.)	G=0.50 Douglas Fir-Larch $Z_{\parallel}$ (lbs.)	$Z_{s\perp}$ (lbs.)	$Z_{m\perp}$ (lbs.)	$Z_{\perp}$ (lbs.)	G=0.49 Douglas Fir-Larch(N) $Z_{\parallel}$ (lbs.)	$Z_{s\perp}$ (lbs.)	$Z_{m\perp}$ (lbs.)	$Z_{\perp}$ (lbs.)	G=0.46 Douglas Fir(S) Hem-Fir(N) $Z_{\parallel}$ (lbs.)	$Z_{s\perp}$ (lbs.)	$Z_{m\perp}$ (lbs.)	$Z_{\perp}$ (lbs.)
1-1/2	1-1/2	1/2	650	420	420	330	530	330	330	250	480	300	300	220	470	290	290	210	440	270	270	190
		5/8	810	500	500	370	660	400	400	280	600	360	360	240	590	350	350	240	560	320	320	220
		3/4	970	580	580	410	800	460	460	310	720	420	420	270	710	400	400	260	670	380	380	240
		7/8	1130	660	660	440	930	520	520	330	850	470	470	290	830	460	460	280	780	420	420	250
		1	1290	740	740	470	1060	580	580	350	970	530	530	310	950	510	510	300	890	480	480	280
1-3/4	1-3/4	1/2	760	490	490	390	620	390	390	290	560	350	350	250	550	340	340	250	520	320	320	230
		5/8	940	590	590	430	770	470	470	330	700	420	420	280	690	410	410	280	650	380	380	250
		3/4	1130	680	680	480	930	540	540	360	850	480	480	310	830	470	470	300	780	440	440	280
		7/8	1320	770	770	510	1080	610	610	390	990	550	550	340	970	530	530	320	910	500	500	300
		1	1510	860	860	550	1240	680	680	410	1130	610	610	360	1110	600	600	350	1040	560	560	320
2-1/2	1-1/2	1/2	770	480	540	440	660	400	420	350	610	370	370	310	610	360	360	300	580	340	330	270
		5/8	1070	660	630	520	930	560	490	390	850	520	430	340	830	520	420	330	780	470	390	300
		3/4	1360	890	720	570	1120	660	560	430	1020	590	500	380	1000	560	480	360	940	520	450	330
		7/8	1590	960	800	620	1300	720	620	470	1190	630	550	410	1170	600	540	390	1090	550	500	360
		1	1820	1020	870	660	1490	770	680	490	1360	680	610	440	1330	650	590	420	1250	600	550	390
3-1/2	1-1/2	1/2	770	480	560	440	660	400	470	360	610	370	430	330	610	360	420	320	580	340	400	310
		5/8	1070	660	760	590	940	560	620	500	880	520	540	460	870	520	530	450	830	470	490	410
		3/4	1450	890	900	770	1270	660	690	580	1200	590	610	510	1190	560	590	490	1140	520	550	450
		7/8	1890	960	990	830	1680	720	770	630	1590	630	680	550	1570	600	650	530	1470	550	600	480
		1	2410	1020	1080	890	2010	770	830	670	1830	680	740	590	1790	650	710	560	1680	600	660	520
	1-3/4	1/2	830	510	590	480	720	420	510	390	670	380	470	350	660	380	460	340	620	360	440	320
		5/8	1160	680	820	620	1000	580	640	520	930	530	560	460	920	530	550	450	880	500	510	410
		3/4	1530	900	940	780	1330	770	720	580	1250	680	640	520	1240	660	620	500	1190	600	580	460
		7/8	1970	1120	1040	840	1730	840	810	640	1620	740	710	550	1590	700	690	530	1490	640	640	490
		1	2480	1190	1130	900	2030	890	880	670	1850	790	780	590	1820	750	760	570	1700	700	700	530
	3-1/2	1/2	830	590	590	530	750	520	520	460	720	490	490	430	710	480	480	420	690	460	460	410
		5/8	1290	880	880	780	1170	780	780	650	1120	700	700	560	1110	690	690	550	1070	650	650	500
		3/4	1860	1190	1190	950	1690	960	960	710	1610	870	870	630	1600	850	850	600	1540	800	800	560
		7/8	2540	1410	1410	1030	2170	1160	1160	780	1970	1060	1060	680	1940	1040	1040	650	1810	980	980	590
		1	3020	1670	1670	1100	2480	1360	1360	820	2260	1230	1230	720	2210	1190	1190	690	2070	1110	1110	640
5-1/4	1-1/2	5/8	1070	660	760	590	940	560	640	500	880	520	590	460	870	520	590	450	830	470	560	430
		3/4	1450	890	990	780	1270	660	850	660	1200	590	790	590	1190	560	780	560	1140	520	740	520
		7/8	1890	960	1260	960	1680	720	1060	720	1590	630	940	630	1570	600	900	600	1520	550	830	550
		1	2410	1020	1500	1020	2150	770	1140	770	2050	680	1010	680	2030	650	970	650	1930	600	910	600
	1-3/4	5/8	1160	680	820	620	1000	580	690	520	930	530	630	470	920	530	630	470	880	500	590	440
		3/4	1530	900	1050	800	1330	770	890	680	1250	680	830	630	1240	660	810	620	1190	600	780	590
		7/8	1970	1120	1320	1020	1730	840	1090	840	1640	740	960	740	1620	700	920	700	1550	640	850	640
		1	2480	1190	1530	1190	2200	890	1170	890	2080	790	1040	790	2060	750	1000	750	1990	700	930	700
	3-1/2	5/8	1290	880	880	780	1170	780	780	680	1120	700	730	630	1110	690	720	620	1070	650	690	580
		3/4	1860	1190	1240	1080	1690	960	1090	850	1610	870	1030	780	1600	850	1010	750	1540	800	970	710
		7/8	2540	1410	1640	1260	2300	1160	1380	1000	2190	1060	1230	870	2170	1040	1190	840	2060	980	1100	770
		1	3310	1670	1940	1420	2870	1390	1520	1060	2660	1290	1360	940	2630	1260	1320	900	2500	1210	1230	830
5-1/2	1-1/2	5/8	1070	660	760	590	940	560	640	500	880	520	590	460	870	520	590	450	830	470	560	430
		3/4	1450	890	990	780	1270	660	850	660	1200	590	790	590	1190	560	780	560	1140	520	740	520
		7/8	1890	960	1260	960	1680	720	1090	720	1590	630	980	630	1570	600	940	600	1520	550	860	550
		1	2410	1020	1560	1020	2150	770	1190	770	2050	680	1060	680	2030	650	1010	650	1930	600	940	600
	3-1/2	5/8	1290	880	880	780	1170	780	780	680	1120	700	730	630	1110	690	720	620	1070	650	690	580
		3/4	1860	1190	1240	1080	1690	960	1090	850	1610	870	1030	780	1600	850	1010	750	1540	800	970	710
		7/8	2540	1410	1640	1260	2300	1160	1410	1020	2190	1060	1260	910	2170	1040	1220	870	2060	980	1130	790
		1	3310	1670	1980	1470	2870	1390	1550	1100	2660	1290	1390	970	2630	1260	1340	930	2500	1210	1250	860
7-1/2	1-1/2	5/8	1070	660	760	590	940	560	640	500	880	520	590	460	870	520	590	450	830	470	560	430
		3/4	1450	890	990	780	1270	660	850	660	1200	590	790	590	1190	560	780	560	1140	520	740	520
		7/8	1890	960	1260	960	1680	720	1090	720	1590	630	1010	630	1570	600	990	600	1520	550	950	550
		1	2410	1020	1560	1020	2150	770	1350	770	2050	680	1270	680	2030	650	1240	650	1930	600	1190	600
	3-1/2	5/8	1290	880	880	780	1170	780	780	680	1120	700	730	630	1110	690	720	620	1070	650	690	580
		3/4	1860	1190	1240	1080	1690	960	1090	850	1610	870	1030	780	1600	850	1010	750	1540	800	970	710
		7/8	2540	1410	1640	1260	2300	1160	1450	1020	2190	1060	1360	930	2170	1040	1340	900	2060	980	1280	850
		1	3310	1670	2090	1470	2870	1390	1830	1210	2660	1290	1630	1110	2630	1260	1570	1080	2500	1210	1470	1030

1. Tabulated lateral design values, Z, for bolted connections shall be multiplied by all applicable adjustment factors (see Table 10.3.1).
2. Tabulated lateral design values, Z, are for "full-body diameter" bolts (see Appendix Table L1) with bolt bending yield strength, F_{yb}, of 45,000 psi.

(continued)

APPENDIX M *(continued)*

Bolts: Reference Lateral Design Values, Z, for Single Shear (Two Member) Connections[1,2] for Sawn Lumber or SCL with Both Members of Identical Specific Gravity [NDS Table 11A]

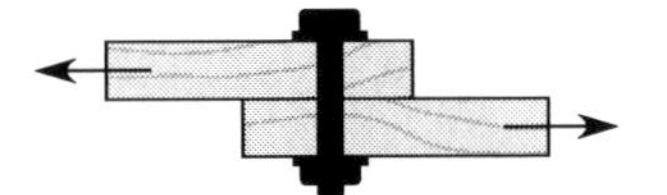

Thickness: Main Member t_m (in.)	Thickness: Side Member t_s (in.)	Bolt Diameter D (in.)	G=0.43 Hem-Fir $Z_{\parallel}$ (lbs.)	$Z_{s\perp}$ (lbs.)	$Z_{m\perp}$ (lbs.)	$Z_{\perp}$ (lbs.)	G=0.42 Spruce-Pine-Fir $Z_{\parallel}$ (lbs.)	$Z_{s\perp}$ (lbs.)	$Z_{m\perp}$ (lbs.)	$Z_{\perp}$ (lbs.)	G=0.37 Redwood (open grain) $Z_{\parallel}$ (lbs.)	$Z_{s\perp}$ (lbs.)	$Z_{m\perp}$ (lbs.)	$Z_{\perp}$ (lbs.)	G=0.36 Eastern Softwoods, Spruce-Pine-Fir(S), Western Cedars, Western Woods $Z_{\parallel}$ (lbs.)	$Z_{s\perp}$ (lbs.)	$Z_{m\perp}$ (lbs.)	$Z_{\perp}$ (lbs.)	G=0.35 Northern Species $Z_{\parallel}$ (lbs.)	$Z_{s\perp}$ (lbs.)	$Z_{m\perp}$ (lbs.)	$Z_{\perp}$ (lbs.)
1-1/2	1-1/2	1/2	410	250	250	180	410	240	240	170	360	210	210	140	350	200	200	130	340	200	200	130
		5/8	520	300	300	190	510	290	290	190	450	250	250	160	440	240	240	150	420	240	240	150
		3/4	620	350	350	210	610	340	340	210	540	290	290	170	520	280	280	170	500	270	270	160
		7/8	720	390	390	230	710	380	380	220	630	330	330	190	610	320	320	180	590	310	310	170
		1	830	440	440	250	810	430	430	240	720	370	370	200	700	360	360	190	670	350	350	190
1-3/4	1-3/4	1/2	480	290	290	210	470	280	280	200	420	250	250	170	410	240	240	160	390	230	230	150
		5/8	600	350	350	230	590	340	340	220	520	290	290	190	510	280	280	180	490	270	270	170
		3/4	720	400	400	250	710	390	390	240	630	340	340	200	610	330	330	190	590	320	320	190
		7/8	850	460	460	270	830	450	450	260	730	390	390	220	710	380	380	210	690	360	360	200
		1	970	510	510	290	950	500	500	280	840	430	430	230	820	420	420	230	790	410	410	220
2-1/2	1-1/2	1/2	550	320	310	250	540	320	300	240	500	290	250	200	490	280	240	190	470	280	240	180
		5/8	730	420	360	270	710	410	350	270	630	350	300	220	610	330	290	210	590	320	280	210
		3/4	870	460	410	300	850	450	400	290	750	370	340	240	740	360	330	230	710	350	320	230
		7/8	1020	500	450	320	1000	490	440	310	860	410	380	260	860	390	370	250	830	370	350	240
		1	1160	540	500	350	1140	530	490	340	1010	440	420	280	980	420	410	270	940	410	390	260
3-1/2	1-1/2	1/2	550	320	380	290	540	320	370	280	500	290	320	250	490	280	300	250	480	280	290	240
		5/8	790	420	440	370	780	410	430	360	720	350	370	300	710	330	350	290	700	320	340	280
		3/4	1100	460	500	400	1080	450	480	390	1010	370	410	320	990	360	400	310	950	350	380	300
		7/8	1370	500	550	430	1340	490	540	420	1180	410	460	350	1160	390	440	340	1110	370	420	320
		1	1570	540	600	470	1530	530	590	460	1350	440	500	380	1320	420	480	370	1270	410	470	350
	1-3/4	1/2	590	340	400	300	580	330	390	290	530	300	330	260	520	290	320	250	510	280	310	250
		5/8	840	480	460	370	820	470	450	360	760	400	390	310	740	380	370	290	730	370	360	280
		3/4	1130	540	520	410	1120	530	510	400	1030	430	430	330	1000	420	420	320	970	410	410	310
		7/8	1390	580	580	440	1360	570	570	430	1200	470	480	360	1170	460	470	350	1130	430	440	320
		1	1590	630	640	480	1550	610	630	460	1370	510	530	380	1340	490	520	370	1290	470	500	360
	3-1/2	1/2	660	440	440	390	660	430	430	380	620	400	400	330	610	390	390	310	600	380	380	310
		5/8	1040	600	600	450	1020	590	590	440	960	520	520	370	950	500	500	350	930	490	490	340
		3/4	1450	740	740	500	1420	730	730	480	1250	650	650	400	1220	630	630	390	1180	620	620	370
		7/8	1690	910	910	540	1660	890	890	520	1460	770	770	440	1430	750	750	420	1370	720	720	390
		1	1930	1030	1030	580	1890	1000	1000	560	1670	870	870	470	1630	840	840	450	1570	810	810	430
5-1/4	1-1/2	5/8	790	420	530	410	780	410	520	400	720	350	470	350	710	330	460	330	700	320	450	320
		3/4	1100	460	690	460	1080	450	670	450	1010	370	560	370	990	360	540	360	970	350	530	350
		7/8	1460	500	750	500	1440	490	730	490	1350	410	620	410	1330	390	600	390	1280	370	560	370
		1	1800	540	820	540	1760	530	800	530	1560	440	670	440	1520	420	650	420	1460	410	630	410
	1-3/4	5/8	840	480	560	410	820	470	550	410	760	400	500	370	740	380	480	360	730	370	470	350
		3/4	1130	540	700	540	1120	530	680	530	1040	430	570	430	1020	420	560	420	1000	410	540	410
		7/8	1490	580	770	580	1470	570	750	570	1370	470	640	470	1350	460	620	460	1320	430	580	430
		1	1910	630	850	630	1890	610	820	610	1760	510	690	510	1740	490	670	490	1700	470	650	470
	3-1/2	5/8	1040	600	660	530	1020	590	650	520	960	520	610	460	950	500	590	440	930	490	580	430
		3/4	1490	740	900	640	1480	730	880	620	1390	650	750	520	1370	630	730	500	1330	620	710	480
		7/8	1950	920	1010	690	1920	910	990	670	1740	820	850	560	1710	800	830	550	1660	770	780	510
		1	2370	1140	1130	750	2330	1120	1100	730	2120	1020	940	600	2080	980	910	580	2030	950	880	560
5-1/2	1-1/2	5/8	790	420	530	410	780	410	520	400	720	350	470	350	710	330	460	330	700	320	450	320
		3/4	1100	460	700	460	1080	450	690	450	1010	370	580	370	990	360	570	360	970	350	550	350
		7/8	1460	500	780	500	1440	490	760	490	1350	410	650	410	1330	390	630	390	1280	370	590	370
		1	1800	540	860	540	1760	530	830	530	1560	440	700	440	1520	420	680	420	1460	410	650	410
	3-1/2	5/8	1040	600	660	530	1020	590	650	520	960	520	610	460	950	500	590	440	930	490	580	430
		3/4	1490	740	920	650	1480	730	900	640	1390	650	770	530	1370	630	750	520	1330	620	720	500
		7/8	1950	920	1030	720	1920	910	1010	700	1740	820	870	590	1710	800	840	570	1660	770	800	530
		1	2370	1140	1150	780	2330	1120	1120	760	2120	1020	960	630	2080	980	930	600	2030	950	890	580
7-1/2	1-1/2	5/8	790	420	530	410	780	410	520	400	720	350	470	350	710	330	460	330	700	320	450	320
		3/4	1100	460	700	460	1080	450	690	450	1010	370	630	370	990	360	620	360	970	350	600	350
		7/8	1460	500	900	500	1440	490	890	490	1350	410	810	410	1330	390	800	390	1280	370	770	370
		1	1800	540	1130	540	1760	530	1110	530	1560	440	920	440	1520	420	890	420	1460	410	860	410
	3-1/2	5/8	1040	600	660	530	1020	590	650	520	960	520	610	460	950	500	590	440	930	490	580	430
		3/4	1490	740	920	650	1480	730	910	640	1390	650	840	560	1370	630	820	550	1330	620	810	540
		7/8	1950	920	1210	790	1920	910	1180	780	1740	820	1010	700	1710	800	980	680	1660	770	920	650
		1	2370	1140	1340	970	2330	1120	1300	950	2120	1020	1100	820	2080	980	1070	790	2030	950	1030	760

1. Tabulated lateral design values, Z, for bolted connections shall be multiplied by all applicable adjustment factors (see Table 10.3.1).
2. Tabulated lateral design values, Z, are for "full-body diameter" bolts (see Appendix Table L1) with bolt bending yield strength, F_{yb}, of 45,000 psi.

APPENDIX N

Bolts: Reference Lateral Design Values, Z,
for Double Shear (Three Member) Connections[1,2]
[NDS Table 11F]

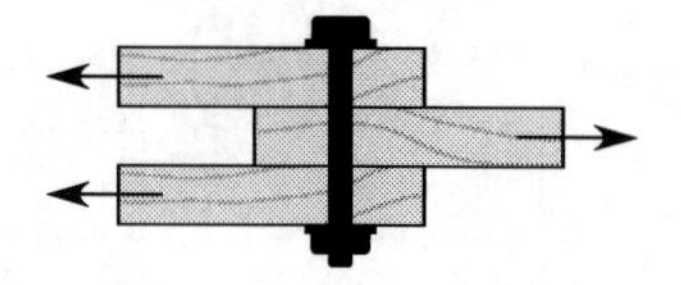

Thickness: Main Member t_m (in.)	Thickness: Side Member t_s (in.)	Bolt Diameter D (in.)	G=0.67 Red Oak			G=0.55 Mixed Maple Southern Pine			G=0.50 Douglas Fir-Larch			G=0.49 Douglas Fir-Larch(N)			G=0.46 Douglas Fir(S) Hem-Fir(N)		
			$Z_{\parallel}$ lbs.	$Z_{s\perp}$ lbs.	$Z_{m\perp}$ lbs.	$Z_{\parallel}$ lbs.	$Z_{s\perp}$ lbs.	$Z_{m\perp}$ lbs.	$Z_{\parallel}$ lbs.	$Z_{s\perp}$ lbs.	$Z_{m\perp}$ lbs.	$Z_{\parallel}$ lbs.	$Z_{s\perp}$ lbs.	$Z_{m\perp}$ lbs.	$Z_{\parallel}$ lbs.	$Z_{s\perp}$ lbs.	$Z_{m\perp}$ lbs.
1-1/2	1-1/2	1/2	1410	960	730	1150	800	550	1050	730	470	1030	720	460	970	680	420
		5/8	1760	1310	810	1440	1130	610	1310	1040	530	1290	1030	520	1210	940	470
		3/4	2110	1690	890	1730	1330	660	1580	1170	590	1550	1130	560	1450	1040	520
		7/8	2460	1920	960	2020	1440	720	1840	1260	630	1800	1210	600	1690	1100	550
		1	2810	2040	1020	2310	1530	770	2100	1350	680	2060	1290	650	1930	1200	600
1-3/4	1-3/4	1/2	1640	1030	850	1350	850	640	1230	770	550	1200	750	530	1130	710	490
		5/8	2050	1370	940	1680	1160	710	1530	1070	610	1500	1060	600	1410	1000	550
		3/4	2460	1810	1040	2020	1550	770	1840	1370	680	1800	1310	660	1690	1210	600
		7/8	2870	2240	1120	2350	1680	840	2140	1470	740	2110	1410	700	1970	1290	640
		1	3280	2380	1190	2690	1790	890	2450	1580	790	2410	1510	750	2250	1400	700
2-1/2	1-1/2	1/2	1530	960	1120	1320	800	910	1230	730	790	1210	720	760	1160	680	700
		5/8	2150	1310	1340	1870	1130	1020	1760	1040	880	1740	1030	860	1660	940	780
		3/4	2890	1770	1480	2550	1330	1110	2400	1170	980	2380	1130	940	2280	1040	860
		7/8	3780	1920	1600	3360	1440	1200	3060	1260	1050	3010	1210	1010	2820	1100	920
		1	4690	2040	1700	3840	1530	1280	3500	1350	1130	3440	1290	1080	3220	1200	1000
3-1/2	1-1/2	1/2	1530	960	1120	1320	800	940	1230	730	860	1210	720	850	1160	680	810
		5/8	2150	1310	1510	1870	1130	1290	1760	1040	1190	1740	1030	1170	1660	940	1090
		3/4	2890	1770	1980	2550	1330	1550	2400	1170	1370	2380	1130	1310	2280	1040	1210
		7/8	3780	1920	2240	3360	1440	1680	3180	1260	1470	3150	1210	1410	3030	1100	1290
		1	4820	2040	2380	4310	1530	1790	4090	1350	1580	4050	1290	1510	3860	1200	1400
	1-3/4	1/2	1660	1030	1180	1430	850	1030	1330	770	940	1310	750	920	1250	710	870
		5/8	2310	1370	1630	1990	1160	1380	1860	1070	1230	1840	1060	1200	1760	1000	1090
		3/4	3060	1810	2070	2670	1550	1550	2510	1370	1370	2480	1310	1310	2370	1210	1210
		7/8	3940	2240	2240	3470	1680	1680	3270	1470	1470	3240	1410	1410	3110	1290	1290
		1	4960	2380	2380	4400	1790	1790	4170	1580	1580	4120	1510	1510	3970	1400	1400
	3-1/2	1/2	1660	1180	1180	1500	1040	1040	1430	970	970	1420	960	960	1370	920	920
		5/8	2590	1770	1770	2340	1560	1420	2240	1410	1230	2220	1390	1200	2150	1290	1090
		3/4	3730	2380	2070	3380	1910	1550	3220	1750	1370	3190	1700	1310	3090	1610	1210
		7/8	5080	2820	2240	4600	2330	1680	4290	2130	1470	4210	2070	1410	3940	1960	1290
		1	6560	3340	2380	5380	2780	1790	4900	2580	1580	4810	2520	1510	4510	2410	1400
5-1/4	1-1/2	5/8	2150	1310	1510	1870	1130	1290	1760	1040	1190	1740	1030	1170	1660	940	1110
		3/4	2890	1770	1980	2550	1330	1690	2400	1170	1580	2380	1130	1550	2280	1040	1480
		7/8	3780	1920	2520	3360	1440	2170	3180	1260	2030	3150	1210	1990	3030	1100	1900
		1	4820	2040	3120	4310	1530	2680	4090	1350	2360	4050	1290	2260	3860	1200	2100
	1-3/4	5/8	2310	1370	1630	1990	1160	1380	1860	1070	1270	1840	1060	1250	1760	1000	1180
		3/4	3060	1810	2110	2670	1550	1790	2510	1370	1660	2480	1310	1630	2370	1210	1550
		7/8	3940	2240	2640	3470	1680	2260	3270	1470	2100	3240	1410	2060	3110	1290	1930
		1	4960	2380	3240	4400	1790	2680	4170	1580	2360	4120	1510	2260	3970	1400	2100
	3-1/2	5/8	2590	1770	1770	2340	1560	1560	2240	1410	1460	2220	1390	1450	2150	1290	1390
		3/4	3730	2380	2480	3380	1910	2180	3220	1750	2050	3190	1700	1970	3090	1610	1810
		7/8	5080	2820	3290	4600	2330	2530	4390	2130	2210	4350	2070	2110	4130	1960	1930
		1	6630	3340	3570	5740	2780	2680	5330	2580	2360	5250	2520	2260	4990	2410	2100
5-1/2	1-1/2	5/8	2150	1310	1510	1870	1130	1290	1760	1040	1190	1740	1030	1170	1660	940	1110
		3/4	2890	1770	1980	2550	1330	1690	2400	1170	1580	2380	1130	1550	2280	1040	1480
		7/8	3780	1920	2520	3360	1440	2170	3180	1260	2030	3150	1210	1990	3030	1100	1900
		1	4820	2040	3120	4310	1530	2700	4090	1350	2480	4050	1290	2370	3860	1200	2200
	3-1/2	5/8	2590	1770	1770	2340	1560	1560	2240	1410	1460	2220	1390	1450	2150	1290	1390
		3/4	3730	2380	2480	3380	1910	2180	3220	1750	2050	3190	1700	2020	3090	1610	1900
		7/8	5080	2820	3290	4600	2330	2650	4390	2130	2310	4350	2070	2210	4130	1960	2020
		1	6630	3340	3740	5740	2780	2810	5330	2580	2480	5250	2520	2370	4990	2410	2200
7-1/2	1-1/2	5/8	2150	1310	1510	1870	1130	1290	1760	1040	1190	1740	1030	1170	1660	940	1110
		3/4	2890	1770	1980	2550	1330	1690	2400	1170	1580	2380	1130	1550	2280	1040	1480
		7/8	3780	1920	2520	3360	1440	2170	3180	1260	2030	3150	1210	1990	3030	1100	1900
		1	4820	2040	3120	4310	1530	2700	4090	1350	2530	4050	1290	2480	3860	1200	2390
	3-1/2	5/8	2590	1770	1770	2340	1560	1560	2240	1410	1460	2220	1390	1450	2150	1290	1390
		3/4	3730	2380	2480	3380	1910	2180	3220	1750	2050	3190	1700	2020	3090	1610	1940
		7/8	5080	2820	3290	4600	2330	2890	4390	2130	2720	4350	2070	2670	4130	1960	2560
		1	6630	3340	4190	5740	2780	3680	5330	2580	3380	5250	2520	3230	4990	2410	3000

1. Tabulated lateral design values, Z, for bolted connections shall be multiplied by all applicable adjustment factors (see Table 10.3.1).
2. Tabulated lateral design values, Z, are for "full-body diameter" bolts (see Appendix Table L1) with bolt bending yield strength, F_{yb}, of 45,000 psi.

Reprinted with courtesy, American Wood Council, Leesburg, VA.

(continued)

APPENDIX N *(continued)*

Bolts: Reference Lateral Design Values, Z, for Double Shear (Three Member) Connections[1,2] [NDS Table 11F]

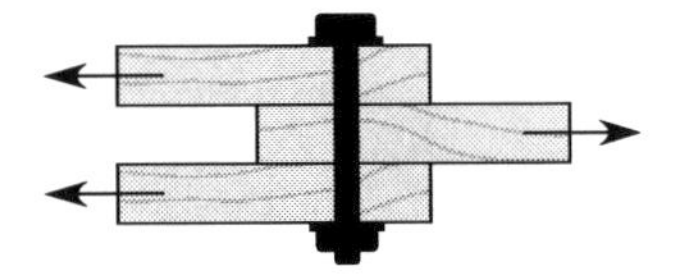

Thickness: Main Member t_m (in.)	Thickness: Side Member t_s (in.)	Bolt Diameter D (in.)	G=0.43 Hem-Fir $Z_{\parallel}$ (lbs.)	$Z_{s\perp}$ (lbs.)	$Z_{m\perp}$ (lbs.)	G=0.42 Spruce-Pine-Fir $Z_{\parallel}$ (lbs.)	$Z_{s\perp}$ (lbs.)	$Z_{m\perp}$ (lbs.)	G=0.37 Redwood (open grain) $Z_{\parallel}$ (lbs.)	$Z_{s\perp}$ (lbs.)	$Z_{m\perp}$ (lbs.)	G=0.36 Eastern Softwoods, Spruce-Pine-Fir(S), Western Cedars, Western Woods $Z_{\parallel}$ (lbs.)	$Z_{s\perp}$ (lbs.)	$Z_{m\perp}$ (lbs.)	G=0.35 Northern Species $Z_{\parallel}$ (lbs.)	$Z_{s\perp}$ (lbs.)	$Z_{m\perp}$ (lbs.)
1-1/2	1-1/2	1/2	900	650	380	880	640	370	780	580	310	760	560	290	730	550	290
		5/8	1130	840	420	1100	830	410	970	690	350	950	660	330	910	640	320
		3/4	1350	920	460	1320	900	450	1170	740	370	1140	720	360	1100	700	350
		7/8	1580	1000	500	1540	970	490	1360	810	410	1330	790	390	1280	740	370
		1	1800	1080	540	1760	1050	530	1560	870	440	1520	840	420	1460	810	410
1-3/4	1-3/4	1/2	1050	670	450	1030	660	430	910	590	360	890	580	340	850	570	330
		5/8	1310	950	490	1290	940	480	1130	810	400	1110	770	380	1070	740	370
		3/4	1580	1080	540	1540	1050	530	1360	870	430	1330	840	420	1280	810	410
		7/8	1840	1160	580	1800	1130	570	1590	950	470	1550	920	460	1490	860	430
		1	2100	1260	630	2060	1230	610	1820	1020	510	1770	980	490	1710	950	470
2-1/2	1-1/2	1/2	1100	650	640	1080	640	610	990	580	510	980	560	490	950	550	480
		5/8	1590	840	700	1570	830	690	1450	690	580	1430	660	550	1390	640	530
		3/4	2190	920	770	2160	900	750	1950	740	620	1900	720	600	1830	700	580
		7/8	2630	1000	830	2570	970	810	2270	810	680	2210	790	660	2130	740	610
		1	3000	1080	900	2940	1050	880	2590	870	730	2530	840	700	2440	810	680
3-1/2	1-1/2	1/2	1100	650	760	1080	640	740	990	580	670	980	560	660	950	550	640
		5/8	1590	840	980	1570	830	960	1450	690	810	1430	660	770	1390	640	740
		3/4	2190	920	1080	2160	900	1050	2010	740	870	1990	720	840	1940	700	810
		7/8	2920	1000	1160	2880	970	1130	2690	810	950	2660	790	920	2560	740	860
		1	3600	1080	1260	3530	1050	1230	3110	870	1020	3040	840	980	2930	810	950
	1-3/4	1/2	1180	670	820	1160	660	800	1060	590	720	1040	580	680	1010	570	670
		5/8	1670	950	980	1650	940	960	1510	810	810	1490	770	770	1450	740	740
		3/4	2270	1080	1080	2240	1050	1050	2070	870	870	2040	840	840	1990	810	810
		7/8	2980	1160	1160	2950	1130	1130	2740	950	950	2700	920	920	2640	860	860
		1	3820	1260	1260	3770	1230	1230	3520	1020	1020	3480	980	980	3410	950	950
	3-1/2	1/2	1330	880	880	1310	870	860	1230	800	720	1220	780	680	1200	760	670
		5/8	2070	1190	980	2050	1170	960	1930	1030	810	1900	1000	770	1870	970	740
		3/4	2980	1490	1080	2950	1460	1050	2720	1290	870	2660	1270	840	2560	1240	810
		7/8	3680	1840	1160	3600	1810	1130	3180	1640	950	3100	1610	920	2990	1550	860
		1	4200	2280	1260	4110	2240	1230	3630	2030	1020	3540	1960	980	3410	1890	950
5-1/4	1-1/2	5/8	1590	840	1050	1570	830	1040	1450	690	940	1430	660	920	1390	640	900
		3/4	2190	920	1400	2160	900	1380	2010	740	1250	1990	720	1230	1940	700	1210
		7/8	2920	1000	1750	2880	970	1700	2690	810	1420	2660	790	1380	2560	740	1290
		1	3600	1080	1890	3530	1050	1840	3110	870	1520	3040	840	1470	2930	810	1420
	1-3/4	5/8	1670	950	1110	1650	940	1100	1510	810	990	1490	770	970	1450	740	940
		3/4	2270	1080	1460	2240	1050	1440	2070	870	1300	2040	840	1260	1990	810	1220
		7/8	2980	1160	1750	2950	1130	1700	2740	950	1420	2700	920	1380	2640	860	1290
		1	3820	1260	1890	3770	1230	1840	3520	1020	1520	3480	980	1470	3410	950	1420
	3-1/2	5/8	2070	1190	1320	2050	1170	1310	1930	1030	1210	1900	1000	1150	1870	970	1120
		3/4	2980	1490	1610	2950	1460	1580	2770	1290	1300	2740	1270	1260	2660	1240	1220
		7/8	3900	1840	1750	3840	1810	1700	3480	1640	1420	3410	1610	1380	3320	1550	1290
		1	4730	2280	1890	4660	2240	1840	4240	2030	1520	4170	1960	1470	4050	1890	1420
5-1/2	1-1/2	5/8	1590	840	1050	1570	830	1040	1450	690	940	1430	660	920	1390	640	900
		3/4	2190	920	1400	2160	900	1380	2010	740	1250	1990	720	1230	1940	700	1210
		7/8	2920	1000	1800	2880	970	1780	2690	810	1490	2660	790	1440	2560	740	1350
		1	3600	1080	1980	3530	1050	1930	3110	870	1600	3040	840	1540	2930	810	1490
	3-1/2	5/8	2070	1190	1320	2050	1170	1310	1930	1030	1210	1900	1000	1180	1870	970	1160
		3/4	2980	1490	1690	2950	1460	1650	2770	1290	1360	2740	1270	1320	2660	1240	1280
		7/8	3900	1840	1830	3840	1810	1780	3480	1640	1490	3410	1610	1440	3320	1550	1350
		1	4730	2280	1980	4660	2240	1930	4240	2030	1600	4170	1960	1540	4050	1890	1490
7-1/2	1-1/2	5/8	1590	840	1050	1570	830	1040	1450	690	940	1430	660	920	1390	640	900
		3/4	2190	920	1400	2160	900	1380	2010	740	1250	1990	720	1230	1940	700	1210
		7/8	2920	1000	1800	2880	970	1760	2690	810	1630	2660	790	1600	2560	740	1550
		1	3600	1080	2270	3530	1050	2240	3110	870	2040	3040	840	2010	2930	810	1970
	3-1/2	5/8	2070	1190	1320	2050	1170	1310	1930	1030	1210	1900	1000	1180	1870	970	1160
		3/4	2980	1490	1850	2950	1460	1820	2770	1290	1670	2740	1270	1650	2660	1240	1620
		7/8	3900	1840	2450	3840	1810	2420	3480	1640	2030	3410	1610	1970	3320	1550	1840
		1	4730	2280	2700	4660	2240	2630	4240	2030	2180	4170	1960	2100	4050	1890	2030

1. Tabulated lateral design values, Z, for bolted connections shall be multiplied by all applicable adjustment factors (see Table 10.3.1).
2. Tabulated lateral design values, Z, are for "full-body diameter" bolts (see Appendix Table L1) with bolt bending yield strength, F_{yb}, of 45,000 psi.

APPENDIX O

Bolts: Reference Lateral Design Values, *Z*, for Single Shear (Two Member) Connections[1,2,3,4] for Sawn Lumber or SCL to Concrete [NDS Table 11E]

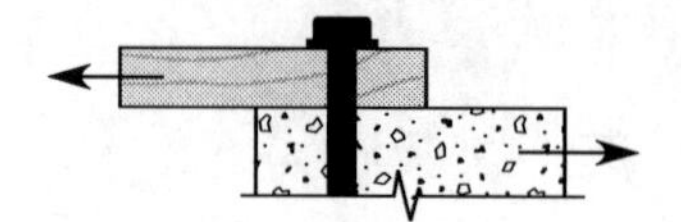

Thickness: Embedment Depth in Concrete t_m in.	Thickness: Side Member t_s in.	Bolt Diameter D in.	G=0.67 Red Oak $Z_{\parallel}$ lbs.	G=0.67 Red Oak $Z_{\perp}$ lbs.	G=0.55 Mixed Maple Southern Pine $Z_{\parallel}$ lbs.	G=0.55 Mixed Maple Southern Pine $Z_{\perp}$ lbs.	G=0.50 Douglas Fir-Larch $Z_{\parallel}$ lbs.	G=0.50 Douglas Fir-Larch $Z_{\perp}$ lbs.	G=0.49 Douglas Fir-Larch(N) $Z_{\parallel}$ lbs.	G=0.49 Douglas Fir-Larch(N) $Z_{\perp}$ lbs.	G=0.46 Douglas Fir(S) Hem-Fir(N) $Z_{\parallel}$ lbs.	G=0.46 Douglas Fir(S) Hem-Fir(N) $Z_{\perp}$ lbs.
6.0 and greater	1-1/2	1/2	770	480	680	410	650	380	640	380	620	360
		5/8	1070	660	970	580	930	530	920	520	890	470
		3/4	1450	890	1330	660	1270	590	1260	560	1230	520
		7/8	1890	960	1750	720	1690	630	1680	600	1640	550
		1	2410	1020	2250	770	2100	680	2060	650	1930	600
	1-3/4	1/2	830	510	740	430	700	400	690	390	670	370
		5/8	1160	680	1030	600	980	550	970	550	940	530
		3/4	1530	900	1390	770	1330	680	1310	660	1270	600
		7/8	1970	1120	1800	840	1730	740	1720	700	1680	640
		1	2480	1190	2290	890	2210	790	2200	750	2150	700
	2-1/2	1/2	830	590	790	520	770	470	760	460	750	440
		5/8	1290	800	1230	670	1180	610	1170	610	1120	570
		3/4	1840	1000	1630	850	1540	800	1520	780	1460	750
		7/8	2290	1240	2050	1080	1940	1020	1920	1000	1860	920
		1	2800	1520	2530	1280	2410	1130	2390	1080	2310	1000
	3-1/2	1/2	830	590	790	540	770	510	760	500	750	490
		5/8	1290	880	1230	810	1200	730	1190	720	1170	670
		3/4	1860	1190	1770	980	1720	900	1720	880	1680	830
		7/8	2540	1410	2410	1190	2320	1100	2290	1070	2200	1020
		1	3310	1670	2970	1420	2800	1330	2770	1300	2660	1260

Thickness: Embedment Depth in Concrete t_m in.	Thickness: Side Member t_s in.	Bolt Diameter D in.	G=0.43 Hem-Fir $Z_{\parallel}$ lbs.	G=0.43 Hem-Fir $Z_{\perp}$ lbs.	G=0.42 Spruce-Pine-Fir $Z_{\parallel}$ lbs.	G=0.42 Spruce-Pine-Fir $Z_{\perp}$ lbs.	G=0.37 Redwood (open grain) $Z_{\parallel}$ lbs.	G=0.37 Redwood (open grain) $Z_{\perp}$ lbs.	G=0.36 Eastern Softwoods Spruce-Pine-Fir(S) Western Cedars Western Woods $Z_{\parallel}$ lbs.	G=0.36 Eastern Softwoods Spruce-Pine-Fir(S) Western Cedars Western Woods $Z_{\perp}$ lbs.	G=0.35 Northern Species $Z_{\parallel}$ lbs.	G=0.35 Northern Species $Z_{\perp}$ lbs.
6.0 and greater	1-1/2	1/2	590	340	590	340	550	310	540	290	530	290
		5/8	860	420	850	410	810	350	800	330	780	320
		3/4	1200	460	1190	450	1130	370	1120	360	1100	350
		7/8	1580	500	1540	490	1360	410	1330	390	1280	370
		1	1800	540	1760	530	1560	440	1520	420	1460	410
	1-3/4	1/2	640	360	630	350	580	320	580	310	560	310
		5/8	910	490	900	480	840	400	830	380	810	370
		3/4	1230	540	1220	530	1160	430	1140	420	1120	410
		7/8	1630	580	1610	570	1540	470	1520	460	1490	430
		1	2090	630	2060	610	1820	510	1770	490	1710	470
	2-1/2	1/2	730	410	730	400	700	360	690	340	680	340
		5/8	1070	540	1060	530	980	480	960	470	940	460
		3/4	1400	710	1380	700	1290	620	1270	600	1240	580
		7/8	1790	830	1770	810	1660	680	1640	660	1600	610
		1	2230	900	2210	880	2080	730	2060	700	2030	680
	3-1/2	1/2	730	470	730	470	700	430	690	410	690	400
		5/8	1140	620	1140	610	1090	550	1080	530	1070	520
		3/4	1650	780	1640	770	1540	680	1510	670	1470	660
		7/8	2100	960	2070	950	1910	870	1880	850	1840	820
		1	2550	1190	2520	1180	2340	1020	2310	980	2260	950

1. Tabulated lateral design values, Z, for bolted connections shall be multiplied by all applicable adjustment factors (see Table 10.3.1).
2. Tabulated lateral design values, Z, are for "full-body diameter" bolts (see Appendix Table L1) with bolt bending yield strength, F_{yb}, of 45,000 psi.
3. Tabulated lateral design values, Z, are based on dowel bearing strength, F_e, of 7,500 psi for concrete with minimum f_c'=2,500 psi.
4. Six inch anchor embedment assumed.

APPENDIX P

1994 Northridge Earthquake:
Performance of Steel Moment Frame Connections

SUMMARY

The special moment-resisting frame (SMRF) beam-to-column connection depicted in Fig. P-1 (the bolted web/welded flange design originally developed by seismic pioneer Egor P. Popov in the early 1970s) is now known to be fundamentally flawed. This connection, although 100% consistent with the 1994 UBC (Sec. 2710(g)1B) and previously considered to be seismically invulnerable, does not have sufficient ductility during multiple inelastic cycling. Although this connection has been the "bread-and-butter" design for new construction for many years, it is no longer used in areas of high seismic activity.

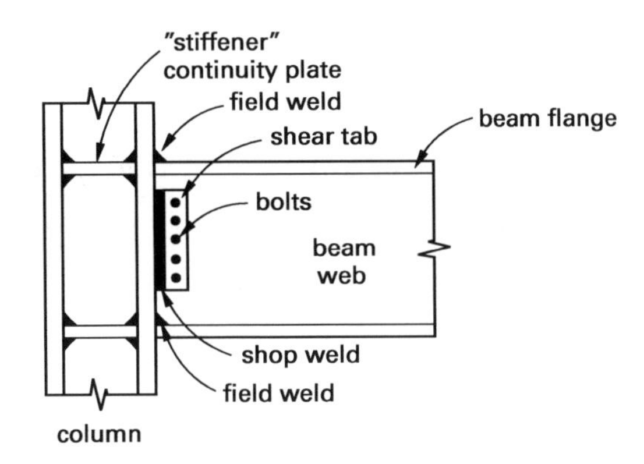

Figure P-1 Connection No Longer Used

NORTHRIDGE EARTHQUAKE

The Northridge earthquake occurred on January 17, 1994. The hypocenter was 11.4 mi (18.4 km) below and 0.6 mi (1 km) to the south of Northridge in Los Angeles. Its magnitude was approximately 6.6. The Northridge thrust fault (previously known as the Pico thrust fault and probably a subsurface extension of the Oak Ridge fault) was involved. This fault was not previously considered to be a seismic danger. Since a thrust fault was involved, the Northridge earthquake had a significant vertical acceleration component.

The duration of the source rupture was 8 sec—longer than the quick jolts of many previous earthquakes. Horizontal ground acceleration was measured as high as 0.93 g at one location. A vertical acceleration of 0.25 g was recorded at one location. Accelerations of 2.3 g horizontally and 1.7 g vertically were recorded at the abutment of the Pacoima Dam. (Horizontally, 2.5 g was measured at the roof of the Olive View Hospital destroyed in the 1971 San Fernando earthquake.)

The Northridge earthquake was the second most expensive U.S. earthquake, with costs and losses exceeding $15 billion. More than 15,000 buildings sustained minor damage, and more than 200 SMRF buildings up to 16 mi (25 km) from the epicenter were damaged. Some of the buildings were brand new, and others were in various stages of completion. (Putting the damage fraction into perspective, there are about 1500 SMRF buildings in Los Angeles.)

There were no building collapses, although a few buildings are said to have been close to collapse. There were 61 fatalities and more than 9000 injuries.

Braced-frame buildings performed well, as did base-isolated structures in distant areas. (There were no base-isolated structures in the vicinity of the actual epicenter.) As expected, reinforced masonry did not perform well. Extensive damage to bridges (most of which were waiting for retrofitting) and infrastructure improvements (e.g., water lines) occurred. Retrofitted bridges fared well, and the viability of carbon-wrapped columns was proven.

Types of Failures

Although there were some failures in column base plates, this was not a widespread occurrence. Local buckling of beams was minimal or nonexistent. Most of the attention has been focused on the welded beam-column connections that failed in record numbers.

Welded connections failed in a variety of ways. There were some laminar tearouts (delamination) where the weld material pulled out of the column face. There were column-flange divots where the weld was strong enough to rip out full-thickness pieces ("nuggets") of the column.

The most common failure involved cracking of the welded connections to various extents. In some cases, small cracks were observed. In other cases, crack propagation was through the column flange and/or beam web.

(continued)

APPENDIX P *(continued)*
1994 Northridge Earthquake:
Performance of Steel Moment Frame Connections

Alarmingly, there was no evidence of inelastic beam deformation occurring before the connection failures. Inelastic failure of the beams is at the core of modern-day seismic protection philosophy.

All of the failures are classified as "brittle failures." It is problematic that the connections failed in a brittle mode, since ductility is the primary goal of modern seismic design.

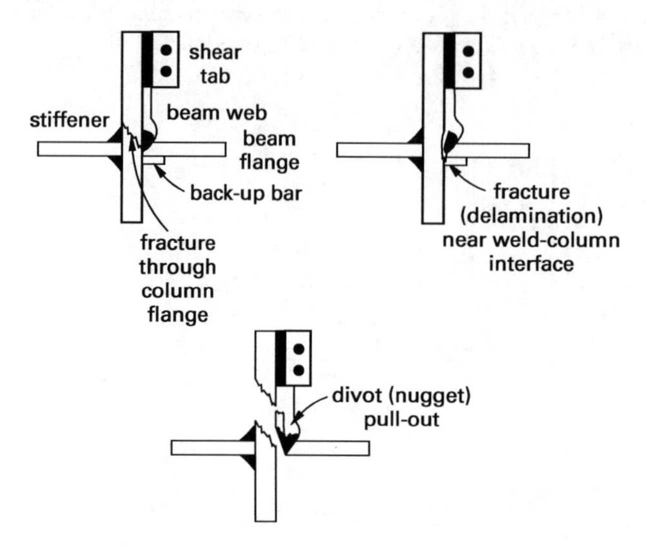

Figure P-2 Types of Connection Failures

Reasons for Failure

The primary reason for the Northridge failures was lack of ductility. The design shown in Fig. P-1 is now known to be "fundamentally flawed" in its ability to perform in a ductile manner. The high degree of ductility that was counted upon to justify the use of an R_w factor of 12 (as specified in the then-current UBC) was not realized. Tests previous and subsequent to the Northridge earthquake indicate that the ductility of this connection design is apparently only 2–4.

Unfortunately, there is no single cause of the connection failures experienced. The poor ductile performance has been blamed on a number of factors: (1) the nature of the earthquake, (2) the materials used, (3) the weld quality, (4) the weld materials, (5) the overall design, and (6) other miscellaneous factors. All of these factors appear to be involved in a complex manner not yet fully understood.

Cause 1: Nature of the Earthquake

The Northridge earthquake was longer, stronger, and more vertical than had been expected. The horizontal acceleration experienced was, in some cases, twice the design acceleration. The vertical acceleration increased the weight supported by the connections. Vertical acceleration was not considered in the original connection designs.

The large vertical acceleration component of this earthquake placed the connections into a state of triaxial stress. Ductile failure is more difficult to achieve under triaxiality, while brittle failure is relatively unaffected by triaxiality. Therefore, fracture is an easier energy-dissipating mechanism than plastic deformation, leading to brittle failure.

Cause 2: Materials

The second suspected factor centers around the beam and column materials. Anisotropy (i.e., different material properties in different directions) in the ductility and through-thickness strength due to the direction of rolling during beam manufacturing has been accused. Some beams had yield strengths higher than 36 ksi material due to statistical variations in the manufacturing processes. This prevented them from yielding as expected and transferred the earthquake energy into the connection.

The buildup of residual stresses during welding, particularly when the beam and column have different characteristics, is also problematic. Typically, designers use grade 50 columns and grade 36 beams to achieve a strong column-weak beam system where beams yield before the columns yield. (Column yielding and story-mechanism failure is considered to be more life-threatening than beam yielding since the column carries the gravity load of the building.)

The intrinsic weldability of grade 50 steel has been questioned by some, though this may not be a significant factor.

(continued)

APPENDIX P *(continued)*
1994 Northridge Earthquake:
Performance of Steel Moment Frame Connections

Cause 3: Quality

It is now known that the original welds were notch-sensitive and particularly susceptible to minor imperfections and flaws. Less than adequate workmanship and inspection were cited in some cases.

In cases involving laminar tearouts, it is fairly certain that the thick base material did not receive sufficient preheating during welding.

Flaws in root weld passes have also been blamed on the difficulty of doing overhead welds on the lower beam flange.

There appears to have been significant notch effects (notch brittleness) caused by the left-in-place backing bars. Backing bars (placed under the flange when welding from the top) can only be removed by flame cutting, followed by grinding of the exposed area. Since backing bars are expensive to remove, they are usually left in place. Most cracks, however, apparently began in the vicinity of the backing bars.

Welding has always been the whipping boy of connection failures. Even the failure of connections during testing at the University of Texas have been blamed on poor welding. However, although poor quality can contribute to the failure mechanism, it is now known that even perfect assemblies will fail with this type of connection.

Cause 4: Weld Material

Initially, little attention was given to the weld electrode material itself. However, a $1 billion class action lawsuit filed against Lincoln Electric Co. in 1997 questioned the use of certain types of welding electrodes. Lincoln Electric is the manufacturer of E70-T4 electrodes used in a semi-automatic flux-core arc welding process. This process is used to weld the beams to the column after the beams have been bolted to the column's shear tab.

The flux-core welding process may produce weld metals with relatively low notch toughness. (However, high notch toughness was not a requirement of prevailing codes or American Welding Societies specifications at the time.) There may be a possible lack of ductility in the E70-T4 electrodes themselves.

Cause 5: Overall Design

The connection design has an intrinsic inability to absorb and dissipate energy in an earthquake with sustained strong motion, particularly with a vertical component to acceleration.

The effect of the presence of thick, strong concrete slabs above the beams probably was not considered in most of the designs. However, these slabs substantially increase the strength and stiffness of the beams. The slabs also shift the neutral axis toward the beam top flanges, increasing the stress at the bottom flange. (Most failures occurred at the bottom flanges.)

Cause 6: Other Factors

Early investigations questioned the use of perimeter framing and the lack of redundancy in some of the damaged buildings. In early years of steel design where riveted construction was used, almost every connection was a moment connection. State-of-the-art design prescribes that not all connections need to be moment-resisting, and just a few selected frames are designed as lateral load-resistance systems.

Current design philosophy requires only a few expansive bays supported by gigantic columns and deep beams. Huge moment connections are required. In some buildings, perhaps only two or three columns around the perimeter might have moment connections. Connections of the traditional bolted seated beam variety are used for interior connections. Thus, in an earthquake, the perimeter connections have to do all the work. In some cases, only six welds are available to supply all the moment resistance for a floor. In order to carry all of the moment load, the columns and beams are made even thicker. This increased thickness just compounds the problem, since full penetration welds become all the more difficult.

However, as indicated in the next section, a lack of redundancy was not observed in all cases. Some buildings with SMRF connections across an entire face were damaged.

Similarities in Damage

Most connection failures had the following common elements: (1) The connections were SMRF. (2) Failures initiated or were located in the bottom flange, near the left-in-place backing bar. Columns were typically grade 50 steel. (3) Failures were limited to connections on the building perimeter, at points of plastic hinging.

(continued)

APPENDIX P *(continued)*
1994 Northridge Earthquake:
Performance of Steel Moment Frame Connections

There has been no statistical correlation between damage and building size, building age, nominal material strength, structural regularity, redundancy, number of bays per frame, frame dimensions, or member sizes. Failures occurred in connections with and without column-flange stiffeners, and with and without return welds on the shear connection plates. Wide flange and box beams were both affected. Interestingly, most buildings were fairly new, with most having been completed in the 1980s.

Most, but not all, buildings were two to four stories in height, and a 22-story building suffered connection failures. The connection failure rate varied from 10% to 100% of the total connections in the building. Also interesting was that most buildings were fairly symmetrical (i.e., rectangular). Building torsion was apparently not an issue.

History

In retrospect, research over more than 20 years has shown that the problem of brittle failure has always existed.

The connection design was originally developed by Egor P. Popov, who released the results of his initial tests in 1972. He had performed tests on both welded-web/welded flange and bolted-web/welded flange connections using a limited number of beam and column sizes. He reported that the all-welded connections showed excellent ductility. The bolted/welded connections performed well, but were less ductile, he reported.

However, the bolted/welded connection was easier and more economical to make in the field, since the beam could be swung into position and bolted to a shear tab already shop-welded to the column. Therefore, the bolted/welded connection was selected, and Popov's parameters were extrapolated to connections much larger than he had actually tested.

In 1986, Popov reported on subsequent testing he had performed, and the news was bad. However, the announcement did not result in any changes in design methodology. Additional testing in the early 1990s, this time at the University of Texas at Austin by Popov's protégé and UT structural engineering professor, Michael Engelhardt, also indicated that the design was not ductile. However, allegedly sloppy workmanship by Texas welders was cited in explaining the test results. In 1993, Engelhardt repeated the tests with specimens welded in California. He announced more bad news just a month before the Northridge earthquake.

Methods of Repair and Remediation

Repair efforts immediately after the earthquake focused on in-kind repair (cleaning, grinding, rewelding, and ultrasonic testing) without any substantial reinforcement. In some cases, the existing bolted shear plates (bolted beam web plates) were also welded. This approach appears to be justified where only a few of the connections failed.

Although the in-kind repair was permitted by the city of Los Angeles, it is now known to be inadequate. Testing at the University of Texas showed that, although strength is returned to normal, ductility is actually reduced slightly.

Various other repair methods were also tested by the University of Texas, and these tests proved the desirability of bringing the forces and moments out from the area of the welded connection. (These tests have also been criticized due to their low budget, use of slowly-applied static loads, absence of axial loading of the columns, and absence of concrete decking.)

The most common repair method used was to add triangular horizontal cover plates ("flange plates") at the top and bottom flanges. This method was the first to be supported by any significant testing. In the University of Texas tests, this method was able to sustain a rotation of 0.0075 radians, which is greater than the 0.005 radians exhibited by the simple-reweld solution, but still far below the requirement of 0.025 radians needed to sustain a magnitude 7 earthquake.

A variation of the flange plate approach is to close the beam sides with vertical plates welded to the column face. After the damage to the original connection has been repaired and the connection has been rewelded, a 1 in thick steel side plate, 5 ft long, is fillet-welded to the column and sides of the beam, creating a 5 ft long "box." (The beam flanges are the top of the box.) This is known as the MNH system, named after Myers, Nelson, Houghton, Inc., which has patented and trademarked a similar "dual strong axes connection" for new construction.

The use of vertical beam-column ribs (gussets), top and bottom, was another connection tested at the University of Texas. However, this method is difficult to apply in situations where access to the top flange of the beam is limited by concrete floor slabs.

A tee-support (haunch support) on the bottom flange only was also tested. Access for welding limits the usefulness of this method for in-field repairs.

Vertical web straps (vertical plates extending from the column face to well into the beam's web) are analogous to flange plates. They also have the ability to move the forces away from the beam-column connection. However, such a repair would be difficult to use where there was a welded shear plate already in place.

(continued)

APPENDIX P *(continued)*
1994 Northridge Earthquake:
Performance of Steel Moment Frame Connections

Another method, which focuses on the use of bolting instead of welding, has been proposed by the Center for Advanced Technology for Large Structural Systems (ATLSS) at Lehigh University. This method, designed by David Bleiman and Kazuhiko Kasai, replaces broken welds with specifically fabricated brackets using high-strength bolts.

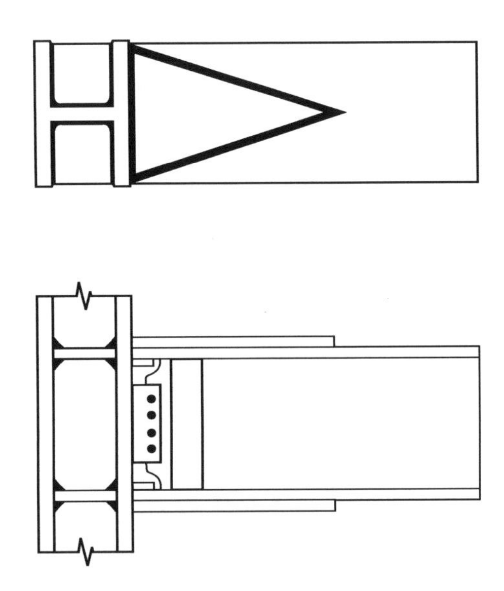

Figure P-3 Types of Connection Repair

Beam Weakening Approaches

It is interesting that beam yielding did not occur in any of the connection failures observed. The connections absorbed all of the earthquake energy. Various beam weakening methods have been proposed in order to ensure that beam yielding occurs in the future. (It is popular to refer to a weakened beam as a "structural fuse," since the beam yields before the connection fails.) These beam weakening methods can be used with old and new construction.

In a "dogbone" beam, the beam flange is shaved to reduce the interior cross section of the beam so that it will yield and buckle locally, away from the connection. This design is known as a *reduced beam section* (RBS) *moment connection.*

Seismic Structural Design Associates (SSDA, Los Angeles) has a different approach for repairing existing connections. The first step is to repair the connection to its original state. The connection is "softened" using horizontal slots in the beam web, top and bottom, and vertical slots in the column web near the beam flanges. The resulting proprietary connections, known as "SSDA slotted web" connections, are prequalified for use in special moment frames.

Connection Design

Until 1988, the UBC did not even specify the type of connection to be used. Rather, the UBC merely specified the force the connection had to support. However, based on the SEAOC's 1985 *Lateral Force Requirements* (the "Blue Book"), the UBC added the design shown in Fig. P-1. For various reasons, including liability to ICBO and SEAOC, the bolted web/welded flange connection detail was stricken from the UBC, which was then in use.

The IBC now prescribes only the seismic force to be resisted, not the connection detail. To ensure ductile behavior in addition to moment transfer, by deference to AISC 341 Sec. K1 and AISC 360, IBC Sec. 1705.12 requires steel beam-column connections used in moment frames to be qualified through testing. AISC has responded by publishing AISC 358, *Prequalified Connections for Special and Intermediate Steel Moment Frames for Seismic Applications*, listing connections and their design procedures. Prequalified designs listed in AISC 358 include reduced beam section (RBS) moment connections, bolted stiffened and unstiffened extended end-plate (BSEEP and BUEEP) moment connections, bolted flange plate (BFP) moment connections, welded unreinforced flange-welded web (WUF-W) moment connections, Kaiser bolted bracket (KBB) moment connections, ConXtech® ConXL™ moment connections, and SidePlate® moment connections. The proprietary SSDA slotted web moment connection is also prequalified.

(continued)

APPENDIX P *(continued)*
1994 Northridge Earthquake:
Performance of Steel Moment Frame Connections

Other Lessons Learned

The Northridge earthquake taught us many important lessons. It may seem obvious, but connections need to be inspected after every major earthquake, regardless of whether there is observed damage. In these inspections, architectural elements and coverings need to be removed. Cracking of the concrete fire-proofing or other telltale signs cannot be counted on to indicate connection failures.

It has also been learned that ultrasonic testing may not be sufficient to detect weld cracks, particularly when the back-up bar has been left in place. Back-up bars are not consistent with the use of flange plates, but there are many thousands of old connections with back-up bars still in place.

Index

D

E

F

G

H

I

J

K

L

M

N

O

Index by Seismic Building Code

I

N

U